IFIP Advances in Information and Communication Technology   478

Editor-in-Chief

IFIP – The International Federation for Information Processing

IFIP was founded in 1960 under the auspices of UNESCO, following the first World Computer Congress held in Paris the previous year. A federation for societies working in information processing, IFIP's aim is two-fold: to support information processing in the countries of its members and to encourage technology transfer to developing nations. As its mission statement clearly states:

> IFIP is the global non-profit federation of societies of ICT professionals that aims at achieving a worldwide professional and socially responsible development and application of information and communication technologies.

IFIP is a non-profit-making organization, run almost solely by 2500 volunteers. It operates through a number of technical committees and working groups, which organize events and publications. IFIP's events range from large international open conferences to working conferences and local seminars.

The flagship event is the IFIP World Computer Congress, at which both invited and contributed papers are presented. Contributed papers are rigorously refereed and the rejection rate is high.

As with the Congress, participation in the open conferences is open to all and papers may be invited or submitted. Again, submitted papers are stringently refereed.

The working conferences are structured differently. They are usually run by a working group and attendance is generally smaller and occasionally by invitation only. Their purpose is to create an atmosphere conducive to innovation and development. Refereeing is also rigorous and papers are subjected to extensive group discussion.

Publications arising from IFIP events vary. The papers presented at the IFIP World Computer Congress and at open conferences are published as conference proceedings, while the results of the working conferences are often published as collections of selected and edited papers.

IFIP distinguishes three types of institutional membership: Country Representative Members, Members at Large, and Associate Members. The type of organization that can apply for membership is a wide variety and includes national or international societies of individual computer scientists/ICT professionals, associations or federations of such societies, government institutions/government related organizations, national or international research institutes or consortia, universities, academies of sciences, companies, national or international associations or federations of companies.

More information about this series at http://www.springer.com/series/6102

Daoliang Li · Zhenbo Li (Eds.)

Computer and Computing Technologies in Agriculture IX

9th IFIP WG 5.14 International Conference, CCTA 2015
Beijing, China, September 27–30, 2015
Revised Selected Papers, Part I

 Springer

Editors
Daoliang Li
China Agricultural University
Beijing
China

Zhenbo Li
China Agricultural University
Beijing
China

ISSN 1868-4238   ISSN 1868-422X (electronic)
IFIP Advances in Information and Communication Technology
ISBN 978-3-319-83920-2   ISBN 978-3-319-48357-3 (eBook)
DOI 10.1007/978-3-319-48357-3

Printed on acid-free paper

This Springer imprint is published by Springer Nature
The registered company is Springer International Publishing AG
The registered company address is: Gewerbestrasse 11, 6330 Cham, Switzerland

Preface

The 9th International Conference on Computer and Computing Technologies in Agriculture (CCTA 2015) was held in Beijing, China, during September 27–30, 2015. The conference was hosted by the China Agricultural University (CAU), Agricultural Information Institute of Chinese Academy of Agricultural Sciences (AIICAAS), China National Engineering Research Center for Information Technology in Agriculture (NERCITA), China National Engineering Research Center of Intelligent Equipment for Agriculture (NERCIEA), International Federation for Information Processing (IFIP), Chinese Society of Agricultural Engineering (CSAE), Chinese Society for Agricultural Machinery (CSAM), Chinese Association for Artificial Intelligence (CAAI), Information Technology Association of China Agro-technological Extension Association, Beijing Technology Innovation Strategic Alliance for Intelligence Internet of Things, Industry in Agriculture, Beijing Society for Information Technology in Agriculture, China (BSITA), Group of Agri-Informatics, Ministry of Agriculture, China, Sino-US Agricultural Aviation Cooperative Technology Center, and the Club of Ossiach. It was sponsored by the National Natural Science Foundation of China (NSFC), Ministry of Agriculture, China (MOA), Ministry of Science and Technology, China (MOST), Ministry of Industry and Information Technology, China (MIIT), State Administration of Foreign Experts Affairs, China (SAFEA), Beijing Administration of Foreign Experts Affairs, China, Beijing Municipal Science and Technology Commission (BMSTC), Beijing Natural Science Foundation, China (BNSF), Beijing Association for Science and Technology, China (BAST), Beijing Academy of Agriculture and Forestry Sciences, China (BAAFS), Dabeinong Education Foundation, China, and the Global Forum on Agricultural Research (GFAR).

In order to promote exchange and cooperation among scientists and professionals from different fields and strengthen international academic exchange, the Joint International Conference on Intelligent Agriculture (ICIA) included the 8th International Symposium on Intelligent Information Technology in Agriculture (8th ISIITA), the 9th IFIP International Conference on Computer and Computing Technologies in Agriculture (9th CCTA), and the AgriFuture Days 2015 International Conference (Agri-Future Days 2015). These events provided a platform, for experts and scholars from all over the world to exchange techniques, ideas, and views on intelligent agricultural innovation. Nine International Conferences on Computer and Computing Technologies in Agriculture have been held since 2007.

The topics of CCTA 2015 covered the theory and applications of all kinds of technology in agriculture, including: intelligent sensing, monitoring, and automatic control technology models; the key technology and model of the Internet of Things; agricultural intelligent equipment technology; computer vision; computer graphics and virtual reality; computer simulation, optimization, and modeling; cloud computing and agricultural applications; agricultural big data; decision support systems and expert system; 3s technology and precision agriculture; the quality and safety of agricultural products; detection and tracing technology; and agricultural electronic commerce technology.

We selected the 122 best papers among the 237 papers submitted to CCTA 2015 for these proceedings. All papers underwent two reviews by two Program Committee members, who are from the Special Interest Group on Advanced Information Processing in Agriculture (AIPA), IFIP. In these proceedings, creative thoughts and inspirations can be discovered, discussed, and disseminated. It is always exciting to have experts, professionals, and scholars with creative contributions getting together to share inspiring ideas and accomplish great developments in the field.

I would like to express my sincere thanks to all authors who submitted research papers to the conference. Finally, I would also like to express my gratitude to all speakers, session chairs, and attendees, both national and international, for their active participation and support of this conference.

September 2016   Daoliang Li

Conference Organization

Organizers

China Agricultural University
China National Engineering Research Center for Information Technology
    in Agriculture (NERCITA)
Agricultural Information Institute of Chinese Academy of Agricultural Sciences
    (AIICAAS)
China National Engineering Research Center of Intelligent Equipment for Agriculture
    (NERCIEA)
International Federation for Information Processing (IFIP)
Chinese Society of Agricultural Engineering (CSAE)
Chinese Society for Agricultural Machinery (CSAM)
Chinese Association for Artificial Intelligence (CAAI)
Information Technology Association of China Agro-technological Extension Association
Beijing Technology Innovation Strategic Alliance for Intelligence Internet Of Things
    Industry in Agriculture
Beijing Society for Information Technology in Agriculture, China (BSITA)
Group of Agri-Informatics, Ministry of Agriculture, China
Sino-US Agricultural Aviation Cooperative Technology Center

Sponsors

National Natural Science Foundation of China (NSFC)
Ministry of Agriculture, China (MOA)
Ministry of Science and Technology, China (MOST)
Ministry of Industry and Information Technology, China (MIIT)
State Administration of Foreign Experts Affairs, China (SAFEA)
Beijing Administration of Foreign Experts Affairs, China
Beijing Municipal Science & Technology Commission (BMSTC)
Beijing Natural Science Foundation, China (BNSF)
Beijing Association for Science and Technology, China (BAST)
Beijing Academy of Agriculture and Forestry Sciences, China (BAAFS)
Dabeinong Education Foundation, China
The Global Forum on Agricultural Research (GFAR)

Organizing Committee

Chunjiang Zhao	China National Engineering Research Center for Information Technology in Agriculture
Daoliang Li	College of Information and Electrical Engineering, China Agricultural University
Ajit Maru	Global Forum on Agricultural Research
Walter H. Mayer	PROGIS Software GmbH
Nick Sigrimis	Department of Agricultural Engineering, Agricultural University of Athens, Greece
Yubin Lan	Texas A&M University, USA
Liping Chen	China National Engineering Research Center of Intelligent Equipment for Agriculture
Xinting Yang	China National Engineering Research Center for Information Technology in Agriculture
Xianxue Meng	Agricultural Information Institute of Chinese Academy of Agricultural Sciences (CAAS)

Chairs

Daoliang Li
Chunjiang Zhao

Conference Secretariat

Xia Li
Fangxu Zhu
Jieying Bi

Contents – Part I

The Growth Analysis in the Wheat Filling Process of the Two Hybrids
and Their Parents Based on Richards Equation 1
  Weiqing Wang

Comparative Study on Metaheuristic-Based Feature Selection for Cotton
Foreign Fibers Recognition 8
  Xuehua Zhao, Xueyan Liu, Daoliang Li, Huiling Chen, Shuangyin Liu,
  Xinbin Yang, Shaobin Zhan, and Wenyong Zhao

Beta Function Law to Model the Dynamics of Fruit's Growth Rate
in Tomato: Parameter Estimation and Evaluation. 19
  Qiaoxue Dong, Lili Yang, Mei Qu, Qinglan Shi, and Shangfeng Du

A Web-Based Cooperation and Retrieval Model of Character Images
for Ancient Chinese Character Research. 27
  Xuedong Tian, Songqiang Yang, Xuesha Jia, Fang Yang,
  and Chong Zhang

Research of Large-Scale and Complex Agricultural Data Classification
Algorithms Based on the Spatial Variability 45
  Hang Chen, Guifen Chen, Lixia Cai, and Yuqin Yang

An Algorithm for the Interactive Calculating of Wheat Plant Surface Point
Coordinates Based on Point Cloud Model 53
  Lei Xi, Guang Zheng, Yanna Ren, and Xinming Ma

Quick and Automatic Generation Method for the Evaluation Report
of the Small Hydropower Substitute Fuel Project 64
  Yingyi Chen, Jiani Xue, Huihui Yu, Jing Xu, Zhumi Zhen, Xingyue Tu,
  Zhijie Ma, Yun Zhao, and Yanzhong Liu

Aquaculture Access Control Model in Intelligent Monitoring and
Management System Based on Group/Role 72
  Qiyu Zhang, Yingyi Chen, Zhumi Zhen, Jing Xu, Ling Zhu,
  Liangliang Gao, and Yanzhong Liu

A Decision Model Forlive Pig Feeding Selection 82
  Xinxin Sun, Longqing Sun, and Yiyang Li

Study on Methods of Extracting New Construction Land Information Based
on SPOT6 94
  Lei Guo, Dongling Zhao, Rui Zhang, Meng Du, Zhixiao Li, Xiang Wang,
  and Yaru Wang

Application of Remote Sensing Technology in Agriculture of the USA   107
Yuechen Liu and Weijie Jiao

Aquatic Animal Disease Diagnosis System Based on Android   115
Min Sun and Daoliang Li

Relationship Between Vegetation Coverage and Rural Settlements
and Anti-desertification Strategies in Horqin Left Back Banner,
Inner Mongolia, China.   125
Jian Zhou, Fengrong Zhang, Yan Xu, Yang Gao, and Xiaoyu Zhao

China Topsoil Stripping Suitability Evaluation Based on Soil Properties   143
Yan Xu, Yuepeng Wang, Xing Liu, Dan Luo, and Hongman Liu

Design of Corn Farmland Monitoring System Based on ZigBee   153
Xiuli Si, Guifen Chen, and Weiwei Li

Simulation of Winter Wheat Yield with WOFOST in County Scale.   161
Shangjie Ma, Zhiyuan Pei, Yajuan He, Lianlin Wang, and Zhiping Ma

Predicting S&P500 Index Using Artificial Neural Network.   173
Shanghong Li, Jiayu Zhang, and Yan Qi

The Extraction Algorithm of Crop Rows Line Based on Machine Vision   190
Zhihua Diao, Beibei Wu, Yuquan Wei, and Yuanyuan Wu

Design of High-Frequency Based Measuring Sensor for Grain Moisture
Content   197
Qinglan Shi, Yunling Liu, and Wen Zhang

Propagation Characteristics of Radio Wave in Plastic Greenhouse   208
Jizhang Wang, Yuli Peng, and Pingping Li

A Review on Leaf Temperature Sensor: Measurement Methods
and Application   216
Lu Yu, Wenli Wang, Xin Zhang, and Wengang Zheng

Distinguish Effect of Cu, Zn and Cd on Wheat's Growth Using
Nondestructive and Rapid Minolta SPAD-502   231
Jinheng Zhang, Yonghong Sun, Lusheng Zeng, Hui Wang,
Qingzeng Guo, Fangli Sun, Jianmei Chen, and Chaoyu Song

Small-Scale Soil Database of Jilin Province, China   239
Xiuli Si, Guifen Chen, and Weiwei Li

Agricultural Environmental Information Collection Device Based
on Raspberry Pi   246
Baofeng Su, Linya Huang, Zhen Gao, and Jiao Guo

Design of Integrated Low-Power Irrigation Monitoring Terminal.   255
  Hongwu Tian, Mingfei Wang, and Jinlei Li

Non-destructive Detection of the pH Value of Cold Fresh Pork
Using Hyperspectral Imaging Technique   266
  Shanmei Liu, Ruifang Zhai, and Hui Peng

Cloud-Based Video Monitoring System Applied in Control of Diseases
and Pests in Orchards   275
  Xue Xia, Yun Qiu, Lin Hu, Jingchao Fan, Xiuming Guo,
  and Guomin Zhou

Research on Coordinated Development Between Animal Husbandry
and Ecological Environment Protection in Australia.   285
  Yiming Zhu and Shasha Li

Determination of Lead (Pb) Content in Vetiver Grass Roots by Raman
Spectroscopy   292
  Yande Liu, Yuxiang Zhang, Lixia Jiang, and Haiyang Wang

Dynamic Analysis of Urban Landscape Patterns of Vegetation Coverage
Based on Multi-temporal Landsat Dataset.   300
  Dong Liang, Ling Teng, Linsheng Huang, Xinhua Xie, Yan Zuo,
  and Jingling Zhao

The Application of the OPTICS Algorithm in the Maize Precise
Fertilization Decision-Making.   317
  Guowei Wang, Yu Chen, Jian Li, and Yunpeng Hao

The Methodology of Monitoring Crops with Remote Sensing
at the National Scale   325
  Quan Wu, Li Sun, Yajuan He, Fei Wang, Danqiong Wang, Weijie Jiao,
  Haijun Wang, and Xue Han

Is Time Series Smoothing Function Necessary for Crop
Mapping? — Evidence from Spectral Angle Mapper After Empirical
Analysis.   335
  Ailian Chen, Hu Zhao, and Zhiyuan Pei

Application Feasibility Analysis of Precision Agriculture in Equipment
for Controlled Traffic Farming System: A Review.   348
  Caiyun Lu, Zhijun Meng, Xiu Wang, Guangwei Wu, Nana Gao,
  and Jianjun Dong

Accurate Inference of Rice Biomass Based on Support Vector Machine.   356
  Lingfeng Duan, Wanneng Yang, Guoxing Chen, Lizhong Xiong,
  and Chenglong Huang

Brazil Soybean Area Estimation Based on Average Samples Change Rate
of Two Years and Official Statistics of a Year Before   366
   Kejian Shen, Weifang Li, Zhiyuan Pei, Fei Wang, Xiaoqian Zhang,
   Guannan Sun, Jiong You, Quan Wu, and Yuechen Liu

Meta-Synthetic Methodology: A New Way to Study Agricultural Rumor
Intervention   375
   Ruya Tian, Lei Wu, Yijun Liu, and Xuefu Zhang

Rapid Identification of Rice Varieties by Grain Shape and Yield-Related
Features Combined with Multi-class SVM   390
   Chenglong Huang, Lingbo Liu, Wanneng Yang, Lizhong Xiong,
   and Lingfeng Duan

The Acquisition of Kiwifruit Feature Point Coordinates Based on the
Spatial Coordinates of Image   399
   Bin Wang, Zixiao Chen, Jianmin Gao, Longsheng Fu, Baofeng Su,
   and Yongjie Cui

The Soil Nutrient Spatial Interpolation Algorithm Based on KNN and IDW . . .   412
   Xin Xu, Hua Yu, Guang Zheng, Hao Zhang, and Lei Xi

Segmentation of Cotton Leaves Based on Improved Watershed Algorithm . . .   425
   Chong Niu, Han Li, Yuguang Niu, Zengchan Zhou, Yunlong Bu,
   and Wengang Zheng

Research on Knowledge Base Construction of Agricultural Ontology Based
on HNC Theory   437
   Hao Xinning, Xie Nengfu, Sunwei, Zhong Xiaochun, and Zhang Xuefu

Method and System of Maize Hybridized Combination Based on Inbred
SSR and Field Test   446
   Zhe Liu, Zhenhong Zhang, Shaolong Fu, Xiaodong Zhang, Dehai Zhu,
   and Shaoming Li

Biomass-Based Leaf Curvilinear Model for Rapeseed (*Brassica napus* L.) . . .   459
   Wenyu Zhang, Weixin Zhang, Daokuo Ge, Hongxin Cao, Yan Liu,
   Kunya Fu, Chunhuan Feng, Weitao Chen, and Chuwei Song

Exploring the Effect Rules of Paddy Drying on a Deep Fixed-Bed   473
   Danyang Wang, Chenghua Li, Benhua Zhang, and Ling Tong

Feature Extraction and Recognition Based on Machine Vision Application
in Lotus Picking Robot   485
   Shuping Tang, Dean Zhao, Weikuan Jia, Yu Chen, Wei Ji,
   and Chengzhi Ruan

Rapeseed (*Brassica napus* L.) Primary Ramification Morphological
Structural Model Based on Biomass   502
    Weixin Zhang, Hongxin Cao, Wenyu Zhang, Yan Liu, Daokuo Ge,
    Chunhuan Feng, Weitao Chen, and Chuwei Song

Effective Wavelengths Selection of Hyperspectral Images of Plastic Films
in Cotton   519
    Hang Zhang, Xi Qiao, Zhenbo Li, and Daoliang Li

Research and Experiment on Precision Seeding Control System of Maize
Planter.   528
    Nana Gao, Weiqiang Fu, Zhijun Meng, Xueli Wei, You Li, and Yue Cong

Study on Time and Space Prediction Model About Rice Yield in Hei
Longjiang Province   536
    Guowei Wang, Hongyan Hu, Hao Zhang, and Yu Chen

Research on the Digital Machine for Killing the Larva of Longicorn Beetle
with Microwave Based on the Arduino   546
    MingXi Shao, XiuMei Zhang, BingGuo Liu, ChangYong Shao,
    AiSheng Ma, ShouSheng Zhang, Sheng Liu, YuJie Liu, LiJing Zhao,
    and Lin Dong

Risk Assessment of Water Resources Shortage in Sanjiang Plain   556
    Qiuxiang Jiang, Yongqi Cao, Ke Zhao, and Zhimei Zhou

Analysis of Soil Fertility Based on FUMF Algorithm   564
    Hang Chen, Guifen Chen, Yating Hu, Liying Cao, Lixia Cai,
    and Sisi Yang

Modeling and Optimization of Agronomic Factors Influencing Yield
and Profit of a Single-Cropping Rice Cultivar.   574
    Weiming Liu and Zuda Bao

The Milk Somatic Cell Image Segmentation Method Based on Dimension
Reduction and Fusion   580
    Jie Bai, Heru Xue, and Yanqing Zhou

Quantitative Detection of Pesticides Based on SERS and Gold Colloid   587
    Yande Liu, Yuxiang Zhang, Haiyang Wang, and Bingbing He

Research on Freshwater Fish Information Service Mode for Modern
Production and Circulation in the Internet + Era   597
    Xinping Fang

Interactive Pruning Simulation of Apple Tree   604
    Lili Yang, JiaFeng Chen, Jing Hua, MengZhen Kang,
    and QiaoXue Dong

Research on Key Technology of Grid Cell Division Method in Rural
Community 612
    Chunlei Shi and Bo Peng

A Review on Optical Measurement Method of Chemical Oxygen Demand
in Water Bodies 619
    Fei Liu, Peichao Zheng, Baichuan Huang, Xiande Zhao, Leizi Jiao,
    and Daming Dong

Analysis of Changes in Agronomic Parameters and Disease Index
of Rapeseed Leaf Leukoplakia Based on Spectra. 637
    Kunya Fu, Hongxin Cao, Wenyu Zhang, Weixin Zhang, Daokuo Ge,
    Yan Liu, Chunhuan Feng, and Weitao Chen

Author Index 655

Contents – Part II

Effects of Waterlogging and Shading at Jointing and Grain-Filling Stages
on Yield Components of Winter Wheat   1
    Yang Liu, Chunlin Shi, Shouli Xuan, Xiufang Wei, Yongle Shi,
    and Zongqiang Luo

The Measurement of Fish Size by Machine Vision - A Review   15
    Mingming Hao, Helong Yu, and Daoliang Li

Study on Growth Regularity of Bacillus Cereus Based on FTIR   33
    Yang Liu, Ruokui Chang, Yong Wei, Yuanhong Wang, and Zizhu Zhao

Soybean Extraction of Brazil Typical Regions Based on Landsat8 Images . . .   41
    Kejian Shen, Xue Han, Haijun Wang, and Weijie Jiao

Study on Landscape Sensitivity and Diversity Analysis in Yucheng City   48
    Xuexia Yuan, Yujian Yang, and Yong Zhang

Application and Implementation of Private Cloud in Agriculture Sensory
Data Platform   60
    Shuwen Jiang, Tian'en Chen, and Jing Dong

Analysis of Differences in Wheat Infected with Powdery Mildew Based
on Fluorescence Imaging System   68
    Shizhou Du, Qinhong Liao, Chengfu Cao, Yuqiang Qiao, Wei Li,
    Xiangqian Zhang, Huan Chen, and Zhu Zhao

Research on Video Image Recognition Technology of Maize Disease Based
on the Fusion of Genetic Algorithm and Simulink Platform   76
    Liying Cao, Ying Meng, Jian Lu, and Guifen Chen

The Design and Implementation of Online Identification of CAPTCHA
Based on the Knowledge Base   92
    Yu'e Song, Chengguo Wang, Ling Zhu, Xiaofeng Chen, and Qiyu Zhang

Research and Application of Monitoring and Simulating System of Soil
Moisture Based on Three-Dimensional GIS   100
    Guifen Chen, Jian Lu, Ying Meng, Liying Cao, and Li Ma

Colorimetric Detection of Mercury in Aqueous Media Based
on Reaction with Dithizone   111
    Zihan Wu, Ming Sun, and Ling Zou

Study on the Prediction Model Based on a Portable Soil TN Detector   117
Xiaofei An, Guangwei Wu, Jianjun Dong, Jianhua Guo,
and Zhijun Meng

A Research on the Task Expression in Pomology Information Retrieval   127
Dingfeng Wu, Jian Wang, Guomin Zhou, and Hua Zhao

Prediction of the Natural Environmental High Temperature Influences
on Mid-Season Rice Seed Setting Rate in the Middle-Lower Yangtze
River Valley   133
Shouli Xuan, Chunlin Shi, Yang Liu, Yanhua Zhao, Wenyu Zhang,
Hongxin Cao, and Changying Xue

Study on the Mutton Freshness Using Multivariate Analysis Based
on Texture Characteristics   143
Xiaojing Tian, Jun Wang, Jutian Yang, Shien Chen, and Zhongren Ma

Research and Application on Protected Vegetables Early Warning
and Control of Mobile Client System   155
Guogang Zhao, Haiye Yu, Lianjun Yu, Guowei Wang, Yuanyuan Sui,
Lei Zhang, Linlin Wang, and Jiao Yang

The Study of Winter Wheat Biomass Estimation Model Based
on Hyperspectral Remote Sensing   163
Xiaowei Teng, Yansheng Dong, and Lumin Meng

Design and Implementation of TD-LTE-Based Real-Time Monitoring
System for Greenhouse Environment Temperature   170
Xin Zhao, Yang Jiao, Lianjun Yu, and Chuanhong Zhang

Research and Design of LVS Cluster Technology in Agricultural
Environment Information Acquisition System   178
Guogang Zhao, Haiye Yu, Lianjun Yu, Guowei Wang, Yuanyuan Sui,
and Lei Zhang

Information Acquisition for Farmland Soil Carbon Sink Impact Factors
Based on ZigBee Wireless Network   185
Bingbing Wang, Dekun Zhai, Lijuan Sun, Dandan Yang, Zhihong Liu,
and Qiulan Wu

Penetration Depth of Near-Infrared Light in Small, Thin-Skin Watermelon . . .   194
Man Qian, Qingyan Wang, Liping Chen, Wenqian Huang,
Shuxiang Fan, and Baohua Zhang

Design and Implementation of an Automatic Grading System of Diced
Potatoes Based on Machine Vision   202
Chaopeng Wang, Wenqian Huang, Baohua Zhang, Jingjing Yang,
Man Qian, Shuxiang Fan, and Liping Chen

A Soil Water Simulation Model for Wheat Field with Temporary Ditches . . .   217
  Chunlin Shi, Yang Liu, Shouli Xuan, and Zhiqing Jin

The Synchronized Updating Technology Research of Spatio-temporal
Supervision Data Model About Organizing of Construction Landuse
Data in Distributed Environment.   225
  Xiaolan Li, Bingbo Gao, Yuchun Pan, Yanbing Zhou, and Xingyao Hao

Comparison of Four Types of Raman Spectroscopy for Noninvasive
Determination of Carotenoids in Agricultural Products.   237
  Chen Liu, Qingyan Wang, Wenqian Huang, Liping Chen,
  Baohua Zhang, and Shuxiang Fan

The Molecular Detection of *Corynespora Cassiicola* on Cucumber
by PCR Assay Using DNAman Software and NCBI   248
  Weiqing Wang

Simulation of Winter Wheat Phenology in Beijing Area
with DSSAT-CERES Model.   259
  Haikuan Feng, Zhenhai Li, Peng He, Xiuliang Jin, Guijun Yang,
  Haiyang Yu, and Fuqin Yang

Design of Monitoring System for Aquaculture Environment   269
  Hua Liu, Liangbing Sa, Yong Wei, Wuji Huang, and Binjie Shi

Research on the Agricultural Skills Training Based on the Motion-Sensing
Technology of the Leap Motion   277
  Peng-fei Zhao, Tian-en Chen, Wei Wang, and Fang-yi Chen

Study of Spatio-temporal Variation of Soil Nutrients in Paddy Rice
Planting Farm.   287
  Cong Wang, Tianen Chen, Jing Dong, Shuwen Jiang, and Chao Li

Path Planning Methods for Auto-Guided Rice-Transplanters   300
  Fangming Zhang, Changhuai Lv, Jie Yang, Caiyu Zhang, Guisen Li,
  and Licheng Fu

Research of the Early Warning Model of Grape Disease and Insect
Based on Rough Neural Network   310
  Dengwei Wang, Tian'en Chen, Chi Zhang, Li Gao, and Li Jiang

Evaluation Model of Tea Industry Information Service Quality.   320
  Xiaohui Shi and Tian'en Chen

Recognition and Localization Method of Overlapping Apples
for Apple Harvesting Robot   330
  Tian Shen, Dean Zhao, Weikuan Jia, and Yu Chen

Retrieval Methods of Natural Language Based on Automatic Indexing   346
  Dan Wang, Xiaorong Yang, Jian Ma, and Liping Zhang

Improving Agricultural Information and Knowledge Transfer in Cambodia
- Adopting Chinese Experience in Using Mobile Internet Technologies   357
  Yanan Hu, Yun Zhang, and Yanqing Duan

Principal Component Analysis Method-Based Research on Agricultural
Science and Technology Website Evaluation   369
  Jian Ma

The Countermeasures of Carrying on Web of the Research Institutions
in the Era of Big Data — Consider the Web of Chinese Academy
of Agricultural Sciences.   382
  Liping Zhang

The Knowledge Structure and Core Journals Analysis of Crop Science
Based on Mapping Knowledge Domains   392
  *Minjuan Liu, Lu Chen, Xue Yuan, Ting Wang, Yun Yan,
  and Yuefei Wang*

Application of Spatial Reasoning in Predicting Rainfall Situation
for Two Disjoint Areas   404
  Jian Li, Yanbo Huang, Rujing Yao, and Yuanyuan Zhang

Simplifying Calculation of Graph Similarity Through Matrices   417
  Xu Wang, Jihong Ouyang, and Guifen Chen

A Systematic Method for Quantitative Diameter Analysis of Sprayed
Pesticide Droplets   429
  Wei Ma, Xiu Wang, Lijun Qi, and Yanbo Huang

Development of Variable Rate System for Disinfection Based
on Injection Technique   437
  Wei Ma, Xiu Wang, Lijun Qi, and Wei Zou

Establishment and Optimization of Model for Detecting Epidermal
Thickness in Newhall Navel Orange   445
  Yande Liu, Yifan Li, and Zhiyuan Gong

Design and Implementation of Greenhouse Remote Monitoring System
Based on 4G and Virtual Network   455
  *Guogang Zhao, Yu Lianjun, Haiye Yu, Guowei Wang, Yuanyuan Sui,
  and Lei Zhang*

The Study of Farmers' Information Perceived Risk in China   463
  Jingjing Zhang

Dynamic Changes of Transverse Diameter of Cucumber Fruit in Solar
Greenhouse Based on No Damage Monitoring   469
   Ruijiang Wei, Xin Wang, and Huiqin Zhu

Study on Laos-China Cross-Border Regional Economic Cooperation Based
on Symbiosis Theory: A Case of Construction of Laos Savan Water
Economic Zone.   479
   Sisavath Thiravong, Jingrong Xu, and Qin Jing

Study on Mode of Laos-China Cross-Border Collaboration Strategy
Facing Symbiosis Relation   487
   Sisavath Thiravong, Jingrong Xu, and Qin Jing

Research of Fractal Compression Algorithm Taking Details in
Consideration in Agriculture Plant Disease and Insect Pests Image   496
   Qiao Deng, Chunhong Liu, and Liting Fu

Commentary on Application of Data Mining in Fruit Quality Evaluation   505
   Jinjian Hou, Dong Wang, Wenshen Jia, and Ligang Pan

Study on Identification of Bacillus cereus in Milk Based
on Two-Dimensional Correlation Infrared Spectroscopy   514
   Zizhu Zhao, Ruokui Chang, Yong Wei, Yuanhong Wang, and Haiyun Wu

Stimulating Effect of Low-Temperature Plasma (LTP) on the Germination
Rate and Vigor of Alfalfa Seed (Medicago Sativa L.)   522
   Xin Tang, Fengchen Liang, Lijing Zhao, Lili Zhang, Jing Shu,
   Huamei Zheng, Xu Qin, Changyong Shao, Jinkui Feng, and Keshuang Du

Evaluation of Timber and Carbon Sequestration Income of Cunninghamia
Lanceolata Timber Forest and Management Decision Support.   530
   Yan Qi, Baoguo Wu, and Shanghong Li

Researches on the Variations of Greenhouse Gas Exchange Flux at Water
Surface Nearby the Small Hydropower Station of Qingshui River, Guizhou . . .   539
   Lei Han, Xuyin Yuan, Jizhou Li, Yun Zhao, Zhijie Ma, and Jing Qin

The Application of Internet of Things in Pig Breeding.   548
   Minghua Shang, Gang Dong, Yuanjie Mu, Fujun Wang,
   and Huaijun Ruan

Research and Exploration of Rural and Agricultural Information
Service – Taking Shandong Province as a Case   557
   Jia Zhao, Jianfei Wang, and Wenjie Feng

Research on Agricultural Development Based on "Internet +".   563
   Wenjie Feng, Lei Wang, Jia Zhao, and Huaijun Ruan

Research and Design of Shandong Province Animal Epidemic Prevention
System Based on GIS   570
    Jiabo Sun, Wenjie Feng, Xiaoyan Zhang, Luyan Niu, and Yanzhong Liu

Research and Design of Wireless Sensor Middleware Based on STM32.   579
    Jiye Zheng, Fengyun Wang, and Lei Wang

Technical Efficiency and Traceability Information Transfer: Evidence
from Grape Producers of Four Provinces in China.   586
    Lei Deng, Ruimei Wang, Weisong Mu, and Jingjie Zhao

Author Index   595

The Growth Analysis in the Wheat Filling Process of the Two Hybrids and Their Parents Based on Richards Equation

Weiqing Wang[✉]

Beijing Vocational College of Agriculture, Beijing, China
weiqingfine@163.com

Abstract. Using Origin software, the grain filling process of hybrid wheats 1–12 and 8-1-54 and their parents was fitted by Richards equation $W = A/(1 + Be^{-kt})^{1/N}$ on computer in order to study the characteristics of grain filling in hybrids wheat. The active grain growth period of hybrid wheats 1–12 and 8-1-54 were longer than that of their parents respectively, and their final grain weight was higher than that of their parents respectively. These results suggest that hybrids 1–12 and 8-1-54 have the higher capacity of graining filling than their parents and that may be associated with longer RGFP and higher maximum grain-filling rate. It was also found that Richards equation was more suitable for fitting the grain filling process of wheat than Logistic equation.

Keywords: Hybrid wheat · Richards equation · Parent plants · Grain-filling

1 Introduction

The grain filling period is one of the most important stages in the life of wheat. During this period, wheat grains are filled and the final yield forms [1, 2]. Researches on this stage have an important significance in revealing the characteristics of grain quality and growth. A large number of reports have been existed on the filling process and filling characteristics of wheat grain [2–4]. Zhang [5] and Ren et al. [6] analyzed the grain filling traits using the third degree polynomial. Gu et al. [7] analyzed the barly grain filling traits using Logistic equation. However the third degree polynomial and Logistic equation are lack of enough plasticity, and difficult to give biological meanings of model parameters. Zhu et al. [8] and Yang et al. [9] once reported that it was much better to use Richards equation to simulate the rice grain filling process than Logistic equation [10]. Application of this equation to wheat grain filling process has not been reported till now.

Using the most available equation for fitting growth analysis (Richards equation) and a set of subsidiary parameters derived from Richards equation [11–13], we propose to simulate and analyze the grain filling data of hybrid wheat 1–12 and its parents (Jing411 and Xiaoyan54), 8–1–54 and its parents (8602 and Xiaoyan54), to compare their filling traits.

© IFIP International Federation for Information Processing 2016
Published by Springer International Publishing AG 2016. All Rights Reserved
D. Li and Z. Li (Eds.): CCTA 2015, Part I, IFIP AICT 478, pp. 1–7, 2016.
DOI: 10.1007/978-3-319-48357-3_1

2 Materials and Methods

2.1 Plant Material

Winter wheat was grown at a farm in Beijing in 2012–2013. The sowing time of wheat seeds was 2 October, 2012. During this time to ensure adequate supply of nutrients and water, so as to avoid the occurrence of stress. Measurements began in May 10th at the stages of anthesis and end in June 13th after flowering. During this period, the weather in Beijing is typical of the spring weather, the average temperature of 19–23 °C, the average daily light intensity (PPFD) of about 1400 μ mol m^{-2} s^{-1} at noon. There are varieties of Jing411, Xiaoyan 54, 1-128602, 8-1-54. 1–12 is the hybrid progeny of Jing411 and Xiaoyan 54, So Jing411, Xiaoyan 54 and 1–12 will be known as the Jing411 series. 8-1-54 is the hybrid progeny of 8602 and Xiaoyan 54, so 8602, Xiaoyan 54 and 8-1-54 will be called the 8602 series.

2.2 Sampling and Harvesting

From anthesis to maturity stage, twenty labeled wheat shikes were selected to sampled every 3 days. The samples were divided into two groups (10 spkies each). Wheat kernels taken out were dried at 75 degrees for 72 h to the constant weight and then weighed.

Using the Origin software, the grain tilling data is fitted by the Richards equation, with the grain weight W as the dependent variable, days after anthesis(t) as the independent variable.

$$W = A/\left(1 + Be^{-kt}\right)^{1/N} \tag{1}$$

W– the grain weight (mg),
A– the final grain weight (mg),
t– the time after anthesis (day),
B, K, and N – the coefficients determined by the regression.

T1 was 5 % of W and t2 was 95 % of W. So the active grain-filling period was defined as the period of t1 to t2. The average rate was calculated by the rate from t1 to t2.

And grain-filling rate (G) was calculated as the derivative of Eq. (1):

$$G = AKBe^{-kt}/N \left(1 + Be^{-kt}\right)^{(N+1)/N} \tag{2}$$

Where G is the growth amount per unit time (g/1000 grains d^{-1})

For further analysis, the following subsidiary parameters can be derived from Eq. (2) to describe the filling traits.

The date of the maximum grain filling rate: $Tmax.G = (lnB-lnN)/k$

The growth quantity in the maximum grain filling rate: $Wmax.G = A(N + 1) - 1/N$

The maximum grain filling rate: $Gmax = (kWmax.G/N)[1 - (Wmax.G/A)N]$

Filling starting potential: R0 = k/N

The active grain-filling phase (about 90 % of total growth quantity): D = 2 (N + 2)/k

The ratio of the growth quantity in the maximum grain-filling rate and the final grain weight: I = Wmax.G/A

3 Results

3.1 Grain Filling Process and Rate

Figures 1 and 2 show the grouting process and grain filling rate of the 5 wheat varieties. In all varieties, the delay and exponential phase can be clearly seen. 31 days of measuring time is sufficient to reveal the rapid filling process. The final grain weight of 1–12 was significantly higher than that of other cultivars. 1–12 into the rapid filling period and dry matter accumulation period was faster and earlier than their parents. The final grain weight of 8-1-54 was significantly higher than that of the parents. Figure 2 shows the curve of the grain filling rate of the five varieties of wheat fitted according to the Richards (1959) equation [13]. From these data, the grain filling rate and the time to reach the maximum filling rate (Tmax) are obtained. In all varieties, the rapid grain filling began at the same days (5 days) after anthesis and ended at the same days (28 days) after anthesis except 1–12 (25 days). 1–12 had the highest maximum grain-filling rate peaking at around 3.0 g 1000-seed day^{-1} that was comparable to its parents jing411 and xiaoyan54, and the maximum grain filling period (MRGFP) of 1–12 occurred earliest (13 days after anthesis) in all varieties, but the grain filling period was shortened (5–25 days). In all varieties, the lowest overall rate of grain filling is 8602 (2.0 g 1000-seed day^{-1}) and this variety showed a longer grain-filling period than its parents.

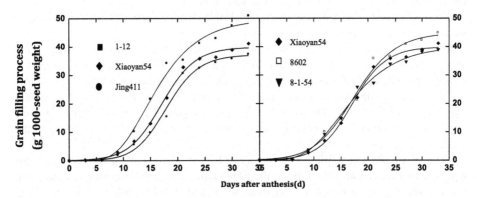

Fig. 1. The grain filling process fitted by Richards equation. Each point represents the mean ± SE of the 1000-seed dry weight, n = 3.

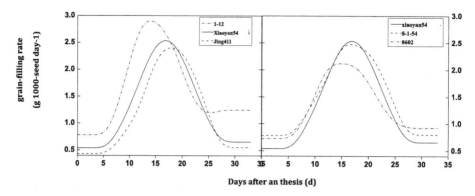

Fig. 2. The grain filling rate of fitting Richards euqation.

These results suggest that hybrids 1–12 and 8-1-54 have the higher capacity of graining filling than their parents and that may be associated with longer RGFP and higher maximum grain-filling rate.

3.2   The Grain Filling Model in Different Varieties and Different Grain Positions

From results of the determination coefficient (R^2) (Table 1), the kernel filling processes in 5 cultivars are better fitting by Richards equation. Hybrid wheat 1–12 and 8-1-54 have a high grain weight, with the final weight about 46.7 g and 44.7 g per 1000 grains, apparently higher than their parents.

From the initial growth potential Ro of the five types of varieties, the lowest initial growth potential (R0) is Xiaoyan54 0.184, followed by jing411 to 0.290, 8-1-54 to 0.322, the higher is 8602 0.839, the highest is 1–12 0.932. So the starting potential of the varieties are: 1–12 > jing411 and xiaoyan54, but 8-1-54 > Xiaoyan54, and <8602. The results showed that the growth potential of 1–12 is greater than its parents. Seed starting potential reflects its growth potential in ovary, Ro value is larger, the endosperm cell division cycle is shorter, split faster and grain filling start earlier; Conversely, Ro value is smaller, the grain filling starts later; If R0 Value is too small, the endosperm cells can not develop normally, grain filling process tends to stop.

The time reaching the maximum grain-filling rate (T_{max}) of the 1–12 is 7.498d, and that of 8-1-54 is 12. 787d. The longest of T_{max} is Xiaoyan54 13.388d. There is obvious regularity: the larger of the initial growth potential Ro of the variety, the shorter of the Tmax.

The maximum grain filling rate (G_{max}) of hybrid wheat 1–12 and 8-1-54 are also higher than their parents. The ratio of the growth quantity in the maximum grain-filling rate and the final grain weight (I): Xiaoyan54 reached 79.94 %, followed by jing411 is 62.63 %, the second is 8-1-54 47.74 %, 1–12 is 40.20 % and 8602 is 39.68 %. To sum up, the maximum grain filling time of 1–12 appeared earlier, the grain filling in active growth is longer. The grain has just completed the final volume of 40 % in maximum graining-filling rate, which showed the later grain filling grain weight gain still can continue to increase if the conditions permit. The maximum filling rate of Xiaoyan54 appeared later, the active growth period is shorter, and when it reached the maximum rate of grain growth, it has been completed around the final growth of 80 %, indicating grain growth is significantly restricted in the later period; To other three varieties, growth characteristics was between the varieties of Jing411 and xiaoyan54. These results suggests that it has heterobeltiosis.

3.3   Division of Grain Filling Type

Richards growth curves are a group of curves determined by N value. From the filling rate curve (Fig. 1 and Table 1) can be seen when $O < N < 1$, the initial growth potential Ro is higher and rate curve is left, the grain is belong to the strong grains. When $N > 1$, rate curve is right and Ro is lower, the grain is belongs to the inferior grains. When $N = 1$, the Richards equation is equal to the Logistic equation, So Richards curve can be used to simulate grain filling types.

Table 1. Parameters of different grain positions in grain filling period

Varieties	A	B	K	N	R_0	T_{max}	G_{max}	$W_{max.G}$	I	D	R^2
1–12	46.7	1.0000	0.215	0.201	0.932	7.498	3.987	18.79	0.4020	20.47	0.996
Xiaoyan54	36.7	2.852×10^{17}	2.085	11.152	0.184	13.388	3.301	29.37	0.7994	12.62	0.995
Jing411	34.7	31396.8	0.8468	2.91886	0.290	10.468	3.479	21.75	0.6263	11.62	0.987
8-1-54	44.8	19.321	0.251	0.779	0.322	12.787	3.502	21.36	0.4774	22.14	0.994
8602	40.6	1.007	0.2	0.168	0.839	9.588	2.784	16.09	0.3968	21.68	0.997

B, K, N: Parameters of the equation; A: The maximum grain weight; R_0: Initial growth potential; Tmax: the time reaching the maximum grain-filling rate; Gmax: The maximum grain-filling rate; $W_{max.G}$: the growth quantity in the maximum grain filling rate; D: The active grain-filling phase (about 90 % of total growth quantity); I: the ratio of the growth quantity in the maximum grain-filling rate and the final grain weight

The N values in the hybrid wheat and 8602 are <1, and the growth rate curves go left. These mean that the kernels grow quickly in the early filling stage, and slow down in the middle and late filling stages. The N value of the male parent (8602) of 8-1-54 was smaller than that in the hybrid, while the growth rate skewed left much more, resulting in a slower growth in the middle and late filling stages (Table 1, Fig. 2).

The N values in jing411 and xiaoyan54 are both larger than 1, and the growth rate curves skewed right, resulting in longer duration to Gmax, Tmax values were 10.468 d and 13.388 d respectively, suggesting slower growth in the early filling stage and fast growth in the middle filling stage (Table 1, Fig. 2).

4   Discussion

Logistic equation and Richards equation can simulate the grain filling process of wheat [14]. There are 3 fitting parameters in logistic equation, but there has an additional parameter N in Richards equation [13, 15]. So the Richards equation has a larger degree of freedom and more flexibility, and in the theory, it should be more fitting than

Logistic equation. The present results showed that Richards equation was fitting well with the filling process for 5 cultivars, resulting in larger determined coefficient (R^2) equation. Furthermore, more information can be achieved according to the value of N. As a result Richards equation is better than Logistic equation in describing the kernel weight increment process, and it could reveal fully the differences in the filling process between varieties.

The results proved in the middle and late stages of graining filling, photosynthetic product and storage capacity of the hybrids of 1–12 and 8-1-54 are adequate. Because their source and sink relationship is coordinated with each other, so they can form a large and full grain; On the contrary, the storage capacity of jing411 and other parents cultivars is sufficient, but supply of photosynthetic products is not insufficient, so grains are big and not full, as the "source" limited type variety. So how to increase the filling duration and fully enhance its production potential to get a higher yield, are the main topic to be done next.

The superior grains varieties were the "storage capacity "limited, so to increase 1000 grain weight, we should pay attention to the promotion of endosperm cell division and expansion. But to the "source" limited variety, we should to improve the nutrition supply to improve grain weight.

Although the results of using polynomial regression and Logistic equation to simulate the process of grain filling are not as good as results of Richards equation, they can be liberalized, and have more mature test equation. They are widely used in agricultural scientific research because of the use of Excel software in computer. Regression equation of Richards equation belongs to nonlinear, cannot use the Excel software to finish, and need to determine the initial values of parameters after using SAS or Matlab software program by Levenberg - Marquardt algorithm and need to establish Richards equation after to select of stable and the minimum Q value. The key point of the whole process is the choice of initial parameter values; but the initial parameters can be determined by crop characteristics and professional knowledge or reference parameters, which is main reason that the Richards equation is not used as the other two equations widely.

References

1. Sali, A., Shukri, F., Udvik, R.: Acta Agric. Slov. **95**(1), 35–41 (2010)
2. Hassan, I.S., Ahmed, M.G.: Field Crops Res. **7**, 61–71 (1983)
3. Yao, S., Kang, Y.H., Lü, G.H., Liu, M.J., Yang, W.P., Li, D.F.: Trans. Chin. Soc. Agric. Eng. **27**(7), 13–17 (2011)
4. Meng, Z.J., Sun, J.S., Duan, A.W., Liu, Z.G., Wang, H.Z.: Trans. Chin. Soc. Agric. Eng. **26**(1), 18–23 (2010)
5. Zhang, X.L.: Acta Agron. Sin. (8) 87–93 (1982)
6. Ren, Z.L., Li, Y.Q.: Sci. Agric. Sin. 12–20 (1981)
7. Gu, Z.F., Feng, C.N., Zhou, M.X., Ji, C.: Acta Agron. Sin. (9) 181–188 (1983)
8. Zhu, Q.S., Cao, X.Z., Luo, Y.Q.: Acta Agron. Sin. **14**, 182–193 (1988)
9. Yang, J., Zhang, J., Wang, Z., Zhu, Q., Wang, W.: Plant Physiol. **127**, 315–323 (2001)
10. Kuraz, M., Mayer, P., Pech, P.: J. Comput. Appl. Math. (7), 02–11 (2014)

11. Confalonieri, R., Rosenumund, A.S., Baruth, B.: Agron. Sustain. Dev. (3), 463–474 (2009)
12. Zhou, J., Liu, F.J., He, J.H.: Comput. Geotech. (54) 69–71 (2013)
13. Richards, F.J.: J. Exp. Bot. **10**, 290–300 (1959)
14. Birch, C.P.D.: Ann. Bot. **83**(6), 713–724 (1999). Oxford University Press Then Academic Press, London
15. Porter, T., Kebreab, E., Kuhi, H.D., Lopez, S., Strathe, A.B., France, J.: Poult. Sci. **89**(2), 371–378 (2010)

Comparative Study on Metaheuristic-Based Feature Selection for Cotton Foreign Fibers Recognition

Xuehua Zhao[1], Xueyan Liu[1,2], Daoliang Li[3(✉)], Huiling Chen[4],
Shuangyin Liu[5], Xinbin Yang[1], Shaobin Zhan[1], and Wenyong Zhao[1]

[1] School of Digital Media, Shenzhen Institute of Information Technology,
Shenzhen 518172, China
lcrlc@sina.com, dyyzlxy@163.com
[2] Key Laboratory of Symbolic Computation and Knowledge
Engineer (Jilin University), Ministry of Education,
Changchun 130012, Jilin, China
[3] College of Information and Electrical Engineering,
China Agricultural University, Beijing 100083, China
dliangl@cau.edu.cn
[4] College of Physics and Electronic Information,
Wenzhou University, Wenzhou 325035, China
chenhuiling.jlu@gmail.com
[5] College of Information, Guangdong Ocean University,
Zhanjiang 524025, Guangdong, China
hdlsyxlq@126.com

Abstract. The excellent feature set or feature combination of cotton foreign fibers is great significant to improve the performance of machine-vision-based recognition system of cotton foreign fibers. To find the excellent feature sets of foreign fibers, in this paper presents three metaheuristic-based feature selection approaches for cotton foreign fibers recognition, which are particle swarm optimization, ant colony optimization and genetic algorithm, respectively. The k-nearest neighbor classifier and support vector machine classifier with k-fold cross validation are used to evaluate the quality of feature subset and identify the cotton foreign fibers. The results show that the metaheuristic-based feature selection methods can efficiently find the optimal feature sets consisting of a few features. It is highly significant to improve the performance of recognition system for cotton foreign fibers.

Keywords: Metaheuristic · Feature selection · Foreign fibers · Recognition system

1 Introduction

The cotton foreign fibers, such as ropes, wrappers, plastic films and so on, are closely related to the quality of the final cotton textile products [1]. In the recent years, the machine-vision-based recognition systems have been widely used to assess the quality of cottons [2, 3], in which classification accuracy is an key measure to validate the

© IFIP International Federation for Information Processing 2016
Published by Springer International Publishing AG 2016. All Rights Reserved
D. Li and Z. Li (Eds.): CCTA 2015, Part I, IFIP AICT 478, pp. 8–18, 2016.
DOI: 10.1007/978-3-319-48357-3_2

performance of recognition systems. To improve the classification accuracy, finding the optimal feature sets with high accuracy is an efficient way due to because it can improve the accuracy and speed of recognition systems.

Feature selection (FS) is a main approach to find the optimal feature sets by reduce the irrelevant or redundant features. Currently, FS has been used to the area of machine learning and data mining [4]. Since to find the optimum feature sets is a NP problem, the researchers begin to turn to find the near optimal feature set and have proposed many algorithms [5, 6].

Currently, metaheuristic algorithms have attracted so much attention, the representive algorithms are particle swarm optimization (PSO for short), ant colony optimization (ACO for short) and genetic algorithm (GA for short) [5, 7, 8]. For metaheuristic algorithms, they are firstly given an evaluating measure of the quality of feature sets, and iteratively improve a specific candidate set. Finally, the excellent feature sets are obtained. The metaheuristic algorithms makes few assumptions on the optimal feature sets and find the optimal feature sets in very large search spaces. This is very suitable for the FS problem.

In this paper, three metaheuristic algorithms for FS are presented to find the optimal feature combination of cotton foreign fibers, which are GA for FS (GAFS for short), ACO for FS (ACOFS for short) and PSO for FS (PSOFS for short). Two classifiers, which are the k-nearest neighbor (KNN for short) [9] and support vector machine (SVM for short) [10], are used to evaluate the quality of subsets and to identify cotton foreign fibers. The aim of our works is applying these algorithms to the data sets of cotton foreign fibers to discover the new and challenging results. The comparison analysis of these algorithms illustrate the excellent search ability of the proposed metaheuristic algorithms and the feature sets obtained by them can efficiently improve the performance of recognition systems of cotton foreign fibers.

The remainder is organized as follows. The applications context and the proposed FS methods are presented in Sect. 2. The results and discussion are described in Sect. 3. Section 4 describes the conclusion.

2 Materials and Methods

2.1 Application Background

The cotton foreign fibers usually fall into six groups, that are feather, hemp rope, plastic film, cloth, hair, polypropylene and, respectively. The cotton foreign fibers induce the quality of the cotton textile products [1]. The grade evaluation using machine-vision-based recognition systems is mainly an approach to solve this problem [2]. These systems in general have three key steps, which are image segmentation, feature extraction and classification of foreign fiber, respectively.

Considering real-time problem, the feature set, which includes a few number of features and has high accuracy, is important due to reduction of the detection time and improvement of classification accuracy. As a result, FS is important to the online recognition systems of cotton foreign fibers. In our work, we have applied three metaheuristic algorithms to this area for obtaining the optimal feature sets of cotton foreign fibers.

2.2 Data Preparation

Firstly, we obtain the foreign fiber images by our test platform, and 1200 representative images including foreign fibers are selected to extract the dataset. The width of the obtained images is 4000 pixels and their height is 500 pixels. Several examples are shown in Fig. 1. These images are divided into six groups in terms of categories of foreign fibers, and every group contains 200 images.

Then, we segment these images into small foreign fiber objects only including cotton foreign fiber. Finally, the 2808 objects are obtained and the number of hair, black plastic film, cloth, rope, polypropylene twines and feather objects is 204, 408, 432, 492, 528 and 744, respectively. The following step is extracting features from these objects.

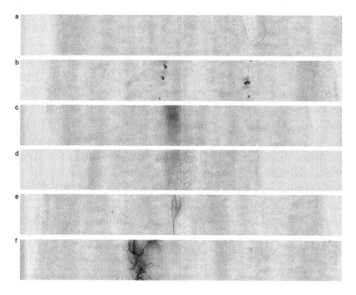

Fig. 1. Images of cotton foreign fibers: (a) hair image, (b) plastic film image, (c) cloth image, (d) hemp rope image, (e) polypropylene image, (f) feather image. (Color figure online)

The color, shape, and texture features can be extracted in cotton foreign fibers. Since accurate classification is difficult in only using one or two features [2]. Therefore we need extraction of all kinds of features including color, shape and texture features, and find the excellent feature combination by FS approaches.

In our experiment, a total of 80 features are extracted from foreign fiber objects, and the number of color, texture and shape features is 28, 42 and 10, respectively. These extracted features are used to build the 80-dimensional feature vector.

After the data is generated, normalization is implemented to reduce the impact of different dimensions.

2.3 Metaheuristic Algorithms for Feature Selection

Fitness Function. Fitness function is important for metaheuristic algorithms, it is used evaluate the quality of each subset. Considering the online classification problem, we expect that the found feature set should has small size and high accuracy. Therefore, fitness evaluation is designed to combine the accuracy of classifier with the length of feature subset. The specific fitness function in this paper is the following Eq. (1):

$$S(X) = \upsilon J(X) + \psi |X| \tag{1}$$

where X denotes the subset, $J(X)$ is the classification accuracy of subset X, $|X|$ denotes the feature number of the subset X, υ and ψ are used to adjust the relative importance of accuracy and size. In this study, two classifiers, KNN and SVM are adopted to evaluate the quality of subset.

Particle Swarm Optimization for Feature Selection. PSO belongs to population-based metaheuristics and is proposed by Kennedy and Eberhart [11]. PSO explores the search space by movements of particles with a velocity and each particle is updated based on its past best position and the current best particle. PSO can efficiently balance the exploration and exploitation and is an efficient optimization algorithm [12, 13]. Supposing the particle i is denoted as $\vec{X}_i = (x_{i,1}, x_{i,2}, \cdots, x_{i,d})$, which has the velocity $\vec{V}_i = (v_{i,1}, v_{i,2}, \cdots, v_{i,d})$. $\vec{P}_i = (p_{i,1}, p_{i,2}, \cdots, p_{i,d})$ denotes the past best position of the particle i. $\vec{P}_g = (p_{g,1}, p_{g,2}, \cdots, p_{g,d})$ denotes the best particle.

Each bit of the particle only lies in one of two states, i.e. zero or one, which will be changed according to probabilities. To change the velocity from continuous space to probability space, the following sigmoid function is using in algorithm:

$$sig(v_{i,j}) = \frac{1}{1 + \exp(-v_{i,j})}, j = 1, 2, \cdots, d \tag{2}$$

The velocity is recalculated in terms of Eq. (3):

$$v_{i,j}^{t+1} = w \times v_{i,j}^t + c_1 \times r_1(p_{i,j}^t - x_{i,j}^t) + c_2 \times r_2(p_{g,j}^t - x_{i,j}^t) \tag{3}$$

where w denotes inertia weight and is updated at the iteration t according to Eq. (4):

$$w_t = w_{min} + (w_{max} - w_{min})\frac{(t_{max} - t)}{t_{max}} \tag{4}$$

where w_{max} denotes respectively the maximal value of the inertia weight, w_{min} denotes the minimum of the inertia weight. The t_{max} is the maximal times of iterations. The parameters c_1 and c_2 denote the acceleration coefficients. The parameters r_1 and r_2 are random numbers varying from 0 to 1. $x_{i,j}$, $p_{i,j}$ and $p_{g,j}$ belong to zero or one. v_{max} is the maximum velocity.

The new particle position is updated by Eq. (5):

$$x_{i,j}^{t+1} = \begin{cases} 1, & \text{if } rnd < sig(v_{i,j}) \\ 0, & \text{if } rnd \geq sig(v_{i,j}) \end{cases}, j = 1, 2, \cdots, d \tag{5}$$

where and rnd denotes the random number in [0, 1] from uniform distribution.

Ant Colony Optimization for Feature Selection. ACO is proposed based on the idea of ants finding food by the shortest path between food source and nest [14]. In ACO, the addressed problem is modelled as a graph, in which the ants search a minimum cost path. The good paths mean the emergent result of the global cooperation among ants. In each iteration finding the optimal solutions, many ants construct their solutions by heuristic information and trail pheromone [15].

In the ACOFS, every candidate solution is mapped into an ant represented by a binary vector where the bit one or zero respectively means that the corresponding feature is selected or not.

Heuristic Information: Heuristic information generally represents the attractiveness of each feature. If heuristic information is not used, the algorithm would be greedy, and the better solution is not found [5]. To evaluate the heuristic information [16], the information gain is calculated in this study.

Feature Selection: In each iteration, the ant k determines whether the feature i is selected or not according to transition probability p_i. The transition probability p_i is given as follows:

$$p_i^k(t) = \begin{cases} \dfrac{[\tau_i(t)]^\alpha [\eta_i(t)]^\beta}{\sum\limits_{u=J^k} [\tau_u]^\alpha [\eta_u]^\beta} & \text{if } i \in J^k \\ 0 & \text{otherwise} \end{cases} \tag{6}$$

where J^k represents the feasible feature set, η_i is the heuristic desirability of the feature i, τ_i is the pheromone value of the feature i. α and β is used to adjust the relative importance of the heuristic information and pheromone.

Pheromone update: After all ants have constructed their feature sets, the algorithm trigger the pheromone evaporation, and according to Eq. (7) each ant k deposits a quantity of pheromone $\Delta \tau_i^k(t)$, which is calculated according the following equation:

$$\Delta \tau_i^k(t) = \begin{cases} \gamma(s^k(t)) & \text{if } i \in s^k(t) \\ 0 & \text{otherwise} \end{cases} \tag{7}$$

where $s^k(t)$ denotes the found feature set by ant k at iteration t. $\gamma(x)$ denotes fitness function.

The pheromone can be updated according to the following rule:

$$\tau_i(t+1) = (1-\rho)\tau_i(t) + \sum_{k=1}^{m} \Delta\tau_i^k(t) + \Delta\tau_i^g(t) \tag{8}$$

where $\rho \in (0,1)$ denote the pheromone decay coefficient which can avoid stagnation. m denotes the number of ants and g is the best ant. For all ants, the pheromone is updated according to Eq. (8).

Genetic Optimization for Feature Selection. GA is proposed by Holland, which is a metaheuristic based on the idea of genetic and natural selection. GA has been used to solve the FS tasks [7]. In evolution, each species has to change its chromosome combination to adapt to the complicated and changing environment for surviving in the world. In GA, a chromosome denotes a potential solution of problem, which is evaluated by fitness function. In GA, a new generation with better survival abilities is generated by crossover and mutation. GA usually includes coding, selection, crossover, mutation.

Encoding: In the GAFS, a chromosome represents a candidate feature set which consist of genes, and a gene is a feature which are encoded in the form of binary strings. If the gene is coded into '1', the corresponding feature is selected, otherwise, the gene is coded into '0'. The bit i in the chromosome denotes the feature i. Each chromosome is initialized randomly.

Selection: The selection is to select the chromosomes into the new generation among the current population. A certain number of chromosomes are probabilistically selected into the next generation, where the probability of selecting chromosomes i is Eq. (9)

$$\Pr(i) = \frac{S(i)}{\sum_{j=1}^{m} S(j)} \tag{9}$$

where $S(i)$ denotes the fitness value of the chromosome i, m is the number of the chromosomes.

Crossover: The crossover operator is used to exchange genes between two chromosomes. The representative crossover operators are multi-point crossover, double-point crossover and single-point crossover [7]. In this study, double-point crossover is performed. After some chromosomes have been selected into the next generation's population, additional chromosomes are generated by using a crossover operation. Crossover takes two parent individuals, which are chosen from the current generation by using the probability function given by Eq. (9), and creates two offspring by recombining portions of both parents.

Mutation: Mutation operator is used to determine the variety of the chromosomes, which let local variations into the chromosomes and keeps the diversity of the population, This can help to find the good solution in search space. Here, the number of chromosomes depends on the mutation rate. Then a gene of the mutating chromosome is selected at random and its value is changed from '1' to '0' or '0' to '1', respectively.

3   Results and Discussion

In our experiments, the configuration of computer is as follows: CPU 2.66 GHz, main memory 4.0 GB and Windows 7 system. All the algorithms are coded and run in the Matlab development environment. Two different classifiers, KNN and SVM, are taken for evaluating the quality of solution. To efficiently evaluate the methods, the 10-fold cross validation is used in our experiments.

The parameters of PSOFS, ACOFS and GAFS is set according to Table 1. These parameters are selected in terms of experiences.

Table 1. Parameter settings for three FS algorithms

Parameters	PSOFS	ACOFS	GAFS
υ	0.9	0.9	0.9
ψ	0.1	0.1	0.1
m	50	50	50
$Iter_{max}$	100	50	100
c_1	2		
c_2	1.5		
w_{max}	0.95		
w_{min}	0.4		
v_{max}	3		
α		1	
β		1.5	
ρ		0.3	
p_r			0.9
p_m			0.05

Table 2. Comparison of performance of the five algorithms

Methods	Classifiers	Average accuracy (%)	Average number of feature
ACOFS	KNN	**92.7**	**20**
PSOFS		92.3	26
GAFS		91.8	27
ACOFS	SVM	91.5	**25**
PSOFS		**91.9**	31
GAFS		91.4	35

Table 2 shows the results of performance comparisons of three algorithms. As we can see, for KNN classifier, ACOFS has the best result among these methods, the selected subset has the smallest size and highest classification accuracy. For SVM, PSOFS can obtain the subset with highest accuracy, but the subset obtained by ACOFS includes the least features.

Figure 2 intuitively shows the accuracy of the different subsets obtained by three algorithms and the original set with KNN and SVM, respectively. As shown in Fig. 2, for KNN and SVM, all the optimal subsets obtained by three metaheuristic-based algorithms achieve much higher accuracy than the original set without feature selection.

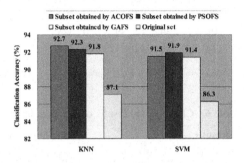

Fig. 2. Results of classification accuracy with KNN and SVM

Table 3. Classification results of the optimal set obtained by ACOFS

Classes	Results						Sample number	Accuracy (%)
	(1)	(2)	(3)	(4)	(5)	(6)		
(1) Plastic film	396	0	0	0	4	8	408	97.06
(2) Cloth	0	408	0	0	0	24	432	**94.44**
(3) Hemp rope	0	8	416	0	28	40	492	84.55
(4) Hair	0	0	0	196	0	8	204	**96.08**
(5) Polypropylene	0	0	4	20	504	0	528	95.45
(6) Feather	8	48	16	0	12	660	744	88.71
Average								**92.72**

Table 4. Classification results of the optimal set obtained by PSOFS

Classes	Results						Sample number	Accuracy (%)
	(1)	(2)	(3)	(4)	(5)	(6)		
(1) Plastic film	396	0	0	0	0	12	408	97.06
(2) Cloth	0	408	0	0	0	24	432	**94.44**
(3) Hemp rope	0	16	412	0	28	36	492	83.74
(4) Hair	0	0	0	192	4	8	204	**94.12**
(5) Polypropylene	0	0	16	8	504	0	528	95.45
(6) Feather	12	44	0	24	0	664	744	89.25
Average								**92.34**

Table 5. Classification results of the optimal set obtained by GAFS

Classes	Results						Sample number	Accuracy (%)
	(1)	(2)	(3)	(4)	(5)	(6)		
(1) Plastic film	396	0	0	0	0	12	408	97.06
(2) Cloth	0	408	0	0	0	24	432	**94.44**
(3) Hemp rope	0	12	408	0	24	48	492	82.93
(4) Hair	0	0	0	192	12	0	204	**94.12**
(5) Polypropylene	0	0	24	12	492	0	528	93.18
(6) Feather	12	48	0	12	12	660	744	88.71
Average								**91.81**

Table 6. Classification results of the original set

Classes	Results						Sample number	Accuracy (%)
	(1)	(2)	(3)	(4)	(5)	(6)		
(1) Plastic film	384	0	0	0	0	24	408	94.12
(2) Cloth	0	372	20	0	0	40	432	**86.11**
(3) Hemp rope	0	12	408	24	12	36	492	82.93
(4) Hair	0	0	32	160	0	12	204	**78.43**
(5) Polypropylene	0	0	28	0	496	4	528	93.94
(6) Feather	24	60	0	12	0	648	744	87.10
Average								**87.10**

Tables 3, 4, 5 and 6 show the detailed classification results of the subset obtained by three meta-heuristic algorithms and the original set using KNN classifier. As shown in Tables 3, 4, 5 and 6, the classification accuracy of the cloth and hair is efficiently improved by using the subset selected by three meta-heuristic algorithms, at least increased by 8 % and 15 %, respectively.

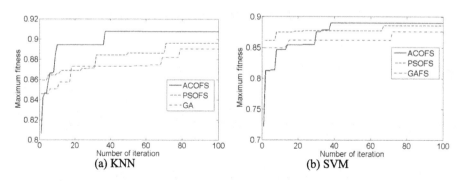

Fig. 3. Fitness of three methods with KNN and SVM

The curves of fitness of the three metaheuristic-based algorithms with KNN, SVM in a certain run are shown in Fig. 3, respectively. The curves shown in Fig. 3 are representative according to our preliminary experiments. As we can see in Fig. 3, for KNN and SVM, ACOFS is more efficient and faster than PSOFS and GAFS. On average, PSOFS and GACOFS need about 70 iterations to find the optimum subset, while ACOFS only needs less than 40 iterations.

4   Conclusions

A key issue in machine-vision-based recognition system of cotton foreign fibers is to find the optimal feature set. In this study, FS based on metaheuristic optimization, namely ACOFS, PSOFS and GAFS, have been proposed to address this FS problem. Two different classifiers, KNN and SVM are taken for evaluating the quality of solution and classification. The experimental results show the presented methods have the excellent ability of finding a reduced set of features with high accuracy in the dataset of cotton foreign fiber. The selected feature set is great significant for machine-vision-based recognition system for cotton foreign fibers. In our future work, we will focus on improving the performance of classifiers used in the recognition rate of recognition systems of cotton foreign fibers.

Acknowledgments. This study is funded by the National Natural Science Foundation of China (61402195, 61471133 and 61571444), Guangdong Natural Science Foundation (2016A030310072), Guangdong Science and Technology Plan Project (2015A070709015 and 2015A020209171), the Science and Technology Plan Project of Wenzhou, China (G20140048), Shenzhen strategic emerging industry development funds (JCYJ20140418100633634).

References

1. Yang, W., Li, D., Zhu, L., Kang, Y., Li, F.: A new approach for image processing in foreign fiber detection. Comput. Electron. Agric. **68**(1), 68–77 (2009)
2. Li, D., Yang, W., Wang, S.: Classification of foreign fibers in cotton lint using machine vision and multi-class support vector machine. Comput. Electron. Agric. **74**(2), 274–279 (2010)
3. Yang, W., Lu, S., Wang, S., Li, D.: Fast recognition of foreign fibers in cotton lint using machine vision. Math. Comput. Model. **54**(3), 877–882 (2011)
4. Lin, J.Y., Ke, H.R., Chien, B.C., Yang, W.P.: Classifier design with feature selection and feature extraction using layered genetic programming. Expert Syst. Appl. **34**(2), 1384–1393 (2008)
5. Bolón-Canedo, V., Sánchez-Maroño, N., Alonso-Betanzos, A.: A review of feature selection methods on synthetic data. Knowl. Inf. Syst. **34**(3), 483–519 (2013)
6. Sun, Z., Bebis, G., Miller, R.: Object detection using feature subset selection. Pattern Recogn. **37**(11), 2165–2176 (2004)
7. Pedergnana, M., Marpu, P.R., Dalla Mura, M., Benediktsson, J.A., Bruzzone, L.: A novel technique for optimal feature selection in attribute profiles based on genetic algorithms. IEEE Trans. Geosci. Remote Sens. **51**(6), 3514–3528 (2013)

8. Chen, H.L., Yang, B., Wang, G., Liu, J., Xu, X., Wang, S.J., Liu, D.Y.: A novel bankruptcy prediction model based on an adaptive fuzzy k-nearest neighbor method. Knowl.-Based Syst. **24**(8), 1348–1359 (2011)

9. Cover, T., Hart, P.: Nearest neighbor pattern classification. IEEE Trans. Inf. Theory **13**(1), 21–27 (1967)

10. Xuegong, Z.: Introduction to statistical learning theory and support vector machines. Acta Automatica Sinica **26**(1), 32–42 (2000)

11. Eberhart, R.C., Kennedy, J.: A new optimizer using particle swarm theory. In: Proceedings of the Sixth International Symposium on Micro Machine and Human Science, vol. 1, pp. 39–43 (1995)

12. Chen, H.L., Yang, B., Wang, G., Wang, S.J., Liu, J., Liu, D.Y.: Support vector machine based diagnostic system for breast cancer using swarm intelligence. J. Med. Syst. **36**(4), 2505–2519 (2012)

13. Kennedy, J., Eberhart, R.C.: A discrete binary version of the particle swarm algorithm. In: IEEE International Conference on Systems, Man, and Cybernetics 1997, vol. 5, pp. 4104–4108 (1997)

14. Dorigo, M., Maniezzo, V., Colorni, A.: Ant system: optimization by a colony of cooperating agents. IEEE Trans. Syst. Man Cybern. B **26**(1), 29–41 (1996)

15. Zhao, X., Li, D., Yang, B., Ma, C., Zhu, Y., Chen, H.: Feature selection based on improved ant colony optimization for online detection of foreign fiber in cotton. Appl. Soft Comput. **24**, 585–596 (2014)

16. Forsati, R., Moayedikia, A., Jensen, R., Shamsfard, M., Meybodi, M.R.: Enriched ant colony optimization and its application in feature selection. Neurocomputing **142**, 354–371 (2014)

Beta Function Law to Model the Dynamics of Fruit's Growth Rate in Tomato: Parameter Estimation and Evaluation

Qiaoxue Dong[1], Lili Yang[2(✉)], Mei Qu[3], Qinglan Shi[1],
and Shangfeng Du[1]

[1] Modern Precision Agriculture System Integration Research Key Laboratory
of Ministry of Education, China Agricultural University,
Beijing 100083, China
[2] Key Laboratory of Agricultural Information Acquisition Technology (Beijing),
Ministry of Agriculture, China Agricultural University, Beijing 100083, China
llyang@cau.edu.cn
[3] College of Agronomy and Biotechnology, Beijing Key Laboratory of Growth
and Development Regulation for Protected Vegetable Crops,
China Agricultural University, Beijing 100193, China

Abstract. The calculation of fruit growth rate (FGR) is the main part of fruit-bearing crop growth models. In this paper, greenhouse tomato was used as a research material. The purpose of this study is to better understand the regulation of single fruit growth response to environmental or genetic factors. A FGR model of tomato fruit was described based on Beta function law because it has flexible ability to generate different curve shape when adjusting its parameters. A field experiments with 4 planting densities in 2012 spring was carried out in China, using the cultivar Weichi. The parameters of Beta-law FGR model were estimated with data from the fruits in the first truss by using the optimization algorithm Nelder-Mead The results showed that the optimization procedure described in this paper found best fitted and robust parameters, which implied that those parameters are cultivar specific, and little environmental dependence. The validation against the data from the second truss and third truss with variations in planting density was found to be acceptable (R2 = 0.9117). This modeling methodology and software program have the potential to become a powerful tool for optimization of ideotype design.

Keywords: Tomato · FGR model · Optimization · Parameter estimation

1 Introduction

Tomato (*solanum lycopersicum*) is one of the most important horticultural crops and food preference in China. With the recent development of greenhouse technology, China has become the first tomato production country in the world as far as total amount of yields, but researches still indicated that the actual yield of the tomato in China is only 7.7 % of the average potential productivity, and there is still a larger space to improve the potential production [1].

© IFIP International Federation for Information Processing 2016
Published by Springer International Publishing AG 2016. All Rights Reserved
D. Li and Z. Li (Eds.): CCTA 2015, Part I, IFIP AICT 478, pp. 19–26, 2016.
DOI: 10.1007/978-3-319-48357-3_3

Crop growth modeling provides a powerful research tool for studying the potential production. [2–5] Calculation of tomato fruit growth rate (FGR) is an important part of tomato simulation models, and the integration of FGR over time gives the accumulated fresh weight along different growth stages. It not only can be used to describe the potential performance in any given environment, but also to describe graded size distribution of final yields. Two kinds of models describing fruit dynamics can be distinguished (1) the empirical models and (2) physiological models. Empirical or regression models are built up based on direct field and experimental observation with some regression functions, for instance the typical S-shaped logistic curves [5].

Because they are descriptive and the parameters of regression models are possible to varies with different level of observation without botanical meaning. Physiological models, on the other hand, explain the observed growth rates from the underlying physiological process and in relation to the environmental factors. So the parameters in such models are usually divided into two categories: cultivar-specific parameters and environment-dependent parameters, where cultivar-specific parameters are assumed constant for some kinds of cultivar and will be estimated and evaluated through optimization technology [6].

Physiological models simulate physiological process by using mathematical equations with different hierarchical level. For instance, in famous tomato growth model of Tomsim and Tomgro, both simulate tomato plant process based on compartment level, and the realization of fruit dynamics depend upon the partitioning of the photosynthesis production into fruit growth phases according to sink strength ratio among compartments, so the final yields is the main simulation objective, while individual fruit growth rate is seldom considered [7, 8].

Tomato functional-structural model (TFSM) provides detailed hierarchical structure and it entails quantitative integration of mechanism at organ level to provide a better modeling frame explaining the behavior related to environment-dependent or genetic-control phenotypic plasticity. However, because of the complexity of the model and the data sets used to estimate model parameter values were limited, the prediction of fruit kinetics was not accurate enough, thus limiting the application into the real production [9, 10].

In practice, management requires well-parameterized predictive models. To accomplish this, a tailored fruit dynamic model from TFSM model for simulating FGR in tomato was presented, which objective is to determine and evaluate those parameters whether they are affected by environment conditions or not. If not, those parameters will be introduced into the physiological models as known invariable cultivar-specific parameters so that the number of parameters to be estimated and the complexity of optimization procedure will be reduced.

2 Materials and Methods

2.1 Field Experiments

An experiment was conducted during spring 2012 (sowing date, 18 January) on tomato (Solanum lycopersicum, WeiChi) at the Xiaotangshan Vegetables Production Zone,

Beijing, China (39.55 N, 116.25E). Plants were planted in a multi-span greenhouse. Rock wool were used as soil-free cultivation medium and four density treatments were applied: high density (HD, 4.2 plants m^{-2}), intermediate density (MD1, 3.5 plants m^{-2} and MD2, 2.8 plants m^{-2}) and low density (LD; 2.1 plants m^{-2}). Irrigation and fertilization were similar for all density treatments. Air temperature, relative humidity and CO_2 were collected every ten minutes by AR5-type data logger. The instrument AV-19Q was used to measure photosynthetic photon flux density.

The first flowering and fruit-set date were recorded. For each treatment, there were three replications and in each replication block, 4 plants were sampled to calculate the average diameter of tomato fruit. The leaf appearance rate was recorded shortly after planting date. After fruit thinning, 8 measurement date were carried out to evaluate the influence of varying density on fruit growth rate (FGR).

2.2 Model Description

The FGR model is executed at time steps corresponding to organogenetic growth cycles (GC), equal to the thermal time (degree days, °Cd) needed to generate a new metamer (or a leaf). According to the measurements and analysis, we found that there is a strong linear relationship between number of leaf and accumulated thermal time (Fig. 1).

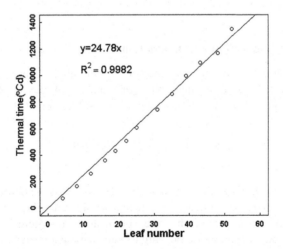

Fig. 1. Method of calculating the growth cycles (GCs) according to the relationship between accumulated thermal time and number of leaves

According to Fig. 1, the definition of one GC was derived as follows:

$$GC = \frac{ATT}{N} \tag{1}$$

where GC is the definition of one growth cycle (about 24.78 degree days); GC is accumulated thermal time ((degree days, °Cd) which base temperature is 10°C. N is the total number of leaves along main stem.

With the definition of time step GC, we formulate a dynamic model for tomato fruit growth rate as the following form:

$$\Delta D(GC_i) = P * f_c(GC_i; par_c) \tag{2}$$

where GC_i means i^{th} growth cycle; ΔD is the real incremental size(mm) at the GC_i;

P is unit cycle expansion strength(mm/GC) determining the potential maximum incremental size for each cycle; $f_c(\ldots)$ will determine the variation shape of growth rate and is assumed as inner law only determined by cultivar type or gene control but not affected by environmental conditions; Here we apply Beta function to model the normalized growth rate for tomato fruit along GCx because it can generate flexible curve shape when adjusting its parameters.

$$f(GC_x) = \frac{(\frac{GC_x+0.5}{T})^{p-1}(1 - \frac{GC_x+0.5}{T})^{q-1}}{\sum\limits_{x=1}^{T} f(GC_x)} \tag{3}$$

where $x = (1, 2, \ldots T)$ and T is maximum expansion cycles and can be observed from field experiments; (p, q) are the parameters of Beta function to be estimated.

2.3   Parameter Calibration and Evaluation

To obtain the parameters of Beta law, a reliable calibration procedure was required. We firstly defined a criteria to evaluate the performance of the FGR model where non-linear least square sum was used as the objective function [14].

$$Z_{NLS}(\theta) = \sum\limits_{i=1}^{N} [y_i - f(x_i; \theta)]^2 \tag{4}$$

Where y_i is the observation ($i = 1, 2, \ldots N = 1$); $f(x_i; \theta)$ is the output of model; θ is the parameter vector to be estimated which final calibrated value will minimize this objective function. To implement this, the optimization algorithm Nelder-Mead was introduced into model calibration procedure to find stable parameters. Finally, The test of calibration results against new data set was the root mean square error (RMSE), calculated as [11].

$$RMSE = \sqrt{\sum (p - o)^2 / n} \tag{5}$$

where p is predicted data, o is observation data and n is the number of observations. The smaller the $RMSE$, the better the prediction.

2.4   Statistical Analysis

The data were analyzed using R package 'stats' (version 3.0.2). R Core Team (2013). R: A language and environment for statistical computing. R Foundation for Statistical Computing, Vienna, Austria. URL http://www.R-project.org/.

3   Results and Discussion

In this section, the FGR model's ability based on beta-law to simulate the expansion rate of tomato fruit was evaluated in two steps. First, the measurements on the fruits' dynamics of the first truss were used for calibration. Second, a validation with parameterization was made using the observations from the fruits in the second and third truss along the main stem.

Those data for calibration were collected every five days from the date of fruit-set (20 Feb. 2012) to the maturity date (11 Apr. 2012). To convert the calendar date to growth cycles which are required in the model, we first compute the accumulated thermal time from fruit-set for each observation date, then use the relationship described in Fig. 1 to calculated the passing growth cycles. So the incremental diameter on each measurement will be divided by number of growth cycles and the calculated value was considered as the average growth rate on the observation date. To capture the essence of the tomato fruit's characteristics and to investigate if the model parameters are environment-dependent or not, the FGR model's parameters were estimated in each condition against target files composed of 4 planting density (LD,MD1,MD2,HD).

The estimated curves for four densities were shown in Fig. 2. Those curves have similar shape and a good agreement between observation and model results was obtained for MD1,MD2,HD (a RMSE of 25 % and below was defined as acceptable agreement), while the RMSE for LD is a little bigger, the possible reason for which is that the date of the first fruit set should be earlier than the others, but in real measurement and calibration procedure, it is recorded the same as the other density. But anyway, we can concluded that the Beta law could be possible to model the growth rate of tomato fruits' dynamics, because it can basically track the variation of tomato fruit's kinetics.

Model parameters were determined for each growing condition by statistical optimization. Table 1 presents the optimized parameter values and their respective standard deviation for each situation. Coefficient of variation (CV%) shows a higher stability of model parameters, and it indicates that increasing densities had little influence on the parameters defining the kinetics of organ fruit (beta law parameters.

Figure 3 shows the iterations process of the optimism algorithm, i.e. Nelder-Mead applied in the non-linear least square calibration procedure. The curve indicated the good performance because the optimization algorithm converged very quickly and found the best fitted values of parameters to minimize the objective function, and it successfully found global minimum while avoiding local one. So this optimization algorithm is suitable for crop growth models.

We made an average on parameters' values for four density and the determined model was validated by the fruits dynamics in the second truss and third truss.

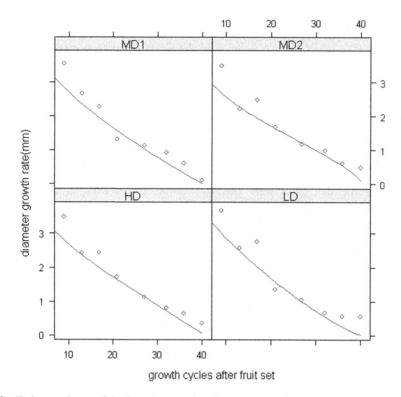

Fig. 2. Fruit growth rate of the first truss as a function growth cycles after fruit-set: measurement verse fitted results with Beta-law FGR model.

Table 1. Value of parameters estimated and evaluation

parameter	LD	MD1	MD2	HD	mean	s.d.	CV%
p	0.94	0.88	0.82	0.89	0.885	0.0487	5.5 %
q	2.23	2.04	1.69	1.92	1.969	0.225	11.4 %
P	4.71	4.87	5.18	4.73	4.872	0.215	4.4 %

The simulation results against the observation were shown in Fig. 4. The simulation of tomato fruit's growth rate generally described the observed field data and the regression of observed data on predicted ones indicated a one to one correspondence ($R^2 = 0.91$). But we still can observe from Fig. 4 that some points are over-predicted or largely under-predicted. One of the possible reasons for inaccurate prediction is that observation is the average diameter of total truss fruits, while in the calibration set, the observation is made from regular date and a single fruit, another reason is that a different nutrient solution was applied in the medium-later growth stages for the second truss and third truss from the one applied in the first truss. However, despite some inaccuracies, the model still provided a reasonable simulation of fruit expansion patterns.

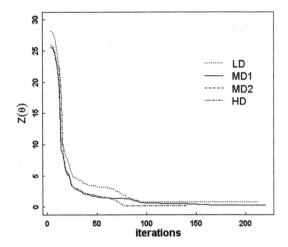

Fig. 3. Convergence paths of the Nelder-Mead optimization algorithm applied to the FGR model.

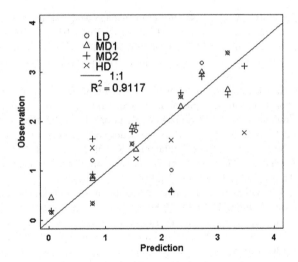

Fig. 4. The validation of model against the data from the second and third truss

4   Conclusion

Form the parameterization of the tomato growth rate model and results' analysis, it is concluded that the FGR beta-law model by using the concepts of growth cycles is able to accurately simulate the expansion rate of greenhouse tomato's fruit. The optimized parameters of model are similar for different planting densities which further strengthened the assumption that those parameters are cultivar- specific and not affected by external environment conditions. The modeling methodology described in this paper will find applications on yield and quality prediction and gene-based ideotype optimization design.

Acknowledgments. This study was supported by the National Natural Science Foundation of China (#61174088 and #31200543), and Special Found for Beijing Common Construction Project. The authors would also like to thank students from College of Agronomy, China Agricultural University involved in the plant measurements.

References

1. Chu, J.X., Sun, Z.F., Du, K.M., Jia, Q.: Spatial distribution of potential productivity for greenhouse tomato variety Zhongza No. 9 Based on TOMSIM. China J. Agrometeorol. **30**(1), 49–53 (2009). (in Chinese)
2. Pogson, M., Hastings, A., Smith, P.: Sensitivity of crop model predictions to entire meteorological and soil input datasets highlights vulnerability to drought. Environ. Model Softw. **29**(1), 37–43 (2012)
3. Luquet, D., Rebolledo, M.C., Soulie, J.C.: Functional-structural plant modeling to support complex trait phenotyping: case of rice early vigour and drought tolerance using ecomeristem model. In: Proceedings - 2012 IEEE 4th International Symposium on Plant Growth Modeling, Simulation, Visualization and Applications, PMA 2012, pp. 270–277 (2012)
4. Soltani, A., Hoogenboom, G.: Assessing crop management options with crop simulation models based on generated weather data. Field Crops Res. **103**, 198–207 (2007)
5. van Ittersum, M.K., Leffelaar, P.A., van Keulen, H., Kropff, M.J., Bastiaansb, L., Goudriaana, J.: On approaches and applications of the Wageningen crop models. Eur. J. Agron. **18**, 201–234 (2003)
6. He, C., Zhang, Z.: Modeling the relationship between tomato fruit growth and the effective accumulated temperature in solar greenhouse. Acta Hortic. **718**, 581–588 (2006)
7. Acutis, M., Confalonieri, R.: Optimization algorithm for calibrating cropping systems simulation models. A case study with simplex-derived methods integrated in the warm simulation environment Acutis M.e Confalonieri R. Ital. J. Agrometeorol. **3**, 26–34 (2006)
8. Jones, J.W., Dayan, E., Allen, L.H., et al.: A dynamic tomato growth and yield model (TOMGRO). Trans. ASAE **34**, 663–672 (1991)
9. Heuvelink, E.: Evaluation of a dynamic simulation model for tomato crop growth and development. Ann. Bot. **83**, 413–422 (1999)
10. Dingkuhn, M., Luquet, D., Quilot, B., de Reffye, P.: Environmental and genetic control of morphogenesis in crops: towards models simulating phenotypic plasticity. Aust. J. Agric. Res. **56**, 1289–1302 (2005)
11. Dong, Q.X., Louarn, G., Wang, Y.M., Barczi, J.F., de Reffye, P.: Does the structure-function model GREENLAB deal with crop phenotypic plasticity induced by plant spacing? A case study on tomato. Ann. Bot. **101**, 1195–1206 (2008)
12. Janssen, P.H.M., Heuberge, P.S.C.: Calibration of process- oriented models. Ecol. Model. **83**, 55–56 (1995)
13. Ma, L.L., Ji, J., He, C.: Studies on tomato leaf area index measurement method based on BP neural network. China Veg. **16**, 45–50 (2009)
14. Wallach, D., Makowski, D., Jones, J.W.: Working with Dynamic Crop Models: Evaluations, Analysis, Parameterization, and Applications. Elsevier, Amsterdam (2006)

A Web-Based Cooperation and Retrieval Model of Character Images for Ancient Chinese Character Research

Xuedong Tian[(✉)], Songqiang Yang, Xuesha Jia, Fang Yang,
and Chong Zhang

College of Mathematics and Computer Science,
Hebei University, Baoding 071002, China
txdinfo@sina.com

Abstract. The character images are the kernel in the research work of ancient Chinese characters. Nowadays, more and more optical sensing device such as scanners and digital cameras are used for transforming ancient Chinese books into digital images for character research instead of paper documents to overcome the difficulties resulted from the rarity and fragility of ancient documents. How to cooperate the images with researchers in web environment becomes a key topic for the higher efficiency and quality of research work. In this paper, considering the requirements in research work of ancient Chinese characters, a web-based cooperation and retrieval model of character images is constructed which consists of several modules including ancient Chinese image management and retrieval, research work cooperation, research conclusion data management and so on. The key techniques employed in this system are discussed. Firstly, the cooperation mechanism of ancient Chinese character images was designed to make character images cooperate with researchers to avoid the occurrence of the situations that a character image is analyzed by several researchers simultaneously or not assigned to any personal for long time. Secondly, the method of integrated image-text arrangement was constructed to adapt to the special situation of ancient Chinese character research that their study conclusion records often contain the mixture of text and images. Thirdly, the conclusion record sorting algorithm by Chinese character radicals was proposed which could arrange the records according to various kinds of dictionary radical orders to meet the needs of research works. Finally, the global and local retrieval method of ancient Chinese character images is proposed for researchers to search the similar character images in database with the global or local area features of character images. The experimental results show that the system constructed with the proposed model has effective assisting function for the research work of ancient Chinese characters.

Keywords: Cooperation and retrieval model · Ancient Chinese character research · Character image · Image retrieval · Image-text integrated arrangement · Conclusion record sorting

© IFIP International Federation for Information Processing 2016
Published by Springer International Publishing AG 2016. All Rights Reserved
D. Li and Z. Li (Eds.): CCTA 2015, Part I, IFIP AICT 478, pp. 27–44, 2016.
DOI: 10.1007/978-3-319-48357-3_4

1 Introduction

The research on ancient Chinese characters is a meaningful work for the digitalization of ancient documents and the promotion of popularization and dissemination of Chinese civilization.

In a large scale research project in this field, researchers are often confronted with large amounts of characters to be studied. To identify the attributes of an ancient Chinese character including its pattern, pronunciation and meaning, researchers must apply their knowledge about ancient Chinese characters sufficiently, read a lot of related references and discuss with other researchers frequently. Meanwhile, it is also necessary for scientific management among researchers, the research task and the research resources including studied character images and related references. No doubt, employing information technology could bring many conveniences and improve the efficiency for these activities.

The number of Chinese characters is hard to be determined. At present, many ancient Chinese characters have still not been included in existing coded character set, which brings many problems when they are processed in computers. As we all know, the characters in computers are stored in code mode for transmitting and processing conveniently. When a character is to be displayed or printed, the corresponding character pattern is loaded by operating system according to its code. So, a character beyond the coded character set could only be displayed or printed in image mode, which will result in the larger memory consuming and compatibility issues with normal text. Moreover, these character images could not be searched with normal text retrieval techniques, they could only be found out with the help of more complex technology of image retrieval. Unfortunately, most of ancient Chinese characters to be researched are not coded and have to be treated as images. These will bring many problems when they need processed together with ordinary text which consists of coded characters.

Through the aforementioned analysis, it is necessary to research and develop a targeted assisting model for ancient Chinese character research.

The related theory and technology of constructing a managing model for assisting the research work of ancient Chinese characters based on network are related to the theory and technology in many fields including CSCW (Computer Supported Cooperative Work), [1] network, database management system, image processing and retrieval. In detail, the model building relates to website construction, ancient Chinese character resource management and cooperation, image-text integrated arrangement, ancient Chinese character sorting according to different dictionary radicals, ancient Chinese character image retrieval, and so on.

The theory of CSCW has been researched for many years which provide a basic support for our model construction. Its object is to construct a computer based system with which people could cooperate with each other to accomplish a common task cooperatively [1]. Rama and Bishop compared the CSCW groupware system including three commercial systems and four academic systems and designed a set of multidimensional criteria for comparing CSCW systems [2]. Penichet et al. proposed a classification method of CSCW system based on logical principles in a flexible and appropriate way [3]. Chen discussed the key problems of cooperative platform system including system hierarchy structure, user interface, consistency maintenance, concurrent control, access control and record management. A prototype system RITIS (Real-time Image and Text Interactive System) composed of clients and servers with centralized and peer-to-peer structure was constructed. It supports the real-time image and text interaction in Internet environment and multi-user interface of WYSIWIS [4].

Image-text integrated arrangement is to organize text and images in a layout concertedly. To realize this object, the spatial information of images and characters must be recorded and utilized [5, 6]. Compared with the technology of text information processing, the image-text integrated arrangement is more complex in sides of input, edition, display and output [7–12]. Yang and Cheng designed a scheme to realize image-text mixed arrangement based on XML. They employed the design pattern of MVC to separate text and its view for minimizing the coupling degree among modules and improving the extensibility of system with java language. Because of the platform independent attribute of java and XML, the proposed prototype system has better flexibility [5]. Lu proposed a method to realize the storage technology of image-text mixed arranging documents and their online edition using the open source server control FreeTextBox and ASPJpeg of ASP.net [6]. Zhang et al. put forward a B/S mode based question bank management system. Through using DSOFramer container and the edition function of Word, they realized the input, edition and composition of test paper containing image-text integrated questions [7]. Fan discussed the method of image-text mixed arrangement of test paper in which not only text but also images frequently appear. He classified the layouts into three types called non-image layouts, single image layouts and multiple images layouts. VB and SQL server are used to process and manage the three types of layouts respectively [8]. Zhang and Chen studied the collection, storage and extraction methods of test paper which contain text, formulas, tables and images. They use Delphi as the developing tool and realize the import and export of test papers [9].

In the field of image clustering and retrieval, two types of strategies called text-based method (TBIR) and content-based method (CBIR) are adopted for obtaining required images from image library [13–20]. Yang researches on image clustering and its application in image retrieval. Through analyzing the existing image clustering features and algorithms, an image retrieval system based on image clustering is designed. The images in library are clustered with AP algorithm and an image index is built firstly. Then, the sample image is searched in the index to find the corresponding class. And the following image matching operation is fulfilled only in this class [13]. Xie studied on the topic of image clustering and retrieval. To solve the problem existing in traditional image clustering algorithms, an image clustering method based on MRF is proposed which transforms the clustering task into energy minimization process. And a local image retrieval method is designed with graph cutting mode [14]. Zhuang et al. proposed a novel method of retrieving Chinese calligraphic characters. The images of Chinese calligraphic characters are matched by the feature of approximate point correspondence algorithm. After the contour points are extracted, the approximate point correspondence is computed and the matching operation of character images is run according to their accumulated matching cost [15]. Zhuang et al. put forward a retrieval method of Chinese calligraphic manuscript images based on probabilistic indexing structure called PMF-Tree (Probabilistic Multiple-Feature-Tree).

Integrated features are used in retrieval such as contour points of character images, character styles and types. The characteristic of this method is that users are allowed to select one of above features as retrieval components [16]. Chen proposed an image retrieval method based on integrated features of global statistic feature and local bitmap feature. The mean-variance of RGB values of images are calculated as the global feature. Then, the image is divided into sub areas to get the mean value with bina-ryzation processing as the local feature. Finally, image retrieval program is run with the combination of the global and local features [17]. Kong et al. design a semi-supervised image retrieval method. The characteristic points of an image are extracted with improved Harris algorithm. The image is divided into the regions of interest and the color and texture features are extracted. Then, the semantic relation between the image and its class is established through semi-supervised learning in image feature space. Finally, the similarity between images and class centers are computed [18].

The theory and technology on the construction of network stations assisting for research work have become mature; the details of them will not be discussed here.

The above work laid the foundation for our research and developing work. In this paper, considering the requirements of ancient Chinese character research, a web-based cooperation and retrieval model of character images for ancient Chinese character research is constructed which is composed of several modules including ancient Chinese image management and retrieval, research work cooperation, research conclusion and reference management. The key techniques employed in this system are discussed such as the character image cooperation mechanism, image-text integrated arrange-ment, conclusion data sorting of ancient Chinese character research according to Chinese characters radicals, global and local retrieval of ancient Chinese character images, and so on.

The paper is organized as follows. Section 2 outlines the overall architecture and functions of the model. In Sect. 3, the key techniques employed in the model are analyzed and introduced. The experimental result is introduced and analyzed in Sect. 4. Finally, conclusions and the further work are discussed.

2   Architecture of the Cooperation and Retrieval Model

The object of the model is to realize the cooperation management among character images to be studied, researchers and the records of research conclusions. Meanwhile, it provides the image and document retrieval service for Chinese character researchers in the process of research work. The architecture of the model is shown in Fig. 1.

The input data of the model is the images of single ancient Chinese character.

Ancient books are digitalized with optical sensing devices (scanners or digital cameras) to form the layout images firstly. Then, page layout analysis and character image segmentation program is employed to segment these layout images into a series of single character images supplied to ancient Chinese character researchers.

The output data of the system is the records of research conclusion data which includes the pattern, pronunciation, meaning and so on of each character image.

Based on the requirement analysis to the ancient Chinese character researchers, the design principles of the model are as follows.

Foreground

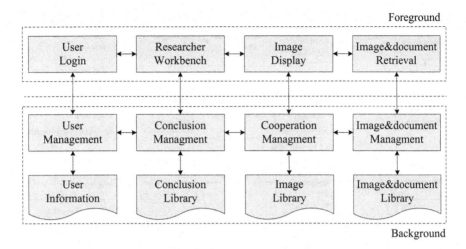

Background

Fig. 1. The architecture of the cooperation and retrieval model

Principle 1. Uniqueness principle. Each image is to be given a unique key code when it is storied into the library of character images and should be allocated to only one researcher for studying.

This could avoid the occurrence of the situation that one character image is assigned to more than one researcher at any time which will result in the confusions of conclusions.

Principle 2. Hierarchy principle. The users of the model are divided into different levels with different authority according to their roles in research work.

Users of the platform with different authorities have different operation scopes, which could effectively avoid the fault operations to the research data.

Principle 3. Independence principle. The research conclusions of a researcher about a character image could only be modified by himself. Other people could give suggestions to him rather than change his research records.

This item is to protect the data of research conclusions from modified by other people rather than original researcher of the character image himself.

Principle 4. Compatibility principle. No matter coded characters or images of no coded characters, the system could organize them normally with the mode of image-text integrated arrangement.

The image-text integrated arrangement problem exists not only in the display operation of research data but also the import and export of the research records in database. So, it is necessary to design a special structure to tackle these problems to ensure the normal use of conclusion records by researchers.

The data flow diagram of the model is shown in Fig. 2.

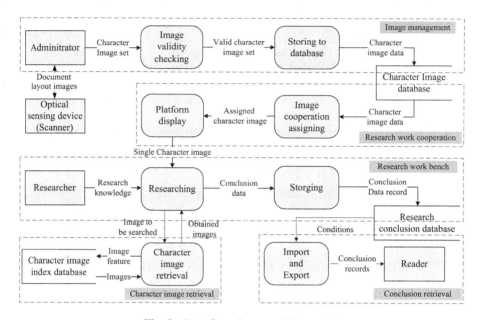

Fig. 2. Data flow diagram of the model

3   Key Techniques in the Model

3.1   Cooperation Mechanism of Research Task

According to the principles of the relationship among researchers, images and research conclusion records, a field controlling operation must be done in data library of different image elements.

Assume ResearchState to be a field in ancient Chinese character image library, SelectID to be a field in researcher library and ExpertNumber to be a field in research conclusion library. The definition of the field value in cooperation is shown in Table 1.

Table 1. Cooperation attributes in library.

Library/Field	Value/Statements
Image/ResearchState	**1**: Current image has not been researched **2**: Current image is being researched **3**: Current image has been researched
Researcher/SelectID	**-1**: The researcher has not allotted an image **Primary Key of image**: The researcher is researching the image
Research conclusion/ExpertNumber	**Primary Key of researcher**: The record could only be modified with the researcher

Through proper setting operations to the semaphores shown in Table 1, the assigning principles of research resource could be abided to ensure the normally running of research work.

3.2   Image-Text Integrated Arrangement

Image-text integrated arrangement mainly consists of three modules: image-text integrated display, edition and their import and export of research data.

3.2.1   Image-Text Integrated Display

In this module, Literal control is employed to realize the image-text integrated display. The coded characters and the tags of image addresses are stored in character strings of Literal. The text attributes of Literal control is linked to Bind ("literal"). The image-text integrated display is shown in Algorithm 1.

Algorithm 1. Image-text integrated display

Input: The content of research conclusions to be displayed in database.

Output: The display data of the content.

  (1) Get the content to be displayed from the field of database table.
    (a) Get the coded characters.
    (b) Get the tags of image addresses.
  (2) Store the characters and the tags of image addresses to be displayed into the Literal control.
  (3) Make the text attributes of Literal control linked to Bind ("literal").

3.2.2   Image-Text Integrated Edition

This module includes the functions of character input and the insertion of character images. Iframe control is utilized to realize the image-text integrated edition which creates an inline frame including a document.

When a character image is inserted, it is stored into database and renamed uniformly. The numbers of character images are recorded and corresponding image labels stored with characters publicly are established.

The algorithm of inserting images into text is shown in Algorithm 2.

Algorithm 2. Image-text integrated edition

Input: The data of character codes and images to be edited.

Output: The image-text integrated data.

  (1) Input the content into iframe control.
    (a) Input characters into iframe control.
    (b) Insert character images into iframe control correspondingly.
  (2) Store the character images into server and rename them uniformly.
  (3) Record the number of character images and store the corresponding image labels.
  (4) Store the characters and the tags of image addresses into the iframe control.

3.2.3   Image-Text Integrated Import and Export

The attribute of the library field in which the labels of characters and images of literal is stored is set to "nvarchar". When we export them into word document, an operation of transforming the html labels to images is fulfilled.

When the data in word documents is imported, the detected images are renamed and stored into database. The locations of the images are replaced with the address links and stored into database. The attributes of the corresponding library fields are set to "nvarchar". The image-text integrated export is shown in Algorithm 3.

Algorithm 3. Image-text integrated export

Input: The research conclusion items.

Output: Word documents including the research conclusion.

(1) Read the research conclusion items.
(2) If the format of the item is image, go to step (3); else go to step (4).
(3) Insert the image. The algorithm terminates.
(4) If the format of the item is not Literal, go to step (5); else go to step (6).
(5) Output the text. The algorithm terminates.
(6) Read the characters of the literal one by one.
(7) If not the end of the string, go to step (8), else the algorithm terminates.
(8) If the character is the tag of the image, go to step (9), otherwise go to step (10).
(9) Convert the tag of the image into the image and output the image, go to step (6).
    (10) Output the text, go to step (6).

The image-text integrated import is shown in Algorithm 4.

Algorithm 4. Image-text integrated import

Input: Word documents including the research conclusion.

Output: The research conclusion item.

(1) Read the data in the imported word document.
(2) If not the end of the string, go to step (3), else go to step (7).
(3) If the format of the character is image, go to step (4), else go to step (6).
(4) Rename the images by the server time and store them into server.
(5) The locations of the images are replaced with the address links. Go to step (1).
(6) Store the text of the word into corresponding field of database. Go to step (1).
(7) Integrate the data and store them into database. The algorithm terminates.

3.3   Research Conclusion Records Sorting by Radical Order

There exist various kinds of radical sets with their corresponding orders related to ancient Chinese characters. So it is necessary to establish a mapping table to realize the sorting function of conclusion records according to different radical sets. The mapping operation of radical sorting of library records is shown in Fig. 3.

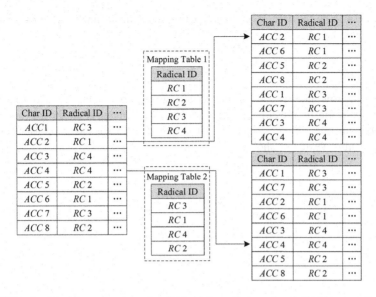

Fig. 3. Mapping operation of radical sorting of research conclusion records

Assume ACC_i to be the key value in conclusion table, RC_j to be the key value of the radical of ACC_i in radical table, $RS(RC_j)$ to be the key value of RC_j in current mapping table.

Then, the series number of record ACC_i in sorted list could be calculated as

$$RS(ACC_i) = \left(\sum_{k=1}^{n} N_{\mathrm{PR}k} \Big|_{RS(RC_k) < RS(RC_j)} \right)$$
$$+ \left(\sum_{l=1}^{m} N_{\mathrm{EQ}l} \Big|_{\substack{RS(RC_l)=RS(RC_j) \\ ACC_l < ACC_j}} \right) + 1 \qquad (1)$$

where $N_{\mathrm{PR}k}\big|_{RS(RC_k)<RS(RC_j)}$ is the number of the records whose $RS(RC_k)$ is less than $RS(RC_j)$, $N_{\mathrm{EQ}l}\big|_{\substack{RS(RC_l)=RS(RC_j) \\ ACC_l<ACC_j}}$ is the number of the records whose $RS(RC_l)$ is equal to $RS(RC_j)$ and $ACC_l < ACC_j$.

3.4   Ancient Chinese Character Image Retrieval

A retrieval algorithm of character images is specially designed in the model to assist researchers to find the local or global similar character images in database. It contains not only the traditional image retrieval functions oriented on the whole area or partial area of a character image, but also a new image searching style called image retrieval in symmetrical areas of character images for searching radicals in Chinese characters.

Assume A to be the area of an ancient Chinese character image which is composed of the sub area a_{ij}

$$A = (a_{ij})^T, (i = 0, 1, \ldots, m - 1; j = 0, 1, \ldots, n - 1) \qquad (2)$$

where m is the row number and n is the column number of meshes divided according to the principle of elastic mesh [19] within the character image A as shown in Fig. 4(a). The directional line elements feature [20] is extracted in sub areas a_{ij} to form corresponding feature vector as shown in Fig. 4(b).

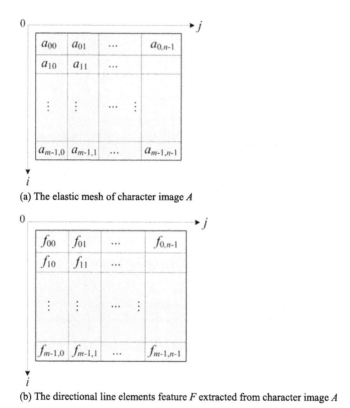

(a) The elastic mesh of character image A

(b) The directional line elements feature F extracted from character image A

Fig. 4. Area division and feature extraction of character image A

$$F = (f_{ij})^T, (i = 0, 1, \ldots, m - 1; j = 0, 1, \ldots, n - 1) \qquad (3)$$

where f_{ij} consists of four directional components.

To improve the efficiency of image retrieval, a hierarchy strategy is employed in which a character image A is clustered into sub clusters previously according to the typical areas A_U, A_D, A_L, A_R, A_C and A_W defined by the structural characteristics of ancient Chinese characters.

The local areas A_U and A_D in vertical are defined as:

$$A_U = \bigcup_{i=0}^{\theta_U} \bigcup_{j=0}^{n-1} a_{ij} \tag{4}$$

$$A_D = \bigcup_{i=\alpha_D}^{m-1} \bigcup_{j=0}^{n-1} a_{ij} \tag{5}$$

where θ_U is the coordinate of the vertical margin of A_U, α_D is the coordinate of the horizontal margin of A_D.

The pixel areas of A_U and A_D in character image is shown in Fig. 5(a) and (b).

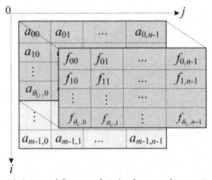

(a) Area and feature of A_U in character image A

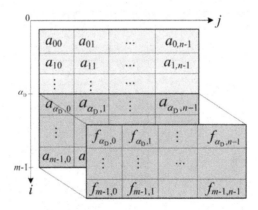

(b) Area and feature of A_D in character image A

Fig. 5. Area division and feature extraction of A_U and A_D in character image A

The local areas A_L and A_R in horizontal are defined as:

$$A_L = \bigcup_{i=0}^{m-1} \bigcup_{j=0}^{\beta_L} a_{ij} \tag{6}$$

$$A_R = \bigcup_{i=0}^{m-1} \bigcup_{j=\delta_R}^{n-1} a_{ij} \tag{7}$$

where β_L is the coordinate of the vertical margin of A_L, δ_R is the coordinate of the horizontal margin of A_R.

The pixel area of A_L and A_R in character image is shown in Fig. 6(a) and (b).

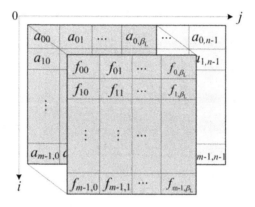

(a) Area and feature of A_L in character image A

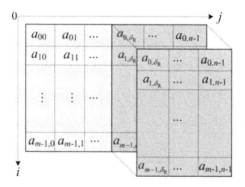

(b) Area and feature of A_R in character image A

Fig. 6. Area division and feature extraction of character image A

The local areas A_C and A_W are defined as:

$$A_C = \bigcup_{i=\gamma_{C1}}^{\gamma_{C2}} \bigcup_{j=\mu_{C1}}^{\mu_{C2}} a_{ij} \tag{8}$$

$$A_W = \bigcup_{i=0;}^{m-1} \bigcup_{i=0}^{m-1} a_{ij} \tag{9}$$

where γ_{C1} and γ_{C2} is the coordinate of the vertical margin of A_C, μ_{C1} and μ_{C2} is the coordinate of the horizontal margin of A_C.

The pixel area of A_C in character image is shown in Fig. 7, while A_W includes the whole area of character image.

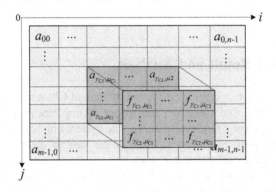

Fig. 7. Area division and feature extraction of A_C in character image A

The clustering operation is fulfilled according to the typical areas. The clustered classes generated by typical areas are C_U, C_D, C_L, C_R, C_C and C_W correspondingly.

When an area A_X is drawn by user within the character image area, it will be matched within the above typical areas A_U, A_D, A_L, A_R, A_C and A_W to find out a best one A_k as the guidance of the following retrieval operation.

$$A_k = \max[(A_X \cap A_i)] \tag{10}$$

where $A_i = A_U$, A_D, A_L, A_R, A_C.

The character image will be searched only in the sub cluster C_k corresponding to A_k based on the feature extracted from A_X. The similarities are measured by the feature distance in $F = (f_{ij})^T$ of A_X and all samples in C_k. The selected character images are returned to the user according to the value of similarities in ascending order.

When a symmetrical retrieval is required, the character image is searched not only in its corresponding classes but also its symmetrical area A_{XS} in character image in vertical direction or horizontal direction such as A_U coupled with A_D, A_L coupled with

A_R. The clustering and retrieval algorithm of ancient Chinese character image is shown in Algorithm 5.

Algorithm 5. Ancient Chinese character image clustering and retrieval

---

Input: Character images.

Output: Matched character images sorted by similarity.

(1) Extracting feature from images and clustering them according to A_U, A_D, A_L, A_R, A_C and A_W correspondingly.
(2) Remark the clustering class ID of each character image.
(3) Obtain A_X.and calculate A_k.
(4) Extracting feature of A_k.
(5) Search the character images in corresponding clustering class C_k.
(6) Sort the retrieval results according to the similarity in descending and display them. The algorithm terminates

---

4   Experimental Result and Analysis

A web-based cooperation and retrieval system of ancient Chinese character images for research with the proposed model in this paper is implemented and utilized in a research project of ancient Chinese characters. The system employs Visual Studio 2008 as the development tool, SQL server 2005 as the storing database, and ASP.NET applied to the system as website frame work.

When a user registers to the system, he will be assigned an authority according to his research task. The setting of authority of the system is shown in Table 2.

Table 2.   Setting of user authority.

No.	Level	Authority
1	I	① Research work assigning ② Character discrimination ③ Character research ④ Character image retrieval ⑤ Research conclusion management ⑥ Document retrieval
2	II	① Character discrimination ② Character research ③ Character image retrieval ④ Document retrieval
3	III	① Character image retrieval ② Document retrieval

A researcher studies the character image assigned to him in the system workbench. He could use the image retrieval function of system to find the similar characters he need locally or globally in database.

The research conclusions of characters could be written by researcher in the corresponding text edition boxes and stored into library. If a researcher needs to write an image of a character, he could use image-text integrated edition mode.

The efficiency of image-text integrated operations is decided by the ratio of the number of coded characters and character images. In our system, there are 1521 records, about 112437 characters in the table of conclusion results. The number of coded characters and character images is shown in Table 3.

Table 3. The number of characters and images in image-text integrated arrangement.

No.	Field ID	Character number	Image number
1	3	654	868
2	4	42080	66
3	6	5739	45
4	7	8403	57

The records of conclusion data of character research stored in database could be displayed in different orders according to the needs of specialists. Table 4 shows several kinds of mapping table for radical sorting.

Table 4. Radical mapping table.

No.	Name	Radical serial number		
		Pinyin	Stroke number	*Kangxi*
1	zhe2	201	6	5
2	bing1	10	22	15
3	shi2	138	8	24
4	da4	29	37	37
5	shan1	132	43	46
6	ya3	176	66	92
7	er3	39	120	128
8	che1	18	150	159
9	qing1	124	168	174
10	ma2	97	200	200

In side of ancient Chinese character image retrieval, we scanned 3176 character images of ancient books for experiment with scanner. Through pre-processing of character images, the feature of clustering and retrieval are extracted. All character images are clustered into six clustering classes according to the definition of A_U, A_D, A_L, A_R, A_C and A_W. Each character image is remark with a clustering class ID.

The distribution of character images in different class is shown in Fig. 8.

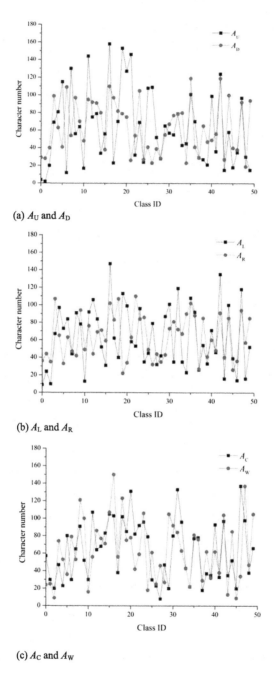

(a) A_U and A_D

(b) A_L and A_R

(c) A_C and A_W

Fig. 8. The distribution of character images in different clusters. (a) The number of character images of every clusters in local area clustering of A_U and A_D. (b) The number of character images of every clusters in local area clustering of A_L and A_R. (c) The number of character images of every clusters in local area clustering of A_C and A_W.

From Fig. 8 we can see that the numbers of character image samples have similar values among classes. This means the high efficiency when image retrieval is done in different classes.

When a searching area is obtained from screen area drawn by user on the character image, the feature of character image is extracted and the retrieval operation is only fulfilled in the corresponding clustering class of the current character image.

5 Conclusion

In this paper, a web-based cooperation and retrieval model of ancient Chinese character images is proposed for assisting ancient Chinese character research work. It includes the functions of ancient Chinese image management and retrieval, research work cooperation, research conclusion data management, and so on. Because of the characteristics of ancient Chinese characters, many problems need to be solved specially in the constructing process of the system. The key techniques developed in this system, such as image cooperation mechanism of ancient Chinese characters, the method of integrated image-text arrangement, the algorithm of character sorting according to Chinese character radicals and the global and local retrieval method of ancient Chinese character images are discussed. The proposed system realized the designing objects of the model including resource management, and research work cooperation. It is helpful for the improvement of the efficiency of character research work.

Our further work is to improve the performance and complete the function of the system. Firstly, in ancient Chinese character image retrieval, we will optimize the algorithm to obtain higher searching speed. This work needs to select more efficient features of character images and corresponding clustering and retrieval algorithms. Secondly, we will develop the recommendation function of research references. This relies on the knowledge arrangement of researchers in ancient Chinese character research work. Thirdly, better user-friendly menus will be developed. No doubt, these improvements will enhance the assisting ability of our model for a better service to the research work of ancient Chinese character research.

Acknowledgments. This work is supported by National Natural Science Foundation of China (No. 61375075) and Natural Science Foundation of Hebei Province (No. F2012201020).

References

1. Shi, M.L., Xiang, Y., Yang, G.X.: Theory and Application of Computer Supported Cooperative Work. Publishing House of Electronics Industry, Beijing (2000)
2. Rama, J., Bishop, J.: 2006 Annual Research Conference of the South African Institute of Computer Scientists and Information Technologists on it Research in Developing Countries, pp. 198–205. South African Institute for Computer Scientists and Information Technologists (2006)
3. Penichet, V.M.R., Marin, I., Gallud, J.A., Lozano, M.D., Tesoriero, R.: Electronic Notes Theor. Comput. Sci. **168** (2007)

4. Chen, X.T.: M D Paper of Harbin University of Science and Technology, Harbin (2004)
5. Yang, Y.G., Cheng, W.Q.: Comput. Appl. Softw. **24** (2007)
6. Lu, Z.Y.: Comput. Program. Skills Maint. (2013)
7. Zhang, J., Fan, Q., Mao, Z.L., Xu, J.Q.: J. Yunnan Univ. (Nat. Sci.) **32** (2010)
8. Fan, J.H.: China Educ. Info (2006)
9. Zhang, W., Chen, S.Q.: Comput. Mod. (2006)
10. Luo, L.X., Zheng, Z.X.: Sci. Technol. Inf. (2007)
11. Huang, K.: Sci. Mosaic (2009)
12. Sun, Q.L., Liu, X.: Comput. Program. Skills Maint. (2009)
13. Yang, C.H.: M D Paper of Nanjing Normal University, Nanjing (2013)
14. Xie, N.Q.: Image clustering and retrieval based on MRF models. M D Paper of Hefei University of Technology, Hefei (2013)
15. Zhuang, Y., Zhang, X., Wu, J., Lu, X.: Retrieval of Chinese calligraphic character image. In: Aizawa, K., Nakamura, Y., Satoh, S. (eds.) PCM 2004. LNCS, vol. 3331, pp. 17–24. Springer, Heidelberg (2004). doi:10.1007/978-3-540-30541-5_3
16. Zhuang, Y., Jiang, N., Hu, H., Hu, H., Jiang, G., Yuan, C.: Probabilistic and interactive retrieval of Chinese calligraphic character images based on multiple features. In: Yu, J.X., Kim, M.H., Unland, R. (eds.) DASFAA 2011. LNCS, vol. 6587, pp. 300–310. Springer, Heidelberg (2011). doi:10.1007/978-3-642-20149-3_23
17. Chen, Y.F., Xue, Q.F., Chen, F.F., Chen, X.D.: J. Changchun Educ. Inst. **29** (2013)
18. Kong, C., Zhang, H.X., Liu, L.: J. Shandong Univ. (Eng. Sci.) **44** (2014)
19. Liu, W., Zhu, N.B., He, H.Z., Li, D.X., Sun, F.J.: J. Chin. Inf. Process. **21** (2007)
20. Wang, H., Ding, X.Q., Halmurat: J. Tsinghua Univ. (Sci. Technol.) **44** (2013)

Research of Large-Scale and Complex Agricultural Data Classification Algorithms Based on the Spatial Variability

Hang Chen[1,2], Guifen Chen[1(✉)], Lixia Cai[1], and Yuqin Yang[1]

[1] College of Information Technology,
Jilin Agricultural University, Changchun 130118, China
chenhang0811@163.com, guifchen@163.com,
419513823@qq.com, 1172126066@qq.com
[2] Institute of Scientific and Technical Information of Jilin,
Beijing 130000, China

Abstract. In the actual classification problems, as a result of lack of clear boundary information between classification objects, that could lead to loss of classification accuracy easily. Therefore, this article from the spatial patterns of the sample properties to proceed, fuzzy clustering algorithm is proposed based on the sensitivity of attribute weights, through using the attribute weights to improve the classification capability between confusing samples, that is for researching and analysing soil nutrient spatial data with consecutive years to collect in Nongan town. Then through the analysis of the visualization technology to realize the visualization of the algorithm. Experimental results show that introducing weights portray attribute information could reduce the objective function value, and effectively alleviate the phenomenon of boundary data that cannot distinguish. Ultimately to improve the classification accuracy. Meanwhile, use of MATLAB to form visualization of three-dimensional image. The results provide a basis for to improve the accuracy of data classification and clustering analysis of large and complex agricultural data.

Keywords: Large-scale and complex data · Spatial variation law · Fuzzy clustering · Soil nutrients · Sensitive attribute weights

1 Introduction

The arrival of the era of precision agriculture [1, 2], makes a variety of complex link relationship between agricultural data features with apparent spatial variability [3] and the correlation. The consequent massive, diverse and dynamic changes, incomplete, uncertain and a series of features, so that each attribute internal link close, but contact between attributes relatively sparse [4]. However, data mining can effectively for data analysis, Wherein the cluster analysis can be used as an independent tool to obtain data distribution situation, so that can observe characteristics of each class, analysis some specific class to move forward a single step, Final extract useful information. But with the rising importance of data structure information and the data on the exponential growth. This shows traditional data mining algorithms have been unable to meet these needs. How to

© IFIP International Federation for Information Processing 2016
Published by Springer International Publishing AG 2016. All Rights Reserved
D. Li and Z. Li (Eds.): CCTA 2015, Part I, IFIP AICT 478, pp. 45–52, 2016.
DOI: 10.1007/978-3-319-48357-3_5

introduce spatial patterns of in large-scale agriculture data [5]. And to strengthen the links between attributes for regional management in order to improve the parallel and distributed implementation strategy of clustering algorithm [6, 7]. All of these are gradually attracted researchers' attention [8]. So, on the basis of K-means algorithm, according to the interdependence of spatial unit location. Li [9], who put forward a new Spatial Contiguous K-Means Cluster algorithm, who removed a lot of debris and isolated cell and taken into account the continuity of the management partition. The actual show that the method is suitable for variable precision agriculture field management operations. Fleming et al. noted that define management zones based on oil properties, terrain and farmers' production experience. Then Appeared feature selection methods which is proposed for large-scale data sets. The purpose is to improve the data processing efficiency and rationality of the decision-making program [10]. While Cui proposed Quick association rules mining algorithm based on a large dense database of vertical data step [11].

Studied the basis of the existing methods, consider the algorithm when dealing with large data sets required scalability and efficiency. Analysis the influence of spatial variability and structural information on the temporal and spatial data, the paper proposed Fuzzy C-Means algorithm that based on attribute weights. In the case of verify its reliability, analysis the algorithm through MATLAB toolbox graphical to further improve the quality of clustering results. The results showed that the introduction of spatial variability and structure information can effectively reduce losses, due to the imbalance caused by the boundary. So, in visual processing, MATLAB played an immediate role.

2 Sensitive Attribute Weights Fuzzy C-Means Analysis

The traditional clustering algorithms in the classification process vulnerable to the sample spatial variability and structure information effect, the existence of boundary data processing hard to demarcation issues, which lead to low accuracy [12]. So, fuzzy c-means cluster algorithm was introduced to analysis of data space structure information [12]. Then constructed fuzzy similar matrix directly after standardization of data. Therefore, this study and master the premise of in soil temporal and spatial variation characteristics and laws, the combination of the spatial variability of attribute weights applied to FCM algorithm. Ultimately improving algorithm's classification capability, reducing the loss of classification precision caused by boundary spatio-temporal data.

2.1 Construction of Sensitive Attribute Weights

Firstly, we analyzed spatial patterns of soil nutrient which according to experience of experts in the field and soil fertility characteristics of test area [13]. The results showed that available p variation coefficient was 31.12 %, and the number of rapidly available potassium was 21.51 %, and available nitrogen was 11.69 % [14]. Secondly, the space coefficient of variation was introduced to the algorithm and AHP could solve the weight coefficients. Now, specific steps are as follows:

(1)   Construct pair wise comparative matrix;
(2)   Selected any n-dimension normalization original vector $\mathbf{w}^{(0)}$;
(3)   Calculate $\tilde{\mathbf{w}}^{(j+1)} = \mathbf{A}\mathbf{w}^{(j)}, j = 1, 2, \cdots$;
(4)   Normalize the $\tilde{\mathbf{w}}^{(j+1)}$;
(5)   For a given precision ε in advance, when $\left| w_i^{(k+1)} - w_i^{(k)} \right| < \varepsilon, j = 1, 2, \cdots, n$,

   Then $\tilde{\mathbf{w}}^{(j+1)}$ is the requirement feature vector; otherwise return (2);
(6)   Compute $\lambda = \frac{1}{n} \sum_{i=1}^{n} \frac{\tilde{w}_i^{(j+1)}}{w_i^{(j)}}$;
(7)   Calculation $CI = (\lambda - n)/(n-1)$;
(8)   Calculated $CR = CI/RI$;
(9)   If the $CR < 0.1$, then through the consistency check; Otherwise, reconstructing paired comparative matrix;
(10)   when all layers are calculated out, to obtain the total target weight vector $A = (a_1, a_2, \ldots, a_m)$; if not, return back to (1).

2.2   Construction of Attribute Weights Fuzzy C-Means Algorithm

When deal with clustering problems with fuzzy concepts [15, 16], each sample is not divided into one class strictly, but belongs to a category at a certain membership. Define:
$$J(U, V) = \sum_{k=1}^{n} \sum_{i=1}^{c} u_{ik}^m d_{ik}^2 (U = (u_{ik})_{c \times n}, d_{ik} = \|x_k - v_i\|),$$ which based on the membership degree matrix [17]. Objective function refers to the sum of weighted square distance between the samples and the center of the cluster. The optimization class refers to make the objective function to take the minimum class. If all points of a class are closed to the center of the class, the value of the goal function is very small.

2.3   MATLAB Modeling Tool

MATLAB is an interactive programming language based on matrix manipulation, the main functions of MATLAB including data analysis, numerical calculation and engineering drawing and so on [18], it can also marked and print the graphics. This paper adopts MATLAB to process and analyze the data, compares this kind of algorithm with the traditional fuzzy c-means, compares the accuracy of the algorithm through the objective function value, and realize the 3d visualization process of the data.

3   Experimental Results and Discussion

3.1   Data Sources

The use of 3S (these are GPS, GIS, RS) and sensor technology acquisition Nongan town soil nutrient information, then based on the geolocation of arable land, then positioning collecting position of the point. Select the main factor affecting soil fertility

[19] (nitrogen, phosphorus, potassium) as the sample data for research, the sampling distribution is as follows:

The plum blossom in Fig. 1 sampling methods is the five sampling method, that is the grid on the four horns and on the center of the grid as the soil samples mixed grid soil sample. Collecting the data content to farmers in 2010 in the town of partial data, for example, as be showed in Table 1.

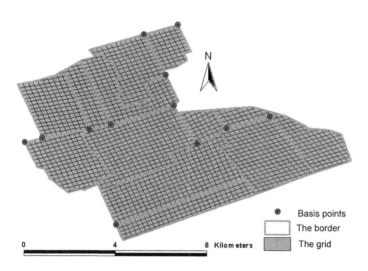

Fig. 1. Soil Sampling Point in Nongan town

Table 1. The sample data

Town name	Alkaline hydrolysis nitrogen (mg/kg)	Available phosphorus (mg/kg)	Available potassium (mg/kg)	Latitude	Longitude
Nongan town	118	30.8	208	44.51535	125.23695
Nongan town	118	36.3	198	44.50502	125.2354
Nongan town	118	42.8	227	44.50268	125.23857
Nongan town	103	32.1	198	44.51077	125.25425
Nongan town	132	15.8	227	44.50702	125.2542
Nongan town	165	10.8	237	44.51245	125.25438
Nongan town	147	30.8	217	44.51385	125.2544
Nongan town	143	21.3	198	44.5057	125.25578

3.2   Data Processing

Processing soil nutrient data in Nongan town from 2008 to 2012, Take on [0,1] evenly distributed random number to determine the initial membership degree matrix. Which determined cluster center by iteratively. Among them, step 1 iteration of the cluster centers is:

$$v^{(l)} = \frac{\sum\limits_{k=1}^{n} (u_{ik}^{(l-1)})^m x_k}{\sum\limits_{k=1}^{n} (u_{ik}^{(l-1)})^m}, \quad i = 1, 2, \ldots c$$

where c is the number of classes, m > 1.

3.3   Application and Analysis of Algorithms

Preprocessing the data after combined with the analysis of the soil nutrient spatial variation. Algorithm through continuous iterative to adjust the size of the objective function value in order to achieve the classification of soil fertility. To objects in the cluster based on the continuity of time and space, after processing the sample data, we use sensitive attribute weights fuzzy C-means algorithm to analysis the data from 2008 to 2012. Experiments show that when taking membership degree exponent 8, clustering result is obvious. In the case of the same power exponent value, compared with the traditional fuzzy C-means clustering algorithm, after repeated experiments and found that the accuracy and operational efficiency of the improved algorithm are both higher traditional algorithm. Wherein the results of 2011 as shown in Table 2.

Table 2. The result

Algorithm	Mean objective function	Average running time (s)
Fuzzy C-Means	15.118689	0.217
Attribute weights Fuzzy C-Means	11.989009	0.180

From Table 2 we can see that under the same conditions, the objective function value is smaller, and the accuracy of the relative increase 21.7 %, also it has a higher operating efficiency. That because the sample edge has no clear demarcation point, and Fuzzy C-Means could improve this problem when dealing with data. On this basis, introducing of spatial variation regularity. Without prejudice to the classification results, the better management area is divided. Combined with the results of the above analysis, using MATLAB visualization toolbox for data processing. The results obtained in Fig. 2 and Table 3.

The Table 3 and Fig. 2 show that after years of continuous precise fertilization, the similar degree of data is improved in gradually, the discrepancy between categories gradually become smaller, the soil fertility difference is leveling off. All above shows that after precise fertilization, the plot of soil in Alkeline-N, Olsen-P and Olsen-K three nutrients data integrated similarity increased year by year; On the other hand also proved that the attribute weights are C clustering algorithm is suitable for evaluation of soil fertility.

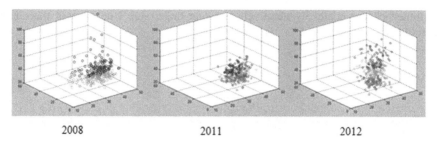

2008   2011   2012

Fig. 2. The three dimensional clustering figure

Table 3. The clustering results

Clustered data	2008	2009	2010	2011	2012
Cluster 0	3	93	44	74	82
Cluster 1	123	99	114	108	108
Cluster 2	174	108	142	118	110

4 Conclusion

The attribute weights fuzzy c-means algorithm is used to analysis and evaluation for Nongan county Nongan town soil nutrient data for five consecutive years (2008–2012). The test results show that after five consecutive years of precise fertilization, soil fertility condition had the obvious change. The attribute weights c-means clustering algorithm is an effective methods of research and evaluation of the soil fertility, in line with the farmers, the change trend of soil fertility.

Firstly, the algorithm consider the spatial variability of soil fertility, combined with AHP to determine the sensitive attribute weights. the original scattered data not only retain the traditional algorithms consider the problem of difficult to deal with the boundary points by using the concept of fuzzy sets, and to overcome the imbalance between the various properties and is sensitive to "noise" and outlier data shortcomings.

Secondly, the paper used the sensitive attribute weights fuzzy c-means algorithm to do the clustering analysis for Nongan soil data in 2011 which included the soil alkaline hydrolysis nitrogen, available phosphorus and available potassium three nutrients data. The results show that the algorithm was 21.7 % higher than that of traditional algorithm of relative accuracy and efficiency increased by 17 %, the improved algorithm clustering effect is better.

Thirdly, using the algorithm to analysis soil nutrient data which precision fertilization consecutive for five years, the results show that the whole plot soil in alkali solution nitrogen, available phosphorus and available potassium in three kinds of nutrient data integrated similarity increased with each passing year. The results of the experiment are consistent with the actual situation of soil fertility, which provides a new reference for the analysis of the status of soil fertility in the future.

Fuzzy clustering is a rather ambiguous concept, the two clustering algorithms should be repeated iterations based on the exponent value of the objective function membership degree, so as to determine the relatively close to the true clustering value. We know MATLAB can handle large-scale data, and the formation of the visual clustering results. Covered in this article the agricultural data mostly a single soil nutrient data. Face of the growing complexity of massive agricultural data, the original matrix processing mode is not enough. In the future, attention should be application testing large data sets, in order to confirm the validity of the algorithm of massive data clustering.

Acknowledgment. The paper was supported by the national "863" project (2006AA10A309), the National spark plan project (2015GA660004), National Spark Plan (2008GA661003) and Shi Hang projects of Jilin province (2011-Z20).

References

1. Zhang, S.: Research of precision agriculture automatic variable fertilization theory and technology based on GPS, GIS, pp. 1–3. Jilin University, Jilin (2003)
2. Chen, L.: Theoretical and experimental studies on variable-rate fertilization in precision farming, pp. 98–104. China Agricultural University, Beijing (2003)
3. Xiang, J., Fu, Q., Wang, Z.: Applied research progress of spatial variability theoretical in soil properties analysis. Soil Water Conserv. Stud. **15**(1), 250–253 (2008)
4. Enlai, Z., Wenning, H., Hang, L., Rong, Yu., Zhu, L.: Compression algorithm of Beidou position redundant data based on time series clustering. Comput. Eng. **2**, 40–42 (2012)
5. Chen, G.: Research and application of spatial data mining technology for precision agriculture. Jilin University, ChangChun (2009)
6. Yun, G.X.F., Xingjie, F.: Incremental - K medoids clustering algorithm. Comput. Eng. **7**(3l), 181–183 (2005)
7. Leung, K.W.T., Ng, W., Lee, D.L.: Personalized concept-based clustering of search engine queries. IEEE Trans. Knowl. Data Eng. **20**(11), 1505–1518 (2008)
8. Deng, M., Liu, Q.L., Wang, J.Q., Shi, Y.: A general method of spatio-temporal clustering analysis. Sci. China Press **42**(1), 111–124 (2012)
9. Li, X., Pan, Y.C., Zhao, C.: Precision agriculture management zones based on spatial continuity of the clustering algorithm. Agric. Eng. J. **21**(8), 78–82 (2005)
10. Li, Z., He, C.: A large data set suitable for feature selection method. Comput. Sci. **33**(4), 184–186 (2007)
11. Cui, J., Li, Q.: Quick association rules mining algorithm based on a large dense database of vertical data step. Comput. Sci. **39**(1), 134–137, 151 (2012)
12. Jing, H., Li, D., Duan, Q., Han, Y., Chen, G.: A fuzzy c-means clustering based algorithm to automatically segment fish disease visual symptoms. Sens. Lett. **10**, 1–8 (2012)
13. Zhao, Y., Han, H., Cao, L., Chen, G.: Study on soil nutrients spatial variability in YuShu City. Comput. Comput. Technol. Agric. **2**, 1–7 (2012)
14. Chen, G.F., Tsao, L.I., Wang, G.: Application of weighted spatial fuzzy clustering algorithm in soil fertility evaluation. Chin. Agric. Sci. **42**(10), 3559–3563 (2009)
15. Backer, E., Jain, A.K.: A clustering performance measure based on fuzzy set decomposition. IEEE Trans. Pattern Anal. Mach. Intell. **3**(1), 66277 (1981)

16. Li, Y., Shi, Z., Cifang, W., Li, F., Cheng, J.: Definition of management zones based on fuzzy clustering analysis in coastal saline land. Sci. Agric. Sin. **40**(1), 114–122 (2007)
17. Helong, Yu., Dayou, L., Guifen, C.: Determination of the soil nutrient management zones based on weighted fuzzy clustering. Trans. Chin. Soc. Agric. Mach. **40**, 177–182 (2009)
18. Bai, X., Xiong, S.: The implementation and application of mathematical experiment system based on MATLAB. Nanchang University (2012)
19. Umeda, M., Kaho, T., Iida, M., Lee, C.K.: Effect of variable rate fertilizing for paddy field. In: 2001 ASAE Annual International Meeting, 2001, Paper Number 01(Part. II)

An Algorithm for the Interactive Calculating of Wheat Plant Surface Point Coordinates Based on Point Cloud Model

Lei Xi, Guang Zheng, Yanna Ren, and Xinming Ma[✉]

College of Information and Management Science,
Agronomy College, Henan Agricultural University, Zhengzhou 450002, China
xinmingma@126.com

Abstract. Employing the 3-dpoint cloud model of the wheat plant as the research object, through the establishment of mapping algorithm from the screen pixel to the coordinate point of the wheat plant body surface, this paper realized the calculating for the point coordinates on the wheat plant surface utilizing the mouse operation. This algorithm was constituted of solving a user view equation of given screen pixels, screening the point cloud near the line of sight, extracting the efficient point sets near the line of sight, surface fitting efficient point set and solving the intersection of the line of sight and the fitting surface five processing flow on the basis of scattered point cloud data preprocessing and rendering. Adopting the mark function of the FastSCAN 3-d digital scanner accessory instrumental software, the comparison validation of the wheat plant body surface coordinates points' selection was conducted. The validation results show that the error less than 2.1 mm and had good precision. Considering the influence of the users' perspective to the calculation of the plant surface points coordinates, the comparison validation was conducted with the 10°, 20° and 0° perspective respectively. The results show that the error is less than 1.8 mm and had good accuracy. The mapping algorithm between the screen pixel and the object surface coordinate point established in the study could map the action of the mouse on the screen window to the operation of the object surface. Moreover, it also provided the technical reference for the establishment of the geometry measurement of the interactive plants based on point cloud data.

Keywords: Wheat plant · Point cloud model · Point coordinates · Interactive

1 Introduction

Virtual plant was always the research hot spot of the interdisciplines such as the digital agriculture and virtual reality. The 3-d geometry measurement of plants was a basic and preliminary work of the virtual-plants modeling. Whether the building of the geometric modeling and the morphogenesis model of plant organs, or the building of the quantitative topology model of the plant, the accurate, complete geometric information was necessary for support. Due to the complex characteristics of the internal morphological structure of the plants, the measurements to the plant geometrical morphology became a time-consuming complex work. With the development of the 3-d information

D. Li and Z. Li (Eds.): CCTA 2015, Part I, IFIP AICT 478, pp. 53–63, 2016.
DOI: 10.1007/978-3-319-48357-3_6

acquisition technology, rapid access to the high-precision point cloud data of the plants became possible. It also provided a new method for obtaining the geometrical morphology data of the plants. The methods of measuring the geometrical morphology data of the plants could be divided into two categories generally: the geometrical morphology measurement to the surface model formed by the reconstruction of point cloud [1–4] and the geometrical morphology measurement using the point cloud data directly [5, 6]. Due to the reconstruction of topological structure between body surface points and establish object surface model in the surface model formed by the reconstruction of point cloud, the design of the measurement algorithm was relatively simple. Moreover, the existing CAD software could be used in the measurement method. However, the shortcomings such as time-consuming and big error (produce redundant triangles) could easily arise during the reconstructing the point cloud data of the whole plant because of the complexity of the plant structure. Therefore, the current study of plant point clouds reconstruction mostly concentrated in the point clouds reconstruction of plant organs [7]. The research of the measurement to the plant geometrical morphology using the point cloud data directly was concentrated in certain fixed shape feature extraction [8] utilizing the point cloud feature extraction technology [9–11]. Although the measurement to the plant geometrical morphology using point cloud feature extraction technology possessed the advantages of the high degree of automation, the interactive measurement to the geometrical morphology of plants such as leaf edge curve, veins curve and space angle with the help of the users' selection to the surface coordinates points had the characteristics of high maneuverability and flexible. Therefore, using the three-dimensional point cloud data of the wheat plant as the object of study, the research in the paper realized the calculation of the plant surface point coordinates using the mouse operation and provided the technical reference for the interactive plant geometrical morphology measurement based on point cloud data.

2   Material and Method

The calculating of the interactive object surface point coordinates aimed to build the mapping between the screen pixels selected by the user mouse operation to the object surface coordinate points. Using mathematical language, it could be described as the calculating the intersection point of the first time intersection between the object's surface and the sight line of users along the line of sight direction. The intersection point was the mapped coordinate point of the screen pixel on the object body surface. Due to the lack of the concept of the "surface" of the object represented by the point cloud data, the intersecting probability of the user's line of sight with any point in the point cloud data all was 0 in theory. Therefore, in this study, the intersection between the line of sight and the object surface was defined as the intersection between the line of sight and the local surface fitted by the point cloud data near the line of sight. Thus, the mapping algorithm from the screen pixel point to the object surface coordinates point on the basis of the preprocessing and rendering to the point cloud data in this paper was constituted by the solving the user sight equation passing the given screen pixel point, screening the point cloud near the line of sight, extracting the efficient point set near the line of sight, surface fitting efficient point set and calculating the intersection between the line of sight and the fitted curve five processing steps and shown in Fig. 1.

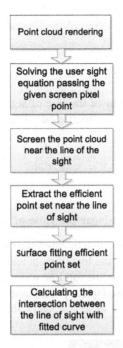

Fig. 1. The algorithm flow of the mapping from the screen pixel point to the object surface coordinate point

2.1 Solving the User Sight Equation Passing the Given Screen Pixel Point

Accounting the coordinate system (object coordinate system) describing the point cloud data was the O system. The coordinate system with the increased depth values describing the window coordinates was named the O' system. The transition from the O system to the O' system could be expressed by the transformational matrix generated by the point cloud rendering, as shown in the formula (1).

$$M = \prod_{i=n}^{1} M_i \tag{1}$$

In the formula, the M was the transformational matrix from the O system to the O' system. The M_i was the transformational matrix generated by the point cloud rendering following the sequence. Suppose that the window coordinate of the depth value of the pixel point passed by the line of sight was the $P'_w(w_x, w_y, w_z)$, the corresponding point of the pixel point P_w in the coordinate system O could be expressed by the formula (2).

$$P_w = M \cdot P'_w \tag{2}$$

The w_x, w_y were expressed by the window coordinate of the mouse during the approach implementation. The w_z could be determined by the projection matrix in the point cloud data rendering and commonly was 0. Suppose that the location of the user's view was P_e, the mathematical equation of the line of the sight could be expressed by the formula (3).

$$P(t) = P_w + v_e \cdot t \tag{3}$$

The v_e was the direction vector of the line of sight and given by the formula (4). The P_e was set during rendering the point cloud data.

$$v_e = \frac{P_e P_w}{|P_e P_w|} \tag{4}$$

2.2   Screening the Point Cloud Near the Line of Sight

As shown in Fig. 2, all the points which the distance to the sight less than the threshold r were regarded as the points near the line of sight and were defined as the point cloud near the line of sight. In the Fig. 2b, the full line was used to express the line of sight of the users. The S_1, S_2, S_3, S_4 was used to the point cloud near the line of sight. Moreover, it also was used to stand for the 4 local surface of the object.

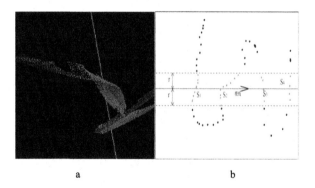

a   b

Fig. 2. The point cloud near the line of sight

Based on the data equation of the line of sight, for any point in the point cloud, the distance to the line of sight could be given by the formula (5). Any point in the point cloud could be conducted the screening and extracting following the formula (5).

$$d = \sqrt{(P_w P_i \cdot v_e)^2 + P_w P_i^2} \tag{5}$$

2.3 Extracting the Efficient Point Set Near the Line of Sight

In general, the point cloud near the line of sight often represented one or more local surfaces. According to the definition of the mapping from the screen pixels to the surface coordinates, only the local surface closest to the viewpoint was used to calculate the surface of the mapping point. That also meant that only the point set which represented the surface was effective. For example, in Fig. 2 b, the point cloud near the line of sight expressed four local surfaces. However, only the point set S_1 was valid.

In the point cloud data near the line of sight, using the clustering based on density and the WaveCluster method [12–15] both could realize the corresponding requirements of identifying effective point set. However, the clustering based on density generally required the users to set the corresponding parameters. Moreover, both the two clustering methods had the shortage that the calculation speed was slow. Considering the requirement to the procedure response speed for the "interactive" design, this research conducted the dimension reduction process to the point cloud data near the line of sight firstly. After that, a linear algorithm was designed to conduct the method for extracting the effective point set.

In the design of the dimension reduction process for the point cloud data near the line of sight, the point cloud located in the different local surface could be found. The projection of the distance from these point cloud to the point of view in the line of sight direction would have bigger difference, as shown in Fig. 3. Hence, J was set as the result for the dimension reduction process to the point cloud data near the line of sight. It could be defined as the formula (6).

$$J_i = |P_e P_i \cdot v_e| \tag{6}$$

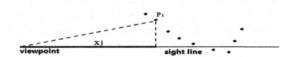

Fig. 3. The projection of the distance from the point to the point of view in the line of sight direction

In the formula, the P_i was the point i in the point cloud data near the line of sight.

Analysis the point cloud near the line of sight, J_i would appear the abnormal increasing from a local surface to another local surface along the line of sight direction usually because the threshold value r was small relatively. Therefore, the efficient point set could be detected through the analysis of the increased abnormal values of adjacent points J_i. The specific algorithm was designed as follows:

The linear table L_p was constructed and saved in the point cloud near the line of sight. After that, it was sort following the J_i of each point. The increment of the J_i of the two neighboring points was calculated and saved in the new linear table L_j. The process could be expressed as the $L_j[i] = L_p[i+1].J - L_p[i].J$. After that, the abnormal value in the L_j was detected to judge whether the point could reach the next part surface.

Considering the interactive response speed, the algorithm adopted the simple abnormal value detection method and determined whether the element in the L_j was the abnormal value by detecting whether the deviation of each element in the L_j was greater than a certain ratio standard deviation.

The two multiple parameters n_1, n_2 were set to be the threshold value for the abnormal value detection in the algorithm. The n_1 was set as the threshold value for the abnormal value-to-be and expressed that the next step detection should to be continue to conduct the abnormal value judgment. The n_2 was the threshold value for the abnormal value and expressed that the abnormal value was detected currently. The algorithm implementation process was shown as follows:

(1) A set S_p was established as a valid point set and initialized to the first three elements of the L_p.
(2) The linear table I_j was established to hold the increment of the J_i in computer and initialized to the first two elements of the L_j. The loop variable was defined i as 4.
(3) At the time of the $i > n$, the algorithm was stopped. n was the length of the L_p.
(4) Calculating the mean I_{avg} and the standard deviation I_{std} of the elements in the I_j.
(5) Executing $tmp = L_j[i-1] - L_j[i-2]$, if $tmp - I_{avg} > n_1 \cdot I_{std}$, go to (8).
(6) Putting the $L_p[i]$ in the set S_p, putting $L_x[i-1]$ in the set I_x, $i = i+1$, go to (3).
(7) If $tmp - I_{avg} > n_2 \cdot I_{std}$, that meant the next surface had been reached, exiting program.
(8) While $i+1 > n$, the algorithm was stopped.
(9) Executing $tmp = L_j[i] - L_j[i-2]$, if $tmp - I_{avg} > n_2 \cdot I_{std}$, exiting program.
(10) Putting $L_p[i-1], L_p[i]$ into the set S_p, putting the $L_x[i-1], L_x[i]$ into I_x, executing $i = i+2$, go to (3).

The set S_p was the efficient point set near the line of sight at the end of the algorithm. The evaluation of the n_1, n_2 were 4, 5 commonly. For the peculiar point cloud data, some small adjustment also could be done. In the Fig. 4, the red point set expressed the result of the efficient point set extraction near the line of sight.

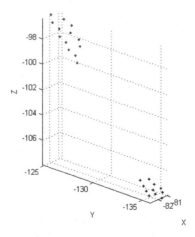

Fig. 4. The result of the efficient point set extraction near the line of sight (Color figure online)

2.4   Surface Fitting Efficient Point Set

According to the interpretation for the intersection of the line of sight and the point cloud, the intersection of the line of sight and the object was actually the intersection of the line of sight and the curved surface expressed by the efficient point set near the line of sight.

The setting of the threshold r was generally smaller during the extracting the point cloud near to the line of sight. At the same time, according to the demand of the rapid response in the user interaction, the least square plain fitting to the efficient point set was adopted to the express the local surface of object in the method. The method was as follows.

The equations of plane was set as $z = -Ax - By + D$. Substituted all the points in the efficient point set into the equation, the least square error formula could be obtained and shown as the formula (7).

$$E = \sum (-Ax_i - By_i + D - z_i) \tag{7}$$

Due to the $\nabla E = 0$, the least square solution of the A, B, D could be obtained and shown as the formula (8).

$$\begin{pmatrix} \Sigma x_i^2 & \Sigma x_i y_i & -\Sigma x_i \\ \Sigma x_i y_i & \Sigma y_i^2 & -\Sigma y_i \\ -\Sigma x_i & -\Sigma y_i & n \end{pmatrix} \begin{pmatrix} A \\ B \\ D \end{pmatrix} = \begin{pmatrix} -\Sigma x_i z_i \\ -\Sigma y_i z_i \\ \Sigma z_i \end{pmatrix} \tag{8}$$

In the formula, n expressed the number of the point sets. x_i, y_i, z_i was the coordinate value of the point i. The least square estimation of the local surface could be obtained through the solving of the equation set. For the convenience of the calculating of the intersection of the line of sight and the fitting plane, the vector V was defined as $v = (A, B, 1)$, the vector P was defined as $P = (x, y, z)$. Thus, the plane equation could be expressed as the vector expression and shown in the formula (9).

$$v \cdot P = D \tag{9}$$

2.5   Calculating the Intersection Between the Line of Sight and the Object Surface

Contact the formula (9) and the formula (3), the parameter t of the point of intersection between the line of sight of users and the fitting local surface of the object could be determined and shown as the formula (10). After that, as shown in Fig. 5, the intersection point could then be found.

$$t = \frac{D - v \cdot P_w}{v \cdot v_e} \tag{10}$$

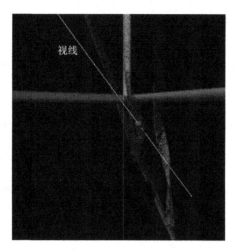

Fig. 5. The intersection point of the line of sight and the object

3 Result and Analysis

Using the point cloud data of the wheat plant collected in Henan agricultural university science park from 2012 to 2013 as the experimental material, adopting the C++ language, with the help of the software library such as OpenGL, PCL [16] and so on, the interactive wheat plant surface pastry coordinates calculating algorithm was realized at the Ubuntu12.04 system environment in the research and shown in Fig. 6. As shown in

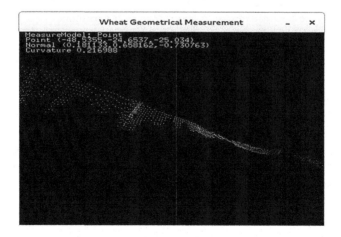

Fig. 6. Interactive measurement of point

Table 1, using the mark function of the supporting tools software of the FastSCAN 3D digital scanner, the mapping from the screen pixels to the object surface coordinates was verified. The results showed that the error of the proposed mapping method in this paper was less than 2.1 mm compared with the FastSCAN and had good accuracy. Because there was no direct intersection between the line of sight and the point cloud in this study, the error of the interactive selection of the points was mainly produced by different user perspective. As shown in Tables 2 and 3, this study conducted the comparison validation with the 10°, 20° and the 0° viewing angle respectively. The results showed that the error of the method was less than 1.8 mm in the normal range of angle and had good accuracy.

Table 1. Experimental verification of screen pixel coordinates point-to-surface point map

Serial number	Fastscan			Mapping method in this paper			Error(mm)
	x(mm)	y(mm)	z(mm)	x(mm)	y(mm)	z(mm)	
1	−93.03	41.06	208.84	−92.9382	42.48706	209.3463	1.51698
2	−72.27	27.55	168.54	−72.2284	27.91212	169.2688	0.814823
3	−52.98	18.29	136.58	−52.4159	19.09847	137.0976	1.113435
4	−49.19	12.52	116.09	−48.1335	13.99432	116.9974	2.028112
5	−37.67	7.37	101.52	−37.2543	7.99218	101.8187	0.805715
6	−24.03	5.63	81.85	−23.5856	6.558495	82.29905	1.123049
7	−33.52	3.68	83.44	−32.4752	4.144433	84.60859	1.634903
8	−13.76	1.7	45.37	−12.7095	1.78187	45.72393	1.111563
9	−0.53	1.58	8.44	0.342988	2.501845	9.016809	1.394495
10	−16.4	−1.71	44.72	−16.0489	−0.42301	45.43838	1.515156

Table 2. The verification of interactive point selection with angle of slight is 0 degrees and 20 degrees

Serial number	0°			10°			
	x(mm)	y(mm)	z(mm)	x(mm)	y(mm)	z(mm)	Error(mm)
1	−92.938	42.487	209.346	−92.083	42.793	209.082	0.948
2	−72.228	27.912	169.269	−72.062	28.466	169.009	0.634
3	−52.416	19.099	137.098	−52.542	18.906	137.786	0.725
4	−48.134	13.994	116.997	−47.982	13.460	116.648	0.656
5	−37.254	7.992	101.819	−36.953	7.862	101.332	0.587
6	−23.586	6.558	82.299	−23.459	6.893	83.078	0.857
7	−32.475	4.144	84.600	−31.813	4.848	83.938	1.176
8	−12.7095	1.782	45.724	−12.049	2.732	44.968	1.382
9	0.343	2.5018	9.0168	−0.240	2.286	8.540	0.784
10	−16.049	−0.423	45.438	−15.319	−0.668	45.156	0.820

Table 3. The verification of interactive point selection with angle of slight is 0 degrees and 20 degrees

Serial number	0°			20°			
	x(mm)	y(mm)	z(mm)	x(mm)	y(mm)	z(mm)	Error(mm)
1	−92.9382	42.48706	209.3463	−93.1475	42.12267	210.6366	1.356964
2	−72.2284	27.91212	169.2688	−72.3269	28.22446	169.8735	0.687665
3	−52.4159	19.09847	137.0976	−51.0175	20.17961	136.6038	1.835253
4	−48.1335	13.99432	116.9974	−48.4386	13.57301	117.768	0.929734
5	−37.2543	7.99218	101.8187	−36.2038	8.447684	100.6524	1.634433
6	−23.5856	6.558495	82.29905	−23.504	6.636894	82.53445	0.261196
7	−32.4752	4.144433	84.60859	−32.5588	3.288491	83.80317	1.178269
8	−12.7095	1.78187	45.72393	−13.1347	2.014106	44.55533	1.26504
9	0.342988	2.501845	9.016809	0.962779	2.533851	9.861952	1.048539
10	−16.0489	−0.42301	45.43838	−15.8339	−0.59462	45.10651	0.43108

4   Conclusion

Using the scattered three-dimensional point cloud data of the wheat plant as the research object, through the procedure including "Solving the user sight equation passing the given screen pixel point–Screening the point cloud near the line of sight–Extracting efficient point set in the point cloud near the line of sight– Surface fitting efficient point set–Calculating the intersection between the line of sight and the object surface", the mapping from the screen pixel point to the object surface coordinate point was formed. It could map the window coordinate of the user's mouse to the object surface which expressed by the point cloud and realize the calculation for the wheat plant surface point coordinate. T This method had certain universality and could to provide technical reference for the construction of the interactive plant to geometry measurement based on point cloud data.

Acknowledgements. This study has been funded by China National 863 Plans Projects (Contract Number: 2008AA10Z220) and key scientific and technological project of Henan Province (Contract Number: 082102140004).

References

1. Boissionnat, J.D.: Geometric structures for three-dimensional shape representation. ACM Trans. Graph. **3**(4), 266–286 (1984)
2. Dey, T.K., Goswami, S.: Tight cocone: a watertight surface reconstructor. J. Comput. Inf. Sc. Eng. **3**(4), 302–307 (2003)
3. Dey, T.K., Goswami, S.: Provable surface reconstruction from noisy samples. Comput. Geom. **25**(1), 124–141 (2005)
4. Yinxi, G., Cheng, H., Zhongke, F., Wenzhao, L., Fei, Y.: Amended Delaunay algorithm for single tree factor extraction using 3-d crown modeling. Trans. Chin. Soc. Agric. Mach. China **44**(2), 192–199 (2013)

5. He, C., Feng, Z., Yuan, J., Wang, J., Dong, Z., Gong, Y.: Three-dimensional volume measurement of trees based on digital elevation model. Trans. Chin. Soc. Agric. Mach. China **28**(8), 195–199 (2012)
6. Paulus, S., Dupuis, J., Mahlein, A.K., Kuhlmann, H.: Surface feature based classification of plant organs from 3D laserscanned point clouds for plant phenotyping. BMC Bioinform. **14**, 238 (2013)
7. Shenglian, L.U., Xinyu, G.U.O., Changfeng, L.I.: Research on techniques for accurate modeling and rendering 3d plant leaf. J. Image Graph. China **14**(4), 731–737 (2009)
8. Zhao, Y., Wen, W., Guo, X., Xiao, B., Lu, S., Sun, Z.: Midvein extraction for 3-d corn leaf model based on parameterization. Trans. Chin. Soc. Agric. Mach. China **42**(4), 183–187 (2012)
9. Takahashi, S.T.: Comperhensible rendering of 3-D shapes. Comput. Graph. **24**(4), 197–206 (1990)
10. Dong, Z., Mingyong, P.: Algorithm for extracting sharp features from point cloud models. Trans. Chin. Soc. Agric. Mach. China **42**(11), 222–227 (2011)
11. Qian, L., Geng, G., Zhou, M., Zhao, L., Li, J.: Algorithm for feature line extraction based on 3D point cloud model. Appl. Res. Comput. China **30**(3), 933–937 (2013)
12. Birant, D., Kut, A.: ST-DBSCAN. An algorithm for clustering spatial-temp oral data. Data & Knowl. Eng. **60**(1), 208–221 (2007)
13. Elsner, G.: A new sequential cluster algorithm for optical lens design. J. Optim. Theory Appl. **59**(2), 165–172 (1988)
14. Hinneburg, A., Gabriel, H.-H.: DENCLUE 2.0: fast clustering based on kernel density estimation. In: R. Berthold, M., Shawe-Taylor, J., Lavrač, N. (eds.) IDA 2007. LNCS, vol. 4723, pp. 70–80. Springer, Heidelberg (2007). doi:10.1007/978-3-540-74825-0_7
15. Zhou, W., Feng, X., Sun, G.: Classification based on feature extraction from cluster of wavelet coefficients. J. Comput. Res. Develop. 1–12, 982–988 (2001)
16. Point Cloud Library (2014). http://www.pclcn.org/study/shownews.php?lang=cn&id=29

Quick and Automatic Generation Method for the Evaluation Report of the Small Hydropower Substitute Fuel Project

Yingyi Chen[1,2,3,4], Jiani Xue[1,2,3,4], Huihui Yu[1,3], Jing Xu[1,3],
Zhumi Zhen[1,3], Xingyue Tu[1,2,3,4], Zhijie Ma[5], Yun Zhao[5],
and Yanzhong Liu[6(✉)]

[1] College of Information and Electrical Engineering,
China Agricultural University, Beijing 100083, People's Republic of China
[2] Key Laboratory of Agricultural Information Acquisition Technology,
Ministry of Agriculture, Beijing 100083, People's Republic of China
[3] Beijing Engineering and Technology Research Center of Internet of Things
in Agriculture, Beijing 100083, People's Republic of China
[4] Key Laboratory of Modern Precision Agriculture System Integration,
Ministry of Education, P.O. Box 121, Beijing 100083, People's Republic of China
[5] Engineering Design and Research Center, IWHR, Beijing 100044, China
[6] Institute of Information for Agriculture, Shandong Academy of Agricultural
Sciences, Jinan 250100, People's Republic of China
lyzl228@163.com

Abstract. The small hydropower substitute fuel (SHSF) project is a commonweal project, give prominence to the ecology benefit. The manager of SHSF need a full and specification report for the evaluation results, rather than a database or a simple data of the result. So, based on the characteristics of the management and evaluation system of SHSF project, this paper presents a multi-indicator comprehensive evaluation module and report template, which includes environmental impacts on climate change, conversion of cropland to forest and natural forest protection project, greenhouse gas emission and the role of social and poverty alleviation. In order to achieve automated building of the comprehensive evaluation report of the small hydropower replacing firewood project, a quick and automatic generation method is constructed for the evaluation report based on C# .Net development environment. It produces evaluation report by reading, traversing and replacing the corresponding position data placeholder of customized multi-indicator evaluation report template. The results showed that the quick and automatic generation method meets the requirements of the management and evaluation system of the small hydropower replacing firewood project successfully, and can be referenced in study and practical application.

Keywords: Small hydropower substitute fuel (SHSF) · Multi-indicator evaluation report template · Quick and automatic generation method

© IFIP International Federation for Information Processing 2016
Published by Springer International Publishing AG 2016. All Rights Reserved
D. Li and Z. Li (Eds.): CCTA 2015, Part I, IFIP AICT 478, pp. 64–71, 2016.
DOI: 10.1007/978-3-319-48357-3_7

1 Introduction

The small hydropower substitute fuel (SHSF) project is a public welfare project that consolidate the achievements of returning farming land to forestry, and solve the problem of fuel for rural life. It gives prominence to the ecology benefit and was named as "projects lightening the hope of mountain area" by local people. From 2009 to 2015, and there are 1022 small hydropower for fuel ecological protection projects in construction in fragile areas. Different from conventional construction of small hydropower energy issues, the small hydropower replacing fuel small hydropower project is to use firewood instead of rural life, so as to achieve the purpose of protecting the forest [1, 2]. At present, the experiment unit work of the small hydropower replacing firewood project is just start, with China's great efforts to developing the small hydropower replacing firewood projects, a large number of project data in urgent need of regulation and evaluation. As an important part of project evaluation and management, efficient and rapid method to generate standardized project evaluation reports is critical.

In order to facilitate the management of the documents generated for further processing and reuse, Word is used as the report output of the application tools can further extend the application's functionality, improve utilization and shareability of data [3–5]. In general, report document formats are relatively fixed [3, 6], so in the most of these reports generated methods, report template is often designed well in advance and the label, or specific keywords are insert into the specific location where need to be filled in by the program. And then the reports are generated by replacing the label, or specific keyword in the build process. The benefit of this approach that the report is product by the program document style is to be designed outside has the advantage of reducing the amount of code, improved implementation efficiency [7].

Facing great development need of small hydropower for fuel ecological protection projects, drawing on the domestic and international water energy management experience as well as the sustainable development of small hydropower management structure, this paper presents a multi-indicator comprehensive evaluation module based on the study of eco-benefit evaluation methods and overall framework of the management and evaluation system of the small hydropower project, it produces evaluation report by using default evaluation report template, which includes environmental impacts on climate change, conversion of cropland to forest and natural forest protection project, greenhouse gas emission and the role of social and poverty alleviation. In order to achieve automated building of the comprehensive evaluation report of the small hydropower project, we construct a quick and automatic generation method for the evaluation report based on C# .Net development environment. It produces evaluation report by reading, traversing and replacing the corresponding position data placeholder of customized multi-indicator evaluation report template. The quick and automatic generation method for the evaluation report of the small hydropower project can provide technical support for benefit analysis of ecological environment.

2   Materials and Methods

2.1   The Structure of Evaluation Report Generation Module

The development objective of the comprehensive evaluation module based on Microsoft Visual Studio integrated development environment and .Net Framework 4.5 platform is to generate Evaluation report quickly and automatically according to the user requirement. The structure of Evaluation report generation module is shown in Fig. 1.

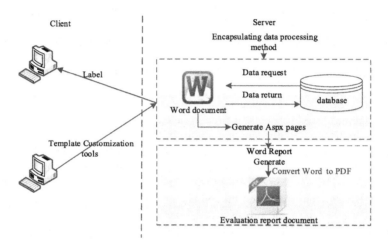

Fig. 1. The structure of evaluation report generation module

The evaluation module design includes three parts: document template customization part, database and report generation part: (1) Document template customization: Document template customization is a tool that runs on a client for the administrator to use for making report document template according to their requirement. Document templates, including fixed and replaceable parts. The replaceable parts is specified using the designated (For example, label in word). (2) Database: Database contains static data and dynamic data tables (3) Report generation: According to user-submitted templates, the report is product by replacing the labels designed in word to specify the content.

2.2   Encapsulating Data Processing Method and Establishing Label

Word report including data recording method and data processing method of data query. (1) the main methods of data entry manual entry of text information. (2) data querying method based on user-selected criteria from the database to retrieve the corresponding data, and graphically displayed. Data-processing methods, including methods and method body, method header indicates the method name, method calls the conditions and method return value type, including method return value type into text strings, images, and tables, one for each input class methods and query class method. Features in each data processing method using the C# language adds a description only shows that it features, the format is a text string [8], and finally through the compiler to encapsulate the data processing method for the dynamic-link library (DLL) files. Technical data processing methods using reflection file description of the analytic methods, sorting packages for labels, and configured service interfaces using Web Service technology release. Labels include tags categories, tags and the corresponding data source [9].

2.3   Report Template Customization

Report template customization is a tool that runs on a client for the administrator to use for making report document template according to their requirement. Report templates, including fixed and replaceable parts, which is specified using the designated (For example, label in word), are stored in database.

2.3.1   Evaluating Indicator of the Small Hydropower Replacing Firewood Project

Report templates makes up of multiple evaluation, which is environmental impacts on climate change, conversion of cropland to forest and natural forest protection project, greenhouse gas emission and the role of social and poverty alleviation.

(1) Environmental impacts on climate change indicators include eco-environment, energy saving and emission reduction, new dam construction technology and capacity to respond to climate change.
(2) Conversion of cropland to forest and natural forest protection project indicator includes the absorption of water conservation, soil conservation, nutrients accumulation, the atmosphere and environment purification, biological diversity protection.
(3) Greenhouse gas emission indicator includes greenhouse gas emissions of the small hydropower replacing firewood station and greenhouse gas emissions of returning farmland to forest.
(4) The role of social and poverty alleviation indicator includes per capita income, Engel's coefficient, increase in employment and social support degree.

2.3.2   Customization of the Multi-indicator Comprehensive Evaluation Report Template

After establishing the corresponding position label is accomplished, the report template can be customize. The steps of creating a custom template as follows: (1) creates a new, blank Word document, in accordance with the changes to the normal document methodology and demand, add reports to be automatically generated by content in the document including fixed static text, header, footer, header, separator(such as symbols) and dynamic text. Among them, the placeholder for dynamic text with the corresponding logo. If the report needs to insert a table, where the paragraph you want to set the Insert table, and then inserts a page break in the next paragraph, make your spreadsheet into a region. If multiple sets of graphs and charts is needed, the appropriate place to insert a text box in Word document template for each size format, such as a text box, and then inserts a placeholder inside a text box. In order to distinguish data placeholders from normal text, you can set a special style for it. (2) Save the word document that contains placeholders for data as a report template (Fig. 2).

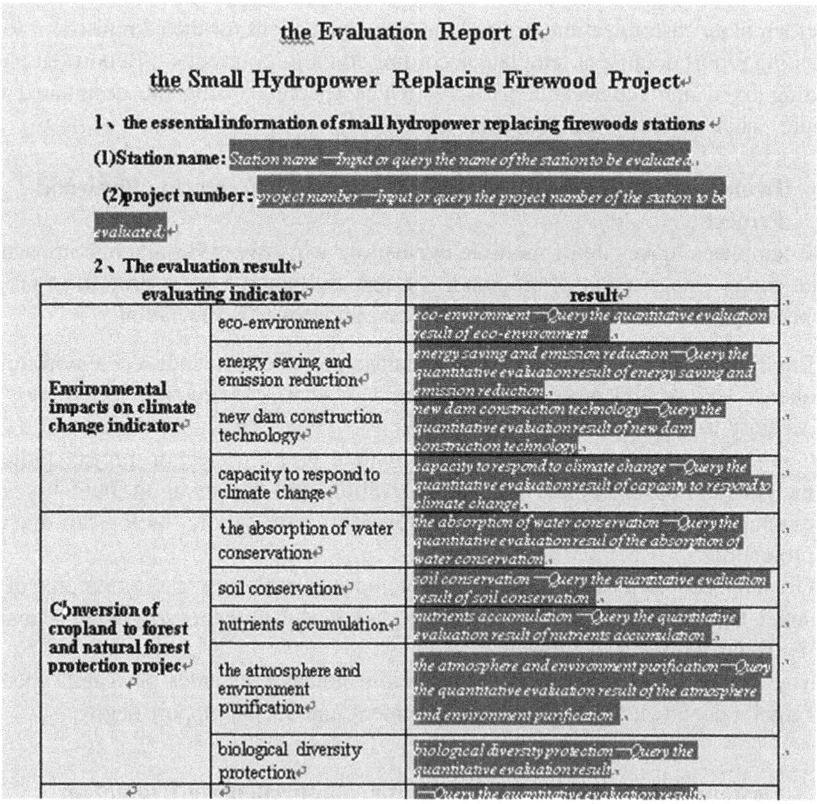

Fig. 2. The partial of the evaluation report template

2.4   Conversation of Dynamic Web Pages

Using report templates to create a report in Word format directly, does not have a function that interacts with a database query based on a condition [10], you also need to convert a Word report templates in the form of dynamic Web pages, follow these steps: (1) The way to traverse the bookmark lookup data placeholders in a template and read the corresponding value, and data processing method of descriptive information. (2) Through reflection technology, according to the description corresponds to the one by one the relationship between information and data processing methods to find data processing methods. (3) the ASPX data display controls is determined depend on the value type of data processing method Reflected. (4) The Word report templates are transformed into Html forms. (5) Using regular expressions to remove comments and style information in the Html file, and add the Aspx pages such as Js and css infor-mation. (6) Through the template engine, using Aspx data components (such as TextBox, PictureBox and GridView) replacing data placeholders in your Html to generate dynamic pages.

3   Evaluation Report Generation

The variety of information in static and dynamic data in the database tables is the input section of the evaluation report, and the evaluation report of the small hydropower replacing fuel project as the output. The generation process of the small hydropower replacing firewood evaluation report is as follows. User finished editing the evaluation report on client under through the data operation of query and inserts by visiting a web page dynamically controls. The Placeholder data and data processing method is one to one in database. The server calls data processing method corresponding to the data placeholders of multi-indicator evaluation report template customized for reading data, and binding Data display components in Dynamic pages. Page structure is similar to final report. In page, queries and modifications operations repeatedly are allowed, and temporarily saved is supported. Then, the contents of all controls are serialized in backed server, and stored to the database.

When the user selects the generating reports Operation, backed server traverse the each data placeholder of multi-indicator evaluation report template customized. At the same time, the corresponding position data placeholder of multi-indicator evaluation report template customized is replaced by the content of the corresponding control that pass through the page. All data placeholders in the template are replaced, and product the word document. Then convert WORD to PDF, and save it as the final evaluation report (Fig. 3).

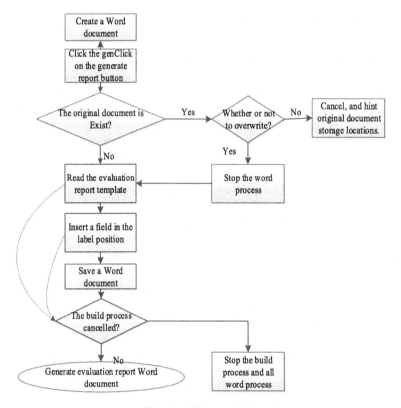

Fig. 3. WO control flow diagram

4   Conclusions

Based on the characteristics of the management and evaluation system of the small hydropower project, in this paper, a quick and automatic generation method based on Microsoft Visual Studio integrated development environment and C# .Net Framework 4.5 platform is proposed. The application results show that this report generation method based on C# .Net language is efficient and rapid, and meets the requirements of the management and evaluation system of the small hydropower replacing firewood project successfully, and can improve the availability and shareability of the material of the management and evaluation system.

Acknowledgements. We acknowledge the financial assistance provided by the Key Program for International S&T Cooperation Projects of China (2012DFA60830).

References

1. Chen, H., Li, R.: Relevant issues at project planning of using small hydropower replacing woody fuel for ecological protection. China Water Resour. **13**, 37–41 (2003)

2. Sun, T., Fei, L., Li, H., et al.: Estimation of the rural electric power consumption in the ecological protection areas with implementing the project of small hydropower for fuel. Trans. CSAE **22**(3), 112–115 (2006). In Chinese with English abstract
3. Lu, B., Yang, X.: Using delphi creates the word report and the dynamical structured form. Comput. Appl. Softw. **4**(3), 180–183 (2007)
4. Zhang, W., Wu, X., Liu, W.: A research on and realization of output of structured documents. J. Spacecr. TT&C Technol. **26**(6), 91–94 (2007)
5. Gao, H.: The Design and Realize of Physical Report Generation System. Dalian University of Technology, Dalian (2008)
6. Ge, F., Wu, N.: A platform for automatically producing word document based on multiple techniques. J. Univ. Electron. Sci. Technol. China **36**(2), 263–266 (2007)
7. Xiong, Y., Chai, Y., Wang, S., et al.: Word document automatically generated technology based on VC. Comput. Era **1**, 52–54 (2010)
8. Troelsen, A.: Pro C# 2010 and the .Net 4. 0 Platform. Posts & Telecom Press, Beijing (2011)
9. Sharp, J.: Microsoft Visual C# 2010 Step by Step. Translated by Zhou Jing. Tsinghua University Press, Beijing (2010)
10. Liu, P., Qin, W., Zhou, Y., Guo, Y.: Automatic generation method of word report based on dynamic web page. Comput. Eng. **38**(5), 279–284 (2012)

Aquaculture Access Control Model in Intelligent Monitoring and Management System Based on Group/Role

Qiyu Zhang[1,2], Yingyi Chen[1,5], Zhumi Zhen[1,5], Jing Xu[1,5],
Ling Zhu[1,4], Liangliang Gao[1,6], and Yanzhong Liu[3(✉)]

[1] College of Information and Electrical Engineering,
China Agricultural University, Beijing 100083, People's Republic of China
[2] Yantai Academy, China Agriculture University, Yantai 264670,
People's Republic of China
[3] Institute of Information for Agriculture,
Shandong Academy of Agricultural Sciences,
Jinan 250100, People's Republic of China
lyzl228@163.com
[4] Shandong Institute of Business and Technology, Yantai 264605,
People's Republic of China
[5] Key Laboratory of Agricultural Information Acquisition Technology,
Ministry of Agriculture, Beijing 100083, People's Republic of China
[6] College of Information Science and Engineering,
Shandong Agricultural University, Taian 271018, China

Abstract. Aquaculture intelligent monitoring and management system (AIMAMS) is a web information system covering businesses, home users, and management of aquaculture information, which have many users. The information security and safety equipment of different users is an important issue that we have to consider. Role-based access control introduces the roles concept into user and access rights, and its basic feature are that divide roles depending on the security policy, assign operating license for each role, and assign roles to each user. The user can access the specified object based on their respective roles. In order to achieve different user needs for security, aquaculture access control model based on group/role is developed which is based on analysis of the role-based access control model. Aquaculture intelligent monitoring and management system consists of system administrators, business users, family farming users and technicians, each category contains a number of other users who have different permissions. For such a complex distribution of competences, the user groups are introduced here on the basis of idea of role-based access control model. Users are divided into four groups which correspond to four categories of users, each group is given the largest collection of operational authority, and the users of each group are assigned different roles, which have all or part of privileges of the current group. Technicians can access business and family farming user's information, but cannot view some sensitive data, such as price, therefore, technicians group is controlled by field-level permissions of data tables, the sensitive fields of business and family farming user data tables is shielded. This model can effectively reduce the overhead of rights management, and can improve system's scalability. Along with the needs of business

Published by Springer International Publishing AG 2016. All Rights Reserved
D. Li and Z. Li (Eds.): CCTA 2015, Part I, IFIP AICT 478, pp. 72–81, 2016.
DOI: 10.1007/978-3-319-48357-3_8

development, the system can easily add new user groups and assign permissions and roles for them. This aquaculture access model is universal, it is not only suitable for aquaculture intelligent monitoring and management system but also has reference value for other systems.

Keywords: Group · Role · Access control · Aquaculture · Permission

Aquaculture intelligent monitoring and management system (AIMAMS) real-time online monitoring the important environmental factors of water quality such as water temperature, dissolved oxygen, PH value in the aquaculture process, by the integration of water quality sensors, Wi-Fi, GPRS and other technologies. This system provides monitoring and management functions which is data query, data collection, curve analysis, and provides more functions which is aquaculture profiles, map browsing, disease prevention and control, feed fed decision, application configuration, provides comprehensive information intelligent automation services for aquaculture users. AIMAMS is a java-based web information system which covers business and many home users. How to ensure information security and the safety equipment of different users? Role-based access control (RBAC) is a good choice. As a security mechanism, RBAC has been widely accepted, which can greatly reduce the cost and complexity of large networked and Web-based systems [1].

1 Role-Based Access Control

The concept of RBAC began in the 1970s, accompanied by multi user and multi application online systems [2]. The basic idea of RBAC is that introduce the concept of role between the user and the permission and link users and roles, and then control user to access system resources by giving permission to role. Role is a set of permission, a user has a role means the user can have all permission owned by the role. A user can have many roles and a role can be assigned to many users; A role can contain multiple permission and a permission can be included more than one role. A user is not directly associated with permission but get permission by role [3].

To facilitate understanding of the various aspects of RBAC, Sandhu et al. [2] defines four models, their relationship are shown in Fig. 1, Fig. 2 describes the basic features.

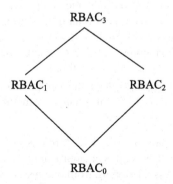

Fig. 1. Relationship among RBAC models

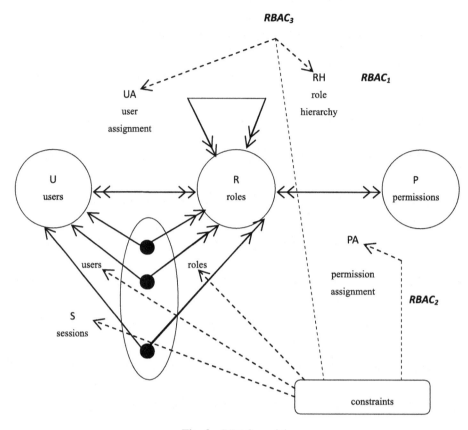

Fig. 2. RBAC models

1.1   RBAC₀ [2]

RBAC₀ is the base model of RBAC, which contains the most basic requiermets. RBAC₀ model is composed of various parts of Fig. 2.

RBAC₀ has the following components:

- U, R, P, and S(users, roles, permissions and sessions),
- $PA \subseteq P \times R$, many-to-many assignment relation to permission and role,
- $UA \subseteq U \times R$, many-to-many assignment relation to user and role,
- user: $S \to U$, function mapping user (s_i) of each session s_i to the single user (user (s_i) is a constant in the session's lifetime)
- roles: $S \to 2^R$, function mapping roles(s_i) of each session s_i to a set of roles, roles (s_i) $\subseteq \{r | (user(s_i), r) \in UA\}$ (which can change with time) and session s_i has the permissions $\cup_{r \in roles(s_i)} \{p | (p, r) \in PA\}$.

In this model, a user is an organization's staff. A role is a function of a task or job title within the organization. An permission is an approval that users access and operate data and resources in the system. As a mapping from one user to many roles, session can be created during the user activates some subset of their roles.

1.2   RBAC$_1$ [2]

Based on RBAC$_0$, RBAC$_1$ increases the concept of role hierarchy that roles can inherit permissions from other roles.

The RBAC$_1$ model has the following components:

- U, R, P, S, PA, UA, and user are unchanged from RBAC$_0$;
- $RH \subseteq R \times R$ is a partial order on R called the role hierarchy or role dominance relation, also written as $\geq$;
- roles: $S \rightarrow 2^R$ is modified from RBAC$_0$ to require roles(s$_i$) $\subseteq \{r | (\exists r' \geq r)$ $[(user(s_i), r') \in UA].\}$ (which can change with time) and session s$_i$ has the permissions $\cup_{r \in roles(s_i)} \{p | (\exists r'' \leq r)[(p, r'') \in PA]\}$.

1.3   RBAC$_2$ [2]

Based on RBAC$_0$, RBAC$_2$ increases constraints mechanism (that is, configuration is limited between different components in RBAC). These constraints are needed to determine the values of various parts in RBAC$_0$ are acceptable or not. Only those values which are acceptable is permitted.

1.4   RBAC$_3$ [2]

RBAC$_3$ combined with RBAC$_1$ and RBAC$_2$, provides both role classifications and hierarchies.

2   Aquaculture Access Control Model Based on Group/Role

The paper studies the role-based access control model, and combines with the characteristics of aquaculture intelligent monitoring and management system, then proposes the aquaculture access control model based on group/role. The model divides into four major groups based on the nature of users, and each large group divides into several different groups by per unit or duties. Each group has a largest collection of identified operational authority. A manager is assigned to a group, who assigns different roles to each user group, the user has all or part of the privileges of the current group.

2.1   User Groups

AIMAMS has many users. Depending on the subordinate units, the users divide into four categories which are system administrators, business users, home users and technicians. System administrators are staff who work in system website operator unit, business users are enterprise who use the system, home users are individual users who use the system, and technicians are users who provide technical guidance to business users and home users, for example, technicians of fishery technical extension station and so on. Except system administrators, each type of user has users coming from different units, these users also contain a number of users who have different functional privileges. For such a complex distribution of competences, the paper introduces the user groups basing on the basic idea of role-based access control model, and divides the four categories into four groups. Each users group divides into different groups by unit.

The group of users $G = \{GA, GE, GF, GT\}$, where $GA = \{GA_0, GA_1, GA_2, ..., GA_m\}$ is system administers group. GA_0 is super administrator group whose function is to manage other system administrators group; $GA_1 \sim GA_m$ is system administrator group whose function is to manage business user, home users and technicians by region, area can be north, northeast, east and other large areas, can also be provincial, municipal and autonomous regions; each region's the administrator user is responsible for the maintenance work which contain managing business users group, home users group and technicians group and initializing the system. $GE = \{GE_1, GE_2, ..., GE_n\}$ is business users group, $GE_1 \sim GE_n$ is specific business user groups. $GF = \{GF_1, GF_2, ..., GF_s\}$ is home users group, $GF_1 \sim GF_s$ is specific home user groups. $GT = \{GT_1, GT_2, ..., GT_t\}$ is technicians group, $GT_1 \sim GT_t$ is specific technician groups who guide business users and home users.

2.2   Users

Because a lot of units involved, the system administrator is impossible to establish an account for each user. A better way is grouped by team management, that is, administrators can specify an administrator user for each group of business users, home users and technicians, and the group administrator can add new user in oriented group.

The user U applies the same packet format as users group G, that is, $U = \{UA, UE, UF, UT\}$, where $UA = \{UA_0, UA_1, UA_2, ..., UA_m\}$, $UE = \{UE_1, UE_2, ..., UE_n\}$, $UF = \{UF_1, UF_2, ..., UF_s\}$, $UT = \{UT_1, UT_2, ..., UT_t\}$. As many user groups, one certain business user group UE_i ($1 \leq i \leq n$) as an example to illustrate users representation. $UE_i = \{ue_{i0}, ue_{i1}, ue_{i2}, ..., ue_{ik}\}$, k is the number of non-administrator users of this group, ue_{i0} is the administrator user of this group.

2.3   Roles

Defining roles is associated with specific applications, these is no fixed pattern, the model defines roles using the "two-step method" [4]. The first step is to determine the class of role, according to the business processes of the business, the commonality of the various stages of the business is abstracted as function terms, in order to define each class of role. The second step is to determine the subclass of role, the model makes sure that the class of user group has permissions (like query, add, update, update, startup, shutdown) to specific data or equipment, according to type of business, users powers, as well as the current state of the system, that is the true sense of role.

They are predefined roles of system, which are determined using above method. For each group, the system pre-defined roles are not directly used, the role of each group must is associated with the data and equipment of current group, and must create a new role by inheriting existing roles or according to the special needs of current group.

The role R applies the same packet format as users group G, that is, R = {RA, RE, RF, RT}, where RE = {RE_1, RE_2,..., RE_n}, RT = {RT_1, RT_2,..., RT_t}. As many user groups, one certain business user group RE_i ($1 \leq i \leq n$) as an example to illustrate roles representation. RE_i = { re_{i0}, re_{i1}, re_{i2},..., re_{ih}}, h is the number of non-administrator roles of this group, re_{i0} is the administrator role of this group.

2.4   Permissions and Representation

2.4.1   Permissions

Depending on the circumstances of AIMAMS, the permissions are divided into six kinds, they are menu permissions, page access permissions, button permissions, access permissions to data area, field-level access permissions and device permissions.

(1) Menu Permission

In order to make different levels of minimum user access to their own user interface, the menu is managed as an authority [5]. The system grants the menu permissions, and organizes flexibly application's menu composition based on the user's permissions, then each role user has a better human-computer interface [6]. When the lack of user permissions, the menu item is grayed out [7] or not displayed. The latter is better.

There are two menus in AIMAMS, they are main menu and tree menu for on-line monitoring. The main menu consists of a menu and submenu, and grants permission for the role by clicking the check box of two-dimensional table composed by a menu and submenu, online monitoring tree menu grants permission for the role by clicking the check box of the tree menu. And these two menus dynamically generated according permission.

(2) Page Access Permission

In order to prevent unauthorized access to the page, the system must set permission for each page, so the page verifies access permission when it is loaded initially to allow or ban the user from accessing the page [6]. The system is modular in design, and creates a file for each module, where stores the relevant page files. IO operations traverse through all folders in the root directory of the system, and get each module JSP pages. The tree structure with check box grants permission for the role, which is composed by module and page.

(3) Button Permission

Button operation is an important operating mode bearer service processing actions. By controlling button permission, applications can achieve transaction processing for multiple users at different stages of the same transaction on the same page, then make enterprise information systems more intuitive and efficient [6]. In fact, the button permissions correspond with the SELECT, INSERT, DELETE, UPDATE operation of the data table. When a user has some permissions, the corresponding button is clicked, otherwise it is gray, that can not be clicked.

(4) Access Permission to Data Area

Each group operating pages of business users and home users are the same, but they can only operate their own data and equipment. According to the business process, the page data are divided by group, so that different user can only handle the data that the user has the permissions on the same page [6].

(5) Field-Level Access Permission

When the different users use the same type of permission, there is a difference that they can access resources and use functions [8]. In this case, the system uses the field-level access control method in data table. Technicians can access information of business users and home users, but can not view sensitive information, such as price and value. During view, each field must determine whether the current role be prohibited to display, only field that is not prohibited can be displayed.

(6) Device Permission

Online monitoring system can control aerators, feeding machines and other equipment according to monitor status of the system. In order to ensure the safety of equipment, only the current group user who have permission can startup and shutdown operating position of the devices.

2.4.2   Representation of Permission

The authorization status of role can be described by access control matrix [9]. In the access control matrix, the row is a role, and each column is an authorized ACL (Access Control List), which is a Boolean value, as shown in Table 1. The matrix value of the i-th row j-th column is represented by RP_{ij}, $RP_{ij} = TRUE$ means that the role of R_i has permission P_j, otherwise it does not have the permission.

Table 1. Access control matrix

	P_1	P_2	...	P_j	...
R_1	TRUE	TRUE	...	FALSE	...
R_2	TRUE	FALSE	...	TRUE	...
...	...	...	...	...	...
R_i	FALSE	FALSE	...	TRUE	...
...	...	...	...	...	...

Role permission can be saved in HashMap, and only the permission of the access control matrix whose value is true can be saved. There are six permissions, the menu permission separates main menu and online monitoring tree menu, so there are seven access matrix. Button permission, access permission to data area and field-level access permission are combined into a HashMap, which all operate the data table. Therefore, the system can devise five HashMap to save all permissions of a role.

2.5   Constraints

In a model, the constraints are divided into relation constraint, precondition constraint, cardinality constraint, time constraint, address constraint and so on.

(1) Relation Constraint

Conflicting role can not be granted to the same user, for example, in order to prevent fraud, the purchaser role and accounting role can not be granted to the same user; conflicting role can not be activated in the same session [10], for example, the administrator role and other role can be granted to a user, but can not be activated at the same time.

(2) Precondition Constraint

A user wants to be granted the role A, it must already have the role B, that is a role wants have a permission, the precondition is that it must have another permission [10]. If a user already has a technician role, it can be granted a technical director role. If a user already has the corresponding menu permissions, it can be granted the submenu permissions.

(3) Cardinality Constraint

Cardinality constraint is a upper bound of users number, these users are granted the same role. n represents the maximum number of users assigned this role [11]. The value of n is specified by the group administrator, and the default value is 1.

(4) Time Constraint

Roles or permissions can be only activated in certain time range [12]. When a role need to grant a user temporary, the role can be setted time constraint, and it can be activated only in the specified time range.

(5) Address Constraint

Address constraint is divided into IP-role [13] and MAC-role constraint, that is, address constraint binds the IP address, MAC address and the role together. The administrator has a very high permission. The extent of damage to the system caused by their operation is far higher than the average user. If the administrator permissions is to be attacked, the damage to the information system is inevitably fatal [14]. So the administrator can be granted the IP-role constraint, MAC-role constraint or IP-MAC-role constraint, and only users whose IP, MAC address is required can activate the administrator roles.

3   Access Control Process

(1) User login, verify the user name and password. If the user passes the verification, the user is prohibited access system. Once authenticated, users get all the role information, and check relationship constraints, time constraints and address constraints, then get the effective role of the user. If a time constraint of a role in time constraint after the current time, this role is an effective role, but there are time constraints.

(2) If the user has multiple valid roles, the six kinds of permissions for each user role represents a seven access control matrix, access control matrix between the

different roles is a logical or operation, the user can get the access control matrix. If the user has a valid role, the user's access control matrix if a role's access control matrix. If the user does not have a valid role, the system gives tips and prohibits the user access.

(3) The access control hash table, which is structured according to the user's access control matrix, is added to the user class, and the user's complete object is added to the session. At the same time, the user name is added to the online user hash table, where the scope is application, the key means user name, value is the flag vale of corresponding seven access control matrix. value= zero means the corresponding access control matrix does not change, value= 1 means change. Each time a user generates access control hash table, the value of the corresponding online user hash table is set to "0000000". If the user is logged, the administrator will determine which access changes when the user roles or permissions changes. The value is set to 1, if the matrix is changed.

(4) When the user has access or operation request, the administrator firstly check the online hash table to see whether role or permission is change. If there is no change, turn (5); if there are changes, the system generates access control matrix again, and generates the hash table.

(5) By querying the corresponding access control hash table, the system judge whether the user has permission. Only the user has permission, its access or operation will be response.

4   Conclusion

Based on the research of role-based access control model, the paper put forward the aquaculture access control model based on group/role, combined with the safety requirements of AIMAMS. The model can simplify the complex permission, reduce the system permissions management overhead effectively, and improve the expansibility of the system. With the business development, the system can easily add new user groups and grant permissions and roles to them. The model is general, it is not only suitable for AIMAMS, but also has reference significance to the other system.

Acknowledgements. This paper was supported by Shandong Province Self-innovation Projects (2012CX90204) and the Shandong Province Key Research & Development Program "Research on aquaculture management and application platform technology based on big data" (No. 2015GGC02066).

References

1. Ferraiolo, D.F., Richard Kuhn, D., Chandramouli, R.: Role-Based Access Control, 2nd edn. Artech House, London (2007)
2. Sandhu, R.S., Coyne, E.J., Feinstein, H.L., Youman, C.E.: Role-based access control models. IEEE Comput. **29**, 38–47 (1996)

3. Shin, M.E., Ahn, G.-J.: UML-based representation of role-based access control. In: IEEE 9th International Workshops on Enabling Technologies: Infrastructure for Collaborative Enterprises (WET ICE 2000), pp. 195–200 (2000)
4. Zhang, W.: Research on role-based access control and its application in court system. Southwest Jiaotong University, Chengdu (2003)
5. Yun, L.: The application of role-based access control technology in union equipment management center system. Xidian University, Xian (2007)
6. Xu, J., Chen, D.: Research of permission management based on role and discretionary access control. Softw. Guide **12**(9), 160–162 (2010)
7. Zhang, H., Liu, Z., Li, Y., et al.: Research and application of role-based access control in privilege management. Microcomput. Inf. **22**(9–3), 29–31 (2006)
8. Yang, F., Zhang, B.: RBAC based classified application system access control method design. In: The Paper Collection of 2012 MIS/S & A Academic Conference, pp. 182–185 (2012)
9. Tang, P., Chen, M., Liu, L., et al.: Design and implementation of a practical role-based access control model. Comput. Appl. **22**(12), 41–43 (2002)
10. Wang, Z., Feng, S.: Specify RBAC constraints using object constraint language. Comput. Eng. Appl. **39**(21), 100–102, 109 (2003)
11. Zhang, H., Zhou, J., Zhang, B.: Research of extension of static constraints mechanism in RBAC Model. J. Beijing Univ. Posts Telecommun. **31**(3), 123–127 (2008)
12. Guang-yu, D., Si-han, Q., Ke-long, L.: Role-based authorization constraint with time character. J. Softw. **13**(8), 1521–1527 (2002)
13. Wu, X.: Web applications of role-based access control. Dalian University of Technology, DaLian (2002)
14. Fan, J., Guan, B., Li, X.: Design of extended role-based access control model and its implementation. Comput. Eng. Des. **29**(18), 4178–4721 (2008)

A Decision Model Forlive Pig Feeding Selection

Xinxin Sun, Longqing Sun[(✉)], and Yiyang Li

College of Information and Electrical Engineering,
China Agricultural University, Beijing 100083, China
{sunxinxin, sunlq}@cau.edu.cn

Abstract. With massive and intensive development of Chinese live pig industry, the imbalance of feeding nutrition turns out to be a primary problem. The main reason is that the nutrition needs to be adaptive to live pig's exact growth conditions in production management process. Regarding this, the paper aims at developing a feeding selection model for pig breeding based on major nutrients predictive model proposed in the Nutrient Requirements of Swine published by U.S. NRC in 1998. The objective of this model is to achieve a minimum cost of feeding stuff under the premise that the nutrients satisfy pig's needs. According to values proposed in NRC and nutritional elements data, a feeding selection model was developed to give proper suggestions on feeding dietary nutrients as well as exact type and quantity of pig feeding stuff. The work presented in this paper would contribute to the optimal precision feeding strategies so as to guarantee live pig growth quality in management process.

Keywords: Live pig breeding · Nutrients prediction · Daily nutrition · Feeding quantity · Precision feeding · Nutrition requirement

1 Introduction

The main cost of live pig feeding comes from the pig feeding stuff cost, accounting for 60–70 % of the total cost of the large-scale pig feeding, and accounting for 70–80 % of the total cost of pig specialized households [1], higher than labor costs (14–22 %) and piglets costs (18–35 %) [2]. There is at least 10 % of the feeding stuff waste every year [3], result in feeding costs continue to rise, limiting our economic development of pig feeding and polluting the environment [4]. In the condition of ensuring the healthy growth of pigs, depending on the growing season, timing and quantitative feeding, fully tapping the feeding stuff nutrient availability, and improving feeding stuff conversion rate, are the key to reduce feeding costs and improve feeding efficiency.

From the live pig industry development status and trends of United States and Europe, there is a large systematic study in precision feeding aspects. In Denmark, pig daily food intake is controlled by computer, depending on the growing season, timing and quantitative feeding, pig breeding achieve information and standardization [5]. The Danish manure normative system (DMNS) provide Danish farmers and authorities with tools for fertilizer planning control, the system includes that dietary nutrient content, nutrient digestibility, feeding stuff intake and nutrient retention in the pig body. According to those information the system included, calculating the standard value of

© IFIP International Federation for Information Processing 2016
Published by Springer International Publishing AG 2016. All Rights Reserved
D. Li and Z. Li (Eds.): CCTA 2015, Part I, IFIP AICT 478, pp. 82–93, 2016.
DOI: 10.1007/978-3-319-48357-3_9

nutrients excreted, then getting the feeding stuff conversion rate, and providing the basis for the next day grain feeding. In European countries, the feeding stuff conversion rate of Danish is the standard to correct their feeding stuff conversion rate [7]. In the Netherlands, the farming management software (Agrovision FARM) and the intelligent sow management system (Velos) are been used in pig feeding [8]. Agrovision FARM can record all important, true and reliable farming information. Velos can afford precise pig feeding in accordance with the feeding curve, avoiding feed wastage caused by artificial feeding and the pig body condition uneven caused by feeding inaccurate, thus ensuring the most accurate for every pig feeding. In the United States, the *Nutrient Requirements of Swine* which was published by U.S. NRC in 1998, providing a reliable basis for pig feeding. According to different stages of pig breeding standards, U.S. researchers designed different feed formulation to reduce nutrient waste and manure nitrogen emissions, save feeding costs [10]. By using the technology of RFID radio frequency identification, wide network video surveillance, reach the intelligent pig feeding [7]. The use of automatically phased system makes pig feeding stuff conversion more efficient, feed fully compatible with the growth rate of pigs, neither because of lacking feeding stuff nutrients affect the growth rate of pigs, not because of excess nutrients in feed result in wastage. In our country, for the differences between China and the U.S. in terms of pig breeding species and breeding conditions, Fu Linsheng, Xiong Benhai and other researchers developed pig nutrition requirements dynamic forecasting system based on the NRC swine dynamic nutrition requirements, through appropriately adjusting the parameters to match the characteristics of Chinese pig breeding [9]. The system provide basic utility for a variety of pigs diet formulation designing under certain condition nutrient requirements, rationally using feed materials, and reducing feed costs and environmental pollution. The study of this system promotes the fine breeding, but the choice of the kinds of feeding stuff and feeding ration is not yet involved.

For the problem of low feeding utilization rate and high breeding cost in the development of aquaculture technology, according to NRC modeling software, the objective of this study was to calculate pig daily nutritional requirements in a growth phase, based on the type of feed and the daily requirement, establishing feeding portfolio optimization model, obtaining the mixed feed and mixed proportions by solving the model, improve feed utilization, reduce breeding costs, achieve precise feeding of pig breeding.

2 Method

2.1 Daily Requirement

Pig daily nutritional requirements is an important basis for the daily design, the pig nutritional requirements the U.S. National Research Council (NRC) recommended is considered the most authoritative pig standards. It uses a mathematical model to estimate the integral energy, protein, amino acids, minerals and vitamins growing pigs needed [9].

NRC is use of three interrelated nutritional needs modes, growth model, pregnancy model and lactation model, to software calculate. In this paper, according to the growth model selecting Growing - Finishing Pigs as the study object, the main two growing stages are 25–38 kg and 38–57 kg. Inputting the data pig's weight, type, energy indicators and so on, outputting the nutritional requirements data, which includes digestible energy (DE)-based energy, amino acid requirements, minerals and vitamins requirement. In the calculation, choosing nutrition requirements model of growth model to software calculate, the data obtained by NRC software model calculating shown in Tables 1, 2 and 3.

Table 1. Daily energy intake for each pig

Body weight (kg)	Total energy (IU)	Digestible energy (IU)	Prediction (g)	Lipid (g)	Crude fiber (g)	Vitamin (IU)	Calcium (g)	Phosphorus (g)
25–38 (kg)*	1.451	4406	119	128	69.65	1926.51	9.00	4.19
38–57 (kg)*	1.904	5901	141	194	89.49	2609.65	11.31	5.26

Notes: *25–38 (kg), *38–57 (kg), mean sub-stages weight of growth pigs.

Table 2. Daily trace elements intake for each pig

Body weight	Sodium (g)	Chloride (g)	Magnesium (g)	Potassium (g)	Copper (mg)	Iodine (mg)	Iron (mg)	Zinc (mg)	Manganese (mg)	Selenium (mg)
25–38 (kg)	1.35	1.06	0.53	3.25	5.78	0.18	89.18	80	2.64	0.269
38–57 (kg)	1.82	1.43	0.71	3.70	6.57	0.25	95.74	98	3.57	0.312

Notes: *25–38 (kg), *38–57 (kg), mean sub-stages weight of growth pigs.

Table 3. Daily vitamin intake for each pig

Body weight	A (IU)	D (IU)	E (IU)	K (mg)	B_6 (mg)	B_{12} (µg)
25–38 (kg)	1714	198	14.51	0.66	1.32	13.7
38–57 (kg)	2322	268	19.65	0.89	1.79	13.35

Notes: *25–38 (kg), *38–57 (kg), mean sub-stages weight of growth pigs.

Data in the table above are the main basis for this paper feeding selection. All selection, recipe selection must meet the growth needs, and the data obtained through the NRC modeling software are the requirements must be meet, that is the constraints in the process of forecast calculation. According to the number of nutritional requirements and cost, select the most appropriate feeds to feed.

2.2 Feeding Decision Model

With the lowest cost as the objective function, the linear programming mathematic model was built [18]. Suppose there are n kinds of feeds to choose from for a growth phase, m_i represents the ith feed, m_i^{max} and m_i^{min} represent the max and min limits of feed, i as the nutrients necessary number for the growth of pigs, n_j represents the jth nutrient element, n_j^{max} and n_j^{min} represent the max intake of nutrient and min intake of nutrients n_j; in unit mass of m_i, the mass percent of nutrients n_j denoted u_{ij}; the market price of feed m_i recorded as c_i, the number of daily feed m_i as x_i, feeding costs can be expressed as the objective function is shown as follows:

$$Z = min \sum_{i=1}^{n} c_i x_i (i = 1, 2, \ldots, n) \tag{1}$$

Where Z represents the cost of feed fed, c_i is the ith feed market price, x_i represents the ith day ration of m_i feed.

Feed usage constraints [(2)–(4)]:

$$m_i^{min} \le x_i \le m_i^{max} (i = 1, 2, \ldots, n) \tag{2}$$

Where x_i represents the ith day ration of m_i feed, m_i^{max} and m_i^{min} represent the max and min limits of feed.

$$n_j^{min} \le \sum_{i=1}^{n} \mu_{ij} x_i \le n_j^{max} (i = 1, 2, \ldots, n) \tag{3}$$

Where n_j^{max} and n_j^{min} represent the max intake of nutrient and min intake of nutrients n_j; n_j represents the jth nutrient element; u_{ij} is the mass percent of nutrients n_j denoted; the number of daily feed m_i as x_i.

$$x_i \ge 0 (i = 1, 2, \ldots, n) \tag{4}$$

Where x_i represents the ith day ration of m_i feed.

Though model computing can obtained x_i as well as get quality for all kinds of feed, while able to calculate the value of Z, which is the minimum cost of feed mix in x_i situations.

2.3 Data Flow

The feeding selection decision process is illustrated in Fig. 1, the model processes are listed below:

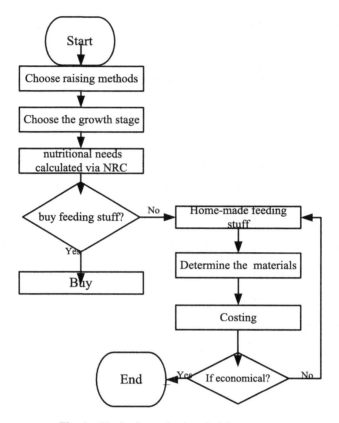

Fig. 1. Pig feeding selection decision process

1. Determine the growth stage of pigs, getting pig weight data.
2. After knowing the weight using NRC modeling software calculate nutritional requirements the pig needed, which includes the digest energy (DE)-based energy, amino acid requirements, minerals and vitamins requirement.
3. Meeting the nutritional needs (get from 2) conditions, can purchased directly competing mixed feeding stuff, can also be mixed in accordance with the existing feeding stuff themselves, if mixed by themselves, determining the type of breeding materials and ingredients contained in the existing ratio.
4. Known nutritional requirements data and raw data feed decision to build a mathematical model to calculate the lowest cost feeding stuff mix, that kind of feed needed, how much needs.
5. If it satisfies the minimum cost and to meet the nutritional needs, you get to choose the end result of the calculation, if not, loop to 3 recalculate until getting the outcome, the calculation ended.

3 Results and Discussion

3.1 Feeding Selection

(1) Data analysis

In this section, through example represent the feeding decision mathematical model simulation process, according to the conditions of meeting the growth needs of pig, with the existing status quo forage species breeding farms, pigs breeding mixed feeding stuff is calculated. There are five different options pig feed, such as A, B, C, D, E, the feed information as shown in Table 4.

Table 4. Proportion of the energy in the feeding stuff

Feeds	Digestible energy (IU)	Prediction (g/kg)	Crude fiber (g/kg)	Vitamin (mg/kg)	Calcium (g/kg)	Phosphorus (g/kg)	Price (yuan/kg)
A	3050	280	100	9.9	45	8	7
B	3060	145	700	10	80	10	6.2
C	3050	160	200	9.5	70	9	6.3
D	3050	310	100	6	50	11	5.5
E	3060	150	100	6.5	10	5	5.2

(2) Modeling

Under the conditions of meeting the nutrients requirements of pig growth, in order to minimize the cost of keeping the goal to establish programming model, such as the ratio of the feed formula (5–10) below.

Objective function:

$$Z_{min} = 7x_1 + 6.2x_2 + 6.3x_3 + 5.5x_4 + 5.2x_5 \tag{5}$$

Where the objective function is to get the cost of the optimal feed ration, Z_{min} is the feeding stuff ration cost; x_1, x_2, x_3, x_4, x_5 represent A, B, C, D, E are five different pig feeding stuff ration, x_i corresponds to the ith feeding stuff ration.

Satisfy the following constraints established:

$$280x_1 + 145x_2 + 160x_3 + 310x_4 + 150x_5 \geq 141 \tag{6}$$

Where the protein are the growth feed pigs needs must be meet of feeding process, x_1, x_2, x_3, x_4, x_5 represent A, B, C, D, E five different pig feeding stuff ration ($x_i \geq 0$, $i = 1, 2, 3, 4, 5;$).

$$100x_1 + 700x_2 + 200x_3 + 100x_4 + 100x_5 \geq 89.49 \tag{7}$$

Where the crude fiber is the growth feed pigs needs must be meet of feeding process, x_1, x_2, x_3, x_4, x_5 represent A, B, C, D, E five different pig feeding stuff ration ($x_i \geq 0$, $i = 1, 2, 3, 4, 5;$).

$$9.9x_1 + 10x_2 + 9.5x_3 + 6x_4 + 6.5x_5 \geq 3.54 \tag{8}$$

Where the vitamins are the growth feed pigs needs must be meet of feeding process, x_1, x_2, x_3, x_4, x_5 represent A, B, C, D, E five different pig feeding stuff ration ($x_i \geq 0$, $i = 1, 2, 3, 4, 5;$).

$$45x_1 + 80x_2 + 70x_3 + 50x_4 + 10x_5 \geq 11.31 \tag{9}$$

Where the calcium are the growth feed pigs needs must be meet of feeding process, x_1, x_2, x_3, x_4, x_5 represent A, B, C, D, E five different pig feeding stuff ration ($x_i \geq 0$, $i = 1, 2, 3, 4, 5;$).

$$8x_1 + 10x_2 + 9x_3 + 11x_4 + 5x_5 \geq 5.26 \tag{10}$$

Where the phosphorus must be meet the growth feeding pigs needs during the feeding process, x_1, x_2, x_3, x_4, x_5 represent A, B, C, D, E five different pig feeding stuff ration ($x_i \geq 0$, $i = 1, 2, 3, 4, 5;$).

(3) **Result**

By decision model formula (5–10) constraint equation knowing that, if breeding farms choose to feed feeding stuff A 0.0586 g per pig per day, feeding stuff B 0.0762 kg, feeding stuff D 0.3663 kg, to meet the growing needs of pigs, and investment the lowest total cost of 2.9 Yuan.

3.2   Analysis

From the simulation knowing that, according to the nutritional requirements data which are shown in Tables 1, 2 and 3, using the mathematical models to calculate to get all kinds of decision-mixed feeding stuff requirements, mixed feeding stuff (Mix) combinations are shown in Table 5. The mixed feeding stuff feeding is better than one feeding stuff feeding, such as A, B, C, D, E are five different methods of feeding pig feeding stuff alone, in the respects of feeding stuff costs and feeding ration., A, B, C, D, E, and Mix feeding volume and feeding cost comparison data are shown in Table 6.

Table 5. Calculated mixed feeding stuff (Mix)

Selected feeds	Feed ration* (kg)	Cost* (yuan)
A	0.0586	0.4
B	0.0762	0.5
D	0.3663	2.0

Note: *Feed ration, feeds a pig need a day; *Cost, feed a pig need spending a day

Table 6. Comparisons of intake and cost between mixed and single feeding stuff

Fed feeds	Feed ration* (kg)	Cost* (yuan)
A	0.89	6.2
B	0.97	6.0
C	0.88	5.5
D	0.87	4.8
E	0.94	4.9
Mix	0.50	2.9

Note: *Feed ration, feeds a pig need a day;
*Cost, feed per pig need spending per day

Analysis showed that, when a single feeding stuff breeding meet all the nutritional needs, some nutritional elements are excess result in feeding stuff waste. When nutrients to meet the requirements of crude fiber, protein and phosphorus ration greater than requirements, indirectly feeding stuff waste and excess nutrients polluting the environment during defecation. Selecting mixed feeding stuff fed to reduce breeding costs and waste of resources, and with a more balanced nutritional needs.

3.2.1   Price Factor
In this paper, the model aims at giving the lowest cost of feeding stuff feeding combination in the breeding process, depending on the kind of prices of feeding stuff. The changing feed prices on the market impact model calculations for changes in feeding stuff prices, using the simulation models in terms of price changes feeding stuff case to get the selection result. For A, B, C, D, E the five kinds of pig feeding stuff, listing three groups changing price (I, II, III). Price I{7,6.2,6.3,5.5,5.2}; Price II {7.0,6.9,6.5,6.5,5.7}; Price III{7.5,6.9,6.5,6.7,4.5}. Calculated by the model, the results obtained 3 Group Mixed Feed Mix I, Mix II, Mix III, as the Table 7 shown.

Table 7. Calculated results of feeding decision model

Mix feeds	Selection feeds	Costs* (yuan)
Mix I	A, B, D	2.9
Mix II	A, B, C, D	3.3
Mix III	B, C, D, E	3.7

Note: *Cost, feed a pig need spending a day

The model results were analyzed, and the three groups of mixed feeding stuff costs were compared with a single feeding costs (Table 6), we find that the mathematical model of mixed feeding stuff to get the lowest cost, shown in Fig. 2. The price I, price II, price III three groups of different prices feeding stuff, mixed feeding stuff Mix cost obtained (shown in black bars shown in block) are minimum by solving the model, the calculation results show that the three groups of computing results, the selection of only one feeding stuff to breed, under the conditions required to meet the nutritional needs of feeding stuff costs were greater than the cost it takes to Mix, price changes in conditions, the model still can calculate the lowest-cost breeding selection results.

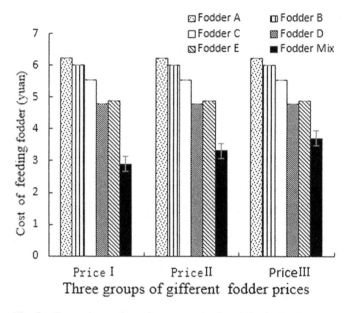

Fig. 2. Comparison of cost between mixed and single feeding stuff

3.2.2   Specified Mixed Feeding Stuff

Under certain circumstances, breeding pig feeding stuff with special requirements, for the same growth stage pig, due to the different growth status the needs of pig feeding stuff different, poor growth conditions may need to be bred a certain or a few specific pig feed feeding stuff, different growth stage with different nutritional requirements, feeding the specific needs of different feeding stuff, in order to meet these conditions, in particular the feeding stuff of feed A and feeding stuff of fed specific B, D, use the model to calculate the breeding selection decision, the results shown in Table 8.

Table 8.   Calculated results of feeding decision model

Mix feeds	Selection feeds	Costs* (yuan)
Specific A	A	3.70
Specific B, D	B, C, D, E	3.71

Note: *Cost, feed a pig need spending a day

The results of the data analysis showed that under certain circumstances the feeding stuff ration, the model calculated the lowest cost of mixed feed, shown in Fig. 3. Feeding stuff A in the feeding of two specific and particular conditions of feeding stuff B of A, B, C, D, E, Mix these six kinds of feeding costs compared to Fig. 3, breeding feeding stuff A group known specific investment Mix Hey cost of 3.70 Yuan, the specific feed fed group B, D Mix feeding cost of 3.71 Yuan, the lowest point are in the fold, the minimum cost of breeding, feeding fixed feeding stuff under specific conditions, to get through the model to calculate the most cost excellent choice of feeding stuff result.

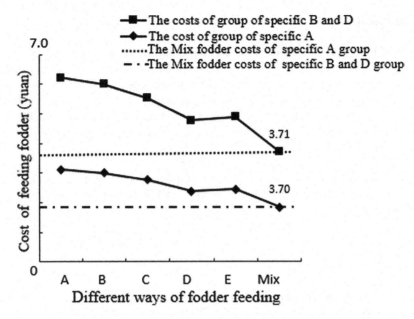

Fig. 3. Comparison of cost between specific feeding A and feeding B, D

3.2.3 Nutritional Needs

In the process of pig breeding, different season with different effects on pig growth, higher temperatures summer and low temperatures winter have different effects on pig feeding stuff intake and nutrient requirements. Summer reduce the needs for appropriate high crude fiber intake of growth pig, and winter low temperatures need to increase energy intake of protein for pig, which is required to change the conditions of the model equation constraint. In summer and winter the two conditions, for example, through a decision-making model for solving the feed mixed feeding stuff to get the results shown in Table 9.

Table 9. Calculated results of feeding decision model

Constrains	Selection feeding stuff	Costs (yuan)
Winner	A, B, D	2.90
Summer	B, D	3.3

Note: *Cost, feed a pig need spending a day

Though the results of the analysis, we know that in the summer and winter both cases, the results calculated by the model are bred in to meet the nutritional needs of growing conditions, day feeding stuff costs 2.9Yuan and 3.3Yuan, comparing with a separate feeding A, B, C, D, E five kinds of feed costs necessary 6.2,6.0,5.5,4.8,4.9 (Yuan), is the optimal cost of the pig breeding, that seasonal changes in nutritional requirements change of pig growth conditions, the model calculated optimal costs mixed feeding stuff.

4 Conclusion

The aim of this paper is to build a pig feeding optimization mathematical model and obtain the lowest-cost feeding mixed results. A model was developed based on optimal cost, considering three elements: price factor, specified feeding stuff and nutritional requirements. The results were analyzed demonstrating its validity according to growing pig conditions. Use of the feeding selection decision model can enable more precise feeding, proper selection of feeding stuff, lower cost and increasing economic benefit of pig breeding.

Acknowledgement. This paper was supported by the Shandong Province innovation special (2014XGA13054). Supported by National Science and Technology Supporting Program (2014BAD08B05).

References

1. Chen, M.: Study on nutrient diagnosis and diet formulation system of swine, no. 6. Chinese Academy of Agricultural Sciences, Beijing (2011)
2. Yu, W., Kong, F., Yu, H.: Chinese farmers backyard pig production cost element analysis. China Swine Ind. (3) (2012)
3. Zhao, X.: How to improve the growing and finishing pigs weight gain and feed efficiency. Swine Prod. (3) (2007)
4. Xiong, Y.: Pig industry technology review and prospects. Farming Feed (10) (2005)
5. Chen, G., Wei, J.: European pig inspection experience. China Anim. Ind. (2013)
6. Jorgensen, H., Prapaspongsa, T.: Models to quantify excretion of dry matter, nitrogen, phosphorus and carbon in growing pigs fed regional daily. J. Anim. Sci. Biotechnol. 4–42 (2013)
7. Gu, Z., Xu, L.: Status quo and development trend of foreign pig industry. Anim. Husb. Vet. Med. **44**(7) (2012)
8. Yi, G.: Intelligent pig management. Glob. Vis. (2013)
9. Fu, L., Xiong, B.: Development of dynamic projection system for nutrient requirements of swine. Agric. Netw. Inf. (7) (2009)
10. Lei, M., Feng, X.: U.S. Swine production technology development of new trends. China Swine Ind. (11) (2012)
11. Xing, Q.: Research and application of intelligent pig-keeping based on association rules. Chongqing University of Technology (2012)
12. Fang, K., Wang, J.: Research on foreign pig breeding information model. Swine Ind. Sci. (5) (2009)
13. Wu, X., Song, Y.: Optimization of the complete feed formula model of Plateau Lean Meat Pig. Ecol. Domest. Anim. **24**(1) (2003)
14. Zhang, Y.: Solving the MATLAB linear programming model in realizing. J. Tonghua Teach. Coll. **30**, 11–31 (2009)
15. Scherer, C., Weiland, S.: Linear matrix inequalities in control. Delft University of Technology (2005)
16. Xu, L., Yu, C., Xing, B., et al.: PDA-based aquaculture feeding decision support system. Trans. CSAE **24**(Supp. 2), 250–254 (2008)

17. Pomar, C., Pomar, J.: The impact of daily multiphase feeding on animal performance, body composition, nitrogen and phosphorus excretions, and feed costs in growing-finishing pigs. Animal **8**(5), 704–713 (2014)
18. Qin, Y.: Design and implement of dairy cow feed stuff prescription decision support system based on .NET. Beijing University of Posts and Telecommunications (2009)
19. Ma, D.: Research and implementation of abalone precision aquaculture decision support system. Shandong Agricultural University (2013)
20. Wang, K.: Analysis on the hog raising scale and their benefit. Southwestern University (2010)
21. Xu, L.: Design and implementation of management system of automatic feeding. Nanjing University of Posts and Telecommunication (2013)
22. Chen, N.: Analysis of feed grain consumption of china and forecasting the demand for feed grain in Chinese hog and poultry husbandry. Southwestern University (2011)
23. Naatjes, M., Susenbeth, A.: Energy requirement of growing pigs under commercial housing conditions. Arch. Anim. Nutr. **68**(2), 93–110 (2014)
24. Li, Y.: Analysis on the hog raising scale and their costs benefit. Northwest Agriculture and Forestry University (2007)

Study on Methods of Extracting New Construction Land Information Based on SPOT6

Lei Guo[1], Dongling Zhao[2(✉)], Rui Zhang[2], Meng Du[2], Zhixiao Li[2], Xiang Wang[1], and Yaru Wang[1]

[1] Hebei Provincial Seismological Bureau, Shijiazhuang 050021, China
guolei0430@126.com, wangx@eq-he.an.cn,
wangyr24@163.com
[2] College of Information and Electrical Engineering,
China Agricultural University, Beijing 100083, China
{zhaodongling, lizhixiao91}@cau.edu.cn,
zhangrui2063@126.com, 1165028791@qq.com

Abstract. SPOT6 is a new remote sensing satellite launched in 2012, with high spatial resolution and strong data acquisition ability. However, a complete data preprocessing technology for the regulation of land resources has not yet been formed. According to the characteristics of SPOT6 satellite images, four different image fusion methods – Gram-Schmidt, HPF, PanSharpand PanSharpening were selected to conduct the comparison experiment by using the software platforms of ENVI, ERDAS and PCI. We evaluate the results' performances from 3 different aspects. First, evaluating the image quality of experiment results qualitatively, then assessed quantitatively by establishing evaluation indexes including mean, standard deviation, information entropy, average gradient and correlation coefficient. Finally, evaluating the applicative effect of fused images based on the classification accuracy. The analysis results shows that the method of PanSharp is best to extract construction land information. Based on the PanSharp fusion image, in order to obtain the texture information under different scales, the authors screened the texture features according to Shannon entropy, and then used distance-based approach J-M to calculate the separation for choosing the optimal texture window. Once got the texture information, combining it with the original image to participate in the multi-scale image classification. The research result showed that multi-window texture participation in classification can improve separation of objects. Finally we extract construction land information with the method of SVM. This study may provide the technical support for application of SPOT6 image in the land resources management.

Keywords: SPOT6 · Image fusion · SVM · Multi-window texture · New construction land

D. Li and Z. Li (Eds.): CCTA 2015, Part I, IFIP AICT 478, pp. 94–106, 2016.
DOI: 10.1007/978-3-319-48357-3_10

1 Introduction

Construction land refers to basic premises of all social and economic activities people engaged in, including residential land, roads, land use, public service area (excluding Greenland and water) [1]. With China's rapid development of industrialization and urbanization construction land most prone to change, especially in the last 30 years, with the rapid expansion and he sprawl phenomenon of construction land, causing decreased utilization of land resources, reduction of arable land resources, environmental pollution and ecological destruction and other issues. Therefore, it has important significance for the rational development of urban planning programs, socio-economic and resource use and sustainable development of ecological environment to obtain new construction information timely and accurately.

For high-resolution image information extraction, it can't rely solely on spectral characteristics, but also greater use of spatial information of remote sensing images. In recent years, the use of auxiliary image texture information of spectral classification in order to enhance the classification accuracy is becoming a hot topic of remote sensing information extraction areas [2]. Based on the research of suitable fusion methods of extracting construction land information, this paper tries to use the object-oriented SVM classification combined with multi-window texture and spectral information to extract new construction land information.

2 Materials and Methods

2.1 Study Area and Data

The study area was located in spa town, Haidian District, Beijing of China, with the controlled area of about 33.32 km^2, with an average elevation of about 50 m. Plain area is 17.79 km^2, accounting for 55 %, mountainous area is 14.53 km^2, accounting for 45 %. The town contains a wealth of land use/land cover types, is an ideal test area of land use information extraction.

The data sources of study area included SPOT 6 satellite data obtained on January 25th, 2013 and a topographic map with the size 1/2000 obtained on 2008. The vegetation coverage of image is low, some places have snow cover.

2.2 The Principle of Support Vector Machine

SVM method is based on the VC dimension theory and structural risk minimization principle of statistical learning theory. Based on the limited sample information, finding the best compromise between complexity of the model and learning ability, in order to obtain the best generalization ability [3, 4].

By learning algorithm, SVM can automatically find the support vector that have a greater ability to distinguish, then construct a classifier that can maximize the interval between classes, so there is a good promotion and a higher classification accuracy.

For simple binary classification problem, SVM is to find an optimal split plane that can be adjusted as correctly as possible to separate two kinds of data with wider distance. The basic idea is illustrated by the following figure (Fig. 1).

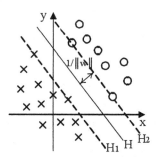

Fig. 1. Optimal hyperplane

Solid point and hollow point of figures represent two categories of training samples; H is classified line; the distance between H1 and H2 are classified intervals. According to the maximum interval method, we require not only the classification lines separate the two classes of samples correctly, but also should make the interval of classification is max; extended to high-dimensional space, the optimal classification line can be extended to the optimal hyperplane.

2.3   Test Flow

The test procedure of this research mainly includes pre-processing of images (orthorectification, image fusion and image clipping), extracting of texture feature, combination of multi-source information, classification of images based on SVM (multi-scale segmentation, parameter setting of the value of C and γ, etc.) and accuracy of evaluation. The test flow chart is expressed as Fig. 2.

2.4   Data Preparation

Data preprocessing including orthorectification, image fusion into and image clipping. In this paper, data preprocessing process is mainly focused on the choice of fusion methods.

Image fusion can compensate for the lack of information in a single image, take advantage of a variety of images, get multi-faceted feature information, is a crucial step of information extraction [5]. According to the characteristics of SPOT6 satellite images, four different image fusion methods of Gram-Schmidt, HPF, PanSharp and PanSharpening were selected to conduct the experiment of comparison. For evaluating the results' performances, we compare them from 3 aspects. The image quality of experiment results was evaluated qualitatively, and also assessed quantitatively by

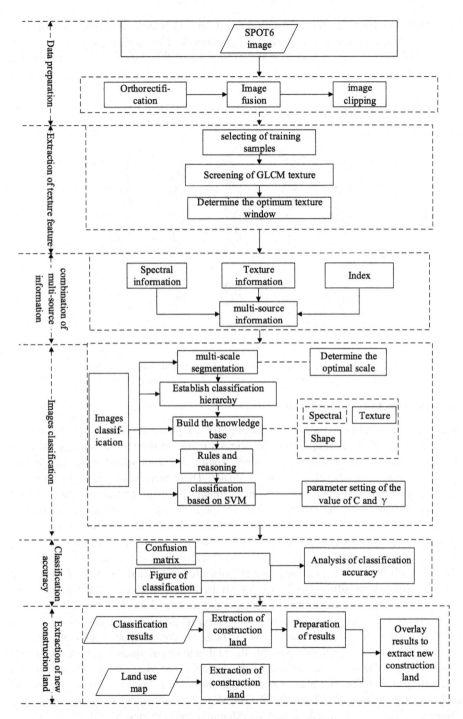

Fig. 2. The flow chart of experimental design

establishing evaluation indexes including mean, standard deviation, information entropy, average gradient and correlation coefficient. The applicative effect of fused images was also evaluated based on the evaluation of the classification accuracy, the results are shown in Fig. 3. The analysis results show that it is better to choose the image fusion methods of PanSharp to extract construction land information.

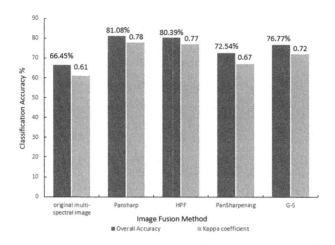

Fig. 3. Image classification overall accuracy and Kappa coefficient

2.5   Texture Feature Selection

Accuracy and precision of analysis will improve coupled with texture information on the basis of spectral information on the original image in the remote sensing thematic information extraction. GLCM can describe spatial distribution and structural characteristics of grayscale of each pixel in the image. There were advantages in improving the effect of ground objects classification in images by using texture of images. Then the texture feature of the first principal component can be extracted according to gray-level co-occurrence matrix (GLCM) which is proposed by Haralick et al. [6]. These texture statistics can be classified combined with multi-spectral remote sensing image. And each texture statistics can be involved in the calculation along with gray value of each band, etc., as a basis for classification [7]. There are 8 GLCM texture statistics which were often used in the 14 kinds, including mean, standard deviation, homogeneity, contrast, dissimilarity, entropy, angle second moment, correlation.

By using principal component analysis (PCA) while extracting the texture feature of SPOT 6 image, the result shows that the cumulative contribution of the first principal component is 95.73 %, which represents the high and low frequency part of 4 bands of these images. The outcome not only has the effect on descending dimension and descending dimension, but also can incarnate the texture feature of land objects and basic tonality feature [8]. Then the 8 GLCM texture statistics of the first principal component can be extracted. Based on Shannon entropy and visual judgment, we

Table 1. Information entropy eight texture features

Mean	Standard deviation	Homogeneity	Contrast	Dissimilarity	Entropy	Angular second moment	Correlation
2.21	0.25	2.18	0.23	1.77	1.99	1.88	0.47

compare and analysis eight kinds of texture features. In this paper, we calculate the information entropy of texture features in 3 × 3 window as example, the results are shown in Table 1.

As can be seen from Table 1, the mean of the highest entropy, provide the most abundant information; homogeneity degree and entropy followed, can also provide a wealth of information. From visual judgment, although the mean entropy is high, but the edge of the texture image is not clear, there is no prominent feature of the border (Fig. 4), therefore excluded. So we chosen homogeneity and entropy involved in classification.

(a) Mean   (b) Homogeneity   (c) Entropy

Fig. 4. Texture features of SPOT6 image (partial)

2.6   The Determination of Texture Window

Research shows that, when the sliding window is larger than 21 × 21, it will be difficult to reflect object properties using the texture features we extracted, so the window size of 21 × 21 is chosen as the largest sliding window [9]. The moving window size is set to 3 × 3, 5 × 5, 7 × 7, 9 × 9, 11 × 11, 13 × 13, 15 × 15, 17 × 17, 19 × 19, 21 × 21 in succession, in order to compare and analyze texture feature at different scales of windows.

As different land objects feature extraction has different optimal texture scale, therefore need to filter the most suitable texture window [10]. In this paper, we choose the index J-M distance to calculate the separability. This paper, we are mainly for extract newly-added construction land, the types of land use in the study area is: vegetation covered area (including vegetation cover cultivated land, woodland), dark place (including water, shadow), construction land (including buildings, roads), bare land and snow all 5 categories. Figures 5 and 6 shows the curve of class separability (J-M distance) with different texture windows.

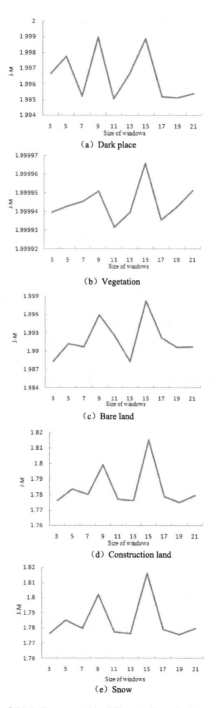

Fig. 5. Curve of J-M distance with different size windows (Homogeneity)

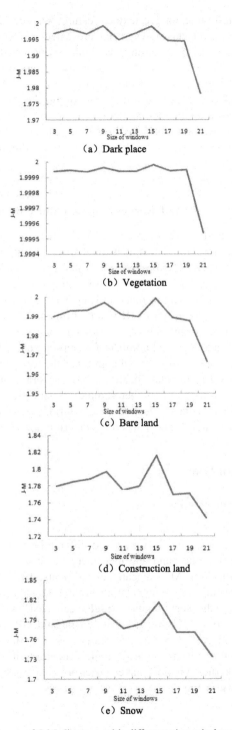

Fig. 6. Curve of J-M distance with different size windows (Entropy)

Figures 5 and 6 show that for the texture features include entropy and homogeneous of SPOT6 images, all classes were not presented in increasing or decreasing trend. For the homogeneous, vegetation, bare land, snow, construction land can be divided good in the 15 × 15 window, the separability of dark place in 9 × 9 window reached the highest. For the texture features entropy, the separability of all classes in 15 × 15 window is highest, and reached the minimum in the window 21 × 21. So for the SPOT6 image, selected texture feature homogeneous in 9 × 9 and 15 × 15 window and entropy in 15 × 15 window take part in the classification. Multi-window texture take part in classification can improve the separability of different types of objects effectively.

2.7 The Determination of SVM Kernel Function and Parameters

This study selected the RBF kernel function as SVM kernel function, because the RBF kernel can map samples into a higher dimensional space, can deal with case when the relationship between class label and feature is nonlinear; and is applied for low dimension, high dimension, small sample, large sample. It has wide convergence domain, is kind of ideal classification function [11]. The kernel parameter and penalty factor C are two necessary parameter of RBF kernel function. This paper we use cross validation algorithm to determine the two parameters, it means the selected training samples will be divided into an equal number of N subset, and make the N − 1 groups as a model of training samples, the rest of the group as the test sample, make use of the test sample to test and verify the classification results accuracy of n − 1 part data, by changing C and γ, we looking for higher sample classification accuracy [12]. This paper we choose the parameters selection model Grid. py which provided by the software libsvm 2.83 to search C and γ. The final determination of C = 32, γ = 0.0001.

3 Results and Analysis

3.1 The Classification Results

Combining the optimal texture characteristics and index characteristics (normalized differential vegetation index, NDVI, normalized differential water index and normalized construction index) as the band and then merge with multi spectral image, participate in image classification. After repeated experiments, determining the parameters of multi-scale segmentation are: segmentation scale 100, shape parameter 0.4 compactness parameter 0.6, the segmentation results can basically ensure most object boundary extraction correctly. Using visual judgment and GPS select site of typical sample. Finally we collected 305 samples altogether, 175 of which were used as training samples, the other 130 as the test sample. According to the training samples, select the SVM RBF kernel function make object-oriented classification, the classification results are shown in Fig. 7.

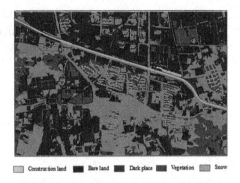

Construction land   Bare land   Dark place   Vegetation   Snow

Fig. 7. SVM classification results of SPOT6 images

To evaluate the classification results accuracy effectively, classification accuracy as shown in the Table 2:

Table 2. The form of SVM classification accuracy evaluation

	Dark place	Vegetation	Bare land	Construction land	Snow	Total	Customer accuracy(%)
Dark place	21	0	1	1	0	22	95.45
Vegetation	0	17	4	0	0	21	80.95
Bare land	0	2	17	6	1	26	65.38
Construction land	4	1	6	45	5	61	73.77
Snow	0	0	0	0	2	2	100.00
Total	25	20	24	51	8		
Producer accuracy(%)	84.00	85.00	70.83	88.24	25.00	84.00	
Overall accuracy: 79.70 %							
Kappa coefficient: 0.717							

Image recognition ability for dark place and vegetation are better than that of bare land and the construction land, construction land and bare land have a certain degree of mixing. SVM classification capability of roads and other linear features is good, even smaller rural road can also be identify effectively.

3.2   The Extraction Results of New Construction Land

Before extracting new construction land, we need deal with classification results. There were some shadows of building existing in class of dark place. Using the neighborhood characteristics of eCognition, set the threshold, those can be classified into construction land; and merge adjacent segmentation map spot. According to the minimum on map spot, those witch area less than 666.67 square meters will be deleted. Finally, make differential operation to classification results which were treated and base land use map. The extraction results are shown below (Fig. 8).

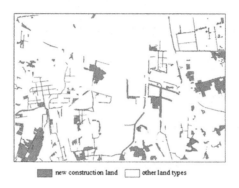

new construction land   ☐ other land types

Fig. 8. New construction extraction results

Statistical results of new construction land are shown in Table 3.

Table 3. Statistics of new construction

Types	The number of polygons	Area of polygons (km²)	Percentage of area (%)
Construction land	38	117.62	88.61
Bare land	12	10.35	7.80
Dark place	5	4.17	3.14
Snow	2	0.60	0.45
Total	57	132.74	
Misclassification rate	11.39		
Leakage rate	3.77		

The accurate rate of extraction is more than 80 %, achieved superior extraction effect, but there is a certain degree of error and omission phenomenon. Among those new construction land which extracted, the bare land divided into construction land had high misclassification rate, which may be related to the image itself, cause the SPOT6 image as the winter image, vegetation coverage is low, may result in the impact on the classification results.

4   Conclusion and Discussion

(1) Image fusion can compensate for the lack of information in a single image, take advantage of a variety of images, get multi-faceted feature information, is a crucial step of information extraction. The most suitable fusion method for the extraction of construction land was selected by quality evaluation system of fusion image. For SPOT6 image, select the fusion method of PanSharp can improve the extraction accuracy of construction land.

(2) Multi-window texture features take part in image classification can improve the class separability and classification accuracy effectively. SPOT6 image, choose the texture features homogeneity of 9×9 window and 15×15 window, choose the texture features entropy of 15×15 window.

(3) The classification method of SVM can solve the small sample, nonlinear and high dimensional problems, obtain global optimal solution, it shown the superiority in the application of road and building.

(4) The SPOT6 image in this paper was winter data, vegetation coverage rate was low, and there are snow covered, this may cause certain effect on the extraction of construction land information. We should select more data to verify the reliability of this method.

Acknowledgment. This research was founded by the Beijing Municipal Science and Technology Projects (Z141100000614001), and the projects of Hebei Provincial Seismological Bureau (DZ20150428102, DZ2015017012).

References

1. Bouziani, M., Goïta, K., He, D.-C.: Automatic change detection of buildings in urban environment from very high spatial resolution images using existing geodatabase and prior knowledge. ISPRS J. Photogrammetry Remote Sens. **65**(1), 143–153 (2010)
2. Huang, Y., Zhang, C., Su, W., Yue, A.-Z.: A study of the optimal scale texture analysis for remote sensing image classification. Remote Sens. Land Resour. **4**, 14–19 (2008)
3. Huang, X., Zhang, L.P., Li, P.X.: Classification of high spatial resolution remotely sensed imagery based upon fusion of multi-scale features and SVM. J. Remote Sens. **11**(1), 48–54 (2007)
4. Zhang, L., Shao, Z.-F., Ding, L.: Semi-supervised collaborative classification for hyperspectral remote sensing image with combination of cluster feature and SVM. Acta Geodaetica Cartogr. Sin. **43**(8), 855–861 (2014)
5. He, H., Yi, F., Meng, Z.: Research on fusion of mapping satellite-1 image and its evaluation. Bull. Surv. Mapp. **2013**(1), 6–9 (2013)
6. Mariz, C., Gianelle, D., Bruzzone, L., et al.: Fusion of multi-spectral SPOT-5 images and very high resolution texture information extracted from digital orthophotos for automatic classification of complex Alpine areas. Int. J. Remote Sens. **30**(11), 2859–2873 (2009)
7. Liu, M.-M., Liu, Y., Sun, G.: SVM land cover classification based on spectral and textural ferture using stratified samples. Remote Sens. Technol. Appl. **29**(2), 315–323 (2014)

8. Li, J.-L., Liu, X., Li, H.: Extraction of texture feature and identification method of land use information from SPOT5 Image. J. Remote Sens. **10**(6), 926–931 (2006)
9. Huang, X., Zhang, L., Li, P.: Classification of high spatial resolution remotely sensed imagery based upon fusion of multiscale features and SVM. J. Remote Sens. **11**(1), 48–54 (2007)
10. Zhang, F., Xue, Y.-L., Li, Y.-C.: Object-oriented building extraction of multi-source remote sensing imagery based on SVM. Remote Sens. Land Resour. **25**(2), 27–29 (2008)
11. Wu, B., Xiong, Z.G., Chen, Y.Z., Zhao, Y.D.: Classification of quickbird image with maximal mutual information feature selection and support vector machine. Procedia Earth Planet. Sci. **1**(1), 1165–1172 (2009)
12. He, L.-M., Shen, Z.-Q., Kong, F.-S., Liu, Z.-K.: Study on multi-source remote sensing images classification with SVM. J. Image Graph. **12**(4), 648–654 (2007)

Application of Remote Sensing Technology in Agriculture of the USA

Yuechen Liu[✉] and Weijie Jiao

Chinese Academy of Agricultural Engineering,
MaiZidian Street 41, Chaoyang District, Beijing 100125, China
liuyuechen@agri.gov.cn

Abstract. With the development both of the information technology and the aeronautics and astronautics technologies, the remote sensing technology has been used in agriculture widely and deeply. After completing the "Agriculture and Resources Inventory Surveys through Aerospace Remote Sensing" plan, remote sensing technology was widely used in agricultural sectors in the USA. Such as the planting area, yield estimation, crop spatial distribution, agricultural insurance, agricultural disaster monitoring, protection of agricultural environment, etc. And meanwhile built a basic database of agricultural remote sensing, based on the data of "Common Land Unit" CLU) and the "Cropland Data Layer" (CDL). The author went to America to visit some officers, professors and specialists, in order to learn the developing situation of remote sensing technology application in agriculture in America. Through communicating and discussing, he got a lot of information and materials and produced some thinking. This paper expounds the present situation of the development of American agricultural remote sensing from data sources, basic data platform and application of remote sensing technology, etc. At the same time, analyze the base about remote sensing technology used widely in American agriculture from based data storage, data sharing mechanism, operation and business system, the degree of automation, etc. Finally, the paper mentioned three points for discussion on the development direction of Chinese agricultural remote sensing. First, strengthen the concepts of universal, sharing, integration of remote sensing technology. Second, lay stress on the goals of guiding production, managing resource, servicing decision by remote sensing technology. Third, grasp the keys of basic data studying, data platform building, technologies advance.

Keywords: Agriculture · Remote sensing technology · Application · Development

Communication, discussion and study is a important method to understand the development of science and technology of the world. We went to America to visit some officers, professors and specialists for learning the developing situation of remote sensing technology application in agriculture of America. It worth learning and using for reference about complete basic data platform and running system, and so on.

The author: Liu Yuechen (1983—), male, engineer, member of Chinese Commission of Agricultural Engineering, the practice of land use and agricultural remote sensing.

D. Li and Z. Li (Eds.): CCTA 2015, Part I, IFIP AICT 478, pp. 107–114, 2016.
DOI: 10.1007/978-3-319-48357-3_11

Agriculture development main met the market competition of domestic in the past, but now it is participating in the market competition from domestic and international. And agriculture technology development will have the newer and higher requirements. [1] The aeronautics and astronautics technologies promoted the development of the technology for earth observation of human and the studying and business application of remote sensing technology. [2] Understanding and mastering the situation of agriculture accurately and timely by agricultural remote sensing technology, which is one of the new sectors of agricultural high technology industrialization at present. [3] The Chinese Ministry of Agriculture Remote Sensing Application Center had been carried on crop monitoring business since 1998, has supplied many a lot of agricultural information to the government for making policy decision of agricultural production. And gain the huge economic and social benefits. [4] However, it is worth attention that some gaps exist in the agricultural remote sensing application, which include the data base preparation, the technology widely useing and the business running system etc., compared with America 's. America's experience is worth learning and using for reference in out country.

In the early 1960s, the agricultural remote sensing laboratory of Prudue University carried out crop acreage monitoring with remote sensing data firstly. It proved that remote sensing data could be used to monitor crops, on the basis of the successful experiments of the corn monitoring. [5] To carry out agricultural monitoring in the large-scale, used remote sensing technology, by American Agriculture ministry from the plan of "Large Area Crop Inventory and Experiment" in the 1970s and the plan of "Agriculture and Resources Inventory Surveys Through Aerospace Remote Sensing" in 1980s. [6] The two plans identified corps, measured area, estimated yield, utilized satellite images. And they completed the growth situation assessment and production forecast for kinds of food crops in America and different regions of the world. Then, the remote sensing technology was used widely in agricultural industry. Such as Crop area statistics, crop yield estimation and crop spatial distribution of cartography, agricultural insurance, disaster monitoring, agricultural subsidies, agricultural environment protection, etc.

1   The Data Resources in the America Agriculture Remote Sensing Monitoring

A lot of data resources are used for America agriculture remote sensing monitoring. The first is aerial image, which is updated in every two or three years by Farm Service Agency. The second is satellite image, which is obtained from satellites of America, such as LantSat and Modis and so on. The third is satellite image, which is obtained from satellites of other countries, such as Spot, IRS, DMC, RapidEye, etc. The remote sensing images can cover the United States and the main agricultural regions of global about five to six times in one year, except aerial image. The plenty of remote sensing data and products provides a strong support to agricultural production management and service in the United States.

2 Two Important Fundamental Spatial Data to Support Agricultural Remote Sensing Business Running Work

The United States Department of Agriculture has established the special agricultural remote sensing database to meet the requirements of agricultural remote sensing monitoring business running work. Both the "Common Land Unit" and the "Cropland Data Layer" are very important data, have been widely used in lots kinds of agricultural remote sensing monitoring works. And they are the bases to agricultural remote sensing application in the special database.

2.1 The Data of Common Land Unit (CLU)

America carry out aerial photography every two or three years. And use the aerial images cover all homeland of the USA. Using fences, rivers, roads and other permanent features to divide the agricultural land plot and other land to make CLU with the main way of computer automatic classification, while combining the method of artificial visual interpretation. Then give each plot a unique number. At the same time, according to multi channel informations coming from farmers' reports and ground survey, etc. to correct it.

2.2 The Data of Cropland Data Layer (CDL)

The USA had drawn the CDL from 2010, covered 48 states and included more than one hundred kinds of crops. The relevant department published data and carried out the information services work through sharing platform. CDL combine CLU may replace the traditional way of the ground quadrat investigation to achieve a new leap forward. The remote sensing monitoring of crop spatial distribution has been changed from the typical sample survey to full coverage survey and the results has been changed from figure to map. It laid a solid foundation for computer automatic interpretation and let the USA became the first country to carry out the business running work of crop spatial distribution drawing of whole nation, and has produced a great influence in the field of international remote sensing.

Mading CDL based on the data of CLU and June Agricultural Survey and ground quadrat investigation, etc. Utilize the intermediate and high resolution remote sensing data and select special land plot, which only planted one crop, as training samples of supervised classification method to identify and divide the different types of agricultural land. In the process, the number of training samples of every kind of crops is more than fifty per cent. It could ensure that the supervised classification accuracy is above ninety percent. In addition, in America, law supervision system is perfect, institutional setting is direct and effective, staff is stabile, professional quality is good, which are the keys to ensure the precision of training samples.

3   The Application of Remote Sensing Technology in America Agriculture

3.1   The Application of Remote Sensing Technology in Agricultural Investigation

Basing on the data of CLU and depending on the tool of Geography Information System (GIS), lay the irregular sampling grids, whose area about 3–4 square miles, then superpose this data layer and the remote sensing images covered the whole country and the CDL to judge the crop's varieties and area in the grid. According to the acreage ratio of crop's variety in grid, with the method of area sampling as the basic sampling frame, merge the same types of grids to the six levels of different ranges. According to a certain proportion of the grid sampling to determine the investigation numbers of each layer. The soft of GIS can help to show the position of grids of each level. On this basis, investigate and verify the samples of agricultural information, and carries on the analysis using the extrapolation model, so as to obtain the national agricultural situation.

3.2   The Application of Remote Sensing Technology in Agricultural Insurance

Establish the complete agricultural information database with the data of CLU, CDL, soil information, meteorological information, agricultural introduction information, and so on. On the one hand, use GIS and many kinds of remote sensing images to monitor the Crop cultivation situation and land plot position, to measure land plot area, and to identify the kinds of crop. In order to judge the authenticity of the insurance investment. On the other hand, after knowing that the occurrence of disasters in some regions, use Normalized Difference Vegetation Index (NDVI), getting from remote sensing data, to analyze the crop growth situation. Estimate the degree of disaster losses through comparing with the many years average case of crop growth situation. Combining with the information of Crop growth period and occurrence time of disaster, judge the situation of crop replant and recovery. Provide the basis for the rapid estimation of insurance amount.

3.3   The Application of Remote Sensing Technology in Flood Disaster Monitoring

Analyzing and comparing multi-temporal satellite images combining with meteorological information to get the spatial distribution situation of water, such as rivers and lakes, in normal years and to measure the area of the water from images. After learning that the occurrence of flood disaster in some regions, superimposed the latest remote sensing images and the water spatial distribution vector data production to monitor the influence scope of the flood disaster. Remote sensing technology can monitor scope of agricultural land flooded, but it unable to probe the deepth of water and to determine whether the formation of flood disaster.

3.4 The Application of Remote Sensing Technology in Drought Monitoring and Its Early Warning

3.4.1 The Application in Drought Monitoring

Different regions have different mean annual precipitation, soil condition and vegetation growth condition, etc. Depend on data of NDVI, meteorological, CDL and CLU to divide some regions with some same attributes. According to the actual situation of the region, formulate the drought judgement index standard and the corresponding levels of drought. Calculate the regional drought index with data of NDVI, basing on some models. Such as growth model, meteorological model, remote sensing model, etc.

3.4.2 The Application in Drought Early Warning

Monitoring temperature of soil and land surface with thermal infrared band of MODIS, high time resolution data, every day. Combining with the data of CDL and CLU, calculate vegetation evapotranspiration in a plot. Wet soil is low soil temperature and large vegetation evapotranspiration. And drought soil is high temperature and small vegetation evapotranspiration. According to the drought judgement index standard and the corresponding levels of drought warn drought occurred early. Improve the spatial resolution through fusing the data of TM to improve the accuracy of monitoring in small scale.

3.5 The Application of Remote Sensing Technology in Groundwater Pollution Monitoring

Early stage field data were gained with a limited number of ground monitoring stations. Calculated the region's situation from some points to the surface by interpolation method, which would formate some mistakes. Nowadays, through monitoring temperature of soil and land surface with thermal infrared band of MODIS, combining with CDL and CLU to identify the kinds of vegetation and the spatial distribution, calculate vegetation evapotranspiration in the polt. Contrast the nearest grassland, has the same attributes and no irrigation, to judge the situation of irrigation of agricultural land. Calculate the water consumption, infiltrate into groundwater, with the vegetation evapotranspiration and the irrigation water capacity recorded by water-meter. Finally, get the data of the total amount of pesticide, fertilizer and other substances into the groundwater. Provide the basis for groundwater pollution control and governance.

3.6 The Application of Remote Sensing Technology in Land Farming Intensity Monitoring

Collect soil spectral information, crop residue (straw) spectral information and other field information by spectrometer to establish interpretation signs information database of remote sensing image, combining with the data of CLU and CDL to identify the kinds of the crop of specific plot. Distinguish and extract area of residue with satellite images for calculating the area percentage about specific plot and the crop residue in it. In order to monitor the land farming intensity and to provide the fundamental data for protecting farmland.

3.7   Summary

There are two prominent features in application of remote sensing technology in agriculture of the USA. The first is to establish a big data platform. Ultilizing a variety of productions of spatial fundamental data, including CDL and CLU and other important data, and combining with other many kinds of data, such as meteorological, hydrology, soil, agricultural production, etc., build a agricultural remote sensing plotform of resource and environment for servicing agricultural production and management. The second is to exploit a variety of models. Exploit and optimize mathematical models pertinently for different application purposes based on big data. A series of remote sensing models, such as corp growth model, vegetation evapotranspiration model, agricultural drought model, meteorological model, agricultural environment evaluation model, and so on, is important tool for broadening the scope of application of remote sensing technology.

4   The Fundament for Wide Application of Agricultural Remote Sensing Technology in the USA

4.1   A Solid Foundation of American Agricultural Remote Sensing Data

Application of agricultural remote sensing need strong support with lots of data and information to support. Especially the CLU and the CDL are necessary and the foundation of basic data for this business. According to these, America achieved standardization and automation to data processing and applied remote sensing technology widely to many fields of agriculture. A lot of intermediate and high resolution remote sensing data covered whole country for five to six times in a year and aerial photography data covered all the land for one time every two years. A plenty of remote sensing image data ensure remote sensing products' accuracy and timing.

4.2   There Is an Open Data Sharing Platform in America

Different department in United States Department of Agriculture (USDA) are responsible to different part of agricultural remote sensing monitoring data. Such as the CLU is produced and managed by Farm Service Agency (FSA), and the CDL is produced and managed by National Agricultural Statistics Service (NASS), and so on. These data are public and shared in the USDA, through an open data sharing platform. The platform makes it possible to talk with data, to formulate agricultural policy with data and to serve with data.

4.3   Agricultural Remote Sensing Running System Is Complete in America

On the one hand of institutional settings, NASS had special remote sensing institution of themselves and developed complete and vertical management work system. It owns independent offices in state and county and both agricultural statistics personnel and ground investigation personnel in it. The staff is stable and specialized. They can operate relevant business softs skillfully and submit results timely, according to the working process. On the other hand of working mechanism, the USDA assigns remote sensing monitoring job and researches of remote sensing application technology to agricultural offices in state and county or entrusts them to the research institutes and universities through the way of government buying products and services. These final results will be shared by the big data platform of USDA.

4.4   There Is High Degree of Automation to Agricultural Remote Sensing Image Interpretation in the USA

Agricultural remote sensing data is complete, open, high degree of standardization in America. Remote sensing image interpretation and information extraction have been achieved intelligentialize and automation. Reasonable institutional settings improve the ability of getting ground investigation data greatly. Stronger coordination ability of spaceflight, aviation and ground monitoring can update the basic information database quickly and provide security for increasing the level of automation of agricultural remote sensing technology.

5   Discussion

Agricultural modernization need modern management technologies. Facing the age of big data, American experience of agricultural remote sensing is benefit to accelerate the business of Chinese agricultural remote sensing and to consolidate the supporting role of agricultural remote sensing. In order to supply stable and reliable services to the level of agricultural modernization upgrade.

5.1   Strengthen the Idea of Universal Suitable for Use, Sharing and Cooperation

Building the idea of data sharing steadily, promoting the creation of big data platform rapidly, enhancing the cooperation of the agricultural remote sensing and agricultural statistics investigation and other ways effectively, are the key to improve the quality of agricultural products.

5.2 Highlighting the Targets of Guiding Production, Managing Resource and Serving Policy Decision

Monitoring the relevant information of crop cultivation and growth, disasters and meterological to guide agricultural production in different regions and to enhance the ability of disaster prevention and mitigation. Monitoring and investigating Cultivated land, grassland, water area and ecological environment, etc. to carry out the agricultural resource management and environmental capacity evaluation and to provide the foundation and basis for promoting agricultural sustainable development. Depend on the important basis of agricultural remote sensing data to promote the integration of depth of remote sensing technology and modern agriculture and to realize scientific management.

5.3 Grasp Well the Priorities of the Basic Data Preparation, the Data Platform Building and the Key Technologies Studying

To start from the food crops, achieve synchronous mapping and annual data updating of a variety of crop. Integrate the data of agricultural remote sensing, agricultural resource and agricultural production to exploit and apply the service function of big data platform effectively. Broaden the field and scope of the application of agricultural remote sensing. Give impetus to the research and innovation of technology and method. Speed up the transformation and the application of the results.

References

1. Xue, L.: The development thinking and application review on the technology of remote sensing agriculture. China Agric. Resour. Reg. Plann. **23**(3), 3–7 (2002)
2. Chen, Z.: Monitoring of agriculture with remote senseing in the age of GEOSS. China Agric. Resour. Reg. Plann. **33**(4), 5–10 (2012)
3. Qinglin, D.: Speeding up technology innovation in the range of agriculture remote sensing and promoting agriculture and countryside economy development. China Agric. Resour. Reg. Plann **23**(3), 1–2 (2002)
4. Huang, Q., Chen, Z., Li, D., et al.: An introduction of BIOMA and the feasibility analysis of application in agriculture monitoring in China. China Agric. Resour. Reg. Plann. **35**(4), 76–80 (2014)
5. Xie, G., Yang, R., Lu, Y.: Advance analysis of agricultural applications of remote sensing techniques. Guangxi Teach. Educ. Univ. (Nat. Sci. Ed.), (2), 88–96 (2014)
6. Yang, B., Pei, Z.: Definition of crop condition and crop monitoring using remote sensing. Trans. Chin. Soc. Agric. Eng. **15**(3), 214–218 (1999)

Aquatic Animal Disease Diagnosis System Based on Android

Min Sun[1,2,3] and Daoliang Li[1,2,3(✉)]

[1] College of Information and Electrical Engineering,
China Agricultural University, Beijing 100083, People's Republic of China
sunnymin100@sina.com, dliangl@cau.edu.cn
[2] Key Laboratory of Agricultural Information Acquisition Technology,
Ministry of Agriculture, Beijing 100083, People's Republic of China
[3] Beijing Engineering and Technology Research Center of Internet
of Things in Agriculture, Beijing 100083, People's Republic of China

Abstract. In recent years, as long as the rapid development of aquaculture, the occurrence of aquatic animal diseases increase year by year, which restrict the sustained, steady, and healthy development of aquaculture. So the rapid diagnosis of aquatic animal disease is particularly important for their prevention and control. In the present work, an Android-based aquatic animal disease diagnostic system has been developed in order to provide more convenient and effective aquatic animal disease diagnostic services to ordinary farmers. The designed system includes the part of the man-machine interface (user interface, expert interface), inference engine, the case base, knowledge base and so on. Interactive interface is a visual display interface. Inference engine was written in the Java programming language. Case base and knowledge base were created using SQLite. The results show that ordinary farmers can easily use this system to realize the convenient, fast, and accurate access to the findings of disease diagnosis and prevention measures as well as aquaculture expert can add, delete, modify, view the case base and knowledge base at any time.

Keywords: Android · Aquatic animals · Disease diagnosis

1 Introduction

Fishing industry has been rapid development since the 1990s in China, and has Fishing is the competitive industries of agriculture economic development, fisheries and aquaculture in the world compared with other edible animal breeding, world aquaculture industry has developed very quickly [1]. However, with the rapid development of aquaculture, aquatic animal disease has become increasingly serious. It is the major factor constraints the sustained and healthy development of aquaculture industry. During April to October in 2004,126 kinds of disease of 74 kinds of aquaculture species are monitored by 30 provinces, autonomous regions and municipalities of China. The results show that from April to October 2004, the direct economic losses due to diseases reached 15.144 billion yuan. 4.56 billion yuan more than 10.584 billion yuan of 2003 [2]. So, how can we make the production of aquaculture on the way to the

© IFIP International Federation for Information Processing 2016
Published by Springer International Publishing AG 2016. All Rights Reserved
D. Li and Z. Li (Eds.): CCTA 2015, Part I, IFIP AICT 478, pp. 115–124, 2016.
DOI: 10.1007/978-3-319-48357-3_12

healthy development has become an issue for aquatic animal experts. The research and promotion of diagnostic methods and prevention of diseases of aquatic animals has become urgent priority of aquaculture development.

In order to reduce the losses caused by diseases in aquaculture, research and develop various aquatic animal disease diagnosis expert system to assist aquaculture farmers has become the main topic among various research institutions. Currently because the aquatic animal disease experts is lack, the personnel of aquaculture are mainly farmers, and the distribution of farms tend to be more dispersed, the technical level of the breeding staff is not high. Thus in the process of diagnosis of aquatic animal disease, losses take places often due to the lack of experts or the experts can not get the scene as soon as possible. In addition, because the disease can not be accurately diagnosed, science prescription and grasp dose, farmers often abused medicines, antibiotics, it also had a significant impact on ecological environment and food safety and the healthy and sustainable development of aquaculture is restricted [3–7]. In order to reduce the economic losses, resolve the conflicts between frequent aquatic animal diseases and lack of experts, many research institutions developed a variety of expert systems for aquatic animal disease diagnosis by using some advanced technology products and information technology. We can use these expert systems to mimic the effect of working off-site, by this way the problem of lacking experts can be resolved.

At present, all kinds of aquatic animal disease diagnostic systems research and development of the country is mainly aimed at the personal computer, Hand personal computer and personal digital assistant applications terminal [8–13]. But for the actual results now, on the one hand, in terms of ordinary farmers, PC machine, PDA, and HPC applications such as intelligent terminal system applications in agriculture, is not only expensive, single function, but also a strong dependence on the network environment, resulting in very limited in its application crowd. On the other hand, the current agricultural intelligent system development platform almost based on Windows operating systems [14], it leading to poor results in terms of free sex, open source, and human-computer interaction. Therefore, it is important to seek one simpler interface and more friendly agricultural intelligent system application terminal and build an open source and free development environment for the popularization of aquatic animal disease prevention and control technology to farmers.

In recent years, with the development of the global smart phone, Google released Android operating system and promoting China's 3G network coverage project, China's 3G smart phone has been rapid development [15]. Especially after thousands of smart phones launched within Android 3G, Android smart phones in China's mobile phone market share has been increasing [16–18]. Android 3G intelligent mobile phone is a set of call, Internet, multimedia and other functions in one of the intelligent terminal, its price is low, and can avoid the "two into". In addition, the Android operating system is a free and open source code operating system based on Linux. It is not only able to provide flexible independent design space for software designers, but also supports voice, touch screen and a series of new interactive technology.

In this paper, an aquatic animal disease diagnosis system based on Android is developed.

2   Design of Aquatic Animal Disease Diagnosis System Based on Android

2.1   Structural Design of Aquatic Animal Disease Diagnosis System Based on Android

The design of the structure of aquatic animal disease diagnosis system based on Android is composed of four modules, human-machine interface, inference engine, case base and knowledge base. Its structure is as shown in Fig. 1.

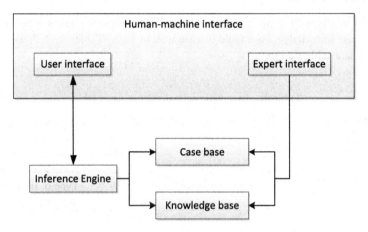

Fig. 1. Aquatic animal disease diagnostic system structure

The man-machine interface includes a user interface and expert interface. The user interface is a visual display interface used by ordinary farmers. Aquatic animal disease expert can use the expert interface to add, delete, modify or view the case database and knowledge base. The inference engine is programmed by JAVA programming language. Case base and knowledge base are developed by SQLite3 and independent of application development.

2.2   Build the Application Development Environment

Android development platform not only supports Windows, Linux also supports Mac OS systems. In this paper, the development environment is built on the Linux operating system and constructed by "JDK (Java development kit) + Eclipse + Android SDK (Software development kit) + ADT (Android development tools)".

2.3   Build the Application Development Environment

SQLite is a compliant relational database management system that embedded lightweight and comply with the ACID. It takes resources are very low, open source and supports most of the SQL statement. This paper uses SQLite database to store large amounts of data, and to achieve the use of data, updating, maintenance and other operations. And install SQLite3 as "aquatic animal disease diagnosis system based on Android" database development platform on Linux operating systems.

3   Design of Aquatic Animal Disease Diagnosis System Based on Android

3.1   Database Design

The system includes three main types of data tables, tables of various types of aquatic animal disease knowledge, case table or case weight table. Tables 1, 2, 3 and 4 are the associated data table design.

Table 1. An aquatic animal disease cases table

Primary key/foreign key	Field name	Field type	Length	Empty	Description
PK	ID	Int	5	N	Case number
	type	Int	5	N	Case type
	Dis_name	varchar	60	N	Names of diseases
	Dis_type	varchar	60	N	Type of disease
	Sym_set	varchar	300		Symptoms collection

Table 2. Weight information table

Primary key/foreign key	Field name	Field type	Length	Empty	Description
PK	ID	Int	5	N	Weight number
	type	Int	5	N	Case type
	Weight_set	varchar	1000	N	Sets of weight

Table 3. An aquatic animal disease information table

Primary key/foreign key	Field name	Field type	Length	Empty	Description
PK	ID	Int	5	N	Disease number
	Dis_name	varchar	60	N	Names of diseases
	Dis_type	varchar	60	N	Type of disease
	cause	varchar	1000		Cause of disease
	prevent	varchar	1000		Precaution
	treatment	varchar	1000		Treatment programs

Table 4. Symptom chart

Primary key/foreign key	Field name	Field type	Length	Empty	Description
PK	ID	Int	5	N	Symptoms number
	Sym_name	varchar	60	N	Symptoms name
	Sym_type	varchar	60		Symptoms type

3.2 System Inference Engine Design and Analysis

This system establish a aquatic animal disease diagnostic reasoning model by using case-based diagnostic reasoning combined with expert symptom scoring method. In order to achieve complementary advantages of the two diagnostic methods, improve the accuracy of diagnosis of aquatic animal diseases. Diagnostic reasoning as follows:

Step 1: Use case diagnosis and numerical diagnosis to do disease diagnosis respectively based on the symptoms information that farmer input, then compare the two results obtained by the two diagnostic methods.

Step 2: If the two diagnostic results are different, the system will show the symptoms information of both diseases and ask farmer input symptoms information again, then diagnosis again.

Step 3: If the two diagnostic results are same, then judge whether the value of Similarity that obtained by case diagnosis is between diagnostic threshold (0.65) and Very similar threshold (0.95), when the condition is satisfied then multiplexed on this case and display diagnostic results.

Step 4: If there is no enough conditions then ask the user whether they are satisfied with the result or not, if they satisfied with it then show the results of the diagnosis. When they are not satisfied with it then back to step 1.

Reasoning process flow chart shown in Fig. 2.:

Because in the implementation process of inference engine need to involve the processing of the database, so SQLiteDatabase categories should be imported, and then create or open the database by the static method of SQLiteDatabase categories.

3.3 System Migration and Updates

Under the development environment of aquatic animal disease diagnostic system based on android, perform "Run As Android Application" can generate a installed package (i.e. APK file) of system, then upload it to the server. Users can download the system to the Android smart phone by 3G wireless network, or download it to the Android smart phone SD by PC, after this the migration of the system is finished.

The system updates is divided into two parts, expert and client. Experts can add, delete, modify or view the case base and knowledge base timely by the end of expert.

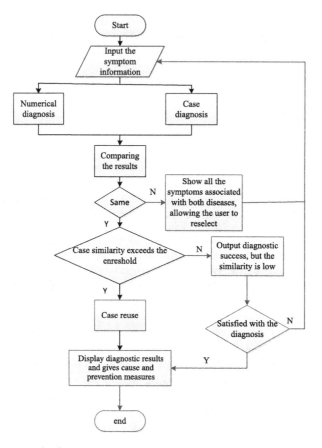

Fig. 2. Aquatic animal disease diagnostic flowchart

Then upload the new database to the server when the condition of network is smooth. Farmers can get the information of system update by the client, they can choose whether to update the system or not.

4　System Integration and Implementation

In this paper, LinearLayout, RelativeLayout classes and AbsoluteLayout class provided by Android SDK are used to design the human-computer interaction of the system. We also use a variety of controls, such as Button, TextView, EditText, etc. to complete the interactive interface design.

After the system is installed on the user's or experts' smart phone, users can enter the login interface and choose breeding objects, then start the operation of disease diagnosis(show as Figs. 3, 4 and 5). Experts enter the login interface and choose the operation such as modify the rules, delete rules(show as Fig. 6).

Fig. 3. Breeding user selects diagnostic object interface

Fig. 4. Disease diagnosis process interface

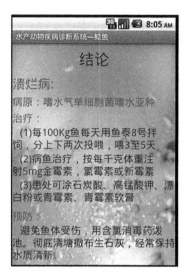

Fig. 5. Disease diagnostic conclusions interface

Fig. 6. Experts function main interface

5   Conclusions

This paper chooses the Android operating system as a development platform and develops an aquatic animal disease diagnosis system based on Android. Allow users to get rid of the constraints and inconvenience caused by computer and network. This system provides a more convenient and effective aquatic animal disease diagnostic services for farmers.

Farmers can use the system put forward in this paper to get the result of disease diagnosis and prevention measures. Experts can operate the case base and knowledge base timely, they also can upload the new data to server when the network is smooth, so that the farmers can get the latest data.

Acknowledgment. This work was supported by the Key Program for International S&T Cooperation Projects of China (2013DFA11320).

References

1. Li, X.X.: Pollution research on aquaculture and agricultural sources. Anhui Agri. Sci. Bull. **13**(11), 61–67 (2007)
2. Yang, Z.G.: Research on prevention and control of aquatic animal diseases. Agric. Technol. Equip. **243**, 50–51 (2012)
3. Zhu, W.: Fish Disease Diagnosis System for Intelligent Call Center. China Agricultural University, Beijing (2006)
4. Wen, J.W.: A Knowledge-Based Fish Disease Diagnosis Reasoning System. China Agricultural University, Beijing (2003)
5. Guo, Y.H.: An Ontology-Based Fish Diseases Knowledge Acquisition and Diagnosis Reasoning Integrating System. China Agricultural University, Beijing (2004)
6. Zhang, J.: A Call Center-Based Fish Diseases Knowledge Acquisition System. China Agricultural University, Beijing (2007)
7. Xing, B.: An Intensive Flounder Farming Disease Early Warning System. China Agricultural University, Beijing (2008)
8. Chen, L.P., Wang, D.H., Zhao, C.J.: Research on development of a development platform on handheld personal computer for agriculture expert system. Trans. CSAE **18**(3), 142–145 (2002)
9. Tu, Y.H., Wang, D.H., Zhao, C.J.: The research on development of a development platform on HPC/PDA for agriculture expert system based on windows CE. High Technol. Lett. **10**, 28–31 (2000)
10. Lei, H.Z.: Windows mobile technology applications in agriculture. Agric. Netw. Inf. **10**, 31–32 (2007)
11. Ouyang, J.Q., Qian, Y.L., Chu, C.Y., Li, J.T.: The design and implement of PDA-oriented expert system in agriculture. Comput. Eng. Appl. **38**(2), 30–31, 114 (2002)
12. Wei, Y.Y., Wang, R.J., Zhang, Y.: A knowledge representation and inference strategy for a development platform of agricultural intelligence system. CAAI Trans. Intell. Syst. **3**(6), 523–528 (2008)
13. Shi, L., Chen, D.Y., Ma, X.Y.: Application and prospects of agriculture expert system. J. Agric. Mechanization Res. **33**(1), 215–218 (2011)
14. Li, Z.D.: Research on the internet-based agricultural expert system. Agric. Sci. Technol. Equip. **6**, 135–136 (2011)
15. Li, W., Lu, D.X., Liu, C.A.: Research on automation testing on windows mobile-based device. Comput. Eng. Des. **27**(21), 4055–4057 (2006)

16. Guan, F.Y., Long, S.T., Huang, J.: Research on rural mobile information service in 3G era. Sci-Tech Inf. Dev. Econ. **21**(4), 134–136, 139 (2011)
17. Sharon, P.H., Anderson, E.: Operating systems for mobile computing. J. Comput. Sci. Coll. Arch. **25**(2), 64–71 (2009)
18. Chen, C.: Android-driven smart phone shipments grew. Commun. World Wkly. **42**, 8 (2010)

Relationship Between Vegetation Coverage and Rural Settlements and Anti-desertification Strategies in Horqin Left Back Banner, Inner Mongolia, China

Jian Zhou[1,2,3], Fengrong Zhang[1,2,3], Yan Xu[1,2,3(✉)], Yang Gao[1,2,3], and Xiaoyu Zhao[1,2,3]

[1] College of Resources and Environmental Sciences,
China Agricultural University, No. 2 Yuanmingyuan West Road,
Haidian, Beijing 100193, China
jzhou2287@163.com, {frzhang,xuyancau}@cau.edu.cn,
gaoyang123@gmail.com, zhaoxiaoyu1992@sina.com
[2] Key Laboratory for Agricultural Land Quality, Monitoring and Control,
The Ministry of Land and Resources, Beijing 100193, China
[3] Research Center of Land Use and Management,
China Agricultural University, Beijing 100193, China

Abstract. This paper investigated the relationship between vegetation coverage and rural settlements in Horqin Left Back Banner. There are 4 spatial patterns about the relationship between vegetation coverage and rural settlements, including high region-low region, high region-low region-high region, high region-low region-high region-low region, and high region. Around rural settlements, land is used as cultivated land and vegetation coverage is high. The overlap effect of overgrazing and extensive cultivation creates the first low region of vegetation coverage. With increasing distance from rural settlements, human disturbance to vegetation growth decreases, vegetation coverage increases, and the high region appears again. The second low region of vegetation coverage is caused by the overgrazing of another rural settlement. At the border region of 2 adjacent villages, 3 different conditions are presented with the vegetation coverage and rural settlements. In the first condition, vegetation coverage is low and is located at the intersection of both low regions of the 2 rural settlements. In the second condition, the vegetation coverage is high and is located at the intersection of both high regions of the 2 rural settlements. In the third condition, vegetation coverage is low and is located at the intersection of both low regions of the 2 rural settlements. According to the changing patterns and reasons for such patterns, we present strategies for anti-desertification, including banning grazing, stopping extensive cultivation, promoting optimal choice of sand-fixation plant, and clarifying land property rights.

Keywords: Vegetation coverage · NDVI · Rural settlements · Anti-desertification · Horqin Left Back Banner (HLBB)

D. Li and Z. Li (Eds.): CCTA 2015, Part I, IFIP AICT 478, pp. 125–142, 2016.
DOI: 10.1007/978-3-319-48357-3_13

1 Introduction

Desertification, defined in the Convention to Combat Desertification and Drought in 1994 as "land degradation in arid, semiarid and dry sub-humid areas resulting from various factors including climatic variability and human activities," will most likely become the greatest threat to humanity in the future because it diminishes the Earth's capacity to support human beings with the steady increase in global population [1]. Desertification is one of the most serious environmental and socioeconomic problems [2, 3]. Thus, many anti-desertification efforts have been made, including determination of the underlying mechanisms of desertification [4–7], risk assessment of desertification [8, 9], monitoring of desertification development [10–13], and investigating the effectiveness of anti-desertification projects [14–16]. China is one of the world's most desertification countries [17–19]. The desertified land in China occupies approximately 13 % of the country's total surface area and is a major source of Asian dust [20]. To address this problem, the Chinese government has launched several projects to fight desertification, such as "Three-North Shelterbelt Project" in 1978, "Grain for Green Project" in 1999, and "Beijing and Tianjin Sandstorm Source Controlling Project" in 2000. Because of these projects, the desertified land has been diminished and environment has become better in China, as well as in Horqin Sandy Land [14].

For desertification mechanism, the following is the general consensus: natural characteristics and human activities were the driving factors of desertification. However, at different locations and different periods, the main driving force of desertification is diversity. For instance, some researchers have proposed that desertification was caused by climate change in the Mu Su Sandy Land and Otindag Sandy Land and their adjacent regions, and this hypothesis was supported by archaeological evidence [21–23], whereas others have argued that human activities were the causes [24, 25]. Meanwhile, previous studies explored the effect of single or multiple factors of human activities on desertification [22, 23]. Otherwise, rural settlement is the place where all human activities occur. The patterns of vegetation coverage around rural settlements are also not clear. So, it is significant to study the relationship from the micro view of rural settlement and from the comprehensive view of taking rural settlement as the place of human activities gathering, between rural settlement and desertification. Vegetation coverage can indicate desertification risk and is an important indicator of the fixation of sand dunes and the restoration of soil fertility. So understanding the patterns of rural settlements and vegetation coverage can offer scientific support for anti-desertification efforts. The following can be determined: the critical potential desertification area, how desertification spreads in area, and what strategies can be adopted to enhance the effects of anti-desertification methods.

As mentioned above, the aims of this study are the following: (I) investigate vegetation coverage patterns (desertification risk degree) around one rural settlement and between 2 adjacent rural settlements, (II) analyze the driving forces of the difference of vegetation coverage in different regions, (III) put forward strategies to anti-desertification by increasing vegetation coverage.

2 Study Area and Study Transect

2.1 Study Area

Horqin Left Back Banner (HLBB) is a county covered 11481 km^2 in Inner Mongolia and located in northern China with a geo-location from 121°30′E–123°42′E and 42°40′N–43°42′N (Fig. 1). HLBB lies in the agro-pastoral fragile ecotone, being a part of Horqin Sandy Land which is the first sandy land in area in China. Being a typical temperate continental climate type, its average annual temperature is 5.3 °C–5.9 °C. Average annual precipitation is 415 mm and is concentrated in June to August, accounting for 70 % of average annual precipitation. The average annual rainfall reduces from east to west and from south to north. Average annual evaporation is 3.9–4.5 times the average annual rainfall. Average annual wind speed is between 3.5 m/s–4.5 m/s and high-speed windy days are concentrated in winter and spring. Generally, the average number of windy days with a speed of >5 m/s (causing sand blowing) reaches 40 in one year. The landform in HLBB includes undulating sand dune, flat sand land, dune slack, and plain. Geomorphic characteristics are connections of fixed-sand dune, semi-fixed sand dune, and moving sand dune and the alternative distribution of undulating sand land and marshy land. Sand dunes are mainly distributed in the north, south, and west of HLBB. The plain is located in the east. Aeolian sandy soil is the main soil type which takes up 68.9 % in all area of HLBB. In general, the physical features of HLBB create conditions for the formation and development of desertification.

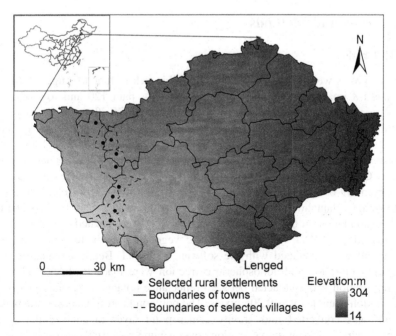

Fig. 1. The location of HLBB and the selected rural settlements

Being an agro-pastoral ecotone, the main animals raised in HLBB are sheep and cattle. Maize is the staple planting crop, and the cropping system is one harvest a year. In 2013, the population was 4.05×10^5, and population density was $35/km^2$. The numbers of sheep and cattle were 4.678×10^5 and 4.013×10^5, respectively. The maize planting area was 1.61×10^5 hm^2 accounting for 86 % in all sown area.

2.2   Study Transect

A total of 262 villages are in HLBB, and analyzing the relationship between vegetation coverage and rural settlement for each village would be difficult. Belt transect method was employed. A belt from north to south, which contains 9 villages (Fig. 1), was chosen. In the belt transect, precipitation ranges from 280 mm in the north to 360 mm in the south. The study transect is located in the sand dune distribution zone, which is vulnerable to land desertification. Elevation drops from 280 m to 40 m. Elevation of HLBB changes from 304 m to 14 m. Agriculture and animal husbandry coexist in the transect area. Cropping system is one harvest a year, and the crop is mainly maize. Thus, the study transect can represent the whole study area in natural conditions and production mode. Villages in the belt transect include Bayantala Gacha (BYTLGC), Momai Gacha (MMGC), Bianjie Gacha (BJGC), Baixingtu Gacha (BXTGC), Genggei Gacha (GGGC), Wuguan Gacha (WGGC), Wudantala Gacha (WDTLGC), Hailasitai Gacha (HLSTGC), and Xinaili Gacha (XALGC) from north to south. Gacha as an administrative unit is equal to village.

3   Materials and Methods

3.1   Materials

Vegetation grows well in August because rainfall is rich in June to August [26]. Thus, one Landsat-8 imagery, acquisition time 30-08-2014, path 120 and row 030, cloud cover 0 %, resolution 30 m, was obtained from USGS (http://glovis.usgs.gov/). Vegetation coverage was presented by calculating vegetation index. Village boundary and rural settlements were extracted from land use database for 2012 (1:10000) which was formed through land use change survey on the base of the Second Land Use Survey of China completed in 2009. In addition, we conducted twice field surveys in HLBB from 09-07-2014 to 19-07-2014 and from 08-10-2014 to 15-10-2014 to investigate the plantation, desertification, vegetation, water, and living conditions. Some questionnaires on the above mentioned aspects were created.

In this paper, NDVI was chosen as the vegetation index to reflect vegetation coverage and was calculated with the software of ENVI. Before calculating NDVI, radiometric calibration and atmospheric correction were made by ENVI. Then, geometric registration was conducted between reflectance imagery and land use map, and the registration accuracy was 0.205 pixels. The polygon layer of 9 rural settlements for 9 villages was converted into gravity center point layer to analyze the spatially changing patterns between the vegetation coverage and rural settlements with the help of geographic information system software (ArcGIS 10.0).

3.2   Methods

3.2.1   Vegetation Index

Desertification is closely related to vegetation coverage. Low level vegetation coverage indicates high risk of desertification, especially in sandy land areas. NDVI is a significant measure that reflects vegetation coverage [27, 28] and was calculated as Formula 1 [29, 30]. The higher NDVI value, the better the vegetation coverage.

$$NDVI = (RED - NIR)/(RED + NIR) \tag{1}$$

RED represents reflectance of red band and *NIR* represents reflectance of near-infrared band.

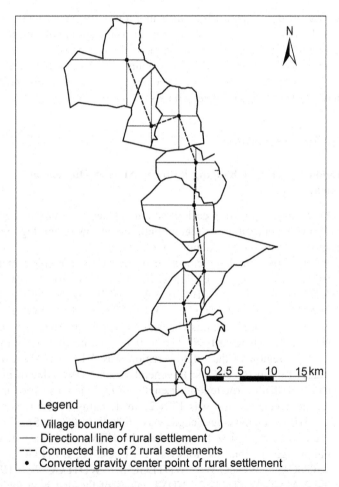

Fig. 2. Directional line of each rural settlement and connected line of 2 adjacent rural settlements

3.2.2   Vegetation Coverage Changing Patterns Around One Rural Settlement and Between 2 Adjacent Rural Settlements

To analyze vegetation coverage changing patterns around one rural settlement and between 2 adjacent rural settlements, four directional lines (east, west, south, and north) were made. These lines started from gravity center points of rural settlements and ended at village boundaries by ArcGIS 10.0. Then, buffering was made for two sides of each line, and the buffering distance was 300 m with the help of ArcGIS 10.0; 300 m is equal to the length of 10 grids. At last, starting from intersection point of the line and the boundary of rural settlement polygon to village boundary, a NDVI value was obtained by averaging NDVI values of 10 grids every 30 m. In this way, vegetation coverage changing condition was obtained in the 4 directions (Fig. 2). The vegetation coverage of 4 directions represents vegetation coverage condition around one rural settlement.

One line was made to connect 2 adjacent rural settlements, and the buffering distance for 2 sides of each line was 300 m (Fig. 2). Starting from the intersection point of connected line and the boundary of one rural settlement polygon and ending at the intersection point of connected line and the boundary of the other rural settlement polygon, the averaged NDVI was calculated every 30 m to determine vegetation coverage condition between 2 rural settlements.

4   Results and Discussions

4.1   Vegetation Coverage Changing Patterns Around One Rural Settlement

In terms of NDVI changing trend in each direction, 4 changing patterns of NDVI were obtained, as follows: high region-low region, high region-low region-high region, high region-low region-high region-low region, and high region.

Sixteen directions of 9 rural settlements had the NDVI changing pattern of high region-low region, as follows: BYTLGC-E, MMGC-S, MMGC-W, BJGC-S, BJGC-W, BXTGC-N, BXTGC-S, BXTGC-W, GGGC-N, WGGC-S, WGGC-E, WDTLGC-N, WDTLGC-E, HLSTGC-N, HLSTGC-E, and XALGC-E (Table 1). Mean NDVI values of 16 high regions changed from 0.58 to 0.81, and mean NDVI values of 16 low regions changed from 0.26 to 0.48. The maximum ratio of average NDVI value of high region to that of low region was 3.12 in XALGC-E, and the minimum ratio was 1.46 in BXTGC-N. In general, mean NDVI value of high regions was twice higher than that of low regions. Lengths of high regions varied from 330 m to 3510 m, and the average length was 1721.25 m. Lengths of low regions changed from 360 m to 3510 m. The average length was 1638.75 m (Table 5).

Eleven directions of 8 rural settlements had the NDVI changing pattern of high region-low region-high region, as follows: BYTLGC-N, BYTLGC-S, BJGC-N, BXTGC-E, GGGC-S, WGGC-W, WDTLGC-S, WDTLGC-W, HLSTGC-S, XALGC-N, and XALGC-W (Table 2). NDVI values of the first high regions varied from 0.49 to 0.86, and average NDVI value of the first high regions was 0.68. NDVI values of the second high regions changed from 0.33 to 0.93, and average NDVI value

Table 1. NDVI changing pattern of high region-low region

Village name and direction	Starting distance/m	Ending distance/m	Average NDVI
BYTLGC-E	0	1050	0.70
	1080	2250	0.41
MMGC-S	0	1830	0.73
	1860	2730	0.30
MMGC-W	0	2190	0.79
	2220	3450	0.47
BJGC-S	0	2880	0.58
	2910	4890	0.29
BJGC-W	0	750	0.70
	780	2190	0.36
BXTGC-N	0	750	0.70
	780	2100	0.48
BXTGC-S	0	330	0.68
	360	4200	0.38
BXTGC-W	0	1140	0.69
	1170	5880	0.34
GGGC-N	0	810	0.59
	840	1740	0.32
WGGC-S	0	3210	0.78
	3240	5070	0.35
WGGC-E	0	3030	0.76
	3060	3360	0.45
WDTLGC-N	0	3510	0.63
	3540	5130	0.33
WDTLGC-E	0	600	0.73
	630	2130	0.35
HLSTGC-N	0	2100	0.63
	2130	4170	0.35
HLSTGC-E	0	2760	0.79
	2790	3750	0.35
XALGC-E	0	600	0.81
	630	1200	0.26

Notes: XXX-X, XXX represents rural settlement name and X represents one direction among east, west, south and north.

of the second high regions was 0.64. NDVI values of low regions changed from 0.26 to 0.71, and average NDVI value of low regions was 0.44. The maximum, average, and minimum ratios of average NDVI value of the first high regions to that of low regions were 2.23, 1.55, and 1.11, respectively. The maximum, average, and minimum ratio of average NDVI value of the second high regions to that of low regions were 2.50, 1.45, and 1.10, respectively. Lengths of the first high regions varied from 480 m to 1590 m, and the average length was 913.64 m. Starting distances of low regions away from the

Table 2. NDVI changing pattern of high region-low region-high region

Village name and direction	Starting distance/m	Ending distance/m	Average NDVI
BYTLGC-N	0	630	0.59
	660	3510	0.40
	3540	5400	0.60
BYTLGC-S	0	870	0.49
	900	4290	0.30
	4320	6600	0.33
BJGC-N	0	1380	0.54
	1410	3630	0.27
	3660	4080	0.47
BXTGC-E	0	480	0.66
	510	3270	0.36
	3300	3390	0.47
GGGC-S	0	990	0.83
	1020	2940	0.42
	2970	5070	0.57
WGGC-W	0	690	0.79
	720	1860	0.71
	1890	2190	0.93
WDTLGC-S	0	1590	0.86
	1620	2670	0.58
	2700	2970	0.84
WDTLGC-W	0	600	0.67
	630	2580	0.55
	2610	3600	0.74
HLSTGC-S	0	1170	0.66
	1200	4350	0.36
	4380	5040	0.56
XALGC-N	0	750	0.78
	780	1110	0.62
	1140	1440	0.86
XALGC-W	0	900	0.58
	930	3270	0.26
	3300	3780	0.65

boundary of rural settlement varied from 510 m to 1620 m. The average starting distance away from the boundary of rural settlement was 943.64 m. Ending distances of low regions away from the boundary of rural settlement ranged from 1110 m to 4350 m. The average distance away from the boundary of rural settlement was 3043.64 m. The average length of low regions was 2100 m. Starting distances of the second high regions away from the boundary of rural settlement varied from 1140 to 6490 m. The average length of the second high regions was 886.36 m (Table 5).

Table 3. NDVI changing pattern of high region-low region-high region-low region

Village name and direction	Starting distance/m	Ending distance/m	Average NDVI
BYTLGC-W	0	330	0.54
	360	3480	0.38
	3510	4800	0.74
	4830	7920	0.42
MMGC-N	0	510	0.74
	540	3990	0.43
	4020	4500	0.73
	4530	8100	0.35
BJGC-E	0	870	0.59
	900	1980	0.30
	2010	2430	0.70
	2460	2700	0.32
WGGC-N	0	870	0.68
	900	2880	0.36
	2910	4620	0.71
	4650	5340	0.45
HLSTGC-W	0	660	0.69
	690	2490	0.36
	2520	6870	0.65
	6900	10410	0.27

Five directions of 5 rural settlements had the NDVI changing pattern of high region-low region-high region-low region, including BYTLGC-W, MMGC-N, BJGC-E, WGGC-N, and HLSTGC-W (Table 3). NDVI ranges of the first high regions, the first low regions, the second high regions, and the second low regions were 0.54 to 0.74, 0.30 to 0.43, 0.65 to 0.74, and 0.27 to 0.45, respectively. The average NDVI values of the first high regions, the first low regions, the second high regions, and the second low regions were 0.65, 0.37, 0.71, and 0.36, respectively. Average length of the first high regions was 648 m. The average starting distance and average ending distance away from the boundary of rural settlement of the first low regions were 678 and 2964 m. The average length of the first low regions was 2286 m. The average starting distance and average ending distance away from the boundary of rural settlement of the second high regions were 2994 and 4644 m. The average length of the second high

Table 4. NDVI changing pattern of high region

Village name and direction	Starting distance/m	Ending distance/m	Average NDVI
MMGC-E	0	660	0.68
GGGC-E	0	2580	0.85
XALGC-S	0	1440	0.90

regions was 1650 m. The average starting distance away from the boundary of rural settlement of the second low regions was 4674 m and the average length of the second low regions was 2220 m (Table 5).

Another NDVI changing pattern for 3 directions of 3 rural settlements was high region, including MMGC-E, GGGC-E, and XALGC-S (Table 4). Average NDVI values of 3 directions were 0.68, 0.85, and 0.90 respectively. Lengths of 3 directions were 660, 2580, and 1440 m.

Generally, the location selection of rural settlement give preference to the place that is rich in water and is fertile in soil resource. So, around one rural settlement, land is exploited into cultivated land and used for food supply and economic income. Furthermore, according to our field research, to keep the sustainable use of the relatively

Table 5. NDVI value and distance of each region

Changing pattern	Region	NDVI			Average distance/m		Length/m
		Max	Min	Average	Starting	Ending	
High region-low region	High region	0.81	0.58	0.71	0	1721.25	1721.25
	Low region	0.48	0.26	0.36	1751.25	3390.00	1638.75
High region-low region-high region	The first high region	0.86	0.49	0.68	0	913.64	913.64
	Low region	0.71	0.26	0.44	943.64	3043.64	2100.00
	The second high region	0.93	0.33	0.64	3073.64	3960.00	886.36
High region-low region-high region-low region	The firs high region	0.74	0.54	0.65	0	648.00	648.00
	The first low region	0.43	0.30	0.37	678.00	2964.00	2286.00
	The second high region	0.74	0.65	0.71	2994.00	4644.00	1650.00
	The second low region	0.45	0.27	0.36	4674.00	6894.00	2220.00
High region	High region	0.90	0.68	0.81	0	1560.00	1560.00

fertile cultivated land, farmers take conservation tillage (for example, crop straw stubble and no-tillage) to fertilize soil and reduce soil wind erosion under windy and less rainfall circumstance. Crops planted in the cultivated land around rural settlements grow well, and NDVI value is high. HLBB is a coexistence zone of farming and grazing and is a vast territory with sparse population. This zone had a population density of 35/km^2 in 2013. Outside cultivated land around rural settlements, the main land use type is grassland. Compared to the limited population, the area of grassland is vast. In case of drought years, the farmers adopt extensive cultivation to get grain. In this way, many grasslands are exploited and planted. However, the exploited land is abandoned in the next few years due to quick decrease in soil fertility after land use. According to the report of grassland resource survey in HLBB, all grassland may be exploited, planted, and abandoned within 8 years. This land use mode destroys land vegetation, and land becomes vulnerable to desertification. At the same time, grazing economy is a significant part of the family income for farmers, which can contribute >50 % of the total family income. Although banning grazing is conducted during 1 May to 1 June every year in HLBB, overgrazing still exists during other times in this zone according to our surveys. Overgrazing and extensive cultivation is devastating to the growth of vegetation. Thus, the NDVI changing pattern of high region-low region is formed, and the low region of NDVI is vulnerable to desertification.

On the basis of NDVI changing pattern of high region-low region, NDVI value increased with increasing distance from rural settlements, thereby indicating that vegetation coverage is improving. Average distance from the starting point of the second high region of high region-low region-high region to the boundary of rural settlement was 3073.64 m. According to our field survey, grazing distance is also about 3000 m. Furthermore, with increasing distance from rural settlement, farmers will not attempt cultivation. Thus, cultivation declines until no cultivation occurs. This finding demonstrated that vegetation coverage can improve in a manner similar to the vegetation coverage of cultivated land around rural settlements if vegetation could grow without disturbance from overgrazing and extensive cultivation in HLBB. The NDVI changing pattern of high region-low region-high region is formed. Relevant literatures and data showed that water consumption of grass is less than 300 mm, and some shrubs consume less than or about 300 mm of water; for example, *Haloxylon ammodendron* and *Caragana microphylla* [31, 32]. The average annual rainfall of HLBB was 328.5 mm during the vegetation growth period. Natural rainfall satisfies the water demands of grass and some shrubs in a large portion of HLBB. Thus, grass and shrub plants rather than trees should be chosen, and overgrazing and extensive cultivation should be forbidden in anti-desertification strategies in HLBB. Thus, shrub and grass can grow well, and vegetation coverage can be high.

Another changing pattern of NDVI was high region-low region-high region-low region. The average distance from starting point of the second low region to the boundary of rural settlement was 4674 m, which was the farthest away from rural settlement and was near the other rural settlement. This changing pattern can be explained in the section on "vegetation coverage between rural settlements."

For MMGC-E, GGGC-E, and XALGC-S, NDVI changing patterns were high region only. Lengths of high regions were 660 and 1440 m for MMGC-E and XALGC-S, respectively. Lengths of the 2 high regions are short relatively, and land is used for cultivation. For GGGC-E, a river passes through from west to east. Water resource is relatively rich, and land is used for cultivation. So, NDVI values are high in the 3 directions. As to GGGC-W, several small scale rural settlements exist along the river, which can affect land use. Thus, no obvious change in trend of NDVI was observed (Fig. 3).

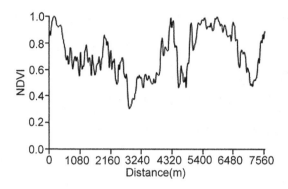

Fig. 3. NDVI changing condition of GGGC-W

In all, a high region of NDVI existed near rural settlements, in which land does not easily undergo desertification. Desertification does not occur easily because land near rural settlements is used for cultivation and is tilled year after year. Outside of cultivated land near rural settlement was a low region of NDVI that comprises mainly grassland. Because overgrazing and extensive cultivation, both of which destroy vegetation growth, coexist in this region, vegetation coverage is poor, and the land is vulnerable to desertification. With increasing distance from rural settlement, a high region of NDVI appears which indicates that vegetation coverage is improving. This phenomenon is mainly due to the decline of effect of human disturbance on vegetation growth, and rainfall can meet the water demand of shrubs and grass. If human disturbance to vegetation growth can be stopped, restoration of vegetation can succeed. As to the second low region of high region-low region-high region-low region, an explanation is given in the section on "vegetation coverage between rural settlements."

4.2   Vegetation Coverage Between Rural Settlements

For vegetation coverage at the border region of 2 adjacent villages, 3 conditions for 8 connected lines existed according to NDVI changing trend along each connected line. First, vegetation coverage at junction area of 2 adjacent villages was at the intersection of low region of one rural settlement and low region of the other rural settlement, including: BYTLGC-MMGC, BJGC-BXTGC, and GGGC-WGGC (Fig. 4a). Second, vegetation coverage at junction area of 2 adjacent villages was at the intersection of high region of one rural settlement and high region of the other rural settlement, including MMGC-BJGC, WGGC-WDTLGC, and HLSTGC-XALGC (Fig. 4b). Third,

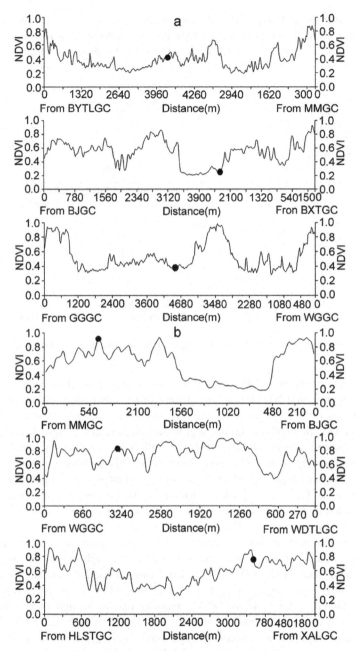

Note: Dot in the curve means the juncture of village boundary and connected line of 2 adjacent rural settlements; a represents the first condition and b represnts the second conditon.

Fig. 4. NDVI changing trend between 2 adjacent rural settlements for the first condition and the second condition

vegetation coverage at junction area of 2 adjacent villages was at the intersection of low region of one rural settlement and low region of the other rural settlement, including BXTGC-GGGC and WDTLGC-HLSTGC (Fig. 5). The difference between the first and third conditions was the NDVI changing pattern of rural settlement.

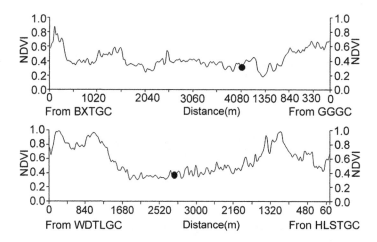

Fig. 5. NDVI changing trend between 2 adjacent rural settlements for the third condition

In the first condition, NDVI changing pattern of one rural settlement was high region-low region, and NDVI changing pattern of the other rural settlement was high region-low region-high region-low region. For the rural settlements with NDVI changing pattern of high region-low region-high region-low region, the second low region does not result from the extensive cultivation and overgrazing itself, but rather from the overgrazing of the other rural settlement. This phenomenon can be demonstrated by distances from the juncture of village boundary and connected line of 2 adjacent rural settlements to 2 rural settlements. The distance from the juncture to BYTL, which had the NDVI changing pattern of high region-low region, was 4350 m. Distance from the juncture to MMGC, which had the NDVI changing pattern of high region-low region-high region-low region, was 5160 m. The distance from the juncture to BXTGC, which had the NDVI changing pattern of high region-low region, was 2340 m. The distance from the juncture to BJGC, which had NDVI changing pattern of high region-low region-high region-low region, was 4410 m. The distance from the juncture to GGGC, which had the NDVI changing pattern of high region-low region, was 4590 m. The distance from the juncture to MMGC, which had NDVI changing pattern of high region-low region-high region-low region, was 4890 m. Human disturbance from rural settlements with the NDVI changing pattern of high region-low region is the reason for the formation of the second low region of high region-low region-high region-low region. According to our field surveys, junction area is closer to the rural settlements, which have NDVI changing pattern of high region-low region. Overgrazing at the second low region mainly comes from the closer rural settlements. At the same time, overgrazing from the farther rural settlements is vanished.

In the second condition, vegetation coverage at junction area of 2 villages was at the intersection of high region of one rural settlement and high region of the other rural settlement. One rural settlement had the NDVI changing pattern of high region-low region-high region, whereas the other had the pattern of high region. Distance from the juncture to MMGC, WGGC, and XALGC, which had NDVI changing pattern of high region, were 630, 1230, and 960 m. Distances from the juncture to BJGC, WDTLGC, and HLSTGC, which had NDVI changing pattern of high region-low region-high region, were 2520, 3300, and 3390 m, respectively. Because distance from the juncture to rural settlements with NDVI changing pattern of high region-low region-high region is farther, human disturbances to the junction zone decreases. At the same time, distance from the juncture to rural settlements with NDVI changing pattern of high region is nearer, and land is used for cultivation. Thus, the vegetation coverage at the junction area improves.

In the third condition, vegetation coverage at junction area of 2 villages was at the intersection of low region of one rural settlement and low region of the other rural settlement. Both rural settlements had the NDVI changing pattern of high region-low region. NDVI at the junction area is much smaller than that of the high region. These results are due to the overgrazing and extensive cultivation from 2 rural settlements. Land in the junction area is more vulnerable to desertification.

5 Conclusions

Four NDVI changing patterns existed around rural settlements, as follows: high region-low region, high region-low region-high region, high region-low region-high region-low region, and high region. Due to natural conditions, for example water and soil, land is used as cultivated land around rural settlements, and vegetation coverage improves. Outside of high region around rural settlements is a low region of NDVI because of the overlapping effects of overgrazing and extensive cultivation; thus, the area is vulnerable to desertification. With the increase of distance far away from rural settlements, human disturbance decreases, vegetation coverage improves, and the high region of NDVI appears again. For the second low region of NDVI of high region-low region-high region-low region, it comes from the overgrazing of another rural settlement.

There were 3 conditions of vegetation coverage at the border region of 2 adjacent villages. First, vegetation coverage at junction area of 2 adjacent villages was at the intersection of low region of one rural settlement and low region of the other rural settlement. One rural settlement had NDVI changing pattern of high region-low region, and the other rural settlement had NDVI changing pattern of high region-low region-high region-low region. The second low region of high region-low region-high region-low region for one rural settlement results from the other rural settlement's overgrazing. Second, vegetation coverage at junction area of 2 adjacent villages was at the intersection of high region of one rural settlement and high region of another rural settlement. One rural settlement had NDVI changing pattern of high region-low region-high region and the other rural settlement had NDVI changing pattern of high region. Third, vegetation coverage at junction area of 2 adjacent villages was at the intersection of low region of one rural settlement and low region of the other rural settlement. Both rural settlements had NDVI changing pattern of high region-low region. The overlap effect of overgrazing and extensive cultivation from 2 rural settlements makes the NDVI at junction area much smaller.

6   Strategies for Anti-desertification

Vegetation coverage condition is closely related to land desertification. Outskirt of cultivated land around rural settlements is the emphasis area of anti-desertification because the vegetation coverage is poor and overgrazing and extensive cultivation are severe in this area. Banning grazing and stopping extensive cultivation should be done to protect vegetation growth and to further promote anti-desertification.

At the same time, far away from rural settlements, another high region of NDVI appears again, indicating that vegetation can grow well in the study area when no human disturbance affects vegetation growth. Integrating with the natural condition of the study area, grasses and shrubs rather than trees should be chosen for anti-desertification. Thus, vegetation can show improved growth, and anti-desertification efforts can succeed further.

The formation of the second low region of NDVI for high region-low region-high region-low region comes from overgrazing of another rural settlement. NDVI value at junction area of 2 adjacent villages, which had the NDVI changing pattern of high region-low region, is much lower. Unclear grassland poverty makes the misuse land resource at the junction area. So, clarification of land property right can affect the rational use of grassland and can help in anti-desertification.

Differences exist among rural settlements in terms of population and number of livestock. Length is different for the same region in the different NDVI changing patterns. So, the determination of emphasis zone range for anti-desertification should be combined with the specific situation of rural settlements.

Acknowledgment. We would like to recognize the National Science Foundation of China (grant 4127111) and the Special Scientific Research of the Ministry of Land and Resource (grant 201411009) for the support of related data and yield survey. We are grateful to the Inner Mongolia Land Surveying and Planning Institute and Inner Mongolia Land Consolidation Institute for support.

References

1. Chen, Y., Tang, H.: Desertification in North China: background, anthropogenic impacts and failures in combating it. Land Degrad. Dev. **16**(4), 367–376 (2005)
2. Kassas, M.: Desertification: a general review. J. Arid Environ. **30**(2), 115–128 (1995)
3. Zhao, H., Li, J., Liu, R., et al.: Effects of desertification on temporal and spatial distribution of soil macro-arthropods in Horqin sandy grassland, Inner Mongolia. Geoderma **223–225**, 62–67 (2014)

4. Fullen, M.A., Mitchell, D.J.: Desertification and reclamation in North-Central China. Ambio **23**(2), 131–135 (1994)
5. Runnström, M.C.: Is northern China winning the battle against desertification? Satellite remote sensing as a tool to study biomass trends on the Ordos Plateau in semiarid China. Ambio: A J. Human Environ. **29**(8), 468–476 (2000)
6. Xue, Z., Qin, Z., Li, H., et al.: Evaluation of aeolian desertification from 1975 to 2010 and its causes in northwest Shanxi province, China. Global Planet. Change **107**, 102–108 (2013)
7. Xu, D., Li, C., Song, X., et al.: The dynamics of desertification in the farming-pastoral region of North China over the past 10 years and their relationship to climate change and human activity. Catena **123**, 11–22 (2014)
8. Santini, M., Caccamo, G., Laurenti, A., et al.: A multi-component GIS framework for desertification risk assessment by an intergrated index. Appl. Geogr. **30**(3), 394–415 (2010)
9. Ladisa, G., Todorovic, M., Trisorio, L.G.: A GIS-based approach for desertification risk assessment in Apulia region, SE Italy. Phys. Chem. Earth **49**, 103–113 (2012)
10. Hill, J., Hostert, P., Tsiourlis, G., et al.: Monitoring 20 years of increased grazing impact on the Greek island of Crete with earth observation satellites. J. Arid Environ. **39**(2), 165–178 (1998)
11. del Barrio, G., Puigdefabregas, J., Sanjuan, M.E., et al.: Assessment and monitoring of land condition in the Iberian Peninsula, 1989–2000. Remote Sens. Environ. **114**(8), 1817–1832 (2010)
12. Dawelbait, M., Morari, F.: Monitoring desertification in a Savannah region in Sudan using Landsat images and spectral mixture analysis. J. Arid Environ. **80**, 45–55 (2012)
13. Li, J., Yang, X., Jin, Y., et al.: Monitoring and analysis of grassland desertification dynamics using Landsat images in Ningxia, China. Remote Sens. Environ. **138**, 19–26 (2013)
14. Li, Y., Cui, J., Zhang, T., et al.: Effectiveness of sand-fixing measures on desert land restoration in Kerqin Sandy Land, northern China. Ecol. Eng. **35**(1), 118–127 (2009)
15. Zhang, G., Dong, J., Xiao, X., et al.: Effectiveness of ecological restoration projects in Horqin Sandy Land, China based on SPOT-VGT NDVI data. Ecol. Eng. **38**(1), 20–29 (2012)
16. Miao, R., Jiang, D., Musa, A., et al.: Effectiveness of shrub planting and grazing exclusion on degraded sandy grassland restoration in Horqin sandy land in Inner Mongolia. Ecol. Eng. **74**, 164–173 (2015)
17. Tao, W., Chen, G., Zhao, H., et al.: Research progress on Aeolian desertification process and controlling in North of China. J. Desert Res. **26**(4), 507–516 (2006)
18. Wang, F., Pan, X., Wang, D., et al.: Combating desertification in China: past, present and future. Land Use Policy **31**, 311–313 (2013)
19. Yang, X., Zhang, K., Jia, B., et al.: Desertification assessment in China: an overview. J. Arid Environ. **63**(2), 517–531 (2005)
20. Song, Z.: A numerical simulation of dust storms in China. Environ. Model. Softw. **19**(2), 141–151 (2004)
21. Weilin, W.: When Maowusu become a desert? - View through new archaeologocal finds. Archaeol. Cult. Relics **5**, 80–85 (2002)
22. Han, Z.: The evolution of the Maowusu desert and the reclamation in the adjacent areas in the Ming Dynasty. Soc. Sci. China **5**, 191–204 (2003)
23. Wang, X., Chen, F.H., Dong, Z., et al.: Evolution of the southern Mu Us Desert in North China over the past 50 years: an analysis using proxies of human activity and climate parameters. Land Degrad. Dev. **16**(4), 351–366 (2005)
24. Hou, R.: Mu Us Desert evolution as indicated by the deserted ancient cities along the Sjara River. Cult. Relics **1**, 35–41 (1973)

25. Wu, W.: Study on process of desertification in Mu Us Sandy Land for last 50 years, China. J. Desert Res. **21**(2), 164–169 (2001)
26. Jia, S., Han, Z., Lv, M., et al.: Extraction of desertification information based on decision tree in northern Liaoning province. Ecol. Environ. Sci. **20**(1), 13–18 (2011)
27. Huang, S., Siegert, F.: Land cover classification optimized to detect areas at risk of desertification in North China based on SPOT VEGETATION imagery. J. Arid Environ. **67** (2), 308–327 (2006)
28. Sternberg, T., Tsolmon, R., Middleton, N., et al.: Tracking desertification on the Mongolian steppe through NDVI and field-survey data. Int. J. Digital Earth **4**(1), 50–64 (2011)
29. Myneni, R.B., Hall, F.G., Sellers, P.J., et al.: The interpretation of spectral vegetation indexes. IEEE Trans. Geosci. Remote Sens. **33**(2), 481–486 (1995)
30. Myneni, R.B., Tucker, C.J., Asrar, G., et al.: Interannual variations in satellite-sensed vegetation index data from 1981 to 1991. J. Geophys. Res. **103**(D6), 6145–6160 (1998)
31. Chang, X., Zhao, W., Zhang, Z.: Water consumption characteristic of Haloxylon ammodendron for sand binding in desert area. Acta Ecologica Sin. **27**(5), 1826–1837 (2007)
32. Yue, G., Zhao, H., Zhang, T., et al.: Characteristics of Caragana microphylla sap flow and water consumption under different weather conditions on Horqin Sandy Land of northeast China. Chin. J. Appl. Ecol. **18**(10), 2173–2178 (2007)

China Topsoil Stripping Suitability Evaluation Based on Soil Properties

Yan Xu[1,2], Yuepeng Wang[1,2], Xing Liu[1,2], Dan Luo[3],
and Hongman Liu[4(✉)]

[1] College of Resources and Environmental Sciences,
China Agricultural University, Beijing 100193, China
xyan@cau.edu.cn, lywangyp@yeah.net,
liuxing90125@163.com
[2] Key Laboratory of Land Quality Ministry of Land and Resources,
Beijing 100193, China
[3] Yantai Municipal Bureau of Land and Resources,
Yantai 265500, China
309255376@qq.com
[4] College of Economics and Management,
China Agricultural University, Beijing 100083, China
liuhm@cau.edu.cn

Abstract. The purpose of topsoil stripping reuse is to provide a good cultivation platform for agricultural site conditions, namely increasing soil thickness and improving soil fertility levels. Based on soil type, using the thickness of the soil and organic matter content as criteria, the natural suitability of stripped topsoil from different regions were evaluated. The evaluation results were divided into five levels: Most suitable, suitable, moderately suitable, less suitable, and unsuitable. The results show that the highest degree of topsoil stripping natural suitability is mainly distributed in the Northeast region, along the middle and lower Yangtze River, and the region south of the Yangtze River. The most unsuitable regions for topsoil stripping natural suitability are mainly located in the hilly regions of the Sichuan Basin and the eastern part of the Northwest. The topsoil stripping process must be based on the soil properties of both the original site and the transplantation location.

Keywords: Soil type · Topsoil stripping · Suitability assessment · Land evaluation

1 Introduction

With socioeconomic development, resource scarcity issues become increasingly prominent, especially the relative shortage of arable land resources, which has created a bottleneck constraint on China's future social and economic development. There is a shortage of arable land resources in China: per capita arable land area is less than 0.094 hm², which is less than 40 % of the world average, and on 60 % to 70 % of the arable land exists erosion, arid and barren conditions, waterlogging, salinization, compaction, gravel, mortar layers, soil gleyization and other factors [1]. Unreasonable

© IFIP International Federation for Information Processing 2016
Published by Springer International Publishing AG 2016. All Rights Reserved
D. Li and Z. Li (Eds.): CCTA 2015, Part I, IFIP AICT 478, pp. 143–152, 2016.
DOI: 10.1007/978-3-319-48357-3_14

fertilization, excessive use of pesticides, industrial waste, and acid rain has significantly worsened soil pollution. Especially in developed areas, around cities and surrounding areas, and along main roads, heavy metals and organic pollutants in soil matter have drastically exceeded standards [2]. Meanwhile, China's arable land reserve resources are scarce; the ecological conditions of the reserve resource regions are fragile and can easily lead to the development of ecological problems. Due to the various needs of construction land occupation, ecological restoration, and restructuring of agriculture, the arable land shortage will continue to be the most critical long-term issue. Therefore, effectively raising the quality of agricultural land and focusing on improving and protecting the ecology and environment has become the key to solving the problem of land use.

Topsoil stripping has become an important way to enhance the quality of agricultural land and improve low-yielding fields. Topsoil stripping refers to removing topsoil, including arable land, humus layers, gardens, woodlands, and grasslands, from construction sites and open-pit mining sites for land restoration, soil improvement, and land reclamation by stripping, storage, transportation, and plough layer construction testing of the topsoil structure and a series of related technologies. Topsoil stripping has the following benefits, such as protecting the surface soil resources, and rapid increase in soil fertility, etc. [3, 4].

Currently, topsoil stripping is common in mining site land reclamation, highway construction site reclamation, soil pollution control, major construction projects and other activities. Topsoil stripping research focuses on exploring the role of stripped topsoil in improving soil [5–15], the technology and process used in topsoil stripping [16–20], and topsoil stripping benefit analysis [21] and so on. Research on the suitability of topsoil stripping is limited, only a few scholars have explored this topic, focusing on the thickness of stripped soil [22, 23], topsoil stripping costs [24–26], as well as fertilization and necessary conditions for topsoil stripping [27]. The above mentioned research focused mainly on the project scale, and topsoil stripping suitability evaluations, especially regional comparisons, have rarely been reported. This article discusses the suitability of topsoil stripping based on the soil type from different regions in China, in order to provide references for future work in topsoil stripping.

2 Materials and Methods

2.1 Data Sources

Research data includes: 1: 100 million China Soil Type Maps, compiled and digitalized by the Institute of Soil Science, Chinese Academy of Sciences, based on the National Soil Survey Office 1995 data. Soil-related property data, achieved depending on typical soil profiles from different soil types, were recorded by the National Soil Survey Office in Chinese Soil [28] and China Soil Type [29].

2.2   Research Design

A variety of factors have a combined effect on topsoil stripping, including soil thickness, surface soil properties, soil organic matter content, soil pollution, etc., which determine the quality of the stripped soil, and also affect the quality of the new farm land or improved farmland. Among these factors, the soil pollution factor has a veto effect. Soil already contaminated can no longer be stripped.

Topsoil stripping increases the amount of new arable land and improves low quality farmland mainly by increasing the thickness of the soil, improving soil structure (including surface soil properties) and increasing nutrient levels. Among these factors, increasing the thickness of the soil is most important because it is the soil nutrient source and a repository for mineral elements in the soil. It is also the determining indicator of soil erosion with significant impact on soil nutritional status. Also, the fertile stripping soil can be used to increase the thickness of the soil layer and provide nutrients for crop growth. So soil thickness and soil organic matter content are chosen as measure indicators for topsoil stripping suitability. The Farmland Grading Regulation GB/T 28407-2012 recommended classification method was used to decide grade and scores. Farmland quality grading regulations were determined by consulting specialists. The weighted average method was used to calculate the topsoil stripping natural suitability index, then suitability levels were divided according to specialists' recommendations.

2.3   Evaluation Indicators

2.3.1   Soil Thickness Index Processing
Soil thickness has a decisive role in topsoil stripping. Soil thickness determines the crop site's conditions, and also affects moisture reserve and root depth. To a certain extent, soil thickness determines the amount of stripping soil, and in turn determines the size and depth of the soil coverage. For mountainous areas, hills and other soil-barren areas, improving soil thickness is particularly important for increasing the amount of arable land.

Based on the national soil classification and index score from Farmland Grading Regulation GB/T 28407-2012, the productive soil thickness of is divided into five levels, and in turn given scores. Results are displayed in the following Table 1:

Table 1. National soil thickness grading and scores

Level	Productive soil thickness (cm)	Score
1	≥150	100
2	100–150	90
3	60–100	70
4	30–60	50
5	<30	10

2.3.2   Indicator Processing of Organic Matter Content

Soil organic matter is an important component of the soil solid phase, which determines the soil fertility level, and also can improve soil physical properties. Typically the agriculture soil layer has the highest organic matter content compared to fast-acting fertilizers, and it needs a long time to form, affects the crop better, and is relatively stable. Topsoil stripping can form nutrient-rich organic layers directly over the transplantation site, shortening the soil maturation process.

According to the national indicators of soil organic matter content classification and score from China Farmland Grading Regulation, the soil organic matter content is divided into six levels, and in turn given scores (Table 2).

Table 2. National soil organic matter content and grading scores

Level	Soil organic matter content (g/kg)	Score
1	≥40	100
2	30–40	90
3	20–30	80
4	10–20	70
5	6–10	60
6	<6	50

2.4   Natural Suitability Score Calculation

The topsoil stripping suitability score can be calculated by the following formula:

$$Z = \sum_{i=1}^{n} x_i p_i$$

In the formula, Z is the impact score for topsoil stripping and re-use. X_i is the indicator score and P_i is the index weight.

In topsoil stripping, soil thickness and the size of the stripping area is limited, so the volume of stripped soil is fixed. Stripping soil thickness is restricted by the volume of stripped soil, but the organic matter content of the soil can be achieved through artificial fertilization, therefore, the thickness of soil is more important than soil organic matter content in topsoil stripping. According to specialists, the correct soil thickness weight is 0.6; organic matter content weight is 0.4.

The impact score Z is calculated based on the soil thickness classification score and the soil organic matter content classification score, which shows the different level of soil thickness and organic matter content's impact on topsoil stripping and re-use. The Z value is divided into five levels:

1. $Z \geq 90$, topsoil stripping suitability level 1, most suitable;
2. $80 \leq Z < 90$, topsoil stripping suitability level 2, suitable;
3. $70 \leq Z < 80$, topsoil stripping suitability level 3, moderately suitable,

4. $60 \leq Z < 70$, topsoil stripping suitability level 4, less suitable;
5. $Z < 60$, topsoil stripping suitability level 5, unsuitable.

3 Results

3.1 Soil Thickness Classification Results

Based on 1: 100 million soil type map, a soil thickness classification map was produced, shown in Fig. 1:

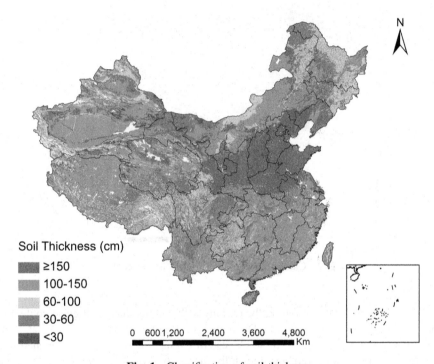

Fig. 1. Classification of soil thickness

As can be seen from the figure, in China the regions with the thickest soil are mainly concentrated in the Northeast, the Loess Plateau and the North China Plain. Black soil and black earth soil of the Northeast, loess soil of the Loess Plateau, and moisture soil of the North China Plain are all thick soils, more than 150 cm thick. The regions with moderate thickness soil are mainly in the south part of China along the middle and lower Yangtze River, the area south of the Yangtze River, the southern China region and the northern part of China, such as the Inner Mongolia Plateau and regions surrounding the Great Wall. In southern China hydrothermal conditions are better, yellow soil, red soil, brick red soil, latosolic red soil, paddy soil, etc. are thicker.

3.2   Classification Results of Soil Organic Matter Content

Based on 1: 100 million soil type map, an organic matter content classification map was produced (Fig. 2).

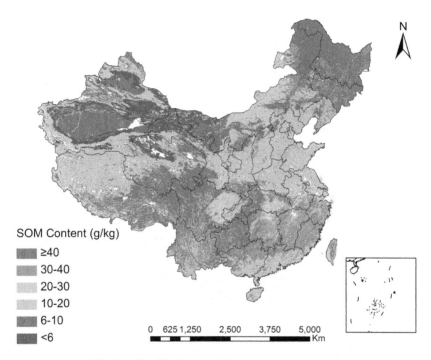

Fig. 2. Classification of soil organic matter content

As can be seen from the figure, the regions with soil that has the highest content of organic matter are mainly in the Daxinganling Region of the Northeast and the Hengduan Mountainous region of southwestern China, where there is untouched natural forest cover, followed by the hilly areas of southern China, with good hydrothermal conditions and rich biomass. The moisture soil of Huang-Huai region and the loess soil of the Loess Plateau region have less soil organic matter content.

3.3   Topsoil Stripping Natural Suitability Assessment

Depending on topsoil stripping impact scores of the different soil types, topsoil stripping suitability evaluation results are displayed according to the different regions in China, shown in Fig. 3:

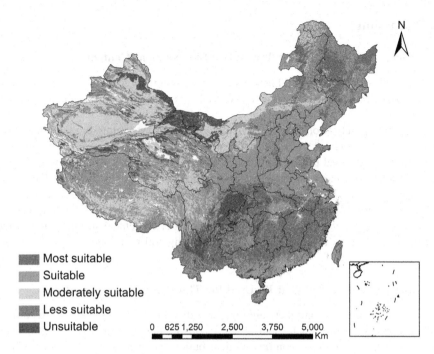

Fig. 3. Classification of topsoil stripping suitability

The most suitable areas for topsoil stripping are mainly in the northeast region, along the middle and lower Yangtze River, and the region south of the Yangtze River. The black soil, black earth soil, meadow Soil, red soil, brick red soil, and yellow-brown soil types, etc., are all rich in organic matter, suitable for topsoil stripping.

There are additional regions with moderate topsoil stripping natural suitability, those soil types are brown soil, cinnamon soil, irrigation-silting soil, heilu soil, paddy soil, moisture soil etc. The thickness of soil and organic matter content of those are not as good as black soil and black earth soil. Therefore the suitability level is lower than the black soil and black earth soil in the Northeast, as well as the red soil and yellow-brown soil in the region south of the Yangtze River.

Moderately suitable topsoil stripping natural suitability areas are mainly located in the northwest regions and the Inner Mongolia Plateau regions. The main soil types are yellow Soil, torrid red soil, red clay soil, brown desert steppe soil, etc. These types of soil have average layer thicknesses and low organic matter content; therefore suitability for topsoil stripping is average.

Less suitable areas for topsoil stripping are mainly scattered in the mountain regions, such as the volcanic ash soil, regosols, cracked soil, etc. These soils have thin layer thicknesses and low organic matter content.

Unsuitable areas for topsoil stripping natural suitability are mainly in the hilly regions of the Sichuan Basin and the eastern regions of northwestern China. These soil types are purple soil, gray-brown desert soil, alpine frost desert soil, cold desert soil, rocky soil, etc.

4   Discussion

4.1   Fertile Topsoil Improves Transplant Site Nutrient Content

Fertility levels of black soil and black earth soil are relatively high, therefore only these types of stripped topsoil can achieve the goal of improving the organic matter and nutrient content.

Therefore, purely from the perspective of topsoil stripping organic matter or nutrients, not all kinds of soils are worth stripping for the purpose of improving soil nutrient levels. Stripped topsoil from black soil and black earth soil regions with high organic matter content are suitable for nutrient enhancement. On the other hand, soil types, such as loessial soil and purple soil, are often subjected to erosion and lose rich nutrients; the use of fertilization and cultivation on these soil types is also adversely subject to erosion. These soils have low organic matter content so stripping the topsoil for soil culture cannot achieve the goal of enhancing soil nutrients.

4.2   Barren Surface Soil Can Be Used for Thickening the Soil Layer

China is vast in territory, and there are regional differences in soil thickness. According to the second soil survey with data of 1627 soil profiles, the country has either significant isolated sections or continuous distribution characteristics: the North China Plain, the Yangtze River Plains, and the Loess Plateau, Inner Mongolia Grassland soils are relatively thick; mountainous and hilly soils are relatively thin; soil layers in Tibet, Qinghai, Xinjiang, Yunnan Plateau and parts of the Sichuan Basin are the thinnest [11]. In these areas the productive thickness of the soil layer is the limiting factor for crop rooting and growth.

Therefore, when considering improving the thickness of the soil and improving site cultivation conditions, even though construction sites are nutrient-barren land, its topsoil and even the soil under the arable layer can be stripped and used for increasing the soil thickness at transplantation areas and improves its water storage drought capacity. For example, hilly and mountainous regions are typically areas where the soil determines the arability and fertilizer determines the productivity. In order for the new site conditions to meet the requirements of increased crop growth and yield productivity depends mainly on whether or not the thickness of the soil can meet the needs of the crop root system, while yield productivity depends on how much soil nutrients are added later, along with other factors. Therefore barren soil areas should emphasize increasing the soil layer thickness with the assistance of the amount of added nutrients.

4.3   Stripping Fertilizing Based on Local Conditions

Stripping topsoil for increasing new arable land and fertilizing barren areas, must based on farmland soil fertility status and productive soil thickness, and soil fertility conditions of the stripped soil to do comparative analysis, then combined with the actual local situation, making the supply of the stripped areas and the needs of the transplant areas match. In general, if soil thickness is a major limiting factor for new arable land, cultivated soil layers below the arable layer can also be used for covering, without considering fertility; if the transplant site itself has enough thickness and soil nutrient content is the only limiting factor, then it must be covered by a soil with a relatively high level of fertility; if the topsoil stripping area has low soil fertility, or the surface fertility is similar with the lower layer, then conducting topsoil stripping is unnecessary.

4.4 Other Noticeable Issues

Factors affecting topsoil stripping include natural, social, economic, geographical, laws and regulations etc. In this article, only the natural factors were chosen for evaluation and the combined impact from social, economic, transportation, laws and regulations factors were not addressed. If these factors are also addressed with comprehensive and quantitative analysis, the results will be more accurate, objective, and easier to guide practical application.

Acknowledgements. Funding from National Natural Science Foundation of China, NO. 41301614.

References

1. Li, Y.: China Land Resources. China Land Press, Beijing (2000)
2. Zhao, Q., Zhou, S., Wu, S., et al.: Cultivated land resources and strategies for its sustainable utilization and protection in China. Acta Pedol. Sin. **43**(4), 662–672 (2006)
3. Yan, S., Wang, T., Dou, S.: Highway borrow pits topsoil stripping engineering technical points. Jilin Agric. (11), 238 (2010)
4. Xu, B., Wang, T., Dou, S.: A preliminary study on topsoil stripping technology. Jilin Agric. (1), 18 (2012)
5. Patzelt, A., Wild, U., Pfadenhauer, J.: Restoration of wet fen meadows by topsoil removal: vegetation development and germination biology of fen species. Restor. Ecol. **9**(2), 127–136 (2001)
6. Tan, Y., Han, C., Wu, C., et al.: Patterns of topsoil stripping for planting use in foreign countries and its enlightenment for China. Trans. Chin. Soc. Agric. Eng. **29**(23), 194–201 (2013)
7. Tallowin, J.R.B., Smith, R.E.N.: Restoration of a Cirsio-Molinietum Fen meadow on an agriculturally improved pasture. Restor. Ecol. **9**(2), 167–178 (2001)
8. Van Diggelen, R., Bakker, J.P., Klooker, J.: Top soil removal: new hope for threatened plant species. Species Dispersal Land Use Process. 257–263 (1997)
9. Verhagen, R., Klooker, J., Bakker, J.P., et al.: Restoration success of low-production plant communities on former agricultural soils after top-soil removal. Appl. Veg. Sci. **4**(1), 75–82 (2001)
10. Hölzel, N., Otte, A.: Restoration of a species-rich flood meadow by topsoil removal and diaspore transfer with plant material. Appl. Veg. Sci. **6**(2), 131–140 (2003)

11. Klimkowska, A., Van Diggelen, R., Bakker, J.P., et al.: Wet meadow restoration in Western Europe: a quantitative assessment of the effectiveness of several techniques. Biol. Conserv. **140**(3), 318–328 (2007)
12. Hu, Z.: Principle and method of soil profile reconstruction for coal mine land reclamation. J. China Coal Soc. **6**, 59–64 (1997)
13. Hu, Z., Wei, Z., Qin, P.: Concept of and methods for soil reconstruction in mined land reclamation. Soils **37**(1), 8–12 (2005)
14. Fu, M., Chen, Q.: Surface soil stripping and its technology in ecological reclamation in mine area. Metal Mine **8**, 63–65 (2004)
15. Jiao, X., Wang, L., Lu, C., et al.: Effects of two reclamation methodologies of coal mining subsidence on soil physical and chemical properties. J. Soil Water Conserv. **4**, 123–125 (2009)
16. Dou, S., Dong, X., Zhang, D., et al.: Technical system of topsoil stripping of songliao plain. J. Jilin Agric. Univ. **36**(2), 127–133 (2014)
17. Sun, H., Ma, Y.: Several measures of soil utilization of highway construction. Heilongjiang Traffic Sci. Technol. (12), 162 (2007)
18. Yang, R., Yang, J.: Research on the reclamation technology for the land temporarily used for railway construction. J. Railway Eng. Soc. (4), 57–61 (2009)
19. Wang, R., Zhang, X., Jiang, W., et al.: Implementation conditions of topsoil stripping of farmland used for construction—taking shifting soil and fertility betterment of three reservoir area as example. J. Hebei Agric. Sci. (1), 90–91 (2011)
20. Wei, A.: The cultivated land fertility of land leveling process to keep the problem. J. Fujian Agric. Sci. Technol. (3), 72–74 (2011)
21. Chen, G., Zhang, X., Wang, R., et al.: Evaluation of topsoil stripping and reuse project performance—taking removed soil fertilization project in three gorges reservoir area as example. Bull. Soil Water Conserv. (5), 239–243 (2012)
22. Grootjans, A.P., Bakker, J.P., Jansen, A.J.M., et al.: Restoration of brook valley meadows in the Netherlands. Hydrobiologia **478**(1–3), 149–170 (2002)
23. Lamers, L.P.M., Smolders, A.J.P., Roelofs, J.G.M.: The restoration of fens in the Netherlands. Hydrobiologia **478**(1–3), 107–130 (2002)
24. Pfadenhauer, J., Grootjans, A.: Wetland restoration in Central Europe: aims and methods. Appl. Veg. Sci. **2**, 95–106 (1999)
25. Ramseier, D.: Why remove the topsoil for fen restoration?—Influence of water table, nutrients and competitors on the establishment of four selected plant species. Bull. Geobotanical Inst. ETH (66), 25–35 (2000)
26. Klimkowska, A., Dzierża, P., Brzezińska, K., et al.: Can we balance the high costs of nature restoration with the method of topsoil removal? Case study from Poland. J. Nat. Conserv. **18**(3), 202–205 (2010)
27. Xu, Y., Zhang, F., Zhao, H., et al.: Prerequisites for preserving the fertility and the soil layer before farmland converted for construction use. China Land Sci. (11), 93–96 (2011)
28. Office of National Soil Survey: Chinese Soil. China Agricultural Press, Beijing (1998)
29. Office of National Soil Survey: Chinese Soil Types · The First Volume—The Sixth Volume. China Agricultural Press, Beijing (1995)

Design of Corn Farmland Monitoring System Based on ZigBee

Xiuli Si[1(✉)], Guifen Chen[1(✉)], and Weiwei Li[2]

[1] JinLin Agricultural University, Changchun 130118, China
adminsxl@163.com, guifchen@163.com
[2] China Greatwall Computer Shenzhen CO., Ltd., Shenzhen 518052, China
ldd604@126.com

Abstract. On basis of ZigBee network theory, a corn farmland monitoring system was designed, integrates free maintenance and low cost ZigBee network, convenient access GPRS network, and strong GIS software with map process and data statistics. It can be used to monitor and warn the real-time state of corn farmland through measuring sensors. Also help technicians and farmers reducing workload on farm management as well as improving corn planting level.

Keywords: ZigBee · GPRS · GIS · Corn · Farmland · Monitoring

1 Introduction

Air humidity, air temperature, soil moisture, and soil nitrogen phosphorus potassium content are vital factors in corn growth. If can be monitored, will reduce the possibility of existence of low fertilizer utilization rate, low yield and not process disease prevention timely while prevent disease and save the cost of agricultural production.

ZigBee network as a wireless network sensor technology based on IEEE 802.15.4 with the characteristics of low power consumption, low cost and low rate, is drawing researchers and enterprise application researchers' attention more and more. It makes network becoming be low complexity, fast, reliable and safe and also supports a large number of network nodes. In which network devices are generally divided into Coordinator node (Coordinator), aggregation node (Router) and sensor nodes (End Device). Each node can be used as the monitoring object, such as a sensor connected directly collect and monitor the original data, automatic transmit data from other nodes of the network [1, 2]. ZigBee network, GPRS wireless transmission and GIS platform, these three makes monitoring air humidity, air temperature, and soil moisture and nitrogen phosphorus potassium content through remote monitoring the target fields become possible.

In ZigBee network, sensor node for the collection of data that is first transmitted into coordinator and then into remote host. This completed via GPRS network so as to achieve remote real-time monitoring [3–5].

D. Li and Z. Li (Eds.): CCTA 2015, Part I, IFIP AICT 478, pp. 153–160, 2016.
DOI: 10.1007/978-3-319-48357-3_15

2 Overall Design of Monitoring System

Monitoring system is mainly composed of two parts, the front data collection sub-system and the terminal early-warning sub-system. The total network consists of several terminals and one monitor center which is composed of more than one computers and a number of coordinators. The overall structure is shown in Fig. 1 [6].

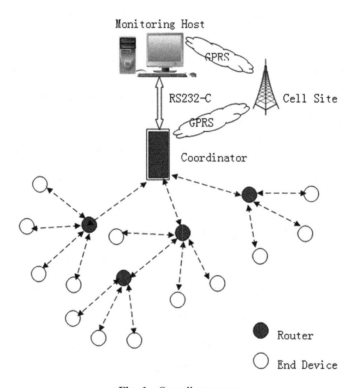

Fig. 1. Overall structure

Front data collection sub-system for the collection of moisture, humidity, temperature, light intensity and nitrogen phosphorus potassium. It includes ZigBee network and relevant application software, consists of coordinator, router and terminal equipments. From the viewpoint of structural network, is composed of coordinator nodes, router nodes and terminal nodes.

Star network is a structure to avoid adding any network nodes arbitrary and constructing poor consumption distribution. A variety of star networks construct ZigBee network in which coordinator is a FFD that achieve a group of services, for the control of sub node communication, data collection and distribution control. A ZigBee network includes at least a coordinator. The information router processed is uploaded by terminal equipment and sent to the coordinator. Terminal equipments, including soil moisture sensor, a soil temperature sensor, air temperature and humidity sensor, a carbon dioxide concentration sensor, light sensor and soil nitrogen phosphorus potassium monitoring sensors. Through which routing data on central nodes and placing on the monitoring points pre arranged. Furthermore, both RS-232C serial port connection and GPRS connection are feasible on coordinator and host. Which means the coordinator can transmit information into remote host by ways of wired or wireless. Therefore real-time data can be analyzed smoothly with GIS software, real data or diagram illustrations under the condition of current growth and early warning of the dangers be given as soon as quickly.

As to terminal early-warning sub-system, includes monitoring hosts and GIS software. The monitoring hosts with GSM module obtain environmental data from the coordinator on farmland via GPRS or RS-232C. Compare to RS-232C, the former indeed has brought convenience in connecting remote. However, when concerns cost down the latter is better.

GIS software installed on monitoring hosts, monitors the sensor data as well as sensor status. On which the sensor's position is displayed, the growth trend of corn is predicted and the working state is adjusted. Meanwhile all sensors' data is monitored, such as farmland local humidity condition, nitrogen phosphorus potassium content, location of high humidity problem determination, and is analyzed especially the original data similar as humidity and nitrogen phosphorus potassium content. In addition, abnormal point information will be displayed on the GIS map if spatial database established.

3 Hardware Components of Monitoring System

The modularization hardware includes microprocessor module, communication module, data collection module and power four modules [7].

3.1 Microprocessor Module

Microprocessor is a central processing unit, to control node operation, routing protocol, synchronous positioning, power management and task management. On which CC2530 has an IEEE 802.15.4 compatible wireless transceiver and RF kernel control and simulation wireless module.

Compare to the previous generation CC2430, CC2530 is often used in the solution system contains second generation chip 2.4 GHz IEEE 802.15.4/ZigBeeTM, can easily set up powerful network with the goal of extremely low cost. It integrates RF transceiver, MCU, programmable flash memory and 8-KB RAM, and provides an interface between MCU and wireless devices which can send commands, read status, automatic operation as well as determine the sequence of events to wireless devices. In addition, CC2530 has a series of unique running modes particularly suitable for ultra low power requirements of the system. Conversion time between operation modes short enough further ensure the low energy consumption.

3.2   Communication Module

This module is used in the communication between control centers, which passes sensor data to monitoring hosts by ways of wired or wireless.

Early warning and remote control are both achieved in wired communication (RS-232C) way or wireless communication (GPRS) way. Wireless communication core devices have SIM900A [5] GPRS module which is a new compact product of SIMCom company using industry standard interface. It belongs to the dual band GSM/GPRS module and packaged with SMT, has advantages of high integration and simple using, and also has stable performance, exquisite appearance and high cost performance. Besides, signal process circuit and receiver/transmitter circuit integrate on SIM900A wireless communication module reduces the difficulties of system development. To meet the requirement of wireless communication, add power supply, SIM and peripheral interface circuit of communication interface. SIM900A can transmit low power realization of voice, data, fax, and SMS. In addition, the size of SIM900A is $24 \times 24 \times 3$ mm, all kinds of design requirements can be applied to M2M applications, particularly suitable for compact product design.

3.3   Data Collection Module

Collecting data in monitoring region and converting analog signals of various sensors (such as signal light, signal, chemical information) into digital signal will be transmitted to the micro processing module. Sensor nodes are powered by two small batteries, on which processor module and data collection module both working in small voltage so as to keep low power consumption and simple peripheral circuit.

Choose SHT11 as the data collection module, is a new intelligent temperature and humidity sensor based on CMOSens technology, has the characteristics of digital output, debugging free, calibration free, peripheral circuit free and whole exchange that are different with traditional temperature and humidity sensor [8, 9]. On CMOS chip integrates temperature and humidity sensors, signal amplifying and conditioning function, AD conversion and two-wire serial interface. This brings benefits of super fast response, strong anti-interference ability and high performance price ratio.

Here soil temperature and humidity, air temperature and humidity sensor, carbon dioxide concentration sensor, light sensor and soil nitrogen phosphorus potassium monitoring sensor have been used. Sensor data are transmitted from terminals to the router through ZigBee network at regular time, consequently are transmitted to the coordinator through the RS-232C or GPRS, these data at last reach at the monitoring center. Obviously, add GPRS wireless communication on serial port connection, which increases the mobility and flexibility.

3.4   Power Module

4.2 V, 5 V and 12 V three voltages will be used. Terminal sensor is powered by two small AA batteries while GPRS wireless communication module SIM900A is powered by 4.2 V. And 12 V is to provide power for other parts of the system. Dormancy mechanism as the main method of saving energy [10] means turning off the wireless communication module or data acquisition module timely when there is no task of data collection or no need to send data again. Test results indicate CC2530's power consumption is less than 1 μA while in sleeping, and more than 99 % of running time of system is in dormant state, thus only 2 AA batteries can maintain communication from six months to two years.

4   Monitoring System Software Design

4.1   Front Data Collection System

Front data collection system software mainly includes three, coordinator node program, routing node program and terminal node program. These three programs working together and accomplish data collection, data transmission and network management [11].

Coordinator node program is composed of main program and subprogram of data collection and processing. Not only is responsible for the network configuration and program management, including the definition of communication channel, the network identifier Profile, and configure the network response node join network requests and binding request, for other node distribution network address, routing table maintenance, but also receives data sensor nodes send, the confluent consolidation back to the computer. Main program actively calls routing nodes and gets data periodically while

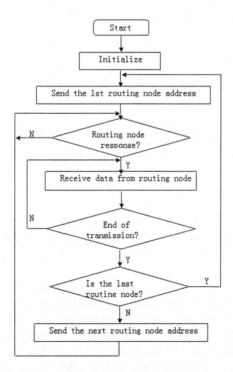

Fig. 2. Main program flowchart of the center node

routing nodes replying center nodes with interrupt way so as to transmit group data (different terminal node data collection to data transmission) smoothly. The main program flowchart shown in Fig. 2.

Since asynchronous transfer mode of routing node and terminal node, a symbol is added before sending effective data in order to improve the reliability of wireless communication. Signals are transmitted to central node from terminal node and then to the next routing node, which to achieve the purpose of circuit detection of terminal node.

Similar to coordinator node, main program of routing nodes sets terminal nodes with different address using the method of calling and receiving field detection data from terminal nodes time-sharing, and transmits data to coordinator node while calling display subroutine. Considering the same basic structure, designs the main program flowchart of routing center code shown in Fig. 3.

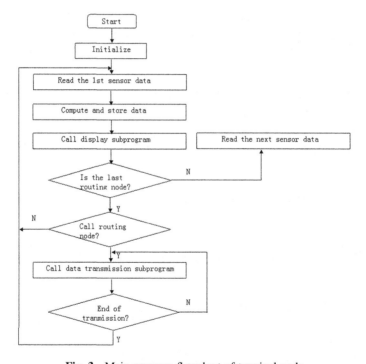

Fig. 3. Main program flowchart of terminal node

Terminal node program is always used to collect and send data. For example, sensor nodes are generally placed in farmland sites. Main program of terminal node begins to initialize after starting, reads parameters of each sensor, and store and display that through compute operation. After collecting all sensor data, checking whether it is necessary send a call to routing node. If not, continue on checking the sensor state on turns. Or else, call the data transmitting subprogram to transmit all data and start to next round. If sensor data overflow or function key was pressed, execute corresponding interrupt service subroutine.

4.2   Terminal Early Warning and Positioning System

Focus on dangerous crossing alarm and issues location of corn farmland so as to achieve monitoring the coordinator node and the routing node, monitoring fixed-points, and monitoring line alarm and source issue localization.

This sub-system runs on the environment of Microsoft.Net4.0 platform combined with WebGIS platform which ArcGIS Server 10.1 undertakes. Farmland GIS maps are created with AcrMap software and deployed on ArcGIS 10.1 server. The current working state of each sensor and real-time state of farmland are listed and displayed on host monitor screen after statistical analyzing. Green, yellow, orange, red, four colors represent the state [12] farmland environment since human eye easily distinguish with and recognize. Among which, green means farmland normal, yellow means minor problems, the orange means affecting the harvest, and red indicates seriously affects the survival of corn crop. According to this, decide whether there is necessary deal with issues on-site. Certainly, warning information and position information are both appeared on screen.

5   Conclusions

The prototype of corn farmland monitoring system is designed in this paper, and simply discussed the characteristics of ZigBee, GPRS and GIS, which make full use of the advantages of these three technologies. If practical system developed, would bring low construction cost, convenient data access and strong data processing function such as data statistics and real-time analysis [13, 14]. Technicians and farmers can easily master the state of their farmland whenever and wherever, also can respond as soon as possible while face to issues early warning system warns. A number of advanced sensors have been used though, there are many issues such as exception handling still need to be solved in practice.

References

1. Jiang, L., Zhuang, W.: Research and application of ZigBee/GPRS technology in precision agriculture. J. Agric. Mechanization Res. (4), 179:182 (2014)
2. Ni, L., Chen, S., et al.: Design and implementation of an information interaction system based on ZigBee technology. Comput. Measur. Control 23(5), 1689–1692 (2015)
3. Sun, Z., Cao, H., Li, H., et al.: A GPRS and web based data acquisition system for greenhouse environment. Trans. Chin. Soc. Agric. Eng. 22(6), 131–134 (2006)
4. Li, X.: Study on paddy field survey information system based on GPS. J. Agric. Mechanization Res. (10), 162–165 (2013)

5. Zhou, Q., Yan, X., Fan, H., Wang, D.: Design of the facility agriculture gas fertilizer monitoring system based on GSM. J. Agric. Mechanization Res. (4), 94–97, 102 (2014)
6. Zhou, G.: Monitoring system of forest fire based on GIS and ZigBee technology. Res. Explor. Lab. (8), 22–24, 28 (2013)
7. Yang, L., Tao, Z.: Design of the environment monitoring network node based on ZigBee. Autom. Instrum. (8), 22–24 (2008)
8. Li, W., Cui, J.: Design of underground personnel positioning system based on ZigBee and GIS. Ind. Mine Autom. (2), 67–69 (2010)
9. Cheng, C., Mao, X., Wu, L.: An online monitoring system of water quality based on ZigBee. Chin. J. Electron Deveices **10**(5), 942–945 (2009)
10. Gao, J.: Study on the energy consumption of the ZigBee node of wireless sensor networks. Electron. Test **2**(2), 1–4 (2008)
11. Bao, C., Li, Z., Zhang, L., et al.: Design of monitoring system for grain depot based on ZigBee technology. Trans. Chin. Soc. Agric. Eng. **25**(9), 197–201 (2009)
12. Yang, G., Zhang, W.: Design of intelligent transportation system based on GPRS/GPS/GIS/ZigBee. Microcomput. Appl. (18), 89–91 (2011)
13. Zhao, S., Zhong, K., Sun, C.: Application of GIS technology in the field of agriculture. J. Agric. Mechanization Res. (4), 234–237 (2014)
14. Sun, T.: Design and implementation of wireless temperature and humidity sensor network monitoring system based on ZigBee. National University of Defense Technology (2009)

Simulation of Winter Wheat Yield
with WOFOST in County Scale

Shangjie Ma[1], Zhiyuan Pei[1(✉)], Yajuan He[1], Lianlin Wang[2],
and Zhiping Ma[2]

[1] Chinese Academy of Agricultural Engineering, Beijing 100125, China
{mashangjie,peizhiyuan}@agri.gov.cn
[2] Yutian Agriculture Planning Office, Yutian 064100, Hebei, China

Abstract. Winter wheat is mainly planted in water shortage area, such as North China and Northwest China. As a key field management measure, irrigation plays an important role in the production of winter wheat. This paper focuses on the improvement of regional winter wheat yield estimation technique in county scale by adjusting the irrigation management measure in crop growth model. The WOFOST (World Food Study) model was used by dividing the whole county into a number of EMUs (Elementary Mapping Units) and then running the model in each unit in sequence. While running, the measured soil moisture and LAI were used to rate the irrigation parameters. Finally, the calibrated irrigation parameters were used to run the model again. The results showed that the simulated winter wheat growth process was normal. During the whole growing period of winter wheat, the change trends of the time series of soil moisture and LAI were basically consistent with that of the measured. The precision of simulated yield was between 87.26 % and 98.68 % among the 5 units, and the average of the precision was 94.56 %. The precision of simulated winter wheat yield was well, and could meet the needs of winter wheat yield estimation in county-wide. This study may provide basis for estimating crop yield in regional area by using the crop growth model.

Keywords: WOFOST · Irrigation parameter · Winter wheat yield

1 Introduction

Winter wheat, one of the main crops in northern China, is mainly planted in water shortage area, such as North China and Northwest China. As a key management measure, irrigation plays an important role in the production of winter wheat. Driving by the light, temperature, water, soil and other environmental variables, the crop growth dynamic model can daily simulate the photosynthesis, respiration, transpiration and evaporation process in crop growth period, and their relationship with meteorological, soil, and other environmental conditions, and can quantitatively describe the dynamic process of crop growth, development, and yield information. WOFOST model originated in the framework of interdisciplinary studies on world food security and on the potential world food production by the Center for World Food Studies in cooperation with the Wageningen Agricultural University [1]. WOFOST is a mechanism

D. Li and Z. Li (Eds.): CCTA 2015, Part I, IFIP AICT 478, pp. 161–172, 2016.
DOI: 10.1007/978-3-319-48357-3_16

model, and had been successfully used in daily business of agro meteorological monitoring and yield forecasting in European Union [2]. Meanwhile, WOFOST was also widely used in board as follows. Wu et al. [3] evaluated the applicability of WOFOST in the North China Plain. Wenxia et al. [4] simulated and validated the rice potential growth process in Zhejiang Province by utilizing WOFOST. According to climate characteristic and ecologic type of winter wheat variety in North China, Ma et al. [1] adjusted the parameters of WOFOST, and simulated phenological phase of overwinter in the region. Chen et al. [5] simulated the integrated impacts of low temperature and drought on maize yield. The abovementioned applications were in field scale. Combining the GIS technology, the application of WOFOST can be extended to regional scale. Ma et al. [6] simulated the winter wheat growth in regional scale in North China Plain by scaling-up WOFOST model. Li et al. [7] evaluated the suitability of meteorological conditions during growing period of Maize in northeast of China. Li et al. [8], Gao et al. [9, 10], Du et al. [11] respectively simulated the growth process of rice, wheat, maize, soybean, beet, potato and so on using WOFOST in Heilongjiang Province. Huang et al. [12] estimated the winter wheat yield in region by the method of optimizing the parameters of WOFOST based on assimilating simulated LAI and MODIS LAI at the re-greening stage of winter wheat. However, there were few reports about the method of simultaneously using the measured LAI and soil moisture to rate the irrigation parameters of model, and to simulate crop growth processing in regional scale.

In this paper, Yutian County, a water shortage region in Hebei Province, was selected as the research area. Both the measured LAI and soil moisture were used to rate the irrigation parameters. The adjusted irrigation parameters were used to run the model to simulate the winter wheat growth processing and to estimate the yield of winter wheat in county scale. All these would provide the basis for regional crop yield estimation by using crop growth model.

2 Materials and Methods

2.1 Study Area

Yutian county locates between 117°31'E ~ 117°56'E and 39°30'N ~ 39°58'N, at the south of Yan Mountains, the center of Eastern Hebei Plain. The terrain of the county slopes gently from northeast to southwest, the altitude in north is about 50 ~ 400 m, in center and south under 25 m, forming the views of the northern hills, central plains, the southern low-lying land pattern. According to the second national soil survey, there are 4 soil types, 9 subtypes in the county. The total area of the main 5 subtypes, leached cinnamon soil, meadow cinnamon soil, swamp alluvial soil, alluvial soil, salted alluvial soil, accounts for nearly 95 %. The county belongs to warm temperate semi humid region in the eastern monsoon region, with the average annual rainfall 655.4 mm, the average temperature 11.6°, and the annual frost free period 196 days. The main crops include winter wheat, spring maize, summer maize. The main varieties of winter wheat are Jingdong NO. 8 and Lunxuan NO. 987, with average yield level $3750 \sim 6000$ kg/hm^2.

2.2   Measured Data

According to the environmental condition and the spatial distribution of winter wheat, 36 sites were chosen to measure ground parameters in 2012. In order to ensure the purity of the remote sensing image pixels, the area of winter wheat must be not less than 200 m × 200 m in each site. The distribution of the sites, shown in Fig. 1, should be representative and homogeneous. Based on the phenological phase of winter wheat, some key stages from re-greening to mature were selected to measure soil moisture, LAI and other crop parameters. The three elements of yield, including the number

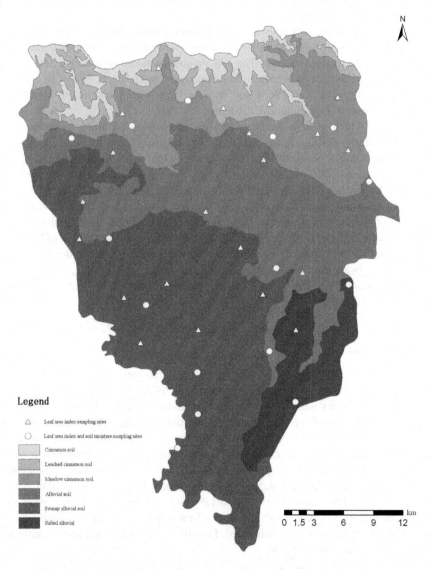

Fig. 1. Spatial distribution of experiment sites and soil types in study area

of heads per hectare, grain number per head, and 1000 grain weight, were collected to estimate the yield from heading stage to mature stage. Dates were respectively April 9–12 re-greening stage, May 2–6 jointing stage, May 16–19 heading stage, June 4–6 grain filling stage, and June 13–17 mature stage. Only soil moisture was measured in grain filling stage, and only crop parameters were measured mature stage.

Method of measuring LAI is as follows. Three subareas about 20 m^2 for each one were homogeneously chosen in each site. Five values of LAI from different locations in each subarea were measured by LAI2000 which was produced by U.S. LI-COR Company. The average of the five values is read as the LAI value of the subarea, and the average of three subareas is written as that of the site. The measurement is done between 09:00 am and 17:00 pm.

Method of measuring soil moisture is as follows. Fifteen representative sites whose area of winter wheat beyond 500 m × 500 m were selected as soil moisture measuring sites from the LAI measuring sites. The soil volume moisture of winter wheat field in the four angles of the site was measured respectively at the depth of $0 \sim 10$, $10 \sim 20$, $20 \sim 40$, $40 \sim 60$ cm by the TSC-I soil moisture measuring instrument which was produced by Chinese Agricultural University.

The winter wheat yield is calculated by the method called three agricultural elements estimating yield algorithm, which is gained from the number of heads per hectare, grain number per head, and 1000 grain weight multiplication [13, 14]. The detailed steps are as follows. In each site, three 1 m^2 rectangular areas were selected randomly. The number of heads in the three rectangular areas during heading stage and grain filling stage are counted respectively. The average of heads in the three rectangles at the same stage is used to calculate the number of heads per hectare at that stage. The average of the number of heads per hectare of the two stages is the number of heads per hectare of the site. At mature stage, total $50 \sim 80$ heads wheat were randomly selected as the sample from the above three rectangular areas in each site. After drying, threshing, counting, weighing, grain number per head, and 1000 grain weight were calculated.

2.3   Study Ideas

Based on the spatial discretization and irrigation parameters rating, the purpose of estimating winter wheat yield in county scale by using the WOFOST model was implemented. The steps were shown as Fig. 2 in follows. First, overlaying the county boundary and the map of soil type spatial distribution, dividing the whole county into several EMUs in which the combination of meteorology and soil type was homogeneous. Then, the parameters of weather, soil, crop, field management measures including initial irrigation management measures extracted from local management habits were prepared in each unit. After that, the model was run. The ground sites in each unit were selected, and the measured LAI and soil moisture in these sites were averaged as that of the unit. With the help of measured LAI and soil moisture, the initial irrigation operations were calibrated. Finally, the calibrated irrigation parameters were used to run the model again.

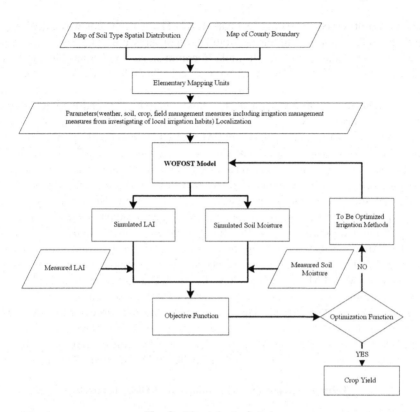

Fig. 2. Flow chart of study

2.4   WOFOST

Model Localization. WOFOST needs parameters of meteorology, soil, crop, management and so on. Both the daily and monthly meteorology parameters can be accepted by the model [15]. The format of daily meteorological parameter is CABO, which contains the maximum and minimum temperature (°C), precipitation (mm), average wind speed (m/s), water vapor pressure at 08:00 am (kPa), solar radiation (MJ/m2•d). The daily measured meteorological data from January 1st, 2010 to June 30th, 2012, including the maximum and minimum temperature, precipitation and sunshine duration, was got from Yutian Meteorology Office (117°44'E, 39°53'N, Altitude 14.4 m). The period covered two whole growth stages of winter wheat, 2010/2011, and 2011/2012. The measured data from neighboring Tangshan Meteorology Observation Station (118° 9'E, 39°40'N) from January 1st, 1957 to June 30th, 2012, including daily average atmosphere pressure, maximum atmosphere pressure, minimum atmosphere pressure, average temperature, maximum temperature, minimum temperature, average relative humidity, minimum relative humidity, precipitation, average wind speed, sunshine duration and so on, was also downloaded on the internet as additional information [16]. The wind speed parameter needed by the model could be replaced by the observed data from Tangshan. The other needed parameters, water vapor pressure and solar radiation, which had no observed data, could be calculated. The water vapor was calculated by relative humidity multiplying the saturation vapor pressure. The saturation vapor pressure was calculated by the formula recommended by China Meteorological Administration [17]. The solar radiation was calculated by sunshine duration, the maximum and minimum temperature [18, 19]. Harmonized World Soil Database from Food and Agricultural Organization of the United Nations [20] and the data from the Second National Soil Survey were also collected. The map of soil spatial distribution was extracted from the soil database of FAO. The cinnamon soil distributed in non cultivated land area, the northern mountainous area, was eliminated firstly. The smaller proportion soil types in cultivated land area were combined to the adjacent bigger types. The main five types, leached cinnamon soil, meadow cinnamon soil, swamp alluvial soil, alluvial soil, salted alluvial soil, were finally extracted. The most physical and chemical properties referred to that in model, and the important parameters, such as field capacity, saturated water content, wilting coefficient, were calculated by SPAM (Soil-Plant-Atmosphere-Water) using the texture, organic matter content from the soil database of FAO [21]. Some crop parameters such as the temperature sum from sowing to emergence (TSUMEM), the temperature sum from emergence to anthesis (TSUM1), and the temperature sum from anthesis to maturity (TSUM2), was calculated by the long time-series observed temperature data in Tangshan and the surveyed long-time phenology of winter wheat in local. The other parameters for crop were confirmed by looking up papers and documents [1, 3, 15, 22, 23] or debugging the model. The detailed crop parameters were listed in Table 1 as bellow. The needed filed management

Table 1. Crop parameters settings for WOFOST model

Parameters	Description	Value
TBASEM	Lower threshold temperature for emergence	0
TEFFMX	Maximum effective temperature for emergence	30.0
TSUMEM	Temperature sum from sowing to emergence	113.0
TSUM1	Temperature sum from emergence to anthesis	1150.4
TSUM2	Temperature sum from anthesis to maturity	900.4
LAIEM	Leaf area at emergence	0.1365
RGRLAI	Maximum relative increase in LAI	0.00917
SPAN	Life span of leaves growing at 35°C	31.3
TBASE	Lower threshold temperature for aging of leaves	0
EFFTB	Light-use efficiency of single leaf	0.45
CVL	Efficiency of conversion into leaves	0.685
CVO	Efficiency of conversion into storage organs	0.709
CVR	Efficiency of conversion into roots	0.694
CVS	Efficiency of conversion into stems	0.662
Q10	Relative increase in respiration rate per 10°C temperature increase	2
RML	Relative maintenance respiration rate of leaves	0.03
RMO	Relative maintenance respiration rate of storage organs	0.01
RMR	Relative maintenance respiration rate of roots	0.015
RMS	Relative maintenance respiration rate of stems	0.015
TBASEM	Maximum relative death rate of leaves due to water stress	0.03

parameters were gained by surveying the planting and managing habit of winter wheat in local. In general, the winter wheat is sowing in early October, harvesting in the second or third ten-days in June. So, when running, the variable sowing date was selected, with the earliest Oct 1st and the latest Oct 10th, the harvesting date was calculated automatically by the model and the longest duration was less than 300 days.

Rating Irrigation Parameters. The irrigation parameters, including irrigation date and irrigation volume, were initially set on the base of surveying local irrigation condition. In general, there is 4 times irrigation during the winter wheat growth period. The volume of irrigation for each time is about $50 \sim 100$ mm. The first time of irrigation often occurs before overwintering period in early December. The second time of irrigation will be implemented after re-greening period in middle ten-days in March next year. The third irrigation may occur before jointing stage, in middle or late April. The fourth time irrigation will be done in heading-anthesis stage in middle May. The times of irrigation may be reduced to three according to the actual weather condition. In the paper, the irrigation parameters were initiated four times, 70 mm for each time, and the dates of which were respectively in November 30th, March 14th the next year, April 14th, and May 14th.

The assumed irrigation parameters were used to run the model firstly. While running, the parameters were rated with the measured LAI and soil moisture by using the trial-and-error method. The initial irrigation volume as the center, 10 mm as the step size, irrigation volume was adjusted between $50 \sim 100$ mm. The initial irrigation date was determined as the center, and 1 day as the step size. The irrigation date was changed from −5 to 5 days, and avoided the rainy day. The parameters which make the error of simulated and measured LAI and soil moisture minimum and the irrigation amount lowest were chosen as the best value. Overlaying the map of soil type distribution and boundary, the whole county was divided into 5 EMUs, which were respectively leached cinnamon soil unit, meadow cinnamon soil unit, swamp alluvial soil unit, alluvial soil unit, and salted alluvial soil unit. The irrigation parameters for each unit were shown as Table 2.

Table 2. Calibrated irrigation parameters

Soil type	First irrigation		Second irrigation		Third irrigation		The 4th irrigation	
	Date	Volume (mm)	Date	Volume (mm)	Date	Volume (mm)	Date	Volume (mm)
Leached cinnamon soil	11-11-30	70	12-3-19	80	12-4-18	80	12-5-17	70
Meadow cinnamon soil	11-11-30	70	12-3-19	70	12-4-18	80	12-5-17	60
Swamp alluvial soil	11-11-30	70	12-3-14	70	12-4-10	70	12-5-13	70
Alluvial soil	11-11-30	70	12-3-14	70	12-4-9	70	12-5-13	70
Salted alluvial soil	11-11-30	70	12-3-19	70	12-4-18	50		

3 Results and Analysis

The WOFOST model was run in each unit in sequence. Both the no irrigation and rated irrigation were run respectively to simulate soil moisture, LAI, and winter wheat yield.

3.1 Soil Moisture Simulation

The simulation results shown as in Fig. 3 showed that the precipitation cannot meet the needs of winter wheat. The soil moisture kept at low level during the whole growth stage. After irrigation, the simulated soil moisture was similar with that of measured. The peak of time-series curve could reflect the irrigation. In general, the average percentage error of the simulated and measured among twenty values, which were measured during the four field experiments for all five soil types, was −17.56 %. For some values, the error of absolute value between the simulated and measured was large. But during the whole growth stage, the trends of simulated and measured soil moisture were basically similar.

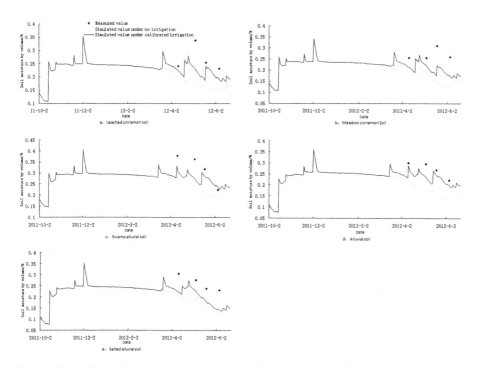

Fig. 3. Comparison of simulated and measured soil moisture under different irrigation methods

3.2 Lai Simulation

Under no irrigation, the results of the LAI simulation (shown as Fig. 4) showed that it is basically normal for the simulated LAI before heading stage of winter wheat. But, after heading stage, the LAI decreased quickly, leading the winter wheat died too early. After irrigation, the change trends of simulated and measured LAI were same. But in the early re-greening stage and lately maturity stage, there were relatively larger deviations between them. In the re-greening stage, the value of simulated was higher than that of measured, which might be caused by ignoring the influence of 'winter loss', a phenomenon that some leaves of winter wheat may be died in overwintering period. The simulated leaf might grow more quickly than the actual leaf in re-greening stage which affected by 'winter loss'. In the lately maturity stage, the simulated LAI was lower than that of measured. The measured LAI might be wrongly added by the dead leaves whose influence were difficulty to be eliminated while measuring. The maximum LAI values in the five units were respectively 4.64, 4.64, 4.46, 4.81, 3.69, and the corresponding simulated value were respectively 4.57, 4.57, 4.59, 4.57, 4.57. The RMSE (Root-Mean-Square Error) of the simulated and measured maximum LAI was 0.41, occurring 1.86 % of the average maximum LAI. The date of the simulated maximum LAI was earlier about 10 days than that of the measured. This difference between them might be influenced by the different time step size of them. LAI would be measured only during the key stage, and the average time interval is about 20 days. But the model was simulated day by day. The different temporal resolution between the simulated and measured LAI might lead the difference of date.

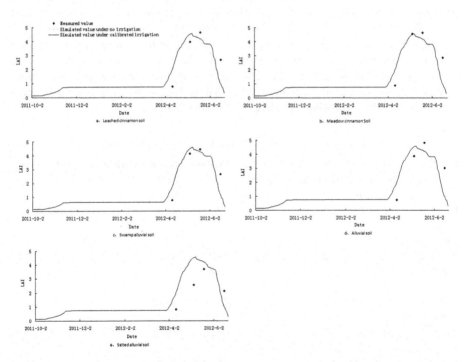

Fig. 4. Comparison of simulated and measured winter wheat LAI under different irrigation methods

3.3 Yield Simulation

While no irrigation, the winter wheat died too early, and the yield was too little. Added irrigation in the model, the simulated growth process was normal, and the yield was also reasonable. The precision equals 1- abs (percentage error), which was used to express the effect of simulation. The measured yields were respectively 4614.29, 4852.65, 5537.62, 5170.72, 4302.24 kg/hm^2 in the five EMUs, and that of simulated were respectively 4943, 4857, 4832, 4970, 4359 kg/hm^2. The precisions for the five EMUs were respectively 92.88 %, 97.85 %, 87.26 %, 96.12 %, 98.68 %. Averaging all the units, the measured yield was 4895.5 kg/hm^2, the simulated yield was 4812.20 kg/hm^2, and the precision was 94.56 %. The difference of simulated yields among the EMUs was obviously lower than that of observed. For example, the minimum measured yield was 4302.24 kg/hm^2, the maximum measured yield was 5537.62 kg/hm^2, the difference between them was 1235.38 kg/hm^2. Simultaneously, the simulated value were respectively 4359 kg/hm^2, 4970 kg/hm^2, 611 kg/hm^2. It had a lot to do with the considered parameters in model and the method of model localization. The considered parameters in model included meteorology parameters, soil parameters, crop parameters and irrigation parameters. But in the farm, the factors that influencing yield were more complex, not only containing the above mentioned factors, but also soil fertility condition, local climate, field terrain, plant diseases and insect pests and so on. During the localization process, being limited by the data acquisition conditions, only soil and irrigation parameters were considered in different EMUs. But the crop and meteorology parameters were same in the whole county. It was not consistent with the actual situation. For example, the precipitation may be different in the whole county, in the region of 55 km from north to south, 36 km from east to west. But there is only one weather observing station in the whole county. Using the same meteorology parameters in all EMUs reduce the difference of simulated yield among the EMUs. But totally, the precision of simulation can meet the estimating yield in county scale (Table 3).

Table 3. Comparison of simulated and measured winter wheat yield

Soil type	Measured yield (kg/ha)	Simulated yield no irrigation (kg/ha)	Precision	Simulated yield calibrated irrigation (kg/ha)	Precision
Leached cinnamon soil	4614.29	1954	42.35 %	4943	92.88 %
Meadow cinnamon soil	4852.65	2352	48.47 %	4957	97.85 %
Swamp alluvial soil	5537.62	2390	43.16 %	4832	87.26 %
Alluvial soil	5170.72	3413	66.01 %	4970	96.12 %
Salted alluvial soil	4302.24	1954	45.42 %	4359	98.68 %
Average	4895.5	2412.60	49.08 %	4812.20	94.56 %

4   Conclusion

As an example, Yutian County in Hebei Province was selected to divide into five EMUs, in which the combination of soil type and meteorology was same. The assumed irrigation parameters from surveying local irrigation condition were used to run the model firstly. Then, run the model, rated the irrigation parameters with the help of measured LAI and soil moisture by using the trial-and-error method. Finally, the rated irrigation parameters were used to run the model again, and completing the estimating yield in county scale. The conclusions were as follows.

(1) The simulated growth process of winter wheat was basically normal. The change trends of the time-series of simulated soil moisture and LAI were basically consistent with the measured in the whole growth period of winter wheat. In the numerical, the difference between the simulated LAI and the measured LAI in the early re-greening stage and lately maturity stage was greater than that of in jointing stage and heading-anthesis stage. The simulated maximum LAI values in the whole stage in all EMUs were close to that of measured. But the date had an error about 10 days which might be affected by the different time scale between simulation and measure.

(2) The simulated results could meet the needs of winter wheat yield estimation in county-wide. The precision of simulated yield was between 87.26 % and 98.68 % among the 5 units, and the average of the precision is 94.56 %.

Because being limited by the difficulty in obtaining parameters in large range, and ignoring some affected factors such as soil fertility condition, local climate, field terrain, plant diseases and insect pests and so on, the simulated difference among the EMUs was obviously lower than that of observed. In the more fine application, in order to improve the precision, the more parameters should be considered in the model and the more detailed input parameters should be used in the simulation.

Acknowledgment. Funds for this research was provided by the Research on the earth observation system by high resolution remote sensing, and the cooperative application technology of high resolution and multi sensors in agriculture (GF13/15-311-003), and was supported by the Innovation Team of Crop Monitoring by Remote Sensing (CMIT), in Chinese Academy of Agricultural Engineering (CAAE).

References

1. Ma, Y., Wang, S., Zhang, L.: Study on improvement of WOFOST against overwinter of wheat in North China. Chin. J. Agrometeorol. **26**(3), 145–149 (2005)
2. Supit, I., Hooijper, A.A., van Diepen, C.A., et al.: System description of WOFOST6.0 crop simulation model implemented in CGMS. In: Theory and Algorithms. The Winand Starting Center for Intergrated Land, Soil and Water Research (SC-DLO), Wagenningen, pp. 1–144 (1994)
3. Dingrong, W., Yangzhu, O., Zhao, X., et al.: The applicability research of WOFOST model in North China plain. Acta Phytoecologica Sinica **27**(5), 594–602 (2003)

4. Wenxia, X., Lijiao, Y., Guanghuo, W.: Simuation and validation of rice potential growth process in Zhejiang by utilizing WOFOST model. Chin. J. Rice Sci. **20**(3), 319–323 (2006)
5. Chen, Z., Zhang, J., Wang, C., et al.: Application of WOFOST model in simulation of integrated impacts of low temperature and drought on maize yield. Chin. J. Agrometeorol. **28**(4), 440–442 (2007)
6. Ma, Y., Wang, S., Zhang, L., et al.: A preliminary study on a regional growth simulation model of winter wheat in North China based on scaling-up approach I potential production level. Acta Agronomica Sinica **31**(6), 697–705 (2005)
7. Li, X., Ma, S., Gong, L., et al.: Evaluation of meteorological suitability degree during maize growth period based on WOFOST in Northeast China. Chin. J. Agrometeorol. **34**(1), 43–49 (2013)
8. Li, X., Wang, Y., Ji, S., et al.: Validation of crop growth monitoring system (CGMS) in Heilongjiang province. Chin. J. Agrometeorol. **26**(3), 155–157 (2005)
9. Gao, Y., Na, J., Gu, H., et al.: Characteristics and compartment of potential climatic productivity of potato (Solanum tuberosum L.) in Heilongjiang province. Chin. J. Agrometeorol. **28**(3), 275–280 (2007)
10. Gao, Y., Nan, R., Hong, G., et al.: Simulation of beta vulgaris climatic productivity and its planting climatic zoning in Heilongjiang province. Chin. J. Ecol. **28**(1), 27–31 (2009)
11. Du, C., Li, J., Wang, C., et al.: Study on dynamic yield forecasting of rice based on WOFOST model in Heilongjiang province. J. Anhui Agric. Sci. **39**(24), 15093–15095, 15122 (2011)
12. Huang, J., Wu, S., Liu, X., et al.: Regional winter wheat yield forecasting based on assimilation of remote sensing data and crop growth model with Ensemble Kalman method. Trans. Chin. Soc. Agric. Eng. (Trans. CASE) **28**(4), 142–148 (2012). (in Chinese with English abstract)
13. Zhang, W., Wu, Y., Yang, Y., et al.: Analysis of interactive relationship among yield combinational factors of winter wheat. J. Henan Vocat.-Tech. Teachers Coll. **20**(3), 1–4 (1992)
14. Zhu, X., Xie, K., Xu, X., et al.: The structure analysis of winter wheat yield and the principal of remote sensing estimation of winter wheat yield. Remote Sens. Environ. China **4**(2), 116–127 (1989)
15. Boogaard, H.L., De Wit, A.J.W., te Roller, J.A., et al.: User's guide for the WOFOST control center 1.8 and WOFOST 7.1.3 crop growth simulation model. Wageningen University & Alterra Research Center, Wageningen (2011). http://www.wofost.wur.nl
16. China Meteorological Data Sharing Service System. http://cdc.cma.gov.cn/home.do
17. Luo, L., Wang, X., Yu, P.: The compare and research of the calculate formula of the saturation water steam pressure. Meteorol. Hydrol. Mar. Instrum. (4), 24–27 (2003)
18. Cao, F., Shen, S.: Estimation daily solar radiation in China. J. Nanjing Inst. Meteorol. **31**(4), 587–591 (2008)
19. Cao, F.: Study on earth surface total solar radiation in China in recent 40 years and daily solar radiation model. Nanjing Institute of Meteorology, Nanjing (2008)
20. Harmonized World Soil Database. http://www.fao.org/nr/land/soils/harmonized-world-soil-database/download-data-only/en/
21. Wei, H., Zhang, Z., Yang, J.: Establishing method for soil database of SWAT model **38**(6), 15–18 (2008)
22. Zhao, H., Pei, Z., Ma, S., et al.: Retrieving LAI by assimilating time series HJ CCD with WOFOST. Trans. Chin. Soc. Agric. Eng. (Trans. CASE) **28**(11), 158–163 (2012). (in Chinese with English abstract)
23. Sijie, W.: Study on Winter Wheat Yield Prediction Based on Assimilating Remote Sensing Data and Crop Growth Model. Central South University, Changsha (2012)

Predicting S&P500 Index Using Artificial Neural Network

Shanghong Li[1], Jiayu Zhang[2], and Yan Qi[1(✉)]

[1] International College Beijing, China Agricultural University, Beijing, China
shanghongli@cau.edu.cn, meganqiyan@sina.cn
[2] Continental Capital, Beijing, China
jiayu.zhang@contiasia.com

Abstract. This paper studies artificial neural network algorithm as a means of modelling and forecasting the financial market data. Such method bypasses traditional statistical method to deal with financial time series data. A recurrent neural network model, Elman network, is implemented to incorporate auto-correlation in time series data. A 3-parameter model is chosen to fit and forecast S&P 500 index. The experimental data is from 2000–2007, to screen out the abnormal market environment after 2008 financial crisis.

Keywords: S&P 500 index · Elman network · Artificial neural network · Time series

1 Introduction

The main purpose of this research is trying to build a preliminary base on how to apply the artificial neural network (ANN) method to model and forecast the financial market activities. Peters (1991) state that it's impossible to predict the stock price in future with the historical data and get a payback in return based on the effective market assumption. However, many traditional study and data shows that the markets do not often following the efficient market hypothesis completely. Fama and Schwert (1977) and Fama and French (1989) stated certain financial and economical time series could be used to predict some variables. Penumadu et al. (1997) showed that several research such as attempting to interpret complex real-world sensor data, ANN are among those most efficient ways available. Trippi and Turban (1996) examined how ANNs could be implemented in financial areas by modeling and predicting the financial time series. It is especially difficult to be modeled by traditional statistical method.

Our work starts from introduction of the fundamental structure of the neural network, till the mostly common used multi-layer model. Then a single-threshold neuron problem as a simple example will be used to illustrate some basic neural network processing features. The special technique of backpropagation and gradient descent algorithm is discussed in detailed to mathematically show how a neural network works. In order to incorporate autocorrelation into time series data, an idea of recurrent networks is introduced to build the time series structure into the neural network internally. As one of the most commonly used recurrent networks, the Elman network is discussed in details as it will be implemented to our real world data.

© IFIP International Federation for Information Processing 2016
Published by Springer International Publishing AG 2016. All Rights Reserved
D. Li and Z. Li (Eds.): CCTA 2015, Part I, IFIP AICT 478, pp. 173–189, 2016.
DOI: 10.1007/978-3-319-48357-3_17

The S&P 500 index is branded as a bellwether for US economy. It is most widely used for large-cap US stocks and often considered as a baseline for comparison in stock and mutual fund Performance charts. As a result we choose S&P 500 as our fitting and forecasting target. As a preliminary base, we would like to keep less parameter, so we choose 3 parameters, one for the real estate market (it contributes 35 %–40 % of the index), one for the yield spread (Huang and Stoll 1998), and the rest for the short-term interest rate.

It is well known that there was a huge financial crisis and turmoil on the financial market in 2008 as the subprime crisis. In order to save the economy and the financial market from the abyss, the Fed started several rounds of quantitative easing (QE) to support the asset prices. It is believed that the crisis and the following QE had a huge impact and noise of the financial market time series data and made some data distorted to the external artificial interference, which is not welcomed in the model. Especially the financial data before 2008 was chosen to train and test the model, for the purpose of avoiding the extreme external noise to impact the model.

2   Artificial Neural Networks

2.1   Overview

ANNs provide an efficient way to approximate various kinds of objective functions such as real-valued, discrete-valued and vector-valued. ANNs are amazing because of their capability to imitate some of the human brain's information process, though in a simplistic way. An ANN is a system of neurons organized into a network in which data can be processed like the learning process in human brain.

A neural network is trained to study from the data set by adjusting the values of the weights between neurons. Basically, a specific input leads to a particular target output in neural networks. When training the networks, the weights are adjusted according to the comparison between the output and the target value, till the error terms fall within a certain error tolerance. In general, a large quantity of data is necessary to obtain a well training network. The following Figure 1 illustrates the training process.

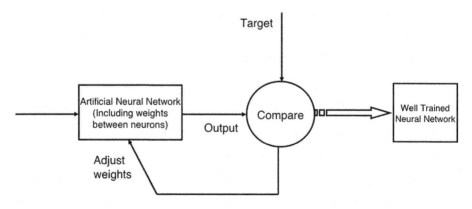

Fig. 1. The process of training neural networks

Neural networks can perform various tasks, including function approximation, pattern recognition, data clustering and trend prediction. Neural networks have a widely application in areas of business and industry, such as: digital process control, code sequence prediction, voice synthesis, speech recognition, market forecasting, portfolio trading, currency price prediction, and so on.

2.2   Network Structures

Neural networks can have various architectures from simplistic to complicate. The simplest one is the single neuron network which only has one neuron in the network. From simplistic to complicate in structure, we divide neural networks into two groups: the single-layer networks and the multi-layer networks. Then we will give a general description of these typical network structures to give readers an insight to neural networks.

2.2.1   Single-Neuron Model
The single-neuron network consists of a scalar input p, a bias input b, a transfer function f and generates an output that is shown in Fig. 2.

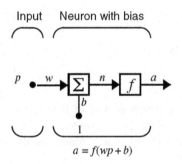

$$a = f(wp + b)$$

Fig. 2. The single-neuron network structure

p is the scalar input and wp is the weighted input. The transfer function input n is the summation of the weighted input and the bias b. a is the output formed by acting the transfer function f on the input. Both w and b are parameters adjustable in the networks. When training the networks the weights and bias parameters will be adjusted based on the training data set to make the networks achieve some desired goal.

2.2.2   Single-Layer Model
A single-layer model with R inputs and S neurons is shown in Fig. 3.

In this network, p is the input vector and W is the weight matrix. n_i is a scalar input generated by combine the bias and weighted input of the ithneron. Finally the transfer function f acts on the net input n to generate a column vector output a. In the single-layer network, the input vector enters the network via the weight matrix,

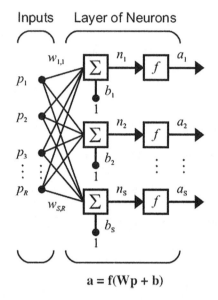

$$a = f(Wp + b)$$

Fig. 3. The single-layer network structure

$$W = \begin{bmatrix} w_{1,1} & w_{1,2} & \cdots & w_{1,R} \\ w_{2,1} & w_{2,2} & \cdots & w_{1,R} \\ \vdots & \vdots & \vdots & \vdots \\ w_{S,1} & w_{S,2} & \cdots & w_{S,R} \end{bmatrix}$$

The row indices indicate which neuron of the weight, while the column indices indicate which input of the weight. For example, $w_{i,j}$ means that weight connected the ith input variable to the jth neuron is $w_{i,j}$. Note that the number of input could be different from that of neurons.

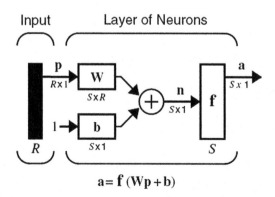

$$a = f(Wp + b)$$

Fig. 4. The single-layer network structure-abbreviated notation

The single-layer network with R inputs and S neurons using abbreviated notation, is shown in Fig. 4.

Here p is an $R \times 1$ input vector, W is an $S \times R$ matrix, and a, b are $S \times 1$ vectors.

2.2.3 Multi-layer Model

A network could have more than one layer and each layer consist of several neurons. The multi-layer model is most common and powerful neural network structure currently used in the application.

To introduce the multi-layer structure clearly, the notation should be extended. First, The weight matrices connected to inputs are called as input weights(IW), while weight matrices connected between layers are called as layer weights(LW).Secondly, the source(second index) and destination (first index) of the weight matrix should be pointed out. For example, the weight matrix connected the input vector and the first layer is noted as $IW^{1,1}$, which means both the source and destination are 1.

A three-layer network is shown in Fig. 5.

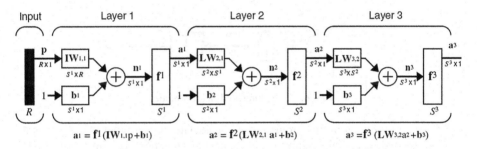

$$a_1 = f^1(IW_{1,1}p + b_1) \qquad a_2 = f^2(LW_{2,1}\, a_1 + b_2) \qquad a_3 = f^3(LW_{3,2}a_2 + b_3)$$

$$a_3 = f^3(LW_{3,2}\, f^2(LW_{2,1}f^1(IW_{1,1}p + b_1) + b_2) + b_3) = y$$

Fig. 5. The three-layer network structure

The network has R inputs, S^1 neurons in the first layer, S^2 neurons in the second layer and S^3 neurons in the third layer. The constant input vector 1 represents the bias for each layer.

The outputs of each layer are automatically the inputs for the next layer so the intermediate layer could be treated as a single-layer network. For instance, layer 2 is a single-layer network with S^1 inputs and S^2 neurons and an $S^2 \times S^1$ weight matrix $LW^{2,1}$. a^1 is the input vector and a^2 is the output vector.

a^3 is the output of the third layer and at the same time is the output of the entire network. The output of the network is always labeled as y. y could be expressed explicitly as:

$$y = f^3 LW^{3,2} f^2 (LW^{2,1} f^1 (IW^{1,1} \cdot p + b_1) + b_2) + b_3)$$

2.3   Network Process

2.3.1   Threshold Neuron Example

We will study a single-threshold neuron problem as a simple example to illustrate some basic neural network processing features.

The problem includes two input variables (x_1 and x_2), one output variable (t) which belongs to one of two categories (0 or 1). The data is given in Table 1. Our task is to correctly divide the input data into two categories (0 or 1). A single-neuron network will be used to achieve this goal, shown in Fig. 6.

Table 1. Classification data

X_1	X_2	t
0.2	0.3	0
0.2	0.8	0
0.8	0.2	0
1.0	0.8	1

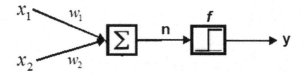

Fig. 6. Liner threshold neuron model

There are two steps in this problem. First, calculate the net input into the transfer function; second, decide the output using a threshold function. To simplify the problem, the bias is ignored, so when calculating the net input, only the input variables and corresponding weights should be taken into account. In this stage, both of the weights will be fixed at 1 which means that there is no learning.

$$\sum = n = \{w_1, w_2\} \cdot \{x_1, x_2\} = x_1 + x_2$$

So we only need to consider the sum of the two inputs, the threshold should be placed anywhere between 1.0 and 1.8. A threshold will be arbitrarily chosen as 1.3 for this problem. Then the threshold function generates the output y such that:

$$f\left(\sum\right) = y = \begin{cases} 0 & u < 1.3 \\ 1 & u \geq 1.3 \end{cases}$$

Using this simple classifier, it is possible to group the data correctly; the results are shown in Table 2.

From above we notice that a specially designed threshold function could classify the data correctly. However, the drawback of the model is that it does not learn from the environment (weights are equal to 1). Later we will discuss how to train the network to learn from the environment, and consequently how to adjust the weights.

Table 2. Performance of the classifier

Input(x_1,x_2)	u	y
(0.2,0.3)	0.5	0
(0.2,0.8)	1.0	0
(0.8,0.2)	1.0	0
(1.0,0.8)	1.8	1

2.3.2 Backpropagation

Standard backpropagation is a gradient descent algorithm, which means that the network weights are adjusted towards the negative direction of the gradient of the network error function. We know that the transfer function is differentiable, and if we define a differentiable error function as the error function of the network, such as the sum-of-squares, then the error function is differentiable with respect to the weights. Therefore we can calculate the derivatives of the errors with respect to the weights and use the derivatives to adjust the weights so as to minimize the error function by using gradient a descent algorithm or some other more powerful methods.

This technique can be divided into two stages. First, errors are propagated backwards through the network so as to calculate the derivatives with respect to weights. Second, weights are adjusted using calculated derivatives by a certain optimization rule.

We now derive the backpropagation algorithm for a general two-layer feed-forward network. It is straightforward to generalize the method from the two-layer model to the multi-layer model. For a general feed-forward network, the net input for each neuron is a weighted sum of the inputs of the form

$$n_j = \sum_i w_{ji} p_i \qquad (1)$$

Where n_j is the net input, p_i is the input, w_{ji} is the corresponding weight.

The net input is transformed by a non-linear differentiable transfer function $g(\cdot)$ in the form

$$a_j = g(n_j) \qquad (2)$$

The error function of the network can be expressed as a differentiable function of the network outputs in the form

$$E = E(y_1, \cdots y_n) \qquad (3)$$

Our goal is to find a method for evaluating the derivatives of the error function with respect to the weight. According to Bishop (1996) the partial derivatives is

$$\frac{\partial E}{\partial w_{ji}} = \frac{\partial E}{\partial n_j} \frac{\partial n_j}{\partial w_{ji}} \qquad (4)$$

Next, we introduce the useful notation

$$\delta_j = \frac{\partial E}{\partial n_j} \tag{5}$$

Using (1) we can get

$$\frac{\partial n_j}{\partial w_{ji}} = p_i \tag{6}$$

Substituting (5) and (6) into (4) we can write

$$\frac{\partial E}{\partial w_{ji}} = \delta_j p_i \tag{7}$$

Therefore so as to calculate the derivatives, the evaluation of δ_j is necessary for each neuron in the hidden layers and output layer.

For the output layer neuron, the calculation of δ_j is straightforward. From the definition (5) we have

$$\delta_k = \frac{\partial E}{\partial n_k} = g'(n_k)\frac{\partial E}{\partial y_k} \tag{8}$$

To calculate δ_j for the hidden layer neurons, we apply the chain rule again,

$$\delta_j = \frac{\partial E}{\partial n_j} = \sum_k \frac{\partial E}{\partial n_k}\frac{\partial n}{\partial n_j} \tag{9}$$

where the sum runs over all the output layer neurons k to which the hidden layer neuron j connects. The structure of neurons and weights is shown in Fig. 7.

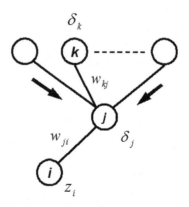

Fig. 7. The evaluation of δ_j for hidden neurons

For the output layer we have,

$$n_k = \sum_j w_{kj} g(n_j) \tag{10}$$

Now substituting the definition of δ_j given by (5) into (9) and use (10), we get

$$\delta_j = g'(n_j) \sum_k w_{kj} \delta_k \tag{11}$$

It's clear that the value of δ_j for the hidden neurons could be calculated by backpropagating δ_k for the neurons higher up in the network.

The backpropagation procedure for evaluating the derivatives of the error function is summarized in the following four steps:

- Apply an input data x_n into the network and forward through the layers using (1) and (2) to calculate the output for all neurons in the hidden layers and output layer.
- Calculate δ_k for all output neurons using (8).
- Backpropagate the δ_k using (11) to obtain δ_j for all hidden layer neurons.
- Use (7) to calculate all required derivatives.

2.3.3   Gradient Descent

In order to complete the learning algorithm, we have to provide a method to update the weights based on the weights. The gradient descent technique will be used here.

We know that the error function can be expressed as a differentiable function of weights in the network

$$E = E(w) \tag{12}$$

We begin with an initial guess for w which can be chosen randomly. Then, we update the weights by moving a small step along the direction in which E decreases most rapidly. By iterating this process, a sequence of weights to minimize the error function is generated,

$$w_{ji}^{\tau+1} = w_{ji}^{\tau} - \eta \left. \frac{\partial E}{\partial w_{ji}} \right|_{w(\tau)} \tag{13}$$

The choice of η (learning rate) is critical in the algorithm, since if it is too small the reduction of the error will be too slow, while if it is too large, the sequence might diverge.

Substituting (7) into (13), we get

$$vw_{ji} = -\eta \delta_j p_i \tag{14}$$

3   Time Series Processing

It seems that it's not appropriate to apply neural networks for time series since they were built mainly for pattern recognition. Another reason is that the initial applications were dealing with detection of patterns in arrays of measurement not change in time (Oeda 2006). Additional structure in the neural network is required for the spatio-temporal property of the time series data. Particularly, memory in time in the structure is necessary for a neural network applied for time series processing. There are two types of nonlinear networks successfully used for time series processing (Box and Tiao 1965). One is the modified backpropagation network and the other one is recurrent network embedded in the structure to capture the long-term memory. We will focus on the latter one as tool to do the time series processing.

3.1   Recurrent Network

The key idea is to incorporate the autocorrelation property of the time series into the network (Samarasinghe 2006). Such networks could figure out the long-term record of the time series by its internal dynamics, instead of depending on external memory filters. Also the recurrent network has feedback loops to memorize the past state of the network and recursively send them back into the network such that the network has an internal long-term-memory property. What's more, memory dynamics and the time lag structure depend on the network itself instead of any other externally designed structure. The way the long-term memory is built into the network is illustrated using a simple model below (Fig. 8):

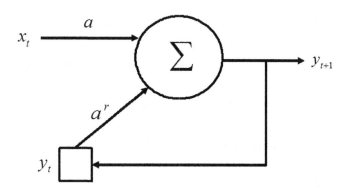

Fig. 8. Recurrent linear neuron

There are two inputs at time t: external input (x_t), internal delayed input (y_t). The neuron forecasts the next state y_{t+1} and becomes a fresh delayed input stored in the square. The whole process is recursive and repeated. At time t, the output could be expressed as

$$y_{t+1} = a^r y_t + a x_t \tag{15}$$

Where a^r is the weight with respect to the delayed input, y_{t+1} then turns into a new delayed input. The output at $t + 1$ is generated in the same manner

$$y_{t+2} = a^r y_{t+1} + a x_{t+1} \tag{16}$$

Substituting (15) into (16), we get

$$y_{t+2} = a^r (a^r y_t + a x_t) + a x_{t+1} = (a^r)^2 y_t + a^{r+1} x_t + a x_{t+1} \tag{17}$$

Suppose that a r = 0.2 and a = 0.3, the output is

$$y_{t+2} = 0.04 y_t + 0.06 x_t + 0.3 x_{t+1} \tag{18}$$

We can see that the network has built in the last two external input and the last predicted output into the structure. This built in structure enables the model to have a long-term memory capability. The recurrent weight determines the weight put on the time lags. Recursively

$$y_{t+n} = a^r (a^r)^n y_t + (a^r)^{n-1} a x_t + (a^r)^{n-2} a x_{t+1} + \cdots + a x_{t+n-1} \tag{19}$$

The weight will be adjusted during the process. When the nonlinear transfer function f is embedded into the model (take time step $t + 2$ as an example), we get

$$y_{t+n} = f[a^r f(a^r y_t + a x_t) + a x_{t+1}] \tag{20}$$

3.2   Elman Networks

Elman and Jordan networks are the two most popular recurrent networks in application. Elman network, mostly a two-layer network, is used in our application and discussed in details here. It's shown as below with feedback loop from the first layer output to the first layer input (Fig. 9).

Elman network contains both *tansig* transfer function and *purelin* transfer function. Therefore it's possible to approximate arbitrary function with any accuracy. To show how the Elman network works, a small network will be studied (the bias is ignored for simplicity).

Here, x_1 and x_2 are two given inputs in the network and y is an internal delayed input. The delayed input at every previous step needs to be stored in a unit and the output result from the hidden layer will be fed back in the next time step. $a_1(t)$, $a_2(t)$ are hidden layer weights and $a^r(t)$ ($a^r(t)$ denotes the recurrent weight). The process from the hidden layer and up is defined as time step $t + 1$. So the hidden layer output weight is $b(t + 1)$ (Fig. 10).

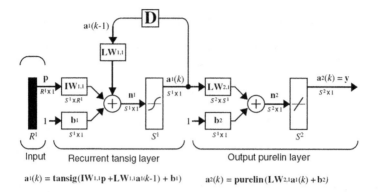

Fig. 9. Two-layer Elman network

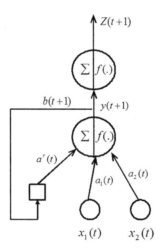

Fig. 10. Simple Elman recurrent network

For each time t, the weighted input $u(t + 1)$ and output $y(t + 1)$ for the hidden layer are given by

$$u(t+1) = a_1(t)x_1(t) + a_2(t)x_2(t) + a^r(t)y(t) \tag{21}$$

$$y(t+1) = f(u(t+1)) \tag{22}$$

Next, $y(t + 1)$ is stored in a unit and will be send back as a delayed input. For the output layer, the weighted input $v(t + 1)$ and output $z(t + 1)$ are given by

$$v(t+1) = b(t+1)y(t+1) \tag{23}$$

$$z(t+1) = f(v(t+1)) \tag{24}$$

where z(t + 1) is the predicted value of the target value T(t + 1). If the error function is a standard sum-of-square function, then the error for the current step is

$$E(t+1) = \frac{1}{2}[z(t+1) - T(t+1)]^2 \tag{25}$$

Then the backpropagation algorithm, introduced in Sect. 2.3, will be used to update the regular weights as well as the recurrent weight. The details won't be expanded on here.

4 S&P 500 IndexForecasting

The S&P 500 is a widely monitored and benchmarked U.S. equity market index, which is consistent of the stocks of top 500 corporations with large market-cap, most of which are based in U.S. All component stocks in the index are large and publicly held companies, which are traded in the top two US equity markets, the NYSE & NASDAQ. As important as the Dow Jones Industrial Average, the S&P 500 is widely referred as the representative index for the US large-cap stocks. As the performance metrics for stocks, equity mutual fund and hedge fund, the S&P 500 index is often chosen as a baseline for performance comparison.

Investors use the index to track the up and down of the U.S equity market and besides that they also trade index based financial products, such as the futures and options for the S&P 500 index. So the future trend of the S&P 500 index is not only the indicator of the future stock market, but also could help the investor make an investment decision and hedge against risk. The historical data of the index can be viewed as time series data. Our task is to try to use appropriate economics indices with the help of the artificial neural network to forecast the future trend of the S&P 500 index several months ahead.

4.1 Input Variables

The appropriate choice of the input variables seriously determines the efficiency and the performance of the model. After consideration and filtering, several economical data become our candidate variables. They are new housing starts, mortgage rates, producer price index, consumer price index, Moody's AAA corporate bond index, Moody's BAA corporate bonds index, 3-month commercial paper and energy price. After testing, the combination of new housing starts, Moody's AAA corporate bond index and 3-month commercial paper gives the best performance among those variables. The underlying explanation of choosing theses variables will be discussed below.

4.1.1 New Housing Starts

The S&P 500 index is a market weighted index, each stock's weight in the index is proportional to its market value. Table 3 illustrates the current top ten sectors contributed to the index

Table 3. Top ten sectors in S&P 500 index

Sector	% of Index
Financial services	20.3
Healthcare	13.4
Industrial material	12.2
Hardware	10.8
Consumer goods	9.7
Consumer services	8.8
Energy	6.5
Software	4.5
Business services	3.9
Media	3.9

The real estate market plays a significant role in the US economy and drives the economy heavily. From the recent mortgage crisis we notice that if the real estate market slumps, the US economy as well as the stock market will be hurt heavily. On the other hand, if the real estate market booms, it will drives the demand for industrial materials, consumer goods and energy consumption. The modern finance industry, closely related with the real estate market, will thrive at the same time.

From Table 3, we notice that these sectors contribute to 35 %–40 % of the index. We pick the new housing starts as an indicator of the real estate market. Since the new housing starts should have a lagging effect on the real estate market and the whole economy, we have good reason to assume that the S&P 500 index will move up over a certain time lag if the new housing starts increases.

Since the above sectors only contribute to 35 %–40 % of the index, only the new housing starts will not be enough to capture the future trend of the S&P 500 index accurately if no other variables introduced into the model. Two other variables are introduced in Sects. (4.1.2) and (4.1.3).

4.1.2 Moody's AAA Corporate Bond

Moody's AAA Corporate Bond, short for "Moody's AAA", is the index of investment grade bond given an AAA rating by Moody's. It is often viewed in macroeconomics as an alternative to the Fed 10-year Treasury Bill as an interest rate indicator.

One parameter that has been proved very effective in predicting the US economy real growth is the interest rate difference between the long-term and short-term debt (*yield spread*). So we pick the Moody's AAA as an alternative indicator to the long-term interest rate.

4.1.3 Commercial Paper

Commercial paper is a kind money-market security issued by large companies or banks. Basically, it use to purchase inventory or to manage working capital instead of long-term financial investments.

We pick the 3-month commercial paper as an alternative indicator to the short-term interest rate into the model.

4.2 Experiment Result

We use monthly historical data to train our model and make a prediction. We pick the Elman network as the network structure to process the time series data. Since it is a nonlinear dynamics network, there is just an output but no explicit function.

100-month historical data is used to train the Elman network before each prediction and then a 10-month ahead prediction is made based on the trained network.

The input variables are sequences of data consisting of the Moody's AAA corporate bond, 3-month commercial paper and the new housing starts in 6 months ago. The target variable is of course the S&P 500 index. Before entering the network, variables have been preprocessed by subtracting the mean and passing the data through a filter to get rid of the noise.

The Elman network we build is a two-layer network with 8 neurons in the hidden layer and the output layer with single neuron. The tan-sigmoid transfer function is used in the hidden layer and the purelin transfer function is used in the output layer. The sum-of-squares is used as the error function. MATLAB R2007a is the working environment. The *trainbr* is picked as the training function.

The simulation runs from Jan-2000 to Jan-2007 semiannually. The following figures show how the model predicts the 10-month ahead trend of the S&P 500 index. The values on the curve show the ratio of change in index from the beginning date.

5 Conclusions

The result shows that with the help of the neural network the new housing starts and long-term, short-term corporate bond could be applied to predict the future trend of the S&P 500 index successfully. When the new housing starts increasing, which means the boom of the real estate market, the whole economy of US will be driven up. S&P 500 index consists of leading companies from widely ranged economics sectors, so the index will go up. On the hand, the index will decrease if the real estate market slumps.

Though we get a good result of the model, there is still a lot to be improved. We need to make our model more robust to contain more parameters and could be more effective for the financial market data after 2008, when there are more noises in the market.

Another problem is that for the nonlinear neural network, there is not an explicit function between the input and output variables. Besides that, the network is not unique either because of its non linearity. It is challenging for us to judge and pick a best model among trained networks.

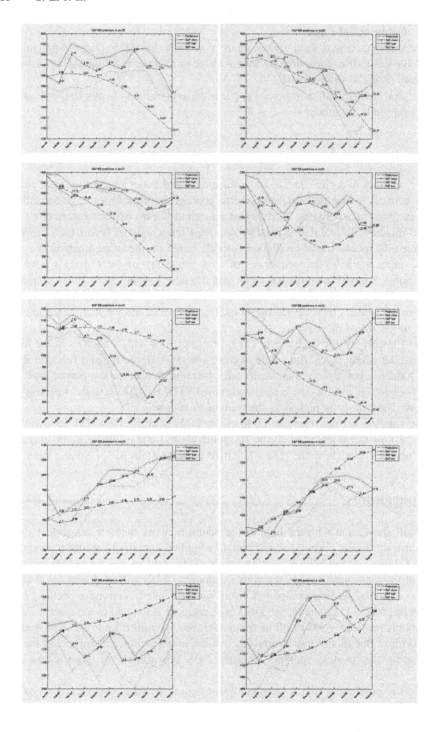

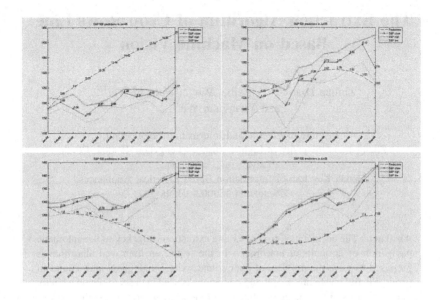

References

Bishop, C.: Neural networks: a pattern recognition perspective. Handbook of neural computation (1996)

Box, G.E.P., Tiao, G.C.: A change in level of a non-stationary time series. Biometrica **53**, 23–35 (1965)

Fama, E.F., French, K.R.: Business conditions and expected returns on stocks and bonds. J. Financ. Econ. **25**, 23–49 (1989)

Fama, E.F., Schwert, W.G.: Assert returns and inflation. J. Financ. Econ. **5**, 115–146 (1977)

Hornick, K., Stinchcombe, M., White, H.: Multilayer feed forward network are universal approximate. Neural Net. **2**, 359–366 (1989)

Huang, R.D., Stoll, H.R.: Is it time to split the S&P 500 future contract. Financ. Anal. J. **52**, 181–192 (1998)

Penumadu, D., Chameau, J.L.: Geomaterial modeling using artificial neural networks. In: Kartam, N., Flood, I., Garrett, J. (eds.) Artificial Neural Networks for Civil Engineers: Fundamentals and Applications, ASCE, pp. 160–184 (1997)

Peters, E.E.: Chaos and Order in capital Markets: a New View of Cycles, Price, and Market Volatility. Wiley, New York (1991)

Samarasinghe, S.: Neural networks for time series forecasting. In: Neural Networks for Applied Sciences and Engineering from Fundamentals to Complex Pattern Recognition (2006)

Oeda, S., Kurimoto, I., Ichimura, T.: Time Series Data Classification Using Recurrent Neural Network with Ensemble Learning. In: Gabrys, B., Howlett, R.J., Jain, L.C. (eds.) KES 2006. LNCS (LNAI), vol. 4253, pp. 742–748. Springer, Heidelberg (2006). doi:10.1007/11893011_94

Trippi, R.R., Turban, E.: Neural networks in finance and investing: using artificial intelligence to improve real-word performance. Irwin Professional Pub, Burr Ridge (1996)

The Extraction Algorithm of Crop Rows Line Based on Machine Vision

Zhihua Diao[1,2(✉)], Beibei Wu[1,2], Yuquan Wei[1,2], and Yuanyuan Wu[1,2]

[1] Institute of Electric and Information Engineering, Zhengzhou University of Light Industry, Zhengzhou 450002, China
dianzhua@163.com
[2] Henan Key Lab of Information Based Electrical Appliances, Zhengzhou 450002, China

Abstract. The accuracy of crop rows line extraction is the key to the automatic navigation of agricultural machinery. In the paper, an improved algorithm is proposed to solve the problem of poor connectivity, single pixel and redundant pixels. Firstly, image pre-processing operations is used in order to obtain a binary image, then the binary image is thinned. In the refinement process, the connectivity of the skeleton is maintained by the introduction of Euclidean distance. Experimental results show that the proposed method has good adaptability to the row crop, and the skeleton lines that are extracted is more accurate. Compared with the traditional algorithms, the error of the navigation line is relatively small by using this algorithm, which could meet the needs of the practical application.

Keywords: Super green algorithm · Crop rows · Skeleton point · Redundant pixels

1 Introduction

Agricultural machine vision navigation plays a key role in automatic picking, irrigation, fertilization and visual navigation is the basis of crop recognition. Precision pesticide technology is a mainstream trend in the development of modern precision agriculture. Its main idea is using an image acquisition device to analyze and process these images. Then a series of processes were used to extract the navigation line of crop to control agricultural machinery for walking.

In the early 20th century, Marchant and Brivot Silsoe research center [1] and the Swedish expert Astrand and Baerveldt [2] had made a contribution to vision navigation algorithm research respectively and achieved some results. Ollis and Stentz [3] used color camera to obtain the field map of crops and the edge of the harvest crop is extracted. Søgaard and Olsen [4] proposed a algorithm with respect to the extraction of crop line based on the Hough transform. Kaizua and Imoub [5] extracted the characteristics of seedling through analysing the spectral characteristics of seedling and the extraction of seedling target line was achieved. In recent years, the domestic also appeared a lot of researches in this aspect. Zhibin et al. [6, 7] who had combined Hough transform and Fisher criterion overcame the insufficiency of the traditional Hough

© IFIP International Federation for Information Processing 2016
Published by Springer International Publishing AG 2016. All Rights Reserved
D. Li and Z. Li (Eds.): CCTA 2015, Part I, IFIP AICT 478, pp. 190–196, 2016.
DOI: 10.1007/978-3-319-48357-3_18

transform and made a model of multiple ridge line. Luo et al. [8] set up a mobile platform for agricultural intelligent operation. Zhao et al. [9] put forward a kind of improvement algorithm based on vertical histogram projection and combined the projection and Hough transform method to detect crop rows. Although these algorithms are valuable, the application of precision pesticide machinery is difficult to achieve because of poor adaptability.

The improved super green algorithm was used greatly reduce the effect of noise. Hough transform reduced the operation time and improved the accuracy. At the same time, an experimental platform was built to conduct different images and a good reconstructions result was obtained, which had good adaptation and accuracy and satisfied the needs of practical application.

2　Segmentation of Crop Rows

2.1　Acquisition of Crop Rows Image

The rows of chinese chives were taked as research object in the experiment. The model of industrial digital camera is MV-VD030SM/SC that was produced by Shanxi Dimensional Image Technology Co., LTD and the model of industrial lens is AFT-0641MP that was produced by AI Feite Photoelectric Technology Co., LTD, which was used to acquire leek images. The output was 8 bit RGB color images. The original image of chinese chives was shown in Fig. 1.

Fig. 1.　Original Chinese chives image

2.2　Gray Processing

Leek images usually contain much color information. It was found that these color information was not wery well for the following process. However, RGB color image was converted to grayscale image had more advantages than using color information processing. Gray processing is the basis of image analysis. The image of leek crop rows taken in farmland had obvious characteristics of green (as shown in Fig. 1). In other words, the G component was higher than the soil background. In order to distinguish crop rows and background better, the traditional super green gray algorithm (2G-R-B) [10] was used to process this image and the effect was shown in Fig. 2a. It was seen that the background noise of the traditional super green gray algorithm was obvious from Fig. 2a, which is not conducive to the follow processing. Therefore, the super green gray algorithm was improved. The improved super green algorithm was as below:

$$Gr = \begin{cases} 1.8G - R - B & G > R \ or \ G > B \\ 255 & others \end{cases} \quad (1)$$

From Fig. 2b, it could be seen that the background noise of chinese chives crop rows were greatly reduced and the crop lines profile were also obvious.

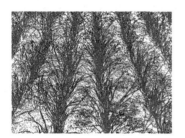

Fig. 2. a. Common gray image b. Improved gray image

2.3   Filtering Denoising

The crop rows after gray processing were still exist noise. In order to get the target of low noise, the grayscale image need to further denoise. The method of combining 3×3 Gaussian template and 3×3 median filter window was used to filter operation. The image after two filtering times was shown in Fig. 3. After filtering, crop rows had been separated clearly.

Fig. 3. Median filtering image   **Fig. 4.** Binary image

2.4   Two Value Processing

In order to further extract the row target, the image was processed by two value operation. The two main methods of two value had threshold and the biggest variance.

The threshold of threshold method was determined according to experiment and experience, but the brightness of different pictures or the same picture were different, which made the application of the method inconvenient. In order to extract the target region, OTSU algorithm was used to process the image automatically. In addition, it was not disturbed by the brightness information and had good treatment result. The binary image was shown in Fig. 4.

3   The Detection of Crop Rows

3.1   Morphological Processing

As there were still many small noise in the crop row image after binarization processing. These noise that was compared with the crop row area was small, so morphological algorithm was used to remove these noise. The main morphological method is erosion and dilation. The crop rows towarded its center gradually and removed the holes that was smaller than structural element by erosion. While, the dilation had the opposite effect, which could filled the void and increased crop line width. In order not to change and eliminate the useful information of crop row, the morphological processing template that would be used must be appropriate. The structure elements of 3*3 template was used in this paper. The times was determined by experiment. The Chinese chives lines contour were shown in Fig. 5.

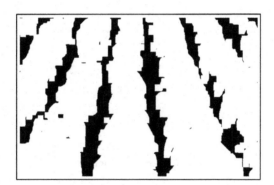

Fig. 5. Morphological processing image

3.2   The Skeleton Extraction of Crop Rows

As the navigation path [11–15] is a linear structure, the skeleton of crop rows should be single pixel width. The outline image after morphological processing could not meet the requirements of machine vision navigation line. Then in order to extract the skeleton information of crop row, thinning algorithm was used(as shown in Fig. 6a). At the same time, the crop rows that were most closely to the center were remained and the redundant skeleton lines was removed to reduce calculation and increase accuracy. The extracted result of the central crop rows was shown in Fig. 6b.

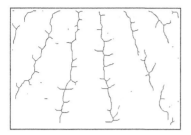 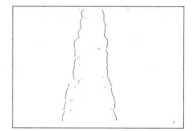

Fig. 6. a. Thinning process b. The skeleton of the center row

3.3   Trunk Line Fitting

The skeleton of central crop rows does not reflect the navigation line. In precision spraying, the skeleton line should be fitted a straight line to realize path planning [16]. The random hough transform [17] was used to fit the crop rows in this paper. The traditional hough transform put the data points of image space map into the parameter space. This was a dispersed mapping of one to many. The random Hough transform improved the velocity of calculation. The result was shown in Fig. 7.

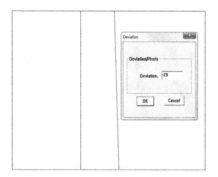

Fig. 7.   Center line image

3.4   The Effect of Extraction of Other Images

In order to verify the algorithm that had been proposed in this paper has a good adaptability to the chinese chives crop rows, the corn crop rows were also carried out the same behavior. The results were shown in Fig. 8. From the figure, it can be seen that the goal lines of the corn crop rows can be extracted clearly, and the skeleton contain less redundant branches. It provided a basis for the analysis of precision pesticide system.

a Original corn crop rows image   b The skeleton image   c Center line fitting image

Fig. 8. The extraction images of corn crop rows

3.5 Deviation Calculation

The deviation that is compared by center line of crop rows after fitting and the actual crop lines was the key technology of precision spraying. In this paper, the deviation of the above two kinds of crop rows were analysed, and it was shown in Table 1. From Table 1, we can see that the algorithm could meet the demand of precision spraying technology in the field.

Table 1. Comparison table of deviation

Experiment object	Deviation/pixels	Running time/ms
Figure 1	−29	215
Figure 8a	18	210

4 Conclusions

(1) The method of gray processing was improved for background segmentation, which had better result and adaptability than before. The median filtering algorithm could reduce the noise effectively.
(2) In this paper, the morphological processing and thinning on binary image were used. The navigation offset was got according to deviation and position information and realized the automatic navigation of agricultural machinery.
(3) The information of corresponding deviation that was got through experiment proved that this method had improved accuracy and adaptability greatly, which achieved desired results on chinese chives. It had certain reference value to the follow-up study.

Acknowledgments. This work was financially supported by the open fund project of NERCITA (KFZN2012W12-012), Zhengzhou Municipal Science and Technology Bureau project (131PPTGG411-13), and backbone teacher plan project of Zhengzhou University of Light Industry, Science and Technology Innovation Fund Project of Zhengzhou University of Light Industry graduate student (2014003).

References

1. Marchant, J.A., Brivot, R.: Real time tracking of plan trows using a Hough transform. Real Time Imag. **1**(5), 363–371 (1995)
2. Astrand, B., Baerveldt, A.-J.: A vision based row-following system for agricultural field machinery. Mechatronics **15**(2), 251–269 (2005)
3. Ollis, M., Stentz, A.: First result in vision-based crop line tracking. In: Proceedings of the 1996 IEEE Conference on Robotics and Automation (ICRA 1996), Minneapolis, MN, pp. 951–956 (1996)
4. Søgaard, H.T., Olsen, H.J.: Determination of crop rows by image analysis without segmentation. Comput. Electron. Agric. **38**(2), 141–158 (2003)
5. Kaizua, Y., Imoub, K.: A dual-spectral camera system for paddy rice seedling row detection. Comput. Electron. Agric. **63**(1), 49–56 (2008)
6. Zhang, Z., Luo, X., Zhou, X., et al.: Crop rows detection based on hough transform and fisher discriminant criterion function. J. Image Graph. **12**(12), 2164–2168 (2007)
7. Zhang, Z., Luo, X., Li, Q., et al.: New algorithm for machine vision navigation of farm machine based on well-ordered set and crop row structure. Trans. Chin. Soc. Agric. Eng. (Trans. CSAE) **23**(7), 122–126 (2007)
8. Luo, X., Ou, Y., Zhao, Z., et al.: Research and development of intelligent flexible chassis for precision farming. Trans. Chin. Soc. Agric. Eng. (Trans. CSAE) **21**(2), 83–85 (2005)
9. Zhao, R., Li, M., et al.: Rapid crop-row detection based on improved hough transformation. Trans. Chin. Soc. Agric. Mach. **40**(7), 163–165, 221 (2009)
10. Zhang, L., Wang, S., Chen, B., et al.: Crop-edge detection based on machine vision. New Zealand J. Agric. Res. **50**(5), 1367–1374 (2007)
11. Nirmal Singh, N., Chatterjee, A., Chatterjee, A., et al.: A two-layered subgoal based mobile robot navigation algorithm with vision system and IR sensor. Measurement **44**(5), 620–641 (2011)
12. Xue, J., Zhang, L., Grift, T.E.: Variable field-of-view machine vision based row guidance of an agricultural robot. Comput. Electron. Agric. **84**, 85–91 (2012)
13. Ji, R., Qi, L.: Crop-row detection algorithm based on random hough transformation. Math. Comput. Model. **54**(3–4), 1016–1020 (2011)
14. Jiang, H., Xiao, Y., Zhang, Y., et al.: Curve path detection of unstructured roads for the outdoor robot navigation. Math. Comput. Model. **58**(3–4), 536–544 (2013)
15. Huang, K.-Y.: Detection and classification of areca nuts with machine vision. Comput. Math. Appl. **64**(5), 739–746 (2012)
16. Ahmed, F., Al-Mamun, H.A., Hossain Bari, A.S.M., et al.: Classification of crops and weeds from digital images: a support vector machine approach. Crop Prot. **40**(10), 98–104 (2012)
17. Ji, J., Chen, G., Sun, L.: A novel Hough transform method for line detection by enhancing accumulator array. Pattern Recogn. Lett. **32**(11), 1503–1510 (2011)

Design of High-Frequency Based Measuring Sensor for Grain Moisture Content

Qinglan Shi[1(✉)], Yunling Liu[1], and Wen Zhang[2]

[1] College of Information and Electric Engineering,
China Agricultural University, Beijing 100083, China
shiql@cau.edu.cn
[2] Department of Electronic Engineering, GuiLin Aerospace Technology,
Guilin 541004, Guangxi, China

Abstract. Accurate measurements of moisture content are indispensable for maintenance of a detecting sensor is designed in this paper to determine the grain moisture content by measuring dielectric constants. To optimize the performance of the designed sensor, electromagnetic waves with suitable frequencies are chosen first followed by deep studies on its transmission characteristic in grain media. Taking wheat as testing samples and applying a total of six frequencies, the network analyzer from Agilent Technologies E5061A is used to measure the dielectric constant and loss tangent of grains with different moisture content. The variation of dielectric constants against moisture under various frequencies is obtained based on which the grain moisture is deduced. According to high frequency transmission line theory, as the impedance of probes wrapped in wet media varies with dielectric constants and is mismatched with transmission lines, standing waves are generated by the composition of reflected waves and incident waves. The strength of reflections depends on probe and characteristic impedances. The relationship of moisture content and dielectric constant can be deduced by analyzing the dynamic variation of reflected and incident waves. Key hardware circuits have been designed involving the moisture detecting circuit, signal generator circuit, high frequency transmission lines, etc. Calibration experiments are carried out, compared with drying method, a cubic polynomial relationship of moisture and output voltage with $R^2 = 0.995$ was obtained by regression analyses. The standard error of predicted results from regression analysis and measured results is 1.09 %. It indicates that the proposed method has a high accuracy.

Keywords: Grain moisture measurement · High-frequency electromagnetic wave · Dielectric constant · Sensor

1 Introduction

The quantity of water in grains usually refers to "moisture content". Detecting and controlling the moisture content of grains play a key role throughout different stages of grain purchase, transportation, storage, processing, commerce, etc. It is essential to employ appropriate methods to measure moisture of grains for the quality maintenance purpose. Many measurement methods have been proposed at home and abroad with

D. Li and Z. Li (Eds.): CCTA 2015, Part I, IFIP AICT 478, pp. 197–207, 2016.
DOI: 10.1007/978-3-319-48357-3_19

various characteristics and a wide scope of application. Generally, they can be divided into two main groups: direct method and indirect method. The former, directly detecting the absolute moisture content by weighting crushed grains before and after drying, is the classic and standard method of grain moisture measurement, such as drying method. Although the direct method is precise and has no effect on the nature of grains, it costs too much time and is not suitable for real-time and on-line measurement. On the other hand, indirect methods such as infrared and microwave methods [1, 2] have high precision, good repeatability but high costs. In addition, physical parameters of media such as dielectric constant, conductance, dielectric loss, etc. may be used to measure grain moisture content [3] and capacitance method [4]. Kandala et al. measured grain moisture by filling grains between plates of parallel capacitors [5]. Mcintosh et al. proposed a sensor available in storerooms to detect grain moisture information [6].

As highly depending on the working environment, the methods mentioned above only apply to the moisture measurement of bulk grains. For packaged grains, non-destructive measurement of grain moisture requires a simply and flexible means of elongate probes inserting into grain packages. Based on high frequency-based electromagnetic wave measurement methods, this paper proposes measures in the design of moisture sensors to enhance its measuring precision.

2 Relationship Between Dielectric Constant and Moisture Content of Wheat

The high frequency electromagnetic measurement method is to measure the dielectric constant of grains by high-frequency electromagnetic waves to indirectly determine grain moisture content [7–9]. Testing samples were collected according to the national standard GBT21305-2007 in China (Determination of moisture content of cereals and cereal products) to study the relation of dielectric constant and moisture content. The samples were harvested in June 2013 at Raoyang town in Heibei province (38°15'N, 115°44'E). After harvest, the samples were stored in a dry and cool place for one year. Red durum wheat in a volume weight of 750 g/L were dried under a constant temperature of 130° ± 2 °C for 90 min ± 5 min and its moisture content was determined. Seven types of tested wheat samples with various moisture content values were then made by mixing water, as listed in Table 1.

Table 1. The moisture content of wheat samples under investigation (bulk density 750 g/L)

Samples	S1	S2	S3	S4	S15	S6	S7
Moisture content (%)	4.52	12.15	14.65	19.94	24.20	32.11	36.54

Dielectric constants of media with different moisture content varies under different frequencies [10]. Nelson studied the frequency characteristics of dielectric constant of different grains [11]. The results indicated that high frequency signals with higher dielectric constants, monotonic variations and less dielectric losses were preferred for moisture measurement. A frequency range of 50 MHz ∼ 300 MHz was recommended [12]. The RF network analyzer Agilent E5061A was used in this study. The dielectric constants of seven samples were measured under six frequencies of 51 MHz, 100 MHz, 150 MHz, 200 MHz, 250 MHz and 300 MHz, respectively. The relationship between dielectric constant and moisture content is plotted in Fig. 1. The relations of real part and imaginary part of permittivity and wheat moisture content are shown in Fig. 1(a) and (b), respectively. It can be seen that the dielectric constant monotonically increases with moisture content. As a dielectric substance, water molecules have strong polar characteristics. The increase in the dielectric constant of dielectric media is due to the increase in its moisture content.

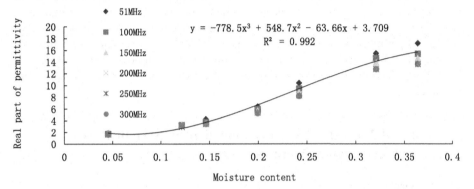

(a)Relation between real part of permittivity and wheat moisture content

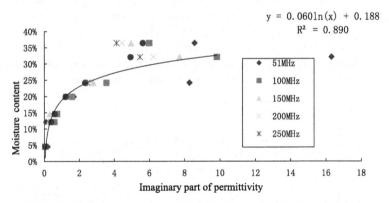

(b)Relation between imaginary part of permittivity and wheat moisture content

Fig. 1. Relation between the dielectric constant and wheat moisture content

From Fig. 6, it can be seen that the measured results are very close for samples with lower moisture content. However it is not the case for saturated samples. Signals in a medium frequency of 100 MHz were taken in the test. The Least Square Method was used to obtain a cubic polynomial relationship between the real part of permittivity and moisture content as follows:

$$\varepsilon' = -778.59^3_{\text{wheat}} + 548.79^2_{\text{wheat}} - 63.669_{\text{wheat}} + 3.709 \tag{1}$$

The coefficient of determination of the real part of permittivity and the fitting curve was calculated as $R_1^2 = 0.9988$.

The relationship between the imaginary part of permittivity and moisture content can be expressed in an exponential function as:

$$\varepsilon'' = 0.063 e^{14.809_{\text{wheat}}} \tag{2}$$

The coefficient of determination of formula (2) $R^2 = 0.890$ was derived in this case.

The two different coefficients of determination in Eqs. (1) and (2) indicate that the real part of permittivity is more related to grain moisture content than the imaginary part of permittivity. In other words, grain moisture content is mainly dependent on the real part of permittivity. Therefore, the moisture of grains can be determined from the real part of permittivity [13]. The dielectric constant in the following sections denotes the real part of permittivity. The high frequency method [14] is proposed based on this principle.

3   Design of Moisture Measurement Sensor System Based on High Frequency Method

3.1   General Design

The block diagram of the sensor system of high frequency-based grain moisture measurements is illustrated in Fig. 2. A high frequency signal is first produced and transferred to probes inserted into tested grains connected through high frequency transmission lines. The dielectric constant measured is larger than 1 (1 for air) since there is water in grains. It varies in a range of $1 \sim 20$ for different moisture content. If the probe in tested grains is regarded as the output load of high frequency signals, the

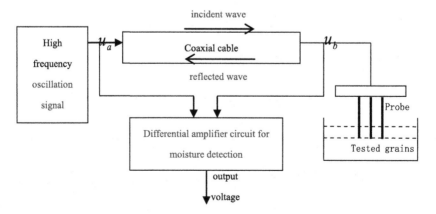

Fig. 2. The block diagram of high frequency sensing circuit system

load impedance varies with grain moisture content. In this way, the feedback on the signal magnitude through output load changes. The moisture content of grains can be determined according to variation in the signal magnitude.

3.2   Signal Generator

According to the frequency characteristics of wheat dielectric constant mentioned above, high frequency signals of 100 MHz were used in the test. Figure 4 illustrates the high-frequency signal generator circuit. The oscillator output frequency is 100 MHz. C1 represents the filter capacitor of power. High frequency signals are output from pin 3 and transmitted to the sensing circuit.

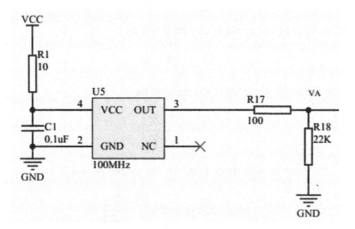

Fig. 3. High-frequency signal generator circuit

3.3   Design of the Transmission Line Circuit

The moisture sensing circuit was designed based on the transmission line theory. From Fig. 3, it can be seen that high a frequency signal generated from high-frequency oscillation sources was transferred to sensor probes by coaxial cables. As each point has its own distributed parameters, the equivalent circuit of the transmission line is depicted in Fig. 5. A high frequency cable of 50 Ω was used for coaxial cables. The two points of the transmission line are represented by A and B. The impedance of probes wrapped in grains is denoted by Z_L.

As wrapped in grains, the impedance of probes changes with its surrounding media which leads to the mismatch of terminal impedance and the signal reflex. The composition of incident waves and reflected waves on transmission lines generates standing waves [15, 16]. If the cable length is taken as one quarter of the signal wavelength (i.e. 0.75 m) and lossless transmission line is assumed, the standing wave ratio can be

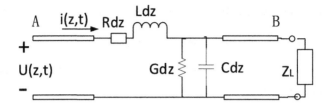

Fig. 4. Distributed parameters of high frequency transmission lines

expressed [17] as $\gamma = \frac{u_{max}}{u_{min}} = \frac{1+\rho}{1+\rho}$, where u_{max} and u_{min} are the peak and trough of u_A and u_B, respectively. The peak voltage of the point A can be calculated as:

$$U_A = A(1+\rho) \tag{3}$$

The peak voltage of the point B can be given as

$$U_B = A(1-\rho) \tag{4}$$

Where ρ is the reflection coefficient defined as

$$\rho = \left|\frac{Z_L - Z_c}{Z_L + Z_c}\right| \tag{5}$$

Therefore, the voltage difference between the points A and B is

$$\Delta U_{AB} = 2A\rho = 2A\left|\frac{Z_L - Z_c}{Z_L + Z_c}\right| \tag{6}$$

where Z_c is the characteristic impedance, A is the voltage constant; the resistance of the cable is taken as 50 Ω.

In order to simplify the circuit and enhance its reliability, the electric cable of 0.75 m was replaced by a 100 µH inductor. The equivalent circuit is shown in Fig. 5.

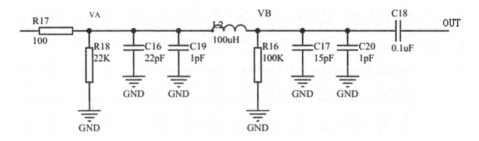

Fig. 5. Equivalent circuit of high frequency transmission lines

The high frequency signal output from signal generators was transmitted to transmission lines by R17 and R18. A π-type filter composed of capacitors C16, C17, C19, C20, was used to filter harmonic components. The 0.75 m cable was replaced by a 100 μH inductor L2 with smaller size and better stability. The voltages at the two ends of the inductor L2 are denoted by VA and VB, respectively.

3.4   Design of Sensor Probes

As shown in Eq. 5, the voltage difference between A and B depends on the impedance of probes. A three-needle type of stainless steel probe was used in this study. This multi-pin probe can be approximated as parallel transmission lines. Its impedance varies between coaxial cables and parallel lines [18, 19]. The probes proposed in references [17, 18] were used for water content measurement of soil. For the same porous media of grains as soil, the impedance is related to the dielectric constant of tested media. The impedance function of soil water measurements was also applied to this study. The function is shown as follows [19, 20].

$$Z_L = -j\frac{Z_0}{\sqrt{\varepsilon}}ctg2\pi f\sqrt{\varepsilon}\frac{l}{C} \tag{7}$$

where Z_0 is the characteristics impedance of probe in air; f is the frequency of high frequency signal source (100 MHz); ε is the dielectric constant of tested grains; l is the length of probes (162 mm); C is the velocity of light, i.e. 3×10^8 m s^{-1}.

3.5   Design of Differential Amplifier Circuit for Moisture Detection

Figure 6 shows the differential amplifier circuit of grain moisture detection. There are dynamic standing waves at u_A, u_B which were transmitted to RF input of LTC5507 chips and transformed into DC voltage. After filter, the DC voltage was sent to the differential amplifier circuit. The AD623, serving as an operational amplifier, is powered by a single supply +5v. Programmable adjustment can be made on its gain by changing resistances. Resistances with error smaller than $0.1 \sim 1$ % were used to obtain precise output voltage gains of AD623. Meanwhile, resistances with low temperature coefficient are preferred to enhance the stability of voltage gains and avoid high gain drift. As the output voltage of AD623 relative to the reference terminal, to obtain a high common mode rejection ratio, an earth REF pins were applied. Transient voltage suppressor SMBJ5.0A was used to prevent the over high instantaneous voltage at the output terminal [21].

$$U_{out} = K_1 \frac{\left|50 + j\frac{Z_0}{\sqrt{\varepsilon}}ctg\frac{2\pi f\sqrt{\varepsilon}l}{\sqrt{\varepsilon}}\right|}{\left|50 - j\frac{Z_0}{\sqrt{\varepsilon}}ctg\frac{2\pi f\sqrt{\varepsilon}l}{\sqrt{\varepsilon}}\right|} \tag{8}$$

Where K_1 is a constant which is always taken as $2 \sim 2.5$ V, this output signal includes the dielectric constant information related to grain moisture. From the relation of dielectric constant and moisture content in Eq. 1, a nonlinear function of output voltages and moisture content can be established. The coefficient of the nonlinear function was determined by calibration experiments.

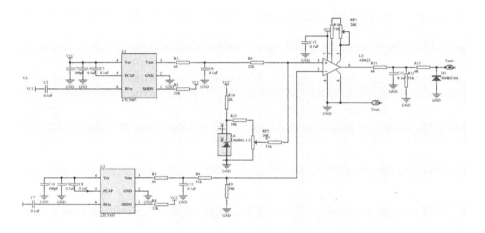

Fig. 6. Differential amplifier for moisture detection

4   Calibration Experiments

The moisture measurements and calibration experiments were carried out on tested samples based on the proposed moisture sensing circuit. Table 2 lists the measured output voltages of the seven samples with different moisture content. Stable readings of voltage meters for each sample were recorded by inserting probes into the tested grains. Repeat the test for five times and take the average value as the output voltage. The variation of measured voltages listed in Table 2 against grain moisture content is plotted in Fig. 7.

Table 2. The output voltage of moisture sensing circuit for wheat with different moisture content

Sample	Moisture content	1st measurement / V	2nd measurement / V	3rd measurement / V	4th measurement / V	5th measurement / V	Average /V
1	4.52 %	0.68	0.69	0.71	0.66	0.7	0.69
2	9.78 %	0.9	0.93	0.91	0.9	0.9	0.91
3	12.15 %	1.02	1.03	1.04	0.98	1.02	1.02
4	14.65 %	1.24	1.17	1.23	1.18	1.18	1.2
5	17.15 %	1.3	1.31	1.29	1.29	1.23	1.28
6	19.94 %	1.33	1.35	1.3	1.34	1.35	1.33
7	36.34 %	1.98	2.01	1.95	2	1.97	1.98

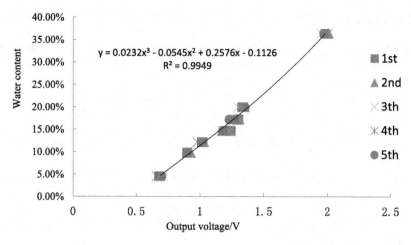

Fig. 7. Relation of output voltage of moisture sensing circuit and wheat moisture content

A cubic polynomial function of moisture and output voltage obtained by the least square method can be given as follows using regression analysis

$$\theta_{wheat} = 0.0232U_{out}^3 - 0.0545U_{out}^2 + 0.2576U_{out} - 0.1126 \tag{9}$$

For the mathematical model derived from the regression analysis, its cross correlation coefficient of predicted and measured results is calculated as

$$R^2 = \frac{\sum_{i=1}^{n} \left(\widehat{\theta}_i - \bar{\theta}\right)^2}{\sum_{i=1}^{n} \left(\theta_i - \bar{\theta}\right)^2} = 0.995 \tag{10}$$

$$RMSE = \sqrt{\frac{1}{n}\sum_{i}^{n} \left(\theta_i - \widehat{\theta}_i\right)^2} = 0.01089 \tag{11}$$

The standard error of predicted results from regression analysis and measured results is 1.09 %. Where n is the total number of samples in the regression analysis and n = 7 in this case; θ_i is the moisture content of the ith sample; $\bar{\theta}_i$ is the average value of measurements of all samples; $\widehat{\theta}_i$ is the moisture content calculated from the output voltage of the ith sample according to the regression equation; $\bar{\widehat{\vartheta}}_i$ is the average value of $\widehat{\theta}_i$.

The result of $R^2 = 0.995$ indicates a good relativity between regression function and measured data.

5   Conclusions

(1) Real and imaginary parts of permittivity of seven groups of wheat samples with various moisture content are measured. The fitting curve shows that the former has a maximum coefficient of determination with moisture content. It indicates that a precise moisture content can be deduced from the real part of permittivity.

(2) The relationship between the moisture and probe impedance can be determined by the relation of moisture and dielectric constant as well as high frequency transmission line theory. A key grain moisture sensing circuit has been designed to measure grain moisture.

(3) Calibration experiments are carried out on the sensor circuit using drying method. A cubic polynomial relationship of moisture and output voltage with $R^2 = 0.995$ was obtained by regression analyses. The standard error of predicted results from regression analysis and measured results is 1.09 %. It indicates that the proposed method has a high accuracy.

It is worth noting that the measuring error of the designed sensor meet the precision demand for wheat. While for other grains, no calibration experiments are conducted and further work should be done.

References

1. Ki-Bok, K., Jong-Heon, K., Seung, S.L., et al.: Measurement of grain moisture content using microwave attenuation at 10.5 GHz and moisture density. IEEE Trans. Instrum. Meas. **1**(51), 72–77 (2002)
2. Jiang, Y.Y., Zhang, Y., Ge, H.Y.: Study on microwave measurement for grain moisture content in grain depot. Comput. Eng. Appl. **46**(29), 239–241 (2010)
3. Zhang, G.Z.: Detector testing of grain moisture based on resistance capacitance method. China Metrol. **10**, 52 (2002)
4. Lu, J.X.: Study on grain's moisture detector based on capacitive sensor. J. Agric. Mech. Res. **6**, 122–123 (2005)
5. Kandala, C.V., et al.: Capacitance sensor for nondestructive measurement of moisture content in NUTS and grain. IEEE Trans. Instrum. Measur. **56**(5), 1809–1813 (2007)
6. Mcintosh, R.B., Casada, M.E.: Fringing field capacitance sensor for measuring the moisture content of agricultural commodities. IEEE Sensors J. **8**(3), 240–247 (2008)
7. You, K.Y., Mun, H.K., You, L.L., et al.: A small and slim coaxial probe for single rice grain moisture sensing. Sensors (Basel, Switzerland) **13**(3), 3652–3663 (2013)
8. Kraszewski, A.W., Trabelsi, S., Nelson, S.O.: Comparison of density-independent expressions for moisture content determination in wheat at microwave frequencies. J. Agric. Eng. Res. **71**(3), 227–237 (1998)
9. Kamil, S., Colak, A.: Determination of dielectric properties of corn seeds from 1 to 100 MHz. Powder Technol. **203**(2), 365–370 (2010)
10. Nelson, S.O.: Microwave dielectric properties of insects and grain kernels. J. Microwave Power **14**(4), 299–303 (1976)
11. Nelson, S.O., Trabelsi, S.: Dielectric spectroscopy of wheat from 10 MHz to 1.8 GHz. Meas. Sci. Technol. **17**(8), 2294–2299 (2006)

12. Stuart, O.N., Samir, T.: Factors influencing the dielectric properties of agricultural and food products factors influencing the dielectric properties of agricultural and food products. Microw. Power Electromagn. Energy **46**(2), 93–107 (2012)
13. Trabelsi, S., Krazsewski, A.W., Nelson, S.O.: New density-independent calibration function for microwave sensing of moisture content in particulate materials. IEEE Trans. Instrum. Meas. **47**(3), 613–622 (1998)
14. Guoa, W., Tiwarib, G., Tangb, J., et al.: Frequency, moisture and temperature-dependent dielectric properties of chickpea flour. Biosyst. Eng. **101**(2), 217–224 (2008)
15. Ansoult, M., Backer, L.D., Declercq, M.: Statistical relationship between apparent dielectric constant and water content in porous media. Soil Sci. Soc. Am. J. **49**(1), 47–50 (1985)
16. Stuart, O.N.: Fundamentals of dielectric properties measurements and agricultural applications. J. Microwave Power Electromagn. Energy **44**(2), 98–113 (2010)
17. Ball, R.J.A.: Characteristic impedance of unbalanced TDR probes. IEEE Trans. Instrum. Meas. **51**(3), 532–536 (2002)
18. Zegelin, S.J., White, I., Jenkins, D.R.: Improved field probes for soil water content and electrical conductivity measurement using time domain reflectometry. Water Resour. Res. **25** (11), 2367–2376 (1989)
19. Sun, Y.R.: Theoretical and experimental approach to calculation of the impedance of soil probe. Acta Pedol. Sin. **1**, 120–126 (2002)
20. Shi, Q.L.: On the transfer function mathematical model and its experimental study for the measurement of soil water content system. J. Agric. Mechanization Res. **2**, 24–26 (2002)
21. Dygas, J.R., Fafilek, G., Breiter, M.W.: Study of grain boundary polarization by two-probe and four-probe impedance spectroscopy. Solid State Ionics **119**(1), 115–125 (1999)

Propagation Characteristics
of Radio Wave in Plastic Greenhouse

Jizhang Wang[1(✉)], Yuli Peng[1], and Pingping Li[2]

[1] Key Laboratory of Modern Agricultural Equipment and Technology,
Ministry of Education and Jiangsu Province, Jiangsu University,
Zhenjiang 212013, China
whxh@ujs.edu.cn, jsujs982@163.com
[2] College of Biology and the Environment, Nanjing Forestry University,
Nanjing 210037, China
lipingping@ujs.edu.cn

Abstract. In order to realize the deployment of wireless sensor network in large areas of plastic greenhouses. The propagation characteristics of radio wave in plastic greenhouse were studied. The Received Signal Strength Index (RSSI) in greenhouse and between greenhouses were studied, and the logarithmic path loss model for RSSI was established. The results show that the radio wave attenuation parameters A and n between greenhouses were 30.785 and 2.89. The attenuation index was larger than the index of free space. In the plastic greenhouse planted with tomatoes, the radio wave attenuation parameters A and n in the ground were 34.99 and 3.64, and in the top of canopy were 35.14 and 2.85, its show that the radio wave transmission has been significantly affected by the crop in the plastic greenhouse.

Keywords: Radio wave · Propagation characteristics · Received Signal Strength Index (RSSI) · Plastic greenhouse

1 Introduction

As a new method for information acquiring and processing, wireless sensor network technology was applied in agricultural environmental monitoring and controlling [1, 2]. In WSNs, it's characterized by realization the transmission of signals by radio waves, while radio waves will be affected by the surrounding environment in the transmission process [3]. So radio propagation is affected by the barrier of crops in the agriculture environment. Therefore, propagation characteristics of wireless channel in agricultural environment has attracted considerable attention, and the research in the propagation characteristics of wireless channel in the typical agricultural environment such as wheat [4–6], apple [7, 8], citrus [9], durian [10] and plum [11] was conducted.

In recent years, as an important content of modern agricultural development, the protected agriculture has developed rapidly, and China has become the largest area country of the protected agriculture. At present, the protected agriculture is developing to the direction of standardization, automation, intelligence and network [12]. The related research has been conducted in the aspect of wireless measurement and control

© IFIP International Federation for Information Processing 2016
Published by Springer International Publishing AG 2016. All Rights Reserved
D. Li and Z. Li (Eds.): CCTA 2015, Part I, IFIP AICT 478, pp. 208–215, 2016.
DOI: 10.1007/978-3-319-48357-3_20

of the greenhouse environment [13–19]. The plastic greenhouse is mainly used for insulation of autumn winter and early spring to realize cultivation in the early spring and late-autumn, which is widely distributed in the south area of china. In the plastic greenhouse, the transmission of radio waves are affected by planting environment and coverings, for the complex environment, so the plastic greenhouse is the typical application environment with obstacles.

In this paper, the plastic greenhouse as the research object, research of propagation characteristics of radio waves between and in plastic greenhouse were study in order to provide basis for the layout of wireless sensor network in the large-scale plastic greenhouse groups.

2 Experiments and Methods

2.1 The Experiment Test Systems

The experiment test systems were shown in Fig. 1, which mainly consisted of receive and send node, USB debugging simulator, PC (Personal Computer), data packet analysis software Packet Sniffer, tripod and tape, etc.

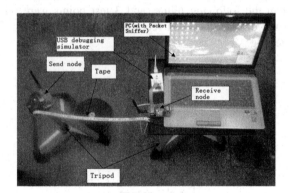

Fig. 1. Experiment test systems

The function of receive and send node was to receive and send radio signals, and the sensor node CC2430 was used as receive and send node. During the testing, the receive node was attached to PC through USB debugging simulator, fixed to the starting point, while the send node moved in sequence according to test points in the direction of test. Radio signals were extracted and analyzed through protocol analyzer software Packet Sniffer. The tripod provided the supporting point in order to keep the receive and send node at the same height, and the tape was used to measure the distance between the receive node and send node.

2.2   Experimental Method

The test place was located in the vegetable planting center of Garden Village, Danyang city Jiangsu province, and linked plastic greenhouse with the same specification were chosen as the experimental subjects, of which the length and width were respectively 70 m and 7.6 m, and the distance between the two sheds was 1.7 m while the distance between the two separated by the road was 12 m; The test of propagation character-istics of radio waves on the crop canopy were respectively conducted between and in the plastic greenhouse.

2.2.1   Experiments Between the Plastic Greenhouse

In order to test the effects of propagation characteristics of radio waves. In plastic greenhouse, this paper designed the experimental scheme as shown in Fig. 2. The plastic greenhouse were numbered, and the left ones were the odd-numbered while the right were the even-numbered. When testing, the receive node was fixed at the plastic greenhouse NO. 1, and the antenna height was 1 m on the ground, and the node was attached to PC computer through USB debugging simulator, and it received signal strength and the number of packet dropout through software, The send node was fixed at the tripod, and the antenna height was also 1 m, which moved respectively in the odd-numbered and even-numbered plastic greenhouse, experiments were divided into 2 groups as shown in the following figure, and 3 test points were set in each plastic greenhouse. And the distance between receive and send node was obtained through calculation after testing.

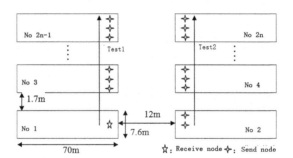

Fig. 2. Experiment scheme

2.2.2   Experiments in the Plastic Greenhouse

Tomatoes planting in the plastic greenhouse were chosen as the experimental subjects, which was in the fruiting period, and the average canopy height was 80 cm while the average row spacing was 60 cm.

In order to analyze the optimum deployment location for nodes, two groups of tests were conducted located in the ground and canopy and the height from the antenna height was 90 cm when the nodes were located in the canopy. During testing, the receive node was fixed at the starting point, while the sender moved in sequence in the middle of the row until there were no signals.

2.3   Test Parameters

(1) Received Signal Strength Index (RSSI)

RSSI (Received Signal Strength Index) was used to judge the link quality, and in this experiments, the protocol analyzer software Packet Sniffer was used for extraction of signals.

(2) Path loss

Path loss is the difference power between effective send radiations and received, which presents attenuation degree of signals, a positive value in dB. The study results [4] show that path loss conform to logarithm path loss model, which can be expressed as:

$$P_r = A - 10n \log d \tag{1}$$

P_r is the average intensity of received signals, A is a parameter of regression analysis model, which presents the effect of application environment on the attenuation of the intensity of radio signals, d is transmission distance of signals; n is attenuation index, which presents decay rate of signal intensity.

3   Results and Discussion

3.1   Propagation Characteristics of Radio Waves Between the Plastic Greenhouses

3.1.1   Effect on Intensity of Received Signals

The RSSI changing with the different distance between receive and send node when radio waves transmitted between the plastic greenhouses was shown as Fig. 3.

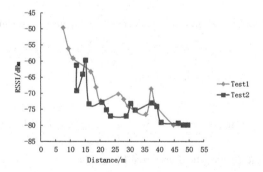

Fig. 3. Received signal strength varies with distance

In the figure, effective propagation path of the two groups of tests was the same, and signal strength of the receive node tended to digression with the increasing distance between the receive and send node, and the strength of received signals decreased sharply when the distance between the receive and send node is smaller while the strength of received signals decreased gently when the distance between the receive and send node increased.

The change of the test 1 and test 2 was the same, and signal strength of both test has the jumping phenomenon, and the data of test 2 was fluctuated obviously. For electromagnetic wave transmitting between the plastic greenhouse is a process of the combination of transmission and reflection, the penetrating power of electromagnetic wave was depended on electric characteristic of the substance, and the insulator will allow electromagnetic wave through with smaller loss, while the good conductor will block the most of radio waves, which accelerated the attenuation of electromagnetic wave. Although the surface of the plastic greenhouse is a layer of plastic with good insulation, the difference of humidity of each plastic greenhouse would change electromagnetic characteristics of the plastic, which leads to the difference of transmission loss in each greenhouse. On the another hand, the reflection of electromagnetic wave can lead to interference phenomenon in which field strength was promoted or down, which will aggravate the jumping phenomenon of the signal strength of electromagnetic wave in short distance. So the combined effect of transmission and reflection leads to fluctuation of signal strength of individual test point in test 1 and test 2.

3.1.2   Result Analysis

A regression analysis was conducted of the value of tested RSSI between the plastic greenhouse in the path loss model by using the least square method, the value of the parameters A and n was obtained as shown in Table 1. The table showed that the relative coefficients R^2 of the test 1 and 2 were 0.943 and 0 916, which means it was reasonable to describe and predict the propagation characteristics between the plastic greenhouse using log-path loss model. The value of parameters of the two groups of tests were basically the same, and the difference of A and were respectively 1.25 and 0.054, which means the propagation characteristics of radio waves was the same in both cases.

Table 1. Regression parameters of log-path loss model

Test	A	N	R2
Test 1	30.16	2.863	0.943
Test 2	31.41	2.917	0.916

3.2   Propagation Characteristics of Radio Waves in the Plastic Greenhouse

3.2.1   Effect on the Strength of Received Signals

The RSSI changing with the distance between the receive and send node in the plastic greenhouse was shown as in Fig. 4. The results showed that the strength of received signals both in the ground and canopy were decreased with the increasing distance between the receive and send node. The strength of received signals decreased sharply when the distance between the receive and send node was shorter, and the decay rate in the ground was sharper than in the canopy. When the strength of received signals declined to about −83 dBm, the transmission distance was 15 m in the ground and 39 m in the canopy. With the increasing distance of the receive and send node, the decay rate of the received signals in the ground and canopy tended to be gentle, and the transmission distance was 15 m in the ground and 39 m in the canopy. The transmission distance in the canopy was obviously longer than that in the ground, and the distance that could be tested was only 30 m in the ground.

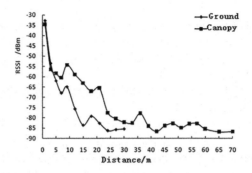

Fig. 4. Received signal strength varies with distance

3.2.2   Result Analysis
A regression analysis was conducted of the value of tested RSSI of tomatoes by using the least square method, and the value of the parameters A and n was obtained as shown in Table 2.

Table 2. Regression parameters of log-path loss model

Location	A	n	R2
Ground	34.99	3.64	0.9699
Canopy	35.14	2.85	0.9091

The results showed that the relative coefficients R^2 in the ground and canopy were respectively 0.9699 and 0.9091, which means it was reasonable to describe and predict the propagation characteristics of tomatoes using log-path loss model. Attenuation index n in the ground was greater than that in the canopy, and the difference between the two location was 0.79, which meant decay rate of radio waves of sensor nodes located in the ground were better than that in the canopy. For the lower location of the tape when the sensor nodes were in the ground, the transmission of the radio waves were affected by leaves of tomatoes greatly, which made attenuation index increased. That also meant the location of nodes in crops decided the attenuation index of radio waves and was related to the transmission distance of sensor nodes.

4   Conclusions

This paper analyzed propagation characteristics of 2.4 GHz radio waves between and in the plastic greenhouse, and conclusions were as follows:

(1) Propagation characteristics of radio waves between the plastic greenhouse conformed to log-path loss model, and the relative coefficients were respectively 0.943 and 0.916. The average value of decay parameters between the plastic greenhouse of A and n were 30.785 and 2.89, and decay index n was greater than that of free-space, which was caused by transmission loss of radio waves.

(2) Propagation characteristics of radio waves in the plastic greenhouse for cultivation of tomatoes conformed to log-path loss model, and the relative coefficients in the ground and canopy were respectively 0.9699 and 0.9091. Decay rate of radio waves was slower when sensor nodes located in the canopy. The sensor nodes should be deployed in the tomato canopy or above of it when the wireless sensor network was deployed in tomatoes.

Acknowledgment. Funds for this research was provided by the prospective research Foundation of Jiangsu province (BY2013065-07), the fund of Jiangsu University (14JDG188) the Priority Academic Program Development of Jiangsu Higher Education Institutions.

References

1. Ning, W., Naiqian, Z., Maohao, W.: Wireless sensors in agriculture and food industry—Recent development and future perspective. Comput. Electron. Agric. **50**, 1–14 (2006)
2. Wei, Y., Minzan, L., Xiu, W.: Status quo and progress of data transmission and communication technology in field information acquisition. Trans. CSAE **24**(5), 297–301 (2008)
3. Yick, J., Mukherjee, B., Ghosal, D.: Wireless sensor network survey. Comput. Netw. **52**(12), 2292–2330 (2008)
4. Liu, H., Wang, M., Zhijun, M., et al.: Performance assessment of short-range radio propagation in crop fields. J. Jiangsu Univ. (Natural Science Edition) **31**(1), 1–5 (2010)
5. Siyu, L., Hongju, G., Jianzhao, J.: Impact of antenna height on propagation characteristics of 2.4 GHz wireless channel in wheat fields. Trans. CSAE **25**(2), 184–189 (2009)
6. Zhen, L., Tiansheng, H., Wang, N., et al.: Path-loss prediction for radio frequency signal of wireless sensor network in field based on artificial neural network. Trans. CSAE **26**(12), 178–181 (2010)
7. Andrade-Sanchez, P., Pierce, F.J., Elliott, T.V.: Performance assessment of wireless sensor networks in agricultural settings. ASABE (2007). Paper No. 073076
8. Xiuming, G., Chunjiang, Z., Xinting, Y., et al.: Propagation characteristics of 2.4 GHz wireless channel at different heights in apple orchard. Trans. CSAE **28**(12), 195–200 (2012)
9. Tao, W., Tiansheng, H., Zhen, L., et al.: Test of wireless sensor network radio frequency signal propagation based on different node deployments in citrus orchards. Trans. CSAE **26**(6), 211–215 (2010)
10. Phaebua, K., Suwalak, R., Phongcharoenpanich, C., et al.: statistical characteristic measurements of propagation in durian orchard for sensor network at 5.8 GHz, communications and information technologies. In: International Symposium on IEEE, ISCIT 2008, pp. 520–523 (2008)
11. Vougioukas, S., Anastassiu, H.T., Regen, C., et al.: Influence of foliage on radio path losses (PLs) for wireless sensor network (WSN) planning in orchards. Biosyst. Eng. **114**(4), 454–465 (2012)

12. Pingping, L., Jizhang, W.: Research progress of intelligent management for greenhouse environment information. Trans. Chin. Soc. Agric. Mach. **45**(4), 236–243 (2014)
13. Zhongfu, S., Hongtai, C., Hongliang, L., et al.: GPRS and WEB based data acquisition system for greenhouse environment. Trans. CSAE **22**(6), 131–134 (2006)
14. Song, Y., Ma, J., Zhang, X., et al.: Design of wireless sensor network-based greenhouse environment monitoring and automatic control system. J. Netw. **7**(5), 838–844 (2012)
15. Park, D.H., Kang, B.J., Cho, K.R., et al.: A study on greenhouse automatic control system based on wireless sensor network. Wirel. Pers. Commun. **56**(1), 117–130 (2011)
16. Park, D.H., Park, J.W.: Wireless sensor network-based greenhouse environment monitoring and automatic control system for dew condensation prevention. Sensors **11**(4), 3640–3651 (2011)
17. Li, L., Gang, L.: Design of greenhouse environment monitoring and controlling system based on Bluetooth technology. Trans. Chin. Soc. Agric. Mach. **37**(6), 97–100 (2006)
18. Li, X., Cheng, X., Yan, K., et al.: A monitoring system for vegetable greenhouses based on a wireless sensor network. Sensors **10**(10), 8963–8980 (2010)
19. Ping, S., Yangyang, G., Pingping, L.: Intelligent measurement and control system of facility agriculture based on ZigBee and 3G. Trans. Chin. Soc. Agric. Mach. **43**(12), 229–233 (2012)

A Review on Leaf Temperature Sensor: Measurement Methods and Application

Lu Yu[1], Wenli Wang[1], Xin Zhang[2,3(✉)], and Wengang Zheng[3]

[1] Institute of Electronic and Informational Engineering,
Hebei University, Baoding 071002, China
15733207576@163.com, 652693254@qq.com
[2] Beijing Research Center for Information Technology in Agriculture,
Beijing 100097, China
zhangx@nercita.org.cn
[3] National Engineering Research Center for Information Technology
in Agriculture, Beijing 100097, China
zhengwg@nercita.org.cn

Abstract. Leaf temperature is the guarantee for the plant to carry out the life activities and closely related to plants' healthy growth and crops' planting management. The accurate measurement of leaf temperature is significant to understand the physiological condition, guide farmland irrigation, select variety and forecast production, etc. The development of plant leaf temperature measurement and requirements of application in recent years at home and abroad were briefly summarized and reviewed in the paper. Firstly, the status of application research and achievements of leaf temperature were introduced from the methods of measurement and scientific experiments. Then it analyzed and compared the principle, advantages and disadvantages and measurement of several common methods in detail including thermal resistance measurement, thermocouple measurement, infrared temperature measurement, infrared thermal imaging measurement and the leaf temperature model. At last, some problems urgently needed to be solved and the development direction of the field were presented, which could provide a reference for the further study of the leaf temperature sensor.

Keywords: Leaf temperature · Sensor · Infrared temperature measurement · Infrared thermal imaging · Leaf temperature model

1 Introduction

Plant leaf temperature is the surface temperature of leaf where is exposed in the atmosphere. The leaf temperature is influenced by the physiological structure of plant itself and the meteorological factors such as solar radiation, air temperature and wind. Leaf temperature varies at different times. In the plants and crops planting, it is more and more seriously to request people to accurately measure plant leaf temperature, so we can understand the relationship between the internal energy, physiological status and environmental factors. Studying the changes of leaf temperature is important for the research on field evapotranspiration, irrigation, variety breeding and yield forecasting.

© IFIP International Federation for Information Processing 2016
Published by Springer International Publishing AG 2016. All Rights Reserved
D. Li and Z. Li (Eds.): CCTA 2015, Part I, IFIP AICT 478, pp. 216–230, 2016.
DOI: 10.1007/978-3-319-48357-3_21

In 1875, the German E. Askenasy explained the concept of leaf temperature, but because of the backward measure technology, the development of leaf temperature application was slow. To measure leaf temperature, thermal resistance could be used [1]. With the development of the thermocouple technology, the leaf temperature was measured by thermocouple. The leaf temperature was measured with thermocouple to study the winter transpiration of greenhouse cucumber in South China [2]. However, due to directly contacting the leaf, the heat conduction is easy to occur, which makes the big error of measurement. It consumes time and the range of temperature is limited [3]. With the rapid development of infrared technology, scientists began to measure the plant leaf temperature with the infrared radiation thermometer, which realized the non-contacting measurement of leaf temperature. Kalyar et al. [4] measured the leaf temperature and used the characteristics of the leaf gas exchange to induce the heat resistance of sunflower. With the infrared technology becoming mature, the thermal infrared imager was produced and applied in the agricultural gradually. The sorghum leaf temperature was measured by infrared imaging instrument to study the water status of the plants under drought stress [5]. The development of leaf temperature measurement technology laid a solid foundation for people to understand the physiological index of plants and the growth of crops.

Starting from the research on leaf temperature measurement, the experiment status and achievements of leaf temperature sensors' application was described and the significance of leaf temperature measurement was showed by the paper. It is summarized that there are several common measurement methods including thermal resistance measurement, thermocouple measurement, infrared temperature measurement, infrared thermal imaging measurement and leaf temperature model. The characteristics and application of five temperature measurement methods were introduced and the problems and development direction of the leaf temperature measurement were pointed out, which provided a reference for the further research on the leaf temperature sensors.

2 Temperature Measurement Theory and Sensing Requirement

Leaf temperature is the surface temperature of the plant leaf, which affects the photosynthesis and transpiration of plants, and it is used to express the temperature of plant and analyze the physiological activities of plants. The degree of plants' drought or water can be indicated and the growth rate and output of the crops can be affected by leaf temperature. There are a lot of methods for leaf temperature measurement such as thermal resistance temperature measurement, thermocouple temperature measurement, infrared temperature measurement, infrared thermal imaging temperature measurement. In addition, the temperature of plant leaves can be obtained by the leaf temperature model. According to different conditions and measurement requirements, the suitable measurement method can be chosen. In the practical application, there are many specific requirements for the leaf temperature sensor. Firstly, the leaf temperature varies from time to time, and the temperature range should be satisfied with the leaf temperature sensor. Secondly, the stability should be guaranteed when the leaf temperature sensor measures in the harsh environment, such as humidity, high temperature and other environment. In addition, to obtain the accurate temperature data in the experiment, the influence of external factors on the leaf temperature sensor should be eliminated as large as possible, which can ensure the measurement accuracy.

3 The Status of the Leaf Temperature Sensor in Application Experiment

The leaf temperature is measured by the leaf temperature sensor. At present, the leaf temperature sensor is widely used in the experiment of plants especially crops. The leaf temperature not only affects the physiological and energy changes of the plants, which reflects the health status of plants, but also reflects the water status of crops, which can guide crop irrigation and drought resistant genotype selection. In addition, the output of crops has a close relationship with the leaf temperature. Here will introduce the experiment status of the specific application of leaf temperature sensor in these aspects.

3.1 Application Experiment of the Leaf Temperature Sensor in the Physiological Health of Plants

Leaf temperature can reflect the physiological activity of plants, and the growth of plants can be monitored by measuring the leaf temperature. Figure 1 shows the leaf temperature is measured by IRT-P5 temperature sensor produced by apogee [6], mounted at the top of plant leaves 4 in., measuring angle is 65°, to measure the

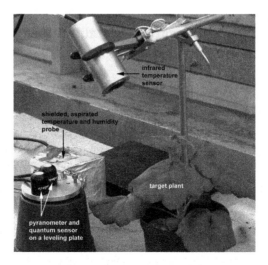

Fig. 1. Geranium plants being measured at the measurement station by Infrared radiometer

temperatures of geranium infected with the pathogen and control group. We could see that the leaf temperature of geranium infected with the pathogen was significantly higher than the temperature of control group. Then it concluded that leaf temperature could predict whether geranium Infective pathogen. In order to study the relationship between temperature and physiological index, the German TFI-50 infrared thermometer was used to measure plant leaf temperature in calm sunny weather [7]. It showed that the higher the stress was, the higher the leaf temperature became, the higher the MDA content was, the lower the content of chlorophyll became, and the physiological characters of the leaves could be judged in a certain degree. The infrared thermography was used to detect physiological parameters of plants in non-contacting and non-destructive manner [8]. Winterhalter et al. [9] used infrared thermometer (Germany, KT15D) to measure plant canopy temperature, which was used to study the high throughput detection and measurement of canopy water quality. Different temperatures were detected by infrared thermal imaging technology between infected tobacco mosaic virus TMV-U1 and uninfected leaves. The results showed that before the visual symptoms appeared the leaf temperature was 0.5–1.3 °C lower than the healthy leaf. Under controlled conditions, the difference of the leaf temperature can be used to distinguish the diseased leaves and healthy leaves. To study the changes of the biomass of three different winter rape infected with A.brassicae and A.brassicicola, the leaf temperature was measured with infrared imaging technology [10]. The results showed that the temperature of diseased leaves was 3–5 °C higher than that of the healthy leaves, in the large area, it could also find the obvious changes of lesion area temperature. Although the research process did not find that the two pathogenic bacteria had any different influence on winter rape, it was found that the thermal imaging technology had great potential in the application of the plant epidemic disease, resistance breeding and crop protection.

3.2 Application Experiment of the Leaf Temperature Sensor in the Irrigation Guidance and Drought Resistant Genotype Screening

The leaf temperature can reflect the water status of crops and guide the crop irrigation and drought resistant genotype selection. The leaf temperature was measured from the beet leaf 1 cm with the infrared thermometer (LaboratoriesC-1600 Linear) [11]. In order to promote the leaf transpiration, tried to select the new leaves when measured, and the time was in the morning 9: 00–10: 00. Study found that under drought conditions the beet leaf temperature difference is not significant between genotype and put forward that using leaf temperature difference between the condition of normal irrigation control and drought stress could better distinguish. A high resolution thermal imager was used to get maize leaves infrared image under conditions of water stress and normal irrigation and found that the leaf temperature obviously changed in two different treatments [12]. The study showed that the temperature of maize seedling could significantly reflect the drought resistance of maize under drought stress and the leaf temperature difference could be regarded as an index of drought tolerance screening of maize at seedling stage. Figure 2 shows the thermal infrared imager of corn canopy temperature is measured and recorded to study the ability of different maize varieties to adapt to drought [13].

Fig. 2. Canopy temperature measured by infrared thermal imager

Using the portable infrared thermometer (the SK-8700) to measure corn leaf temperature [14], the visual angle is 45° and the distance from the leaf is10 cm. It studied the daily difference of corn leaf-air temperature and the relationship between the temperature and the environment factors in detail. The experiment showed that corn leaf-air temperature difference are relatively stable in full water supplied conditions, basically maintained at the range of −4.0–1.0 °C. The leaf-air temperature difference showed an increasing trend under drought stress and the difference was 5 °C between severe stress and adequate water supply. The leaf-air temperature difference can not only well reflect the moisture content of the corn plant but also be used as an important indicator of the drought state of corn (as shown in Fig. 3).

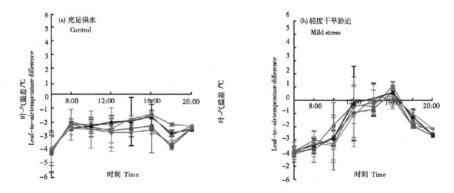

Fig. 3. Characteristic of diurnal variation of leaf-air temperature difference for maize hybrids at the stage of 13th leaf expanded under different drought stress

3.3   Application Experiment of the Leaf Temperature Sensor in Output Forecasting

The leaf temperature is also important in the output forecasting. Han et al. [15] using the BAU-I infrared thermometer to measure leaf temperature, from the side away from the rice leaf 1 cm determined, and studied the relationship between leaf temperature and water conditions at the booting stage and yield through the experiment. A hand-held infrared thermometer (domestic M312216) was used to observe leaf temperature and study the relationship between leaf temperature of early rice and late rice and production [16]. From the Table 1, the rice varieties and the cultivation measure are reasonably selected, leaf temperature of late growth period is reduced and leaf-air temperature difference are increased, rice yields will be raised. In addition, at the same growth conditions of planting density, irrigation and fertilization, genotypic differences among varieties led to the leaf-air temperature difference. Therefore, leaf-air temperature difference can be the selection index of high yield breeding for rice in different growth stages. There were many researchers who used infrared thermometer to measure crop canopy temperature, and to study the relationship between canopy temperature and canopy-air temperature difference and crop yield [17, 18].

Table 1. The correlation between leaf temperature and yield of 14:00 during rice growth period

Growth period	No-super rice		Super rice			Correlation coefficient r
	JinYou463	JinYou402	JinYou458	GanXin203	LuLiangYou996	
Leaf stage	2.2	1.8	2.1	1.7	2.3	0.5748
Differentiation	2.5	2.4	2.8	2.6	2.6	0.6734*
Heading stage	2.8	2.6	2.9	2.7	2.9	0.8329**
After heading	2.4	2.3	2.6	3.0	2.7	0.5727
Plot output/ (t/hm2)	6.71	6.22	7.45	7.12	8.11	

Note: * and ** respectively, expressed at 0.05 and 0.01 levels of significant difference.

4   The Methods of Leaf Temperature Measurement

Since modern times, scientists have studied the method of measuring plant temperatures and until now it has developed into a very mature technology. The methods include contacting measurement of temperature, non-contacting measurement of temperature and leaf temperature model. This paper mainly introduces thermal resistance temperature measurement, thermocouple temperature measurement, infrared temperature measurement, infrared thermal imaging measurement and leaf temperature model. Each method has its own characteristics, the Table 2 shows the technical parameters, advantages and disadvantages and the specific application of these methods in detail.

Table 2. Technology parameters, advantages and disadvantages of four methods of temperature measurement

Method	Product	Range	Precision	Advantage	Disadvantage
Thermal resistance	LT-1 M leaf temperature sensor	0–50 °C	0.15 °C	High precision Simple installation	resistance easily disturbed by outside environment
Thermocouple	CB-0231 thermocouple thermometer	−50–50 °C	0.05 °C	Quick reaction High precision Simple operation	Response a long time Influence leaf environment
Infrared	MI series infrared radiometer	−30–65 °C	0.3 °C	8–14 μm germanium window reduces the absorption of water vapor With the radiation shielding, minimize interference	Influenced by emissivity, distance, environment temperature and atmospheric absorption
Infrared thermal imaging	FLIR A300 infrared camera	−20–120°C	2°C	built-in analysis, alarm functionality and independent communication technology	Expensive price Poor precision

4.1   Contacting Measurement of Leaf Temperature

4.1.1   Thermal Resistance Measurement

The thermal resistance is a temperature sensor which is made of known resistance varying with temperature, such as platinum resistance temperature measurement. Thermal resistance temperature measurement often uses constant current source of three wires to drive the resistance sensor in order to obtain the voltage of thermal resistance. After voltage amplification, A/D conversion and SCM processing, the temperature of the object will be obtained. The thermal resistance is the method of contact temperature measurement, which has a lot of advantage including high accuracy, simple operation and low cost. At present, the research on using thermal resistance to measure leaf temperature is not too much. The changes of carbon dioxide concentration in winter wheat canopy were studied using TL series glass thermal resistor temperature measurement [19]. The LT-1 M leaf temperature sensor [20] (as showed in Fig. 4) has a subminiature touch probe that measures leaf temperature. The lightweight stainless steel wire clip holds high precision glass packaging thermistor, and the probe is very small and specially designed, which almost makes no effect on the natural temperature of the plant leaves. LT-1M leaf temperature sensor has high

Fig. 4. The LT-1M leaf temperature sensor

measurement precision, installed simply and user can customize the length of water-proof cable, which can avoid the effect of humidity environment when measuring the leaf temperature. It can also be used in plant physiological and ecological monitoring system and photosynthesis measuring instrument.

4.1.2 Thermocouple Measurement

Thermocouple is a thermoelectric type of temperature sensor, which converts temperature signal into electric potential signal. The thermocouple temperature measurement instrument amplifies electrical signal and converts analog signal into digital signal. Then it sends data to the microprocessor, displays the results and gets the object temperature.

Thermocouple is also a method of contact temperature measurement which can be used to directly measure the crops leaf temperature. It is easily carried and operated, and its price is cheap. At present, the CB-0231 thermocouple temperature measurement instrument produced by CID eco scientific instrument co., ltd is often applied in plant leaf temperature measurement. CB-0231 [21] is a precision temperature measurement instrument which is first applied in physiological ecology and teaching in our country. According to the temperature difference and electromotive force principle, the thermocouple probe has the advantages of small volume, small heat capacity and fast response to measure the continuous trace changing temperature of the organisms. It is not only widely used in plant physiological ecology freezing injury, fruit storage and refrigeration but also used to measure the surface temperature. The CB-0231 thermocouple thermometer was used to measure the leaf temperature to study the effects of soil water stress on quercus variables leaf temperature [22].

4.2 Non-contacting Measurement of Leaf Temperature

4.2.1 Infrared Temperature Measurement

All objects above zero degree are constantly emitting infrared radiation energy to the surrounding. It is the basis of the infrared radiation temperature measurement to obtain the object surface temperature accurately by measuring its infrared energy [23]. According to the infrared radiation of the object, infrared thermometer relies on its internal optical system to gather the infrared radiation energy on the infrared sensor and converts into electrical signal. After the amplification circuit, the compensation circuit and the linear processing, the detected object temperature is displayed in the terminal. The infrared temperature measurement system includes optical system, infrared sensor, signal amplifier and signal processing and display output. The core is the infrared sensor which converts the incident radiation into a measurable electrical signal [24].

The first infrared thermometer appeared in 1931 and its strong advantage attracted the world's attention. As early as in the 1960s he used infrared thermometer to measure the plant temperature [25]. Later researchers and ecological scholars constantly tried to measured plant leaf temperature by infrared thermometer and analyzed the physiological status of plant, and studied the growth of plant crops. Infrared temperature measurement is non-contacted and different from the traditional temperature measurement. It has advantages of quick response, high accuracy and reliability, wide range and it is not easily damaged [26]. In recent years, with the rapid and smooth development of China's infrared products market, a few foreign companies have increased the market share quickly and owned considerable advantages. The infrared temperature measurement instrument has broad market prospect and economic benefit so that its application will be more and more extensive. The MI series infrared radiometer [27] (as showed in Fig. 5) produced by the Apogee company has been widely used in plant leaf temperature measurement. Infrared radiometer receives object infrared radiant energy of probe field of view and converts it into electrical signal so that the object surface temperature will be measured of. It can quickly measure the application environment and its response time is only about 0.6 s. The recording of the germanium probe is equipped with a radiation shield, which can effectively reduce the error of the measurement data caused by abrupt changes of environment temperature. Because of high accuracy, high sensitivity and a variety of field of view, infrared radiometer is very suitable for the measurement of plant leaf temperature.

Fig. 5. Apogee MI series infrared radiometer

4.2.2 Infrared Thermal Imaging Measurement

All objects whose nature temperature is higher than the absolute zero are emitting radiated energy in the form of electromagnetic wave, including 0.7–1 μm of infrared light wave. The infrared temperature has a high temperature effect, which is the basis of infrared thermal imaging. Infrared thermal imaging is the technology that receives infrared radiation by infrared detectors and changes to video of thermal image by signal processing system. It transforms the thermal distribution of the object into visual images and displays the gray or pseudo color on the monitor, and then gets the temperature distribution field information [28].

Since early 1980, far infrared imaging technology has firstly applied to agriculture and environmental detection. The real-time observation of plant leaves is possible with the features of multi-function, high accuracy and resolution. The application of far infrared imaging technology in plant research has caused a boom [29–31]. The results of the study were remarkable and the object of the study was also widened. The infrared thermal image (as showed in Fig. 6) is recorded with the FLIR A310 infrared camera (as showed in Fig. 7) which is produced by FLIR Systems [32], with a spectral infrared range of wavelength λ from 7.5 to 13 μm, a temperature range of −20 to +120 °C and an accuracy of ±2 %. The FLIR A300 infrared camera comes with the 18 mm standard lens providing a 25° × 19° field of view and 320 × 240 resolution. With its composite video output, the camera is a good choice for measuring leaf temperature. The FLIR A310 owns its built-in analysis, alarm functionality and independent communication technology. It has multi-camera utility software including the FLIR IP config and FLIR IR monitor.

Fig. 6. The infrared thermal image of leaf

Fig. 7. FLIR A300 infrared camera

4.3   Leaf Temperature Model

The leaf temperature model is a simulation model in the greenhouse, which regards the environment conditions (air temperature, air humidity and other parameters) as the driving variable. This paper mainly introduces the model between environment temperature and leaf temperature. The model takes leaf temperature as the dependent variable and environment temperature as the independent variable, building their fitted equation. The model method has good forecast effect and can guide actual production. In foreign countries, it was feasible to control the greenhouse environment according to the leaf temperature and establish the mechanism model of the leaf temperature [33]. Yao et al. [34] established the leaf temperature simulation model based on the mechanism model of the blade energy balance.

Using the common cucumber as the experimental material in the simple sunlight greenhouse, the model of plant leaf temperature was established [35]. The study showed the change regulation of the environment temperature to leaf temperature in a day and analyzed the physiological adaptive mechanism of the leaf temperature. From the Fig. 8, the environment temperature and leaf temperature are fitted and the equation is: $y = 1.188a - 7.6662$. The 'y' was the leaf temperature and the 'a' was environment temperature, coefficient of determination was $R2 = 0.9128$. The equation was significantly associated with 0.01 levels and verified good correlation between leaf temperature and environment temperature.

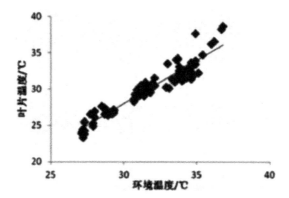

Fig. 8. The fitting of leaf temperature and environment temperature

5   Problems and Development Direction

5.1   Problems

There are many researches on the application of leaf temperature. Although the technology of leaf temperature measurement is developing and many advanced products are produced, there are some problems in the application of leaf temperature and measurement methods that are mainly reflected in the following aspects:

(1) The leading wire of thermal resistance sensor and the connecting wire of the resistance have a great influence on the measurement results. The leading wire of thermal resistance is often in the measured temperature environment which fluctuates seriously and varies with temperature. It is so difficult to estimate and correct that causes larger measurement error [36]. Because the platinum is expensive metals, it is generally used for high precision industrial measurement and rarely applied in plant temperature measurement.

(2) Because the thermocouple is belong to contacting measurement [37], it is necessary to absorb heat from the leaves and bring heat conduction, which will influence on the surface environment of leaves. Because of the disadvantages of hard workload and longtime response, the thermocouple temperature measurement is not suitable for large area measurement. In addition, contacting with the leaves easily leads to the change of the leaf environment and affects the growth of the crops and the measurement precision. Many experts and scholars measures leaf temperature by thermocouple measurements, but it is difficult for thermocouple to overcome the limitations of radiation and heat conduction. So thermocouple temperature measurement is not too widely used in plant leaf measurement.

(3) Infrared thermometer is affected by the leaf emission rate which is a very complex parameter. So it is difficult to obtain the real temperature. The result needs to be corrected for material emissivity and temperature measurement data is processed difficultly. In addition, the infrared thermometer is greatly influenced by the environment, such as dust, steam, carbon dioxide, etc. [38].

(4) There are some problems in the infrared thermal imaging technology [39]. Firstly, the infrared imaging camera depends on the temperature difference to form images, but leaf-air temperature difference is generally small, so the contrast of infrared thermal image is low and the ability of resolving details is poor. Secondly, the infrared imaging camera could not realize the target through transparent barrier, such as the window glass because it can't make the infrared imaging camera detect temperature difference. Thirdly, high cost and price is the biggest factor that limits the thermal imaging camera products widely used.

5.2 Development Direction

The research on the leaf temperature in our country started relatively late and it has not reached to the world leader level in the application and measurement of leaf temperature, so there is a lot of room for developing. Firstly, research on the application of leaf temperature is still in the primary stage and using leaf temperature to study evapotranspiration of crop, irrigation, variety breeding and yield forecast should be further improved. The application of leaf temperature should be developed toward the direction of precision agriculture. Secondly, when measuring the crop leaf temperature we should study the effects on temperature at different moments and different parts of leaves and the degree of young and old leaves may make leaf temperature different. It is the key for application of leaf temperature to measure temperature accurately. We should pay attention to the development of leaf temperature measuring instrument and the key technology and improving the accuracy of measurement, which is the guarantee for the healthy and rapid development of our country agriculture.

6   Conclusion

Now more and more experts and scholars pay more attention to the study of crop leaf temperature, and leaf temperature has become one of the most important factors in the process of crop growth. Plant leaf temperature is one of the most important parameters of plant affecting its healthy growth. It is significant for the study on crop evapo-transpiration, irrigation, variety breeding and yield forecast. In order to measure leaf temperature, thermal resistance measurement, thermocouple measurement, infrared temperature measurement, infrared thermal imaging measurement and the leaf temperature model can be used. Each method has its own characteristics in measuring the leaf temperature. Obviously, infrared temperature measurement has more advantages in leaf temperature measurement. This non-contact type of measurement has advantages of wide range and fast response and it is suitable for large area measurement. However, the factors that affect the accuracy of infrared temperature measurement, such as emissivity, distance, environment temperature and atmospheric absorption must be noticed. The accuracy of infrared sensors and the key technology must be improved. Reducing the cost of infrared instrument production and the use of infrared temperature measurement will be more common.

Acknowledgement. The research was supported by the National High Technology Research and Development Program ("863" Program) of China (2013AA103005).

References

1. Hackl, H., Baresel, J.P., Mistele, B., Hu, Y., Schmidhalter, U.: A comparison of plant temperature as measured by thermal imaging and infrared thermometry. J. Agron. Crop Sci. **198**(6), 415–429 (2012)
2. Weihong, L., Xiaohan, W., Jianfeng, D., et al.: Measurement and simulation of cucumber canopy transpiration in a subtropical modern greenhouse under winter. Acta Phytoecologica Sinica **28**(1), 59–65 (2004)
3. Gaoming, J.: Plant temperature and measurement. J. Plant, 31–32 (1998)
4. Kalyar, T., et al.: Utilization of leaf temperature for the selection of leaf gas-exchange traits to induce heat resistance in sunflower (*Helianthus annuus* L.). Photosynthetica **51**(3), 419–428 (2013)
5. Yitao, W., Yufei, Z., Fengxian, L., et al.: Relationship between leaf temperature and water status in sorghum under drought stress. Agric. Res. Arid Areas **31**(6), 146–151 (2013)
6. Omer, M., James, C.L., Frantz, J.M.: Using leaf temperature as a nondestructive procedure to detect root rot stress in geranium. Horttechnology **17**(4), 532–536 (2007)
7. Xiaotong, Z., Yadong, H., Jiping, G., et al.: Leaf temperature and physiological traits of panicle under different soil water potential. Hubei Agric. Sci. **50**(1), 33–36 (2011)

8. Xu, H., et al.: Early detection of plant disease using infrared thermal imaging. Opt. East, 638110–638110-7 (2006). International Society for Optics and Photonics
9. Winterhalter, L., et al.: High throughput phenotyping of canopy water mass and canopy temperature in well-watered and drought stressed tropical maize hybrids in the vegetative stage. Eur. J. Agron. **35**, 22–32 (2011)
10. Baranowski, P., Mazurek, W., Jedryczl, M.: Temperature changes of oilseed rape (Brassica napus) leaves infected by fungi of Altemaria sp. Roliny Oleiste **30**, 21–33 (2009)
11. Shaw, B., Thomas, T.H., Cooke, D.T.: Responses of sugar beet (Beta vulgaris L.) to drought and nutrient deficiency stress. Plant Growth Regul. 77–83 (2002)
12. Liu, Ya., Subhash, C., Yan, J., Song, C., Zhao, J., Li, J.: Maize leaf temperature responses to drought: thermal imaging and quantitative trait loci (QTL) mapping. Environ. Exp. Bot. **71** (2), 158–165 (2010)
13. Atherton, J.J., Rosamond, M.C., Zeze, D.A.: Phenotyping maize for adaptation to drought. Front. Physiol. **3**, 305 (2012)
14. Yunpeng, L., Si, S., Yuqiang, P., et al.: Diurnal variation in leaf-air temperature difference and the hybrid difference in maize under different drought stress. J. China Agric. Univ. **19** (1), 13–21 (2014)
15. Han, et al.: The relation between leaf temperature and water condition in boot stage on rice. Chin. Agric. Sci. Bull. **22**(2), 214–216 (2006)
16. Ziming, W., Wei, Z., Qinghua, S., Xiaohua, P.: The study on double-season rice leaves temperature change law and its relationship with yield. Chin. Agric. Sci. Bull. **28**(18), 86–92 (2012)
17. Shan, H., et al.: Genetic differences in canopy temperature of different early-rice varieties and its relationship with yield. Acta Agriculturae Universitatis Jiangxiensis **36**(6), 1179–1184 (2014)
18. Xuemin, R., et al.: Relationships between yield characteristics and canopy temperature of peanut. J. Northwest Agric. For. Univ. **42**(12), 39–45 (2014)
19. Baohua, D., Chengf eng, T., Yangping, D.: Studies on the changing patterns of CO_2 concentration in winter wheat canopy. Chin. J. Agrometeorology **06**, 26–31 (1996)
20. LT-1M leaf temperature sensor. Sensors and systems for monitoring growing plants. www.phyto-sensor.com
21. CB-0231 thermocouple thermometer. China Chemical Instrument Network, 20 January 2009. http://www.chem17.com/Offer_sale/detail/53822.html
22. Xiaoming, S., Guo, D., Yong, Z., et al.: Influence of soil water stress on Quercus variabilis leaf temperature and characteristics of chlorophyll fluorescence. J. Henan Agric. Univ. **06**, 691–697 (2013)
23. Zifei, Z.: Infrared thermometers outline. Metrol. Measur. Tech. **33**(10), 22–23 (2006)
24. Liu, J., Wang, H., Fan, L.: Principle of infrared thermometer and problem of application. Mod. Instrum. 50–51 (2007)
25. Tanner, C.B.: Plant temperature. Agron. J. **55**(2), 210–211 (1963)
26. Xia, D., Wenjuan, S., Fuming, W.: Design of infrared thermometer based on single-chip microcomputer. Shanxi Electron. Technol. 21–23 (2011)
27. Infrared Radiometer Meters MI-200 Series: http://www.apogeeinstruments.com
28. Li, Y., Xiaogang, S., Jihong, L.: Research on temperature measurement technology and application of infrared thermal imaging. Mod. Electron. Tech. 112–115 (2009)
29. Kitaya, Y., Kawai, M.: The effect of gravity on surface temperatures of plant leaves. Plant Cell Environ. **26**, 497–503 (2003)
30. Mustilli, A.C., Merlot, S.: Arabidopsis OST1 protein kinase mediates the regulation of stomatal aperture by abscisic acid and acts upstream of reactive oxygen species production. Plant Cell **14**, 3089–3099 (2002)

31. Merlot, S., Mustilli, A.-C.: Use of infrared thermal imaging to isolate Arabidopsis mutants defective in stomatal regulation. Plant J. **30**, 601–609 (2002)
32. FLIR A310. http://www.flircameras.com/flir_a-series_a310.htm
33. Rohmy, et al.: Environmental control in greenhouse based on phytomonitoring-leaf temperature as a factor controlling greenhouse environments. Acta Hortic. **761**, 71–76 (2007)
34. Zhenkun, Y., et al.: A simulation model of the relationship between tomato leaf temperature and ambient factors in solar greenhouse. Jiangsu J. Agric. **26**(3), 587–592 (2010)
35. Li, T.: Solar greenhouse in northern crop leaf temperature and physiological adaptability mechanism model. J. Inner Mongolia Agric. Univ. **01**, 25–29 (2014)
36. Ting, L., et al.: A new method of lead compensation for three-wire resistance temperature detectors. Electron. Measur. Technol. **01**, 38–39 (2001)
37. Xueli, C., et al.: Analysis and correction method of thermocouple temperature measurement error. Sci. Technol. **11**, 303–304 (2008)
38. Jilin, L., et al.: Radiation temperature measurement and calibration technology. Measuring Press, Beijing (2009)
39. Enliang, P.: Application and development of infrared thermal imaging technology in intelligent monitoring. China Public Secur. **22**, 66–68 (2014)

Distinguish Effect of Cu, Zn and Cd on Wheat's Growth Using Nondestructive and Rapid Minolta SPAD-502

Jinheng Zhang[1(✉)], Yonghong Sun[3], Lusheng Zeng[2], Hui Wang[1,2],
Qingzeng Guo[1], Fangli Sun[1], Jianmei Chen[3], and Chaoyu Song[3]

[1] Institute of Eco-Environment and Agriculture Information,
College of Environment and Safety Engineering,
Qingdao University of Science and Technology,
Qingdao 266042, Shandong, China
zhangjinheng1973@gmail.com

[2] College of Resource and Environment, Qingdao Agricultural University,
Qingdao 266109, Shandong, China

[3] Qingdao Agricultural Science Research Institute,
Qingdao 266100, Shandong, China

Abstract. This paper focused on the SPAD values response to different treatment levels of heavy metal Cu, Zn and Cd stress on growth of wheat. Random blocks design experiment has been carried out to simulate five concentration levels' heavy metal Cu, Zn and Cd respectively. The in-situ SPAD values were measured for each treatment on December 8th 2014 (seedling stage), December 13th 2014 (tillering stage), January 8th 2015 (tillering stage), April 18th 2015 (elongation stage), April 30th 2015 (booting stage), May 12th 2015 (heading stage) and June 3rd 2015 (filling stage) respectively using Minolta SPAD-502. The standard deviation of SPAD in different between check plot (CK) and level i, Normalized SPAD values and their variation coefficients among five heavy metal treatments for Cd, Zn and Cu have been analyzed at different measurement time. SPAD values of CK treatment decreased slightly from seedling stage to tillering stage, followed by rapid increase to elongation stage, keeping a shoulder to heading stage, then rapid decrease to filling stage. With the increase of heavy metal treatment levels, standard deviations of SPAD values between CK and level i increased gradually, and the minimum of standard deviation appeared at booting and heading stage. The Normalized SPAD had been calculated by the SPAD ratio of heavy level i to CK. The higher Normalized SPAD, the lower Normalized SPAD value with the minimum at booting and heading stage. Variance coefficients of Normalized SPAD of Cd treatments at seedling stage and tillering stages lower than Zn and Cu treatments. Normalized SPAD among Cd, Cu and Zn treatments had lowest and similar variance at booting stage and heading stages.

Keywords: Cu · Zn · Cd · SPAD-502 · Wheat

© IFIP International Federation for Information Processing 2016
Published by Springer International Publishing AG 2016. All Rights Reserved
D. Li and Z. Li (Eds.): CCTA 2015, Part I, IFIP AICT 478, pp. 231–238, 2016.
DOI: 10.1007/978-3-319-48357-3_22

1 Introduction

Heavy metals (HM) means that some elements which have specific weights of more than 5 g/cm^3 (Lasat 2000). Industrial revolution causes high anthropogenic emission of heavy metals into soil and water (Ayres 1992). Plants' rootsare the primary contact site for the metal ions (Eisenreich et al. 1996). As for plant metabolism, HMs such as Cu and Zn are essential microelement. But when present in excess, Cu and Zn also can become extremely toxic just like low levels of Cd, Hg, and Pb, (Gupta 2013). Incorporated various techniques of analysis have been used to quantify the effects of HMs on plant including leaf injury with visual assessment, biochemical assessment and spectral measurements.

The SPAD-502 m uses two light-emitting diodes (650 nm and 940 nm) and a photodiode detector. The light-emitting diodes and photodiode detector measure transmission sequentiallythrough leaves (Markwell et al. 1987). SPAD-502 has been utilized for determining the nitrogen status, chlorophyll content, thickness of leaf, succulence of leaf, specific leaf area, leaf mass, leaf water content, photosynthesis, and so on. (Campbell et al. 1990; Castelli et al. 1996; Azia and Stewart 2001; Balasubramanian et al. 2000; Richardson et al. 2002; Giunta et al. 2002; Wang et al. 2004; Jifon et al. 2005; Netto et al. 2005; Uddling et al. 2007; Marenco et al. 2009). SPAD readings may represent a useful screening criterion of crop growth status. And SPAD-502's utilization saves much more time and resources than traditional destructive methods (Netto et al. 2005). SPAD readings come from the leaves' transmitted spectra at 650 nm and 940 nm. It has been recognized as a simple and portable diagnostic tool to measure the greenness or chlorophyll content of leaves, thus it may become useful tool being used to assess the stress effects on plant growth.

The plant leaves had very strong enrichment ability for some elements and plant trace elements had a strong influence growth status of leaves, which can response to transmitted spectra of leaves at red and infrared bands used by SPAD. Few publications reported on SPAD as a tool to detect heavy metal stress of wheat. This research was undertaken to quantify the responses of SPAD readings of wheat among different heavy metal treatments. The objective of this present study was to find the potential of SPAD-502 assessing heavy metal Cu, Zn and Cd stress on wheat.

2 Materials and Methods

The field plots (genotype of wheat is called Jimai 22) was treated with heavy metal Cu (CuSO$_4$·5H$_2$O), Zn (ZnSO$_4$·7H$_2$O)and Cd (3CdSO$_4$.8H$_2$O) at five-level treatments (see Table 1) and three repeated trials. Seed on 17th October 2014. Plots with area of 2×3 m^2 had been arranged in a randomized block consisting of same fertilizer applications (Fig. 1). *In-situ* nondestructive SPAD values were measured by Minolta SPAD-502 on December 8th 2014 (seedling stage), December 13th 2014

Table 1. The concentration treatment levels of Cd^{2+} Zn^{2+} Cu^{2+} (mg/kg)

	Ck	L1	L2	L3	L4
Cd^{2+}	0.00	1.00	3.00	5.00	8.00
Zn^{2+}	0.00	250.00	500.00	750.00	1000.00
Cu^{2+}	0.00	100.00	300.00	600.00	900.00

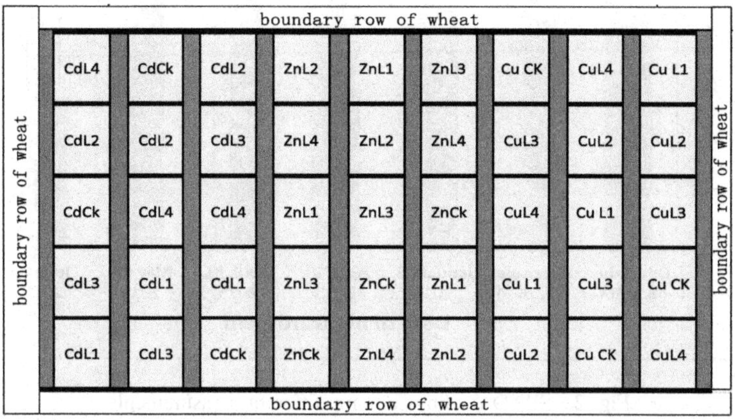

Fig. 1. Distribution diagram of field plots (PVC sheets and earthen rows had been put among different blocks in order to block heavy metals diffuse among adjacent blocks)

(seedling stage), January 8th 2015(tillering stage), April 18th 2015 (elongation stage), April 30th 2015 (booting stage), May 12th 2015 (heading stage) and June 3rd 2015 (filling stage) respectively. Middle of the upper full expanded leaves had been chosen to measure the SPAD values. *In-situ* Minolta SPAD-502 was taken 20–30 values per block.

3 Results and Discussion

SPAD values of check plot (CK) for Cd, Zn and Cu treatments changed as follow: slight decrease from seeding stage on Dec. 8th, 2014 to tillering stage on Jan 13th, 2015. Then the SPAD values increase rapidly from tillering stage on Jan 13th, 2015 to elongation stage on April 18th, 2015. There is a shoulder from elongation stage on April 18th, 2015 to heading stage on May 12th, 2015, then decrease rapidly again, which indicated that different SPAD values appeared different growth stages of wheat under normal growing status. The highest SPAD values appeared at elongation and heading stages (Fig. 2).

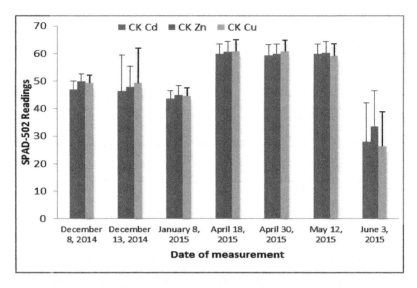

Fig. 2. SPAD readings among different measurements

Compare the different between SPAD readings measured from check plot (CK) and heavy metal treatment level 1 (L1), level 2 (L2), level 3 (L3) and level 4 (L4) respectively at the whole growth stages. As the heavy metal treatment levels increased from L1 through L4, the dispersion degree increase gradually according to the sequence between CK and L1, CK and L2, CK and L3, CK and L4 (Fig. 3a–d). The dispersion degree also could be indicated by standard deviations of SPAD readings between CK and Level i (i = 1, 2, 3, 4 respectively). With the increase of heavy metal treatment levels, standard deviations of SPAD values increased step by step followed by the sequence between CK and L1 < CK and L2 < CK and L3 < CK and L4 (Fig. 4), which also can be proved by the SPAD value ratio of different level to CK (Fig. 5).

CK-Li means standard deviation between CK and heavy metal treatment level i.

SPAD reading had been normalized using the ratio of different level to CK, which called Normalized SPAD. The higher heavy metal level, the lower ratio value appeared at whole growth stages. Minimum difference of both the ratio and the standard deviation appeared on April 30th 2015 (booting stage) and May 12th 2015 (heading stage) (Fig. 5), which indicated that heavy metals have the least influence on booting stage and heading stage of wheat reflected by the changes of SPAD readings (Fig. 5). The phenomenon also can be proved by variance coefficients of Normalized SPAD.

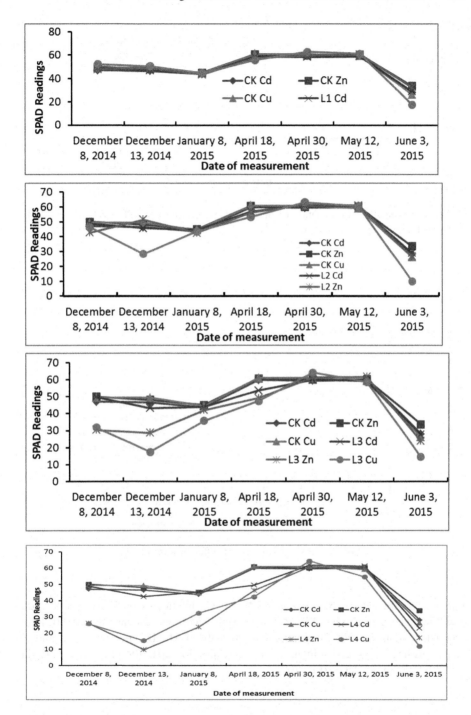

Fig. 3. The dispersion degree increase gradually according to the sequence between CK and L1, CK and L2, CK and L3, CK and L4

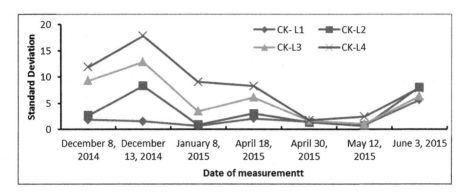

Fig. 4. Compare standard deviation of SPAD values between CK and Level i (i = 1, 2, 3, 4 respectively)

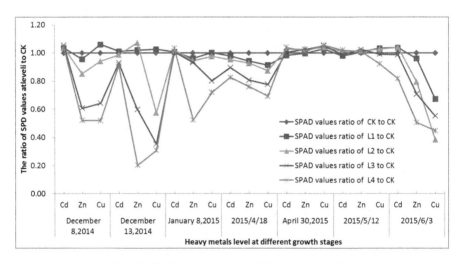

Fig. 5. SPAD value ratio of different level to CK

Variance coefficients of Normalized SPAD of Cd treatments on December 8th 2014 (seedling stage), December 13th 2014 (tillering stage) and January 8th 2015(tillering stage) less than Zn and Cu treatments. Normalized SPAD among Cd, Cu and Zn treatments had lowest and similar variance on April 30th 2015 (booting stage) and May 12th 2015 (heading stage) (Fig. 6).

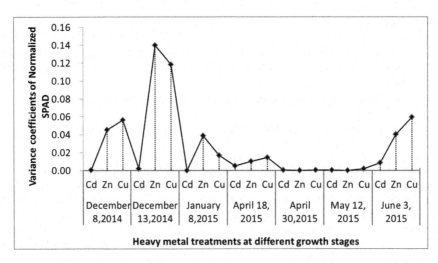

Fig. 6. Variance coefficients of Normalized SPAD of heavy metal treatments at different growth stages

4 Conclusion

Minolta SPAD-502 can be utilized to distinguish the effects of Cd, Cu, Zn on wheat as a nondestructive and rapid tool. The key findings from this research as follow:

(1) Different SPAD values appeared different growth stages of wheat under normal growing status. The highest SPAD values appeared at elongation and heading stages, which indicated the most vigorous growth stages of wheat are elongation and heading stages.

(2) The SPAD readings different and their standard deviations between CK and heavy metal treatment level i (CK and L1, CK and L2, CK and L3, CK and L4) showed that with the increased heavy metal treatment levels, the dispersion degree and standard deviations increased gradually, which indicated that the higher heavy metal concentration treatment, the more influence over wheat's growth.

(3) Higher heavy metal level had lower Normalized SPAD values. Minimum difference of Normalized SPAD and standard deviation between CK and level i appeared at booting stage and heading stage. And Normalized SPAD among Cd, Cu and Zn treatments had lowest and similar variance at booting and heading stage. These results indicated that heavy metals have the least influence over wheat's growth at booting stage and heading stage.

(4) Variance coefficients of Normalized SPAD of Cd treatments at seedling stage and tillering stage lower than Zn and Cu treatments, which indicated the influence of Cd over wheat's growth is less than heavy metal CU and Zn.

Acknowledgement. This paper was supported by the National Natural Science Foundation of China [41471279].

References

Netto, A.T., Campostrini, E., de Oliveira, J.G., Bressan-Smith, R.E.: Photosynthetic pigments, nitrogen, chlorophyll a fluorescence and SPAD-502 readings in coffee leaves. Sci. Hortic. **104** (2), 199–209 (2005)

Ayres, R.U.: Toxic heavy metals: materials cycle optimization. Proc. Natl. Acad. Sci. USA **89**, 815–820 (1992)

Azia, F., Stewart, K.A.: Relationships between extractable chlorophyll and SPAD values in muskmelon leaves. J. Plant Nutr. **24**, 961–966 (2001)

Balasubramanian, V., Morales, A.C., Cruz, R.T., Thiyagarajan, T.M., Nagarajan, R., Babu, M., Abdulrachman, S., Hai, L.H.: Adaptation of the chlorophyll meter (SPAD) technology for real-time N management in rice: a review. Int. Rice Res. Notes **25**(1), 4–8 (2000)

Campbell, R.J., Mobley, K.N., Marini, R.P., Pfeiffer, D.G.: Growing conditions alter the relationship between SPAD-501 values and apple leaf chlorophyll. HortScience **25**, 330–331 (1990)

Castelli, F., Contillo, R., Miceli, F.: Non-destructive determination of leaf chlorophyll content in four crop species. J. Agron. Crop Sci. **177**, 275–283 (1996)

Giunta, F., Motzo, R., Deidda, M.: SPAD readings and associated leaf traits in durum wheat, barley and triticale cultivars. Euphytica **125**(2), 197–205 (2002)

Jifon, J.L., Syvertsen, J.P., Whaley, E.: Growth environment and leaf anatomy affect nondestructive estimates of chlorophyll and nitrogen in Citrus sp. leaves. J. Am. Soc. Hortic. Sci. **130**, 152–158 (2005)

Markwell, J., Osterman, J.C., Mitchell, J.L.: Calibration of the Minolta SPAD-502 leaf chlorophyll meter. Photosynth. Res. **46**(3), 467–472 (1987)

Lasat, M.M.: Phytoextraction of metals from contaminated soil: a review of plant/soil/metal interaction and assessment of pertinent agronomic issues. J. Hazard. Subst. Res. **2**(5), 1–25 (2000)

Dharmendra K. Gupta (2013)

Eisenreich, W., Menhard, B., Hylands, P.J., Zenk, M.H., Bacher, A.: Studies on the biosynthesis of taxol: the taxane carbon skeleton is not of mevalonoid origin. Proc. Natl. Acad. Sci. USA **93**, 6431–6436 (1996)

Marenco, R.A., Antezana-Vera, S.A., Nascimento, H.C.S.: Relationship between specific leaf area, leaf thickness, leaf water content and SPAD-502 readings in six Amazonian tree species. Photosynthetica **47**(2), 184–190 (2009)

Richardson, A.D., Duigan, S.P., Berlyn, G.P.: An evaluation of nonivasive methods to estimate foliar chlorophyll content. New Phytol. **153**, 185–194 (2002)

Uddling, J., Gelang-Alfredsson, J., Piikki, K., Pleijel, H.: Evaluating the relationship between leaf chlorophyll concentration and SPAD-502 chlorophyll meter readings. Photosynth. Res. **91**(1), 37–46 (2007)

Wang, Q.B., Chen, M.J., Li, Y.C.: Nondestructive and rapid estimation of leaf chlorophyll and nitrogen status of peace lily using a chlorophyll meter. J. Plant Nutr. **27**, 557–569 (2004)

Small-Scale Soil Database
of Jilin Province, China

Xiuli Si[1(✉)], Guifen Chen[1(✉)], and Weiwei Li[2]

[1] JinLin Agricultural University, Changchun 130118, China
adminsxl@163.com, guifchen@163.com
[2] China Greatwall Computer Shenzhen CO., Ltd., Shenzhen 518052, China
1dd604@126.com

Abstract. Choosing the example of Jilin province, the construction method of small scale soil database is discussed; the materials gather from the second soil survey projects are digitized; Jilin province 1:1000000 small scale soil database are established. Commonly, soil database includes the three parts of the spatial database, the attribute database and the metadata database. In the end of this paper, the application prospect of soil database is forecasted; the shortage of existing research results is discussed. The method would provide reference for the construction of small scale soil database in other provinces and regions.

Keywords: Small scale · Soil · Database

1 Introduction

The study of soil resource database has been performing abroad since 1960s. In 1966, US soil protection division (SCS) established a new, comprehensive national soil information system NASIS. Later, in 1971, Canada developed CanSIS which starting from 1963. Then United Kingdom, Australia and other developed countries established their individual soil database.

Although launch researching soil database relatively late in China, it develops rapidly in recent years. Soil workers began to attempt the construction of soil database from the mid-1980s. In 1986, soil erosion information system was developed by remote Sensing Center of Peking University, which firstly completing the construction of regional soil erosion information database in domestic. Besides this, under the aid of Chinese Academy of Sciences *Tenth Five-Year Plan*, Nanjing Soil Research Institute built China soil database and established China soil information system SISChina [1, 2].

In this paper taking Jilin small scale soil database as an example, the construction method is introduced, the building process of soil database is explored, and the problems existed are analyzed. Expect providing reference for relevant departments and fields while using soil database.

2 Overview of Research Area

Jilin province lies in the central part of northesten China, is located in the east longitude 122–131, 41–46 degrees north latitude. Jilin province consists of eight prefecture-level cities(including Changchun, Jilin, Siping, Songyuan, Baicheng, Tonghua, Baishan,

D. Li and Z. Li (Eds.): CCTA 2015, Part I, IFIP AICT 478, pp. 239–245, 2016.
DOI: 10.1007/978-3-319-48357-3_23

Liaoyuan) and one autonomous prefecture(Yanbian). It has a total area of 190,000 km². In which, agricultural land accounted for 21.1 %, forestry land accounted for 48.6 %, animal husbandry land accounted for 8.1 %, fisheries land accounted for 3.4 %, and other land accounted for 18.8 %. And has a total population of 27.3 million, agricultural population occupies 50.34 % above, 13.49 million.

Jilin has the temperate monsoon climate and the obvious continental nature. Southeast is high, northwest is low and the Midwest is broad plain. The terrain is tilted from southeast to northwest, is characterized by obvious southeast and low northwest.

The total of province is divided into two first grade landform areas, the west is the subsidence based Songliao plain depression area, the eastern part of the uplift and erosion based Laoyeling Changbai Mountain uplift zone including medium mountains, low mountains, hills, platform and plain 5 kinds of landforms. In 1979 the second soil survey, the interim plan was formulated, and the total soil was classified into 19 soil categories, 59 sub categories [3].

3   Soil Data Analysis

The soil database is composed of spatial database, attribute database and metadata database, which concerns a variety of data. These can be divided into spatial data, attributes data and metadata by type [4], also can be divided into the census data and research data by source. There are the second Jilin province soil survey *soil Zhi* 55 copies (city, county), small scale maps and Jilin province second soil survey thematic research bulletin that China science and technology press in 1991.

3.1   Spatial Data

From the Table 1, it is clear that spatial data includes 1:500000 scale geological maps, 1:500000 scale topographic map and small scale other thematic maps of the second soil survey. This covers special soil type, parent material, nutrient index, trace element, and modified partition [5].

Table 1. Spatial data

Name	Scale	Create time	Source	Format
Jilin geological map	1:500000	1989	Jilin Bureau of Surveying and Mapping	Paper
Jilin topographic map	1:500000	1989	Jilin Department of Geology and Mineral	Paper
City/Country soil maps	1:500000	1983–1985	City or country soil survey office	Paper

3.2   Attribute Data

Refer to *Jilin Soil* and other 55 soil books (province, city, district and county), with the final classification system of the second Jilin soil survey as the standard, original census classification and code name are retained in database, and the national standard of soil classification is added (Table 2). Moreover, the topography, geology and other types of small and medium scale spatial database has been completed in China. In order to avoid data duplication of collection we should pay attention to priority on using of digital products.

Table 2. Attribute data

Name	Scale	Create time	Source	Format
Jilin soil map	1:1000000	1991	Soil and Fertilizer Station of Jilin Province	Paper

3.3   Spatial Metadata

It describes the contents, quality, representation, spatial reference and management of data, and achieves the dual purpose of effective management and reasonable sharing via providing geospatial information. For data producers, on the one hand, could release the geographic spatial data efficiently, maximize play the value of existing geospatial information, and facilitate understanding of the relevant information. On the other hand, according to the contents of the spatial metadata could effectively manage and maintain the existing geospatial data, so as to prevent the influence caused by staff changing and improve the data availability. For the data exchange center at all levels, realizing the fast searching and accurate positioning of the geospatial information based on the spatial metadata that data producer provided, and achieving data sharing last [6].

Metadata carries maximum convenience for data users querying and retrieving effective geospatial data, understanding the existing geospatial data, choosing the appropriate geospatial data, and promoting the reuse of geospatial data. Spatial metadata refers to national geographic information metadata standard.

4   Construction of Soil Database

Establish soil spatial database, property database and metadata database with ArcGIS 10 under the condition of soil data acquisition. Connect property data and metadata data with the key fields, and finally form soil database.

4.1   Construction of Spatial Database

Choose ArcGIS 10 as the database and spatial information processing platform. Image maps are vectored after scanning with the built-in ArcScan and edited with ArcMap, and metadata is processed (inputting and editing) by ArcCatalog, spatial data was finally stored in Geodatabase format.

4.1.1   Map Pretreatment and Scanning

Choose well preserved paper maps, if there is deformation, to preprocess the paper maps prior to scanning. It depends on the clean condition of the surface, remove the stains, clean the elements of the map, paint, which all to reduce the error of the later processing and avoid rework or repeat construction [6]. Select appropriate scanning methods and parameters on map medium and map element while scanning, province, city, county, district, a total of 55 soil diagram sequentially scanned into the computer. While scanning must pay attention to keep supply level of paper angle, maps and a horizontal line angle does not exceed 0.2 degrees, the scanning resolution shall not be less than 300 dpi, image scanning should be clearly and correctly reflect the design elements, the deformation is larger amplitude block scanning, unqualified image should be re scanned.

4.1.2   Map Mosaic Registration

Mosaic and register maps using ArcGIS after scanning, formed soil maps have the unified parameters. Subsequently, open the soil maps and see the loading image content, register images using georeferencing function of ArcMap and add control points. Increase the number of control points when the deformation errors exceed the limit, Then click the Auto adjust button to correct the images, and click the save button to save the control points to the file, and load the control point from the file. Finally save and generate the sampling data, click the rectify toolbar to correct and raster to generate a new raster file.

4.1.3   Element Vector

According to the contents and characteristics of soil type map, the various factors are divided into point thematic elements, linear thematic elements and areal thematic elements (Table 3). The elements are vector based on the layers with the interactive method. Vector elements in the process of constructing the spatial database, soil map digitalization is tedious, time-consuming, and is also the most important, mainly including the extraction of soil boundary, soil type note notes and profile position coordinates information. In order to meet the requirement of precision, errors are corrected by the automatic generation of standard frame theory module on ArcGIS, also are eliminated in spatial data acquisition and record.

Table 3. Vector

Type	Content
Point thematic elements	Section point, height point
Linear thematic elements	Soil classification wiring, highway, railway, river
Areal thematic elements	Soil type maps, reservoirs, lakes, and residential areas

4.1.4   Data Check

In order to guarantee the quality of database, the vector results must be checked. Data checking is divided into two parts, routine checking and topological checking. Routine checking is mainly checking the vectorization results and scanning the original maps.

And checking wrong painting automatic digitization brought, screen vectorization leakage phenomenon, confirm the original fuzzy elements, and fine correction of picture mosaic, correct figure line missing, note record does not match the other errors, added an important feature information and each element attribute values are evaluated to confirm whether the error. Topology check is to check and revise several of topological errors of vector data, and finish related face shape element topology construction, and check the consistency of pattern spot attributes, at last forms Jilin province's small scale soil spatial database.

4.2 Construction of Attribute Database

Soil attribute data mainly from the soil records of counties and districts and reference to the soil data in the provinces, the municipal soil annals and other documents. The attribute data is stored in two ways of the pattern spot attribute and attribute table. Each map layer corresponds to a particular two-dimensional attribute table for the storage of spatial and non spatial attribute information of this layer entity. As to the information that map surface difficult to express, such as layered soil, bulk density, texture and mechanical components. These information all obtained from the soil records and other related materials. Attribute data table is established [7, 8].

4.2.1 Soil Attributes Data

Soil attribute data are supposedly based on the second soil survey of Jilin province, mainly including soil profile data (soil profile of traits), nutrients (soil and fertilizer) data, background data (in the study area over the years of agricultural production, study area in recent 20 years meteorological data and society, depending upon the type of economic, demographic information). It will be seen that the basic mapping units are soil properties and soil species on Jilin 1:100000 soil maps, soil properties and soil analysis data are classified which refer to Jilin province local records and regional soil records. Preserving the original census classification name and coding, and increasing the national standard soil classification in database. Considering the possibility of errors in soil annals and soil map, attribute data must be tested first.

4.2.2 Attribute Data Coding

In order to achieve the efficient management of data information, the soil type name and other information must be coded while establishing the soil database. Attribute data can provide three kinds of classification query, that is, the county census classification, the provincial classification and the national standard classification, so as to establish the traceability relationship of different soil classification system. The coding of profile attributes is composed of two parts, that is, the coding of the section sample points and the information coding of the physical and chemical properties of the profile. Other statistical information is coded according to relevant national requirements.

4.2.3 Construction of Attribute Database

According to the types and features of attribute data, design field name, field type, build attribute table structure, and organize prepared attribute data into tables.

Build attribute data's E-R model according to connections and restricted conditions between various entities and their attributes on the attribute data while building the attribute database, design the organizational structure of tables for each entity, include field names, field type, field length, etc. Reduce data redundancy as possible, maintaining the consistency of data in database usage and maintenance. The organization of soil attribute data is of the same importance as the soil map digitization. Because of the attribute of the section and the data entry is more complicated, which easily bring out errors while inputting data. Therefore checking process is very important. The soil attribute data entry includes the profile data and the data in table. Among them, the typical profile is mainly from the county and the district soil.

4.3   Construction of Metadata Database

Metadata is the data describes data, mainly for describing data, data quality, data source, spatial reference, access and access mode, etc. Soil database metadata can be for data producers to provide the appropriate data, to facilitate understanding of the relevant information, so as to effectively save, management and maintenance of the data and prevent the influence caused by the change of data production staff, improve the data availability; on the other hand, soil metadata is conducive to user data correctly and quickly to query the data retrieval, better understanding of the data content, promote soil data sharing and exchange. In the construction of small scale soil database, the spatial database and attribute metadata database completed by referring to national geographic information metadata standard.

4.4   Construction of Soil Database

Establishment of soil database, spatial database and attribute database are indispensable. Without attribute data, the soil database established can only show soil types in the province's spatial distribution, and each physical and chemical properties of soil type not are stored in soil database.

Regional differences and complexity of data type leads to various kinds of corresponding relations between soil mapping unit and attribute data, which bring new difficulties realizing the connection according to a key field. With the increase of the scale, soil patches number also showed orders of magnitude increase.

Metadata database, adopt a "county (city, district) + soil type + digital" approach to spatial association pattern and attribute data.

First, analysis the layer metadata, obtain area ranges and properties field, classify mapping unit and property data via field searching according to the county (city or district). Then search corresponding map spot number and attribute table respectively in the spatial database and attribute database according to the soil type, and record the corresponding relations, so as to realize the one-to-one correspondence of the attribute list attribute record and spatial pattern unit. The tables of Jilin soil database includes surface tillage bedding characters statistics, typical profiles of attribute tables, statistical profile attribute data table, table query classification, distribution of soil, and taxonomic classification of reference.

5 Conclusions

The data might be out of date since the second soil survey carried out more than thirty years ago and great changes have taken place in the current land use and soil quality. In addition, the errors on paper diagram under the influence of drawing level and preservation methods are unavoidable, for example, pattern line deletion, pattern note omission, different drawing boundaries dislocation and other.

The errors on paper soil map were corrected in process of building database, the quality of digital map compared to the original soil maps have greatly improved. Meanwhile, supplementary terrain, geology, land use and other information also makes database practicality has been enhanced. However, in the current database and the results of the application level, there are still some problems worthy of discussion and analysis. Soil database as a basic achievement, its application prospect is broad and related to agricultural, forestry, ecology and other disciplines and fields have a greater demand [9–11]. It is necessary that carry out the new soil survey and update soil patches and attribute data, both based on the new soil database, and refer to the second land survey and the cultivated land fertility survey results.

References

1. Lei, Q., Zhang, R., Xu, A., et al.: Construction and development direction of digital soil in China (in Chinese). J. Soil Sci. **41**(5), 1246–1251 (2010)
2. Zhang, W., Zhang, R., Xu, A., et al.: Development of China digital soil maps (CDSM) at 1:50 000 scale (in Chinese). Sci. Agric. Sinica **47**(16), 3195–3213 (2014)
3. Soil, J.: Soil and Fertilizer Station of Jilin Province. China Agriculture Press, Beijing (1998). (in Chinese)
4. Chen, G., Li, X., Ren, H., et al.: Establishment of the soil database of Laixi City based on MAPGIS (in Chinese). Bull. Soil Water Conserv. **4**, 168–171 (2009)
5. Wu, J., Hu, Y., Zhi, J., et al.: A 1:50000 scale soil database of Zhejiang Province, China (in Chinese). Acta Pedol. Sinica (1), 30–39 (2013)
6. Jing, C., Zhi, J., Zhang, C., et al.: Construction of medium and small scale soil geographic data bases, Zhejiang Province, China (in Chinese). Bull. Sci. Techn. (11), 99–105 (2012)
7. Shi, X.Z., Yu, D.S., Warner, E.D., et al.: Soil database of 1:1000000 digital soil survey and reference system of the Chinese genetic soil classification system. Soil Surv. Horiz. **45**(4), 129–136 (2004)
8. Kening, W., Yang, F., Qiaoling, L., et al.: Construction and application of 1:200000 soil data base in Henan Province (in Chinese). J. Henan Agric. Sci. **5**, 77–80 (2007)
9. Shi, X., Yu, D., Gao, P., et al.: Soil information system of china (SISChina) and its application (in Chinese). Soils **39**(3), 329–333 (2007)
10. Panagos, P.J., Liedekerke, M.V., Montanarella, L.: Multi-scale European soil information system (MEUSIS): a multi-scale method to derive soil indicators. Comput. Geosci. **15**(3), 463–475 (2011)
11. Zhou, H.: Sharing of soil information data distributed inquiry data base of 1:4 M soil information of China (in Chinese). Acta Pedol. Sin. **39**(4), 483–489 (2002)

Agricultural Environmental Information Collection Device Based on Raspberry Pi

Baofeng Su[1(⊠)], Linya Huang[1], Zhen Gao[2], and Jiao Guo[1]

[1] College of Mechanical and Electronic Engineering,
Northwest A&F University, Yangling 712100, China
bfs@nwsuaf.edu.cn, hlyl552187701@126.com,
jiao.g@163.com
[2] School of Mechanical Engineering, Shandong University,
Jinan 250061, Shandong, China
sdugaozhen@gmail.com

Abstract. Integrated agricultural environmental information real-time collection, transmission and management is extremely critical for precision agriculture (PA). This paper describes the basic knowledge and primary principles used in an agricultural environmental information collection device. The device is based on Raspberry Pi, combining with GPS module, some digital sensors and analog sensors to measure environmental temperature and humidity, barometric pressure, light intensity, soil moisture and other environmental information accurate collection. All collected data was real-time transmitted to a remote server specified database for management. In the experiment process, the device has been proved to have the ability to catch the distributed information and also can get different centralized data management. The results show the integration of operational information collection, transmission and management, as well as the use of open source software to make it easy to collect multiple types of data parallelism, which provides an important solution for the rapid transformation of agricultural production.

Keywords: Agricultural environmental information · Integration · Open source · Raspberry Pi

1 Introduction

Efficient collection and scientific management of farmland environmental information farmland for information technology has great significance [1]. Efficient collection of environmental information is beneficial to have timely access to agricultural information, to meet the diverse needs of farm data information, and more comprehensive and accurate grasp of environmental development. And scientific management is essential for post-processing of information. Through scientific management of information, and we can normalize all kinds of scattered and mixed messages, which is conducive to statistics and data classification. Comprehensive use of agricultural information technology will become an important means of twenty-first century rational use of agricultural resources, improve crop yields, reduce production costs and

D. Li and Z. Li (Eds.): CCTA 2015, Part I, IFIP AICT 478, pp. 246–254, 2016.
DOI: 10.1007/978-3-319-48357-3_24

improve competitiveness in international markets of agricultural products. However, our positioning on the use of GPS technology to collect field information technology than in other countries started late, most information acquisition devices' detection and management technology are decentralized, while the collection of data immutable, which restricts the acquisition of real-time equipment, effectiveness and flexibility. Meanwhile, most parts of China are still using manual collection of agricultural information. This method has many limitations that cannot meet the sophisticated needs and the development of agriculture, such as the poor timeliness, information collection and intelligent digital low, poor visualization, and low level of shared management of the effects of data.

For agriculture, farm environmental information collected intelligence requires integrated information collection, transmission and management in order to achieve real-time and efficient gathering information and science, improve the management of data [2]. There have been some international research institutions and researchers with experience of successful experiments. Vieira et al., on the Amazon River Basin for environmental information, the development of a monitoring node, and completed by WSN transmission of environmental information [3]. Yang et al. studied the measurement and transmission test for soil moisture [4]. Xin et al. studied on monitoring network for crop growing environment [5]. Throughout the study of these scholars we find that their research focused on a microcontroller as the core function of simple farmland information collection device, such as GPS positioning technology combined with measurement of soil moisture, soil nutrients, planting density and other parameters. They finished just one-sided information collection work, such as a single with a measurement and use manual records or indirect transfer remote storage after collection. Most of the study only focused on access to information unilaterally, and the integration of information collection, transmission and management did not materialize. This type of data acquisition equipment is characterized by the development of a low cost, single function and the need for complex post-processing data.

Currently, with the development of open source hardware equipment and electronic technology, Raspberry Pi increasingly showing its superiority. The Raspberry Pi's appearance and behavior provides a new stage for the research platform of environment monitoring and management. Raspberry Pi with its simple operation, high processing speed and a variety of interface get a lot of attention to research staff. Slawomir introduced the Raspberry Pi as a measure to control the use of cell [6], Sheikh studied the environmental monitoring system based on the Raspberry Pi and Arduino [7], Vladimir and other researchers used the Raspberry Pi to achieve a smart home network node design [8]. However, on the use of agricultural information collection Raspberry Pi research are rare.

This paper proposes a farm environment information collection device, and designs an agricultural information collection device based on Raspberry Pi. Raspberry Pi equipped with a variety of sensors for many types of parallel acquisition and transmission of information, data acquired by the built-in memory card Raspberry Pi for online storage or sent via GPRS module. In practice, offline maps can be saved in the Raspberry Pi, so users can watch the current environmental information in real time, improving flexibility. The multi-class data collection procedures was controlled by python that open source software. The device can store information or online transmission, according to the needs of users, for database management information, namely, the achievement of the integrated operation of information collection, transmission and management. These features improve the timeliness, reliability, convenience and flexibility of the device, so that it can solve most of the current information collection problems encountered in the process of agricultural production.

2   Experiments and Methods

For the agricultural information collection device, which through digital sensors for measuring air temperature and humidity, barometric pressure, soil moisture and light intensity and other environmental information accurate collection. The information collected after treatment Raspberry Pi, the user can select the data storage as needed or sent online. Data is stored in the processor's built-in selectable storage memory or external memory card, online delivery is to the C/S model will be sent to the remote server that stores data via GPRS module. Meanwhile, the device can achieve human-computer interaction, user-friendly command displays the input and farmland environmental information. The entire device use Raspberry Pi achieve environmental information farmland acquisition, processing, storage and delivery of integrated operations, as well as store location map. The main hardware components of the device are Raspberry Pi, information collection module, a wireless network communication module and the auxiliary module. Wherein the auxiliary module is touch input and display modules, scalable data storage module and power supply module.

2.1   System Hardware Design

Raspberry Pi. "Raspberry Pi Foundation" developed the Raspberry Pi and which is a registered charity in the United Kingdom. Raspberry Pi is a card-type computer, and its size is only 85.6 × 53.98 × 17 mm [9]. As of April 2015, Raspberry Pi Foundation has released a total of four Raspberry Pi, respectively, Raspberry Pi Model A, Model B, Model B + and 2 Model B. The type B Raspberry Pi is the choice of this study, the core module of this system is produced Broadcom's BCM2835, and its memory is 512 MB. SD card can be used as the storage medium. It integrated GPIO, I2C, UART and SPI interfaces, and it can connect a variety of sensors to collect environmental information farmland [10]. The type B Raspberry Pi can run on many Adopt SupNIR-5700 NIRS (Focused Photonics (Hangzhou), Inc.) to collect NIR spectra of all samples. Spectral measurement of samples uses random RIMP software and its testing method is: transmission, measurement range: $1000 \sim 1800$ nm, scanning speed: 10 times/s, spectral resolution: 6 nm, temperature of sample cell: 60 °C, testing method: load the sample into the three-quarters of sample bottle, and then place the sample bottle into the sample cell. Stabilized in constant temperature for 5 min, the bottle is taken out to check if there exist bubbles. It starts to collect spectrogram if there is no bubble, and each sample averages out three times. Use NIRS random RIMP software and MATLAB7.8 to collect spectra and convert data format, use chemo metrics software

Unscrambler X 10.1 to pretreatment the spectral data and analyze principal component, and use SVM pattern recognition and regression software package designed by a prof. Lin from National Taiwan University to build SVM models in MATLAB7.8 and parameters optimization.

Farmland Environmental Information Collection. What the collection point of object farm setting information collection module collected mainly include latitude and longitude, altitude, barometric pressure, air temperature and humidity, soil moisture, light intensity collection points and other factors. The latitude and longitude, altitude information collection point by Ubox NEO-6 M GPS module connected via UART interface; The GPS location information module serial NMEA-0183 output format uses standard communication format. The output data is in ASCII code, contains the latitude, longitude, altitude, speed, date and other information.

The device collects air temperature and humidity information through the AM2301 digital temperature and humidity sensor. Its transmission distance is up to 20 m or more. The device collects pneumatic pressure through the GY-65 digital pressure sensor, and is connected via I2C interface to communicate with the Raspberry Pi. GY-65 sensor is a high-precision, low-power digital pressure sensor. It built-in chip is BMP085, and its measurement range of $300 \sim 1100$ hPa. It does not require an external clock circuit, and it built-in temperature compensation, it is possible to reduce the influence of environmental factors. In this study, we collected light intensity by GY-30 sensor, which is connected via I2C interface to communicate with the Raspberry Pi. GY-30 light intensity detection sensor uses BH1750FVI chip, and the illumination range is $0 \sim 65535$ Lx. The sensor does not distinguish between ambient light, which has close to the visual sensitivity of the spectral characteristic. Soil moisture information collected using YL-69 sensors that standard single-bus interface. Soil moisture data detection module uses a D/A dual output mode. Its sensitivity is adjustable, and its comparators work stable LM393 chip. Agricultural Environmental Information collected for processing information in a central Raspberry Pi of the device, and the information in real time by GPRS module is sent to the user.

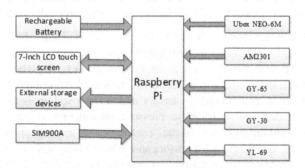

Fig. 1. Hardware framework

Auxiliary module of the agricultural information collection device mainly includes a touch input display module, scalable data storage module and power module. Touch input display module as a 7-inch LCD touch screen, and to support their daily work by

5 V supply. Help users operate and collect information collection terminal display. Scalable data storage module for Raspberry Pi can add removable storage devices, and users can capture information stored in the internal memory or a Raspberry Pi terminals need to select the external memory. 5 V voltage power supply module provides rechargeable battery or mobile power, providing power to support the normal operation of the device. The farmland hardware environment information collection device overall block diagram shown in Fig. 1.

Wireless Network Communication. For the wireless network communication module, GPRS module can be used. GPRS is the abbreviation of General Packet Radio Service. It is a global mobile phone system (GSM) based data transmission technology. GPRS support and directly connected to the IP network, it is possible to provide SMS, MMS and other services. GPRS network short access time, transmission efficiency is higher than the GSM, theoretical bandwidth can reach 171.2 Kbit/s. GPRS network supports standard data communication protocol applications that can interconnect with IP networks, X.25 networks, point to point and point to multipoint service support. Users can freely move their distribution and network points, for IP communication desired location. This device is mainly applied to Internet-enabled GPRS module. In this device, we use SIM900A module to achieve the requirements of the Internet by APN.

2.2   Application System Developing

The central processor of the present study is the type B Raspberry Pi, built a Linux system to support python, Java, C and other programming languages. This study is based on Python programming. Python is an interactive, object-oriented, dynamic semantics and syntax beautiful scripting language. Python is an efficient development tool that supports multiple operating systems, with a free open-source, high-level language, portability, object-oriented, embeddable scalability characteristics. In this study, the device uses the python programming language, to achieve a sensor of farmland environmental information collection, collection terminals to communicate with a remote server socket based C/S mode, the operation of the database.

Information Collection Sensor System. In recent years, Dallas Semiconductor has introduced a unique single-bus technology. Single bus widely used for single-host system, which can control one or more slave devices, data exchange between them only through a signal line. For sensor collection of farmland environmental information, AM2301 temperature and humidity sensors and YL-69 soil moisture sensor to air temperature, humidity and barometric pressure information via a single bus and Raspberry Pi communication with the control easy, expansion of convenience advantages. GY-65 GY-30 pressure sensors and light intensity sensor used to collect air pressure and light intensity information. For parameters characteristic of the collection, the device uses an I2C interface to communicate with the Raspberry Pi. This bus developed by the PHILIPS two-wire serial bus is widely used in the microelectronics field communication control a bus standard. It is used to connect the microcontroller and its peripherals for the 2-wire synchronous serial communication with the interface cable less, simple control mode, a smaller device package, higher communication speed and so on. For Ubox NEO-6 M GPS module, it is sent via UART communication interfaces and raspberries. A serial interface allows data to order a transfer, which is characterized by simple communication lines, and as long as the pair of transmission lines can be two-way communication, thereby reducing the cost of the equipment. Information collected scheme is shown in Fig. 2.

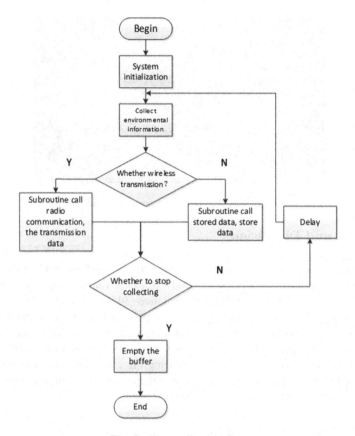

Fig. 2. Data collection flow

Database Operation Programming. This study was based on PostgreSQL database. PostgreSQL is California, Berkeley, developed with PostgreSQL version 4.2-based object-relational database management system. PostgreSQL uses a more classic C/ S (Client/ Server) structure, with good scalability, is open source free software. Python PostgreSQL database for the operation is done by Python psycopg2 module.

Data Collection Terminal Socket Development. The device implementation study based wireless network communication C/S (Client/Server) architecture. In the application the client/server model, the client generally based embedded computer processor core units; the server is generally user-created data center. In this mode, the data center server as a user, you need to have a fixed IP address or domain name address using the fixed conversion dynamic DNS, so the client can easily obtain the IP address of the server in order to establish a TCP connection. The system to collect terminal for the client to the remote server to server-side, to achieve collection terminals communicate with a remote server using python programming socket. The type B Raspberry Pi and sim900A wiring shown in Fig. 3.

Fig. 3. The type B Raspberry Pi connection with sim900A

3   Results and Discussion

Experimental testing of the system is carried out in a wheat field, located in the neighborhood of college of Mechanical and Electronic Engineering, Northwest A&F University, Yangling District of Shaanxi Province (34°17'33.22"N, 108°04'11.8"E). Before the experiment make Yangling district maps stored into the Raspberry Pi. Using collection terminal operator in the field of environmental information collected and sent via GPRS module. The operator chose a more open testing ground for the trial subjects, and placed the sensor in the fields. After starting the collection terminal, initialize the system, manual collection touch button, select the collection of information stored or transmitted. Remote server data store as shown in Fig. 4.

	east_longitude numeric(9,6)	north_latitude numeric(8,6)	altitude numeric(4,1)	atmospheric_pressure numeric(5,2)	air_temperature numeric(3,1)	air_humidity_percentage numeric(3,1)	illmination_intensity_Lux numeric(5,2)	soil_humidity_percentage numeric(3,1)
1	108.070018	34.291790	524.4	959.09	19.9	66.9	297.50	66.5
2	108.070107	34.291563	525.7	959.10	19.8	66.8	297.65	66.7
3	108.070025	34.292012	526.4	959.08	19.9	67.1	298.32	66.5
4	108.070053	34.291852	524.5	959.11	19.9	66.5	297.60	66.5
5	108.070020	34.291821	527.1	959.09	19.8	66.5	297.61	66.6
6	108.070031	34.291780	525.3	959.08	19.8	67.0	297.65	66.7

Fig. 4. Data storage on remote server

Results show that agricultural environmental information collection device in this paper can effectively collect real-time environmental information farmland and the ability to select the storage or sent to the database server's designated according to user needs. The device has collected information accuracy and short cycle advantages. Each of which will perform data collection, transfer and management, to achieve the integration of agricultural information collection operation.

4   Conclusions

This paper designed a farm environment information collection device. The Raspberry Pi system as the core, combined with multi-sensor and GPS module for real-time acquisition of farmland environmental information. It does this by using GPRS module socket communication mechanisms to achieve the wireless transmission and storage of information. It gives Python operation of each sensor, GPRS module and PostgreSQL database program implementation. The device realizes the accuracy of the information collection, scientific management and dynamic distribution requirements. The device uses the connection Raspberry Pi and information collection, transmission and transmit modules to achieve the integration of information collection, transmission and management. It meets the accuracy of the information collection, scientific management, dynamic distribution requirements, and have real-time high, convenient and reliable, low power consumption advantages.

Acknowledgment. The authors would like to thank the anonymous reviewers for their critical and constructive review of the manuscript. This study was supported by the National Natural Science Foundation of China (No. 41401391), the Fundamental Research Funds for the Central Universities of China (No. 2014YB071, No. Z109021423), and the Exclusive Talent Funds of Northwest A&F University of China (No. 2013BSJJ017, No. Z109021108).

References

1. Yuan, W.S., Zhang, H.L.: The development status, existing problems and countermeasures of agricultural information construction in China. In: Proceedings of 2008 International Conference on Information, Automation and Electrification in Agriculture, pp. 46–51 (2008)
2. Wu, J., et al.: Environmental issues in China: monitoring, assessment and management. Ecol. Ind. **51**, 1–2 (2015)
3. Vieira, R.G., et al.: An energy management method of sensor nodes for environmental monitoring in Amazonian Basin. Wirel. Netw. **21**(3), 793–807 (2015)
4. Yang, J., et al.: Integration of wireless sensor networks in environmental monitoring cyber infrastructure. Wirel. Netw. **16**(4), 1091–1108 (2010)
5. Xin, Z., et al.: Study on wireless-monitoring technology for crop growing environment. In: 2008 Proceedings of Information Technology and Environmental System Sciences: ITESS 2008, vol 3, pp. 953–959 (2008)
6. Michalak, S.: Raspberry Pi as a measurement system control unit. In: 2014 International Conference on Signals and Electronic Systems (ICSES) (2014)
7. Ferdoush, S., Li, X.R.: Wireless sensor network system design using Raspberry Pi and Arduino for environmental monitoring applications. In: 9th International Conference on Future Networks and Communications (FNC 2014)/The 11th International Conference on Mobile Systems and Pervasive Computing (MobiSPC 2014)/Affiliated Workshops, vol. 34, pp. 103–110 (2014)
8. Vujovic, V., Maksimovic, M.: Raspberry Pi as a sensor web node for home automation. Comput. Electr. Eng. **44**, 153–171 (2015)
9. Mcmanus, S.: Introducing the Raspberry Pi. Electron. World **121**(1945), 8–9 (2015)

10. Banerjee, S., et al.: Secure sensor node with Raspberry Pi. In: 2013 International Conference on Multimedia, Signal Processing and Communication Technologies, pp. 26–30 (2013)
11. Mcmanus, S.: Downloading the Raspberry Pi's operating system. Electron. World **121** (1946), 8–9 (2015)

Design of Integrated Low-Power Irrigation Monitoring Terminal

Hongwu Tian[(⊠)], Mingfei Wang, and Jinlei Li

Beijing Research Center of Intelligent Equipment for Agriculture,
Beijing 100097, China
{tianhw,wangmf}@nercita.org.cn,
lijl@cerctia.org.cn

Abstract. In order to realize low-cost automatic control of field irrigation and water metering, an integrated low-power irrigation monitoring terminal based on wireless data communication was designed. Powered by battery or solar, the terminal could acquire and store data of sensors and water meters in real-time. Taking valve as control object, the terminal can provide safe irrigation strategy based on multiple control logic, such as data overrun, time and manual operation. The detection of fault on-site and reasonable judgment of irrigation can be carried out by using self-check function, querying of local data and remote data transmission based on GPRS and Modbus Protocol. Experimental results showed that the terminal could work stably, and control irrigation water usage accurately with low power cost.

Keywords: Field irrigation · Low-power · Multi-control · GPRS · Protect · Fault detection

1 Introduction

With continuous growth of population and economic in China, increasing pressure has been placed on the water management and conservation, and continues to drive growing demand for improving water use efficiency for agricultural irrigation and developing water saving technology in China. Irrigation automatic control technology has been widely used, and water-saving agriculture also has become agricultural development strategy for many countries [1]. China tops the list of farmland effective irrigation area in the world. In terms of water-saving automatic control, in the past few decades, China has introduced many advanced equipment and irrigation control system from abroad enterprises, including Eldar-Shanny, RainBird, Hunter, Toro, etc. However, high cost, poor compatibility with the characteristics of domestic irrigation and different user habits restricted its large-scale application [2–4]. Agricultural water-saving information technology research was started in the 1950s and great progress was achieved in China.

Fund Project. National High-tech R&D Program of China (863 Program) (2011AA100509). The Ministry of Agriculture of China 948 Program (2011-G33). Special Fund for Agro-scientific Research in the Public Interest (201203012-4-1).

D. Li and Z. Li (Eds.): CCTA 2015, Part I, IFIP AICT 478, pp. 255–265, 2016.
DOI: 10.1007/978-3-319-48357-3_25

Intelligent decision control model and control system were constructed based on information fusion technology. Jiangsu University developed winter wheat precise irrigation intelligent system based on fuzzy decision theory; Tianjin Normal University of engineering constructed soil water potential intelligent decision model and support system; National Engineering Technology Research Center for Information Technology in Agriculture developed water-saving irrigation automatic control system, which could realize water pump and valve automatic control [1]. Those systems and applications mentioned above required the users to be professional, which hindered the promoting of information system. The No. 1 central file released by GOV.CN in 2011 put forward "Three Red Line" management system, which made specific limits on water use quantity, water use efficiency and in water pollution in function area. However, the actual situation is that stations for agricultural water information monitoring are concentrated and lack in number, which results in difficulties to accomplish real-time monitoring [5–9]. To some extent, it has limited the development of information technology applied to the agricultural water-saving management.

A low power cost field irrigation control terminal was designed in this paper. Different application mode could be built, and a variety of irrigation control mode, information collection, storage and download function were also developed. Meanwhile, the introduction of simplified menu design, online data query and equipment local online self-check function could greatly help users with the controller used improve the efficiency of equipment maintenance, which made the developed terminal easier to be promoted.

2 Hardware Structure

Considering that the equipment would be practically used in open field, where the stable utility power lines are not available, the terminal was powered by battery or solar. MCU C8051F964 was selected as its core chip, which was a high rate and low power cost single chip microcomputer of high-performance. The sleep-mode power consumption of this chip could be as low as 700 nA. A built-in ultra-low power consumption clock provided the equipment with timing awaken resource. Two LDO served as independent power for main chip systems and peripherals. Independent external clock calendar served for equipment basic time and equipment operation. All the peripheral devices, including sensor, solenoid valve control, USB and Bluetooth, GPRS module, LCD display were controlled separately by power switch. Communication part adopted the design of plug and play mode, and also reserved interfaces for standard RS485 BUS and GPRS module, which guaranteed the extending ability for networking functions, data remote transmission and local download function. The framework of the proposed system hardware and picture are shown in Fig. 1.

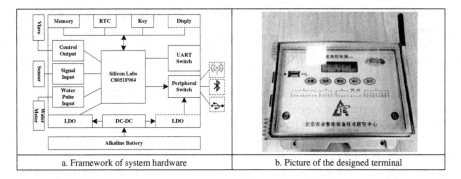

| a. Framework of system hardware | b. Picture of the designed terminal |

Fig. 1. Framework of system hardware and picture

3 System Function Design

The function was designed based on the requirements of end user, equipment practical applications and the possible problems that might arise in practical use. The reliability, convenience, protection and fault tolerance were all taken into consideration, in order to improve the performance of the equipment.

3.1 Data Acquisition Function

3.1.1 Sensor Data Acquisition

Sensor information acquisition, especially the soil moisture [10], is indispensable function in irrigation monitoring and control system. The obtained data can not only reflect real-time measurement of soil moisture, but also can be used as a reference index to guide irrigation control. Considering the particular occasions, there would be a sensor damage or malfunction, especially the short circuit fault. Therefore, in the sensor interface design, power independent control and OCP (Over Current Protection) were used, which meant power was activated only when data sampling happens through each channel and shut down for the rest of time. If voltage or current value over range was detected, the device would automatically cancel the measurements, shut off the power, and prompt error message, such as display hints or text message.

Using the proposed design, it can not only reduce the power consumption of the equipment running status, but also can avoid long time short-circuit fault which might lead to paralysis.

3.1.2 Water Meter Data Collection

In view of the increasingly serious situation of water resources and the present situation of the agricultural water inefficient use, in addition to irrigation control, accurate measurement of water consumption is equally important. The water use information can be used to help with water management. The irrigation water use information, including timing of data acquisition, water use accumulation in a certain area, water use accumulation of a specific valve, and the density of water use in different period of time could be analyzed, which could provide the basis for realizing effective water management [11].

The equipment has 4 channels of remote meter signal acquisition interface of impulse type. Using MCU's pulse detection and level matching function, it achieved the real time measurement of water use. The operating principle is: two continuously impulse generated by water meter separately means the preset smallest unit. In practical use, external interference signal jitter could produce continuous single pulse on one signal wire. Low pass filtering in MCU could effectively prevent the happening of such error. Port match level flip function could also prevent the water meter signal from deadlock, which might cause program to enter infinite loop. The combination of both protection procedures could ensure water pulse measurement accurate and reliable.

3.2   Record Store and Download

Combined with Sect. 2.1, using nonvolatile memory of external expansion, data record, store and download functions were developed for the logging system. The record types were divided into S (sensor data), T (time water record), and Q (fixed quantity of the water). In order to guarantee the consistency of the data format, fixed-length data package structure was used. Figure 2 shows the data store format.

History data could be checked locally, and downloaded through USB or Bluetooth. APP based on Android was developed to accomplish data download using smart phone. The maximum data records capacity was 8000, which meant the storage of half a year with saving interval by 30 min. All the data could be downloaded within 2 min with TXT or EXCEL file format.

Type	Year	Month	Day	Hour	Min.	Data1	Data2	Data3	Data4
1Byte	1Byte	1Byte	1Byte	1Byte	1Byte	HSB&LSB	HSB&LSB	HSB&LSB	HSB&LSB
W	0x0E	0x07	0x0F	0x0C	0x20	0x00 0x20	0x10 0x20	0x01 0x12	0x00 0x13

Fig. 2. Data store format

3.3   Data Communication

Data communication methods included: (1) RS485 bus communication; (2) GPRS wireless network communication.

RS485 communication mode was mainly used for multi-point network by external 433 MHz or 2.4 GHz wireless data transmission module. Data remote transmission was accomplished based on GPRS or SMS using embedded SIM900A module. In actual use, due to high current consumption GPRS at both starting and running period, SIM900A would be timed to start and connect. At the same time, many problems might lead to net connection failure, including module failure, SIM card error, network instability, default network interruption, server without electricity, and so on. In order to ensure the wireless communication reset function without frequent restarting the module, method by resetting times and reconnection time management was used to save power and ensure the GPRS general communication. The two communication modes both followed industry standard Modbus Protocol [12].

4   The Software Design

4.1   Task Partitioning

The main task for the software was to accomplish valve control and water management based on different logic. According to the characteristic of time benchmark for the system, the main task was divided into several subtasks, including data acquisition, communication response, key press, calendar update, data storage, updating of operation parameters, and equipment status information display. Time slice polling and software timer mechanism accomplished the main function. Figure 3 shows the main program structure.

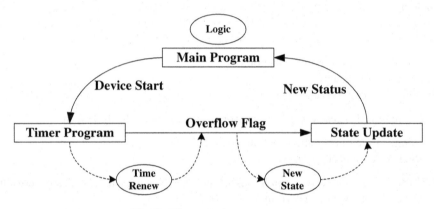

Fig. 3. The main program structure

The terms in Fig. 3 are explained as following:

Control logic: Logical judgment and mission task started according to the user preset mode.

Timing of equipment running: In the main loop, the equipment start flag would be checked regularly and was used to start the counting down timer. Equipment running status would be updated when overflow flag was detected.

Update of equipment status: According to the countdown overflow, equipment running status would be updated, and the latest status of the equipment would be used as the factor for main program logic judgment.

4.2   Control Logic

In control logic design, mainstream irrigations control methods were referenced, such as Hunter, Netafim, etc. Meanwhile, combining with the target user demand and domestic use, menu was simplified in order to make less information input by user and reduce the incidence of parameter errors in practical use. Four types of control logic, including transfinite control, timing control, quantitative control and manual control, were accomplished in the design, and protection mechanism was also taken into consideration as supplementary.

4.2.1 Sensor Limit Value Control

Control mode based on soil moisture sensor measurements has been widely used in actual irrigation control system and it could help to fill water in time while crop in water stress status. Its disadvantage is that, firstly unrepresentative install position of the sensor will greatly affect the reasonable irrigation. Secondly, sensor fault or damage would lead to decision errors, and serious consequences would happen under the extreme condition.

Aiming at the problems above, we focus on the analysis of sensor failure factors, which might lead to the measured value to be continuous high or low. Reasonable supplement irrigation and, limited irrigation mechanism were designed to avoid excessive and insufficient irrigation happening. According to empirical analysis, relatively low measured value might be caused by the following three factors:

(1) Illogical measure point selection;
(2) Pulling out of sensor;
(3) Sensor damage or disconnection.

The reason that leads to larger measured value may probably lied in:

(1) Illogical measure point selection, such as low-lying place;
(2) Sensor h connection error.

If one of the above two conditions occurs, the changing trends of soil moisture would be like the charts shown in Figs. 4 and 5.

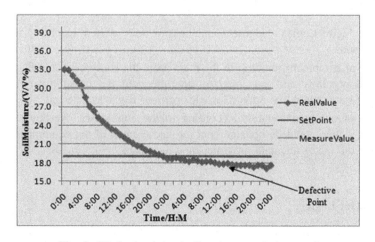

Fig. 4. "Defective irrigation" moisture variation trend

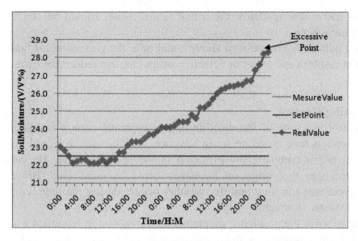

Fig. 5. "Excessive irrigation" moisture variation trend

To solve the problems above, overrun control setting parameters were designed as shown in Table 1. Besides the commonly used preset, two additional parameters, 'maximum delay time' and 'daily maximum irrigation quantity', were added. They had higher priority than other parameters. The overrun control parameters are shown in Table 1.

Table 1. Overrun control parameters setup list

Controlling condition	Parameter setup
The reference number of sensor (1–4)	S1
The reference value (%)	15.0
Returned difference (%)	3.0
Entry condition	Ultra lower limit
Maximum delay time (DD)	02
The maximum daily amount of irrigation (m³)	20

The terms in Table 1 are explained as following.

Maximum delay time: day was the smallest unit, after overrun irrigation program starting, the timer begins to countdown. If irrigation started again before the overflow of the countdown timer, the timer would reset and restart. If there was no restart during the maximum delay time, forced irrigation would be activated. Through this way, irrigation would start at least one time during the maximum delay time interval and ineffective irrigation caused by larger soil moisture value could be avoided.

The maximum daily amount of irrigation: If the sensor measurement was continuous too low to start irrigate and no action is taken, flood irrigation would happen due to the unchanging soil moisture. The maximum irrigation quantity could control the upper limit of daily irrigation water accumulation, which could avoid over irrigation.

If no water meter was installed, the actual largest water should be converted to the largest irrigation time.

The two parameters introduced above could offer the protection for fault condition which might result in excessive or defective irrigation, and reduce the extreme damage to the crops.

4.2.2   Timing Control

Timing control was one of the most common automatic control irrigation methods, efficient irrigation time could be set in advance, and the system would automatically start the valve and complete the irrigation in the specified time. The advantage of timing control was that irrigation frequency could be guaranteed. However, if the irrigation cycle was too long and dry weather occurs during the time, crop would be withered for water shortage.

In order to solve the problem, 'supplementary irrigation' mechanism was applied to ensure crop hydrate properly in the occasional extremely dry weather conditions, and to reduce the resulting losses. While 'supplementary irrigation' happening, 'the largest irrigation quantity' was also set to avoid over-irrigation in the condition of sensor error occurred.

4.2.3   Manual Control Mode

Valve control failure often happens in practical use, and human intervention irrigation is a common practice. Also users do not want the original parameters and scheme to be affected after manual control operation, such as a supplementary irrigation. For example, the user would start manual supplementary irrigation one time in dry weather. Manual control is an effective solution to solve the problems above. It was designed as a "mandatory" irrigation method which had the highest priority. All of the control logic wouldn't be unlocked until the manual control completed. Thus, it could be used in the case of valve failure. Valve could be opened for a short time and left enough time for the user to observe the valve status. Additionally, a forced irrigation time was set in advance for complementary irrigation in a prospective special time. Users do not need to close the valve manually, and the valve would be closed automatically after finishing the preset running time.

The completeness and flexibility of the system would be greatly improved through the combination of manual control, normal control logic and control mode based on priority.

4.3   Power Consumption Analysis

Internal timer and external events trigger mode were designed for the system's wake-up function. MCU could be waken-up once per minute by internal real time clock, logic judgment, valve control, data sampling and data transmission task. Key push and water pulse input were two main external wakeup sources. Sleep mode, idle mode, sensors data sampling, valve control, key pressing and GPRS data transmission were the main working statuses of the terminal. The average current consumption calculating equation is written as:

$$I_{av} = I_{mcu} + I_{sam} + I_{ctr} + I_{comm} \tag{1}$$

Where, I_{av} is Average current; I_{mcu} is MCU working current; I_{sam} is Sensor sampling current; I_{ctr} is Valve activating current; I_{comm} is Data transfer current.

$$I_{mcu} = (I_{sl}T_{sl} + I_{act}T_{act})/(T_{sl} + T_{act}) \tag{2}$$

Where, MCU's sleep current I_{sl} is 45 μA; Working current I_{act} is 3 mA; System was waken up per minute; Work time is 1 s; Sleep time is 59 s; Calculation is 110 μA.

$$I_{sam} = (N_{times}N_{num}T_{run}I_{run})/(24 \times 3600) \tag{3}$$

Where, N_{times} is Sampling frequency; N_{num} is number of sensors; T_{run} is sampling time; I_{run} is sampling current.

The parameters were set as following: the number of sensors was four; sampling cycle was 500 ms; sampling current was 30 mA; sampling interval was in accordance with 30 min; the average sampling current sensor was about 34 μA.

Assume the number of sensor is four, valve active current is 30 mA and switching impulse width is 100 ms. If each valve open and closes 10 times a day, according to Eq. (3), the average current of valve is 3 μA. Also assume that the GPRS data transmits twice a day, working current is 60 mA, active time is 10 s, the average current is 10 μA.

Therefore, according to Eq. (1), the system's average current cost I_{av} is 157 μA. If a 9 V battery with 1200 mA h is used, considering battery capacity loss and other factors, the actual utilization was calculated as 50 %, the equipment continuous working time would be 160 days which means it could be normally used for a complete irrigation season.

4.4 Self-test Mode

Self-test mode was integrated in software, in order to provide the users with intuitive online test information through relatively simple means, and quickly locate fault. The following options show the common test mode options:

(1) Channel data sampling test;
(2) Valve ON-OFF test;
(3) Channel power output test;
(4) GPRS connection test.

Users can enter test mode through menu options, select the test option, then the device enters test procedure automatically, test results or prompt information will be displayed in the LCD. User can preliminary judge fault according to the test result, and communicate with remote staff about the related message with quick fault-locating and propose solutions. It can help to improve the efficiency of device maintenance.

5 Conclusion

A low cost field irrigation control terminal is designed in this paper, in order to suit with the agriculture water use situation in China. The highlights of the proposed design are as following.

(1) Low power design based on power management and timing wakeup technology made it possible for the equipment to work for a whole irrigation season with a 9 V alkaline battery.
(2) It could not only be used as a water recorder to install on the pipe cooperating with GPRS for data remote transmission, but also as a small irrigation control equipment for automatic irrigation control in a small scale. The provided communication interfaces can be extended to multi-point wireless measurement and control system.
(3) By using simplified control logic based on 'excessive' and 'defective' protection mechanism, the reliability of equipment can be greatly improved. The developed fault location online test function, has greatly improved the efficiency of equipment maintenance.
(4) The records of water use can provide the data base for precise irrigation decision making and water management decision-making.

In general, the developed terminal is simple to install, low cost and with a good practicability and application prospect. However, when using battery as its power, the reliability and stability also need to be improved.

References

1. Chunjiang, Z., Wengang, Z.: Applications of Information Technologies in Agricultural Water Saving, pp. 1–3. Science Press, Beijing (2012)
2. Jinqiang, A., Kai, W., Liqian, W., Zengfang, S.: Internet of things based accurate automation irrigation. J. Northwest A&F Univ. **41**(12), 220–221 (2013)
3. Kranz, W.L., Evans, R.G., Lamm, F.R., et al.: A review of mechanical move sprinkler irrigation control and automation technologies. Appl. Eng. Agric. **28**(3), 389–397 (2012)
4. Cardenas-Laihacar, B., Dukes, M.D.: Soil moisture sensor landscape irrigation controller: a review of multi-study results and future implications. Trans. ASABE **55**(2), 581–590 (2012)
5. Zhu Huanli, R., Zhengbo, R.X.: Development of automatic control system for irrigation. J. Irrig. Drainage **28**(4), 124–126 (2009). (in Chinese with English abstract)
6. Shouyong, X., Xiwen, L., Shuzi, Y., et al.: Design and implementation of fuzzy control for irrigating system with PLC. Trans. CASE **23**(6), 208–210 (2007)
7. Ming, J., Qigong, C., Xingfang, Y.: Precision irrigation system based on fuzzy control. Trans. CASE **21**(10), 17–20 (2005). (in Chinese with English abstract)
8. Zhifang, C., Ni, S., Jinglei, W.: Water-saving irrigation management and decision support system
9. Zenglin, Z., Wenting, H.: Application of automatic control in water saving irrigation system. Water Saving Irrig. (10), 65–68 (2012). (in Chinese with English abstract)

10. Hengwen, W., Xidong, C., Mingyu, Y.: Burying depth of soil moisture sensors in intelligent drip irrigation system. J. Irrig. Drainage **29**(4), 16–20 (2010). (in Chinese with English abstract)
11. Haorui, C., Jiesheng, H., Gejun, Z., et al.: Research on WUA-based & total water quantity controlled management mode of irrigation scheme. J. Irrig. Drainage **28**(4), 14–17 (2009). (in Chinese with English abstract)
12. Gang, S., Wengang, Z., Chunjiang, Z., et al.: A wireless irrigation control system based on modbus protocol. Water Saving Irrig. (10), 60–63 (2009). (In Chinese with English abstract)
13. Junfu, Z., Junqi, C., Jianfei, H., et al.: Control system of urban green land precision irrigation based on GPRS/GSM and µC/OS embedded technology. Trans. CASE **25**(9), 1–6 (2009). (in Chinese with English abstract)
14. Hongy, R., Xiaoyi, M., Xxiangyang, L., et al.: Research and application of information transmission protocol for irrigation area based on GPRS. J. Agric. Mechanization Res. **32**(4), 143–146 (2010). (in Chinese with English abstract)
15. Wei, Z., Yong, H., Zhengjun, Q., et al.: Design of precision irrigation system based on wireless sensor network and fuzzy control. Trans. CASE **25**(S2), 7–11 (2009). (in Chinese with English abstract)

Non-destructive Detection of the pH Value of Cold Fresh Pork Using Hyperspectral Imaging Technique

Shanmei Liu[✉], Ruifang Zhai, and Hui Peng

College of Informatics, Huazhong Agricultural University,
Wuhan 430070, China
{lsmei,zhairuifang,moonbird}@mail.hzau.edu.cn

Abstract. In this paper, the pH value of cold fresh pork was non-destructively detected based on hyperspectral imaging (HSI) technique, and some useful data processing methods were discussed. After some sample set partition methods, some spectral pretreatment methods, and some optimum wavelength selection methods were compared respectively, the most suitable data processing method was chosen and the robust hyperspectral model for predicting the pH value of cold fresh pork was established. The results indicated that the pH value hyperspectral model of cold fresh pork established by using the whole wavelengths after the sample set was divided by using concentration gradient (CG) algorithm, and the spectral data was pretreated by using normalization combined with mean center(MC) had the best prediction abilities, with the determination coefficients R_{cv}^2 equaled to 0.768, R_p^2 equaled to 0.694, RMSECV equaled to 0.1113, and RMSEP equaled to 0.1204. The results also indicated that the model established by using the characteristic wavelengths which were selected by using CARS algorithm had better prediction abilities, with R_{cv}^2 equaled to 0.8581, R_p^2 equaled to 0.8668, RMSECV equaled to 0.0858, and RMSEP equaled to 0.0772. All the results showed that suitable data processing methods was advantageous to the prediction ability of the model, and that HSI technique can be utilized to measure the pH value of cold fresh pork in a rapid and non-destructive way.

Keywords: Cold fresh pork · pH value · Hyperspectral imaging technique · Data processing

1 Introduction

Pork is one of China's staple meats. In recent years more and more people have shown increasing concern with pork quality and safety. So some reliable, quick and non-destructive detection methods for pork quality control become more and more important. However, traditional pork quality detection methods mostly depend on physical or chemical analysis, sensory evaluation which are not conductive to rapid detection of pork products in circulation. As one of the important evaluation indexes of pork quality, the pH value is closely associated with meat's color [1, 2], water-holding, tenderness [3], shelf life [4], soluble protein concentration, after-cooking flavor [5], and

© IFIP International Federation for Information Processing 2016
Published by Springer International Publishing AG 2016. All Rights Reserved
D. Li and Z. Li (Eds.): CCTA 2015, Part I, IFIP AICT 478, pp. 266–274, 2016.
DOI: 10.1007/978-3-319-48357-3_26

some other quality attribute properties [6]. The pH value of cold fresh pork is generally measured by utilizing the pH meter, which is time consuming, unsuitable and invasive when large amounts of samples are measured [7]. So it is essential to develop a rapid and accurate technique for pork pH value detection.

Recently, hyperspectral imaging technology provides an alternative methodology in agricultural and livestock product (pork, beef and some other meats) nondestructive detection field. Some works that adopt HSI for evaluating the quality of meat characteristics, such as tenderness, water-holding, and marbling level, have been done [8–16]. However, only a few studies have been reported that focus on the use of HSI for evaluating the pH value of cold fresh pork. In this study, the HSI technique was tried to detect the pH value of cold fresh pork by a non-destructive way. Furthermore, some data processing methods were discussed to find out the most suitable one, and then improve the prediction ability of the model of pH value. After discussing the sample set partition methods and spectral pretreatment methods, a quantitative PLSR prediction model is built with the whole wavelengths. Furthermore, the characteristic wavelengths are selected by utilizing competitive adaptive re-weighted algorithm, after which a new PLSR model is established by using the characteristic wavelengths.

2 Experiments and Methods

2.1 Experimental Samples

A total of 161 samples of cold fresh pork were collected from several local supermarkets in Wuhan, Hubei, China. The samples came from 79 No. 0 indigenous pigs and 82 Enshi mountain pigs. Each sample was taken from the back of an individual pig with the size of 6 cm × 6 cm × 3 cm.

2.2 Experimental Equipment

The HSI system used in this study was the HyperSIS HSI system developed by Beijing Zhuo Li Han Guang Instrument Co. Ltd. The HSI system is mainly made up of a hyperspectral imager, a CCD camera with an objective lens, an illumination source, a sample mobile station, and a computer which had software SpectraSENS to acquire images and control the camera.

The main experimental instruments and equipment required in measuring the reference pH value of cold fresh pork included a laboratory pH meter (FE20) developed by the Swiss company Mettler- Toledo, meat grinder, beaker, and others.

2.3 Acquisition of Hyperspectral Data

Thirty minutes before the experiment, the HSI system was turned on for preheating to ensure a stable performance. The hyperspectral data acquisition software SpectraSENS in the computer was opened for parameter settings. In this experiment, the CCD parameter settings were as follows: image size of 1392 × 500 pixels, wavelength range of 391 ~ 1100 nm, Bining X and Bining Y values of 2, and exposure time of 0.1 ns.

The motor parameter settings were as follows: the scanning rate of 40 nm/s, scanning distance started at 80 nm and ended at 220 nm. At last, the focal length of the CCD camera was regulated.

Before hyperspectral data acquisition, black and white calibration was conducted to remove the influence of black and white background on the hyperspectral images. The black calibration image was acquired after all the lights were switched off and the len of the CCD camera was covered, and the white image was obtained by using a white tile. The black calibration image and white calibration image were imported into the software SpectraSENS, which would automatically complete the black and white calibration of the hyperspectral images acquired by the HyperSIS HSI system and output the calibrated hyperspectral images of the measured samples.

2.4   Measurement of the pH Value

After hyperspectral data acquisition, the reference pH value of each sample of cold fresh pork was measured on the basis of the Chinese agriculture industry standard NY/T 821-2004 [17]. First, each sample was minced and divided into four copies of solution. The pH values of the cold fresh pork were measured by a lab pH meter (FE20), and the mean value was calculated and considered the reference pH value of the sample. The statistical results of the reference pH values of the 161 pork samples are as follows: the minimum of 5.2400, the maximum of 6.4100, the mean of 5.5861, and the standard deviation of 0.2221.

2.5   Data Processing Methods

Data processing in this study mainly included spectral extraction, outlier detection, sample set division, spectral pretreatment, model establishment, and optimum wavelength selection. Spectral extraction was completed in ENVI4.7, and the remaining steps were performed in MATLAB R2010a.

The average spectrum of every sample was calculated and considered as the spectrum of this sample. The average spectrum of every pork sample was extracted from the region of interest (ROI) selected by the polygon tool in ENVI4.7. Through the polygon tool, the edges and seriously reflective regions of the sample were efficiently excluded from the ROI. The extracted spectral data from the cold fresh pork samples were put in a matrix named X (161 rows, 520 columns), where the rows represent 161 pork samples and the columns represent 520 wavelengths.

After spectral extraction, the Monte Carlo sampling algorithm (MCS) [18] was utilized for outlier detection. Several methods of sample set partition, such as Kennard-Stone (KS) [19], sample set partitioning based on joint X-Y distances (SPXY) [20], random sampling (RS), duplex, CG, were utilized to divide the sample set of cold fresh pork into the calibration and test sets. The most appropriate sample set partition method was determined based on the statistical results of the calibration and test sets obtained by different methods.

Mean centering (MC) can significantly improve the performance of the PLSR model. Thus, this method should generally always be used. The spectra of the samples of cold fresh pork were processed by several spectral preprocessing methods: MC, auto-scale (AS), multiplicative scattering correction (MSC) combined with MC, standard normalized variate (SNV) combined with MC, first derivative (FD) combined with MC, second derivative (SD) combined with MC, orthogonal signal correction (OSC) combined with MC, and normalization combined with MC. The PLSR models were then established using the whole wavelengths, and the most suitable spectral preprocessing method was determined based on the performance parameters of the PLSR models.

Hyperspectral data has large amount of redundant information because some different wavelengths have the same spectral information, which will cause some adverse effects on the prediction accuracy and speed of the hyperspectral model. Some characteristic wavelength selection algorithm can solve this problem well. In this paper, the competitive adaptive reweighted sampling algorithm [21], which has remarkable ability in selecting characteristic wavelengths, was utilized to select the characteristic wavelengths of the pH value of cold fresh pork. A new model (CARS-PLSR) was established with the selected optimum wavelengths.

2.6 Model Performance Parameters

Model performance was evaluated by utilizing leave one out validation with four parameters (R_{cv}^2, R_p^2, RMSECV, RMSEP). The values of R_{cv}^2 and R_p^2 are larger, and the values of RMSECV and RMSEP are lower. Hence, the performance of the corresponding model is better.

3   Results and Discussions

3.1   Spectral Profiles

The spectra which were obtained from the hyperspectral images of the 161 samples of cold fresh pork are shown in Fig. 1.

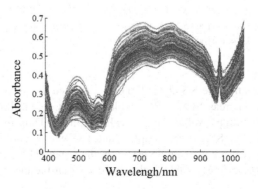

Fig. 1. Reflectance spectra of the samples of cold fresh pork

3.2   Outliers of Pork Samples

The outliers of the sample set of cold fresh pork were detected by MCS method, and the results are presented in Fig. 2.

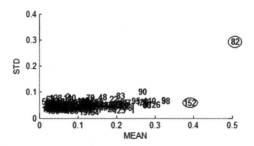

Fig. 2. Results of the outlier detection by MCS

Figure 2 indicates that sample no. 82 and sample no. 152 were outliers. After deleting the two outliers, the remaining 159 normal samples were used for subsequent analysis.

3.3   Selection of the Sample Set Partition Method

The statistical results of the calibration set (120 samples) and test set (39 samples) obtained by KS, SPXY, RS, duplex, and CG are shown in Table 1.

Table 1. Statistical results for the pH values of cold fresh pork

Sample partition method	The calibration set			The test set		
	Range	Mean	Standard deviation	Range	Mean	Standard deviation
KS	5.2425 ~ 6.4100	5.5885	0.2245	5.2400 ~ 6.2100	5.5713	0.2166
SPXY	5.2400 ~ 6.4100	5.5997	0.2372	5.2825 ~ 6.0375	5.5367	0.1605
RS	5.2400 ~ 6.4100	5.5818	0.2272	5.2425 ~ 6.2325	5.5918	0.2079
Duplex	5.2400 ~ 6.4100	5.5727	0.2053	5.2950 ~ 6.3525	5.6199	0.2670
CG	5.2400 ~ 6.4100	5.5863	0.2281	5.2825 ~ 6.2325	5.5779	0.2050

Table 1 indicates that the KS algorithm only included the maximum pH value into the calibration set, whereas SPXY, RS, duplex, and CG algorithm all integrated the minimum and maximum pH values into the calibration set. Thus, the pH value range of the test set obtained by SPXY, RS, duplex, and CG was within their corresponding calibration set only. Before the sample partition, the mean pH value of all the samples was determined to be 5.5861. Table 1 also indicates that the mean value of 5.5863 in the calibration set and the mean value of 5.5779 in the test set obtained by CG were the closest values to the mean value of 5.5861 of all the samples. In other words, the calibration and test sets obtained by the CG algorithm were well-distributed. Hence, the CG algorithm was determined to be the most suitable sample set partition method in this study.

3.4   PLSR Models Using the Whole Wavelengths

The PLSR models were established using the whole wavelengths after preprocessing the spectra using different methods, the results are presented in Table 2.

Table 2. Results of PLSR models built by using the whole wavelengths after different spectral pre-treatments

Spectral pretreatment method	Principal component numbers	R_{cv}^2	RMSECV	R_p^2	RMSEP
MC	20	0.7417	0.1184	0.6809	0.1286
AS	20	0.7311	0.1208	0.6788	0.1269
MSC + MC	18	0.7549	0.1148	0.6685	0.1235
SNV + MC	19	0.7488	0.1161	0.6802	0.1278
FD + MC	20	0.7606	0.1130	0.5772	0.1442
SD + MC	20	0.6969	0.1274	0.5809	0.1492
OSC + MC	20	0.7445	0.1186	0.6810	0.1315
Normalization + MC	**18**	**0.7680**	**0.1113**	**0.6940**	**0.1204**

Table 2 indicates that, after the spectra were preprocessed by normalization combined with MC, the performance of the pH value PLSR model built using the whole wavelengths was optimal, with R_{cv}^2 equaled to 0.7680 and R_p^2 equaled to 0.6940, which were the maximum values, RMSECV equaled to 0.1113 and RMSEP equaled to 0.1204, which were the minimum values. Thus, normalization combined with MC was determined to be the most suitable spectral preprocessing method when the PLSR model for predicting the pH value of cold fresh pork was established using the whole wavelengths.

3.5   Selection of Optimum Wavelengths

By CARS algorithm, 35 optimum wavelengths closely related to the pH value of cold fresh pork, were selected. The distribution of the characteristic wavelengths is presented in Fig. 3.

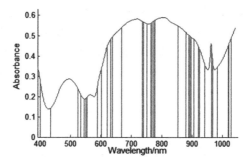

Fig. 3. Distribution of the characteristic wavelengths selected by CARS

3.6   PLSR Models Using Optimum Wavelengths

New PLSR models were built by using the characteristic wavelengths after the spectra were preprocessed by different methods. The results are shown in Table 3.

Table 3. Results of the PLSR models built by using the selected wavelengths after different spectral pre-treatments

Spectral pretreatment method	Principal component numbers	R^2_{cv}	RMSECV	R^2_p	RMSEP
none	20	0.8434	0.090	0.828	0.0946
MC	**18**	**0.8581**	**0.085**	**0.866**	**0.0772**
AS	19	0.8553	0.086	0.879	0.0727
MSC + MC	20	0.8276	0.094	0.832	0.0858
SNV + MC	20	0.8230	0.096	0.831	0.0859
OSC + MC	17	0.8563	0.086	0.842	0.0855
Normalization + MC	20	0.8557	0.086	0.839	0.0866

Table 3 clearly indicates that, after the spectra were preprocessed by MC, the new PLSR model built by using the characteristic wavelengths had very good prediction ability, with R^2_{cv} equaled to 0.8581, R^2_p equaled to 0.8668, RMSECV equaled to 0.0858, and RMSEP equaled to 0.0772. Table 2 and Table 3 also show that, better results were achieved by the new PLSR model than by the PLSR model built using the whole wavelengths. Moreover, more than 93 % of the wavelengths were eliminated (i.e., 35 vs. 520). The prediction ability of the new PLSR model and the number of characteristic wavelengths all indicated that the CARS algorithm was efficient in selecting the optimum wavelength in this study.

4   Conclusions

In this study, the pH value of cold fresh pork was non-destructively detected based on hyperspectral imaging (HSI) technique, and some useful data processing methods were discussed. The PLSR prediction model was built by using the whole wavelengths and the characteristic wavelengths, respectively. Different methods of sample set division and spectral preprocessing were compared. The CG algorithm was determined to be the most suitable method of sample set division for cold fresh pork samples, and normalization combined with MC method was identified as the optimal spectral preprocessing method for the PLSR model established using the whole wavelengths, which resulted in R_{cv}^2 equaled to 0.7680, R_p^2 equaled to 0.6940, RMSECV equaled to 0.1113, and RMSEP equaled to 0.1204. The characteristic wavelengths were selected by using CARS method to improve the performance of the PLSR model. Out of 520 wavelengths, only 35 optimum wavelengths were considered in developing a new PLSR prediction model, which performed effectively, with R_{cv}^2 equaled to 0.8581, R_p^2 equaled to 0.8668, and RMSECV equaled to 0.0858, and RMSEP equaled to 0.0772, after preprocessing the spectra by MC. The overall results indicated that suitable data processing method can effectively improve the performance of the PLSR model for predicting the pH value of cold fresh pork and the HSI technique is a powerful method for non-destructive detection of the pH value of cold fresh pork.

Acknowledgment. Fund for this research was provided by the project 2662015QC024 supported by the Fundamental Research funds for the Central Universities.

References

1. Brewer, M.S., Novakofski, J., Freise, K.: Instrumental evaluation of pH effects on ability of pork chops to bloom. Meat Sci. **72**(4), 596–602 (2006)
2. Andrews, B.S., Hutchison, S., Unruh, J.A., et al.: Influence of pH at 24H postmortem on quality characteristics of pork loins aged 45 days postmortem. J. Muscle Foods **18**(4), 401–419 (2007)
3. van Laack, R.L., Stevens, S.G., Stalder, K.J.: The influence of ultimate pH and intramuscular fat content on pork tenderness and tenderization. J. Anim. Sci. **79**(2), 392–397 (2001)
4. Knox, B.L., Van, Laack R.L.J.M., Davidson, P.M.: Relationships between ultimate pH and microbial, chemical, and physical characteristics of vacuum-packaged pork loins. J. Food Sci. **73**(3), M104–M110 (2008)
5. Bryhni, E.A., Byrne, D.V., Rodbotten, M., et al.: Consumer and sensory investigations in relation to physical/chemical aspects of cooked pork in scandinavia. Meat Sci. **65**(2), 737–748 (2003)
6. Yitao, L., Yuxia, F., Xueqian, W., et al.: Online determination of pH in fresh pork by visible/near-infrared spectroscopy. Guang Pu Xue Yu Guang Pu Fen Xi **30**(3), 681–684 (2010)
7. He, H.J., Wu, D., Sun, D.W.: Application of hyperspectral imaging technique for non-destructive pH prediction in salmon fillets. In: Proceedings of the 3rd CIGR International Conference of Agricultural Engineering (CIGR-AgEng2012), Valencia, Spain
8. Naganathan, G.K., Grimes, L.M., Subbiah, J., et al.: Visible/near-infrared hyperspectral imaging for beef tenderness prediction. Comput. Electron. Agric. **64**(2), 225–233 (2008)
9. Naganathan, G.K., Grimes, L.M., Subbiah, J., et al.: Partial least squares analysis of near-infrared hyperspectral images for beef tenderness prediction. Sens. Instrum. Food Qual. Saf. **2**(3), 178–188 (2008)

10. Elmasry, G., Sun, D.W., Allen, P.: Non-destructive determination of water-holding capacity in fresh beef by using NIR hyperspectral imaging. Food Res. Int. **44**(9), 2624–2633 (2011)
11. Elmasry, G., Sun, D.W., Allen, P.: Near-infrared hyperspectral imaging for predicting colour, pH and tenderness of fresh beef. J. Food Eng. **110**(1), 127–140 (2012)
12. Kamruzzaman, M., ElMasry, G., Sun, D.W., et al.: Prediction of some quality attributes of lamb meat using near-infrared hyperspectral imaging and multivariate analysis. Anal. Chim. Acta **714**(3), 57–67 (2012)
13. Qiao, J., Wang, N., Ngadi, M.O., et al.: Prediction of drip-loss, pH, and color for pork using a hyperspectral imaging technique. Meat Sci. **76**(1), 1–8 (2007)
14. Qiao, J., Ngadi, M.O., Wang, N., et al.: Pork quality and marbling level assessment using a hyperspectral imaging system. J. Food Eng. **83**(1), 10–16 (2007)
15. Yankun, P., Jing, Z., Wei, W., et al.: Potential prediction of the microbial spoilage of beef using spatially resolved hyperspectral scattering profiles. J. Food Eng. **102**(2), 163–169 (2011)
16. Feifei, T., Yankun, P., Yongyu, L., et al.: Simultaneous determination of tenderness and escherichia coli contamination of pork using hyperspectral scattering technique. Meat Sci. **90**(3), 851–857 (2012)
17. NY/T 821-2004. Agriculture standard: technical specification for determination of meat quality of pigs
18. Dongsheng, C., Yizeng, L., Qingsong, X., et al.: A new strategy of outlier detection for QSAR/QSPR. J. Comput. Chem. **31**(3), 592–602 (2010)
19. Wu, W., Walczak, B., Massart, D.L., et al.: Artificial neural networks in classification of NIR spectral data: design of the training set. Chemometr. Intell. Lab. Syst. **33**(95), 35–46 (1996)
20. Galvao, R.K.H., Araujo, M.C.U., Jose, G.E., et al.: A method for calibration and validation subset partitioning. Talanta **67**(4), 736–740 (2005)
21. Hongdong, L., Yizeng, L., Qingsong, X., et al.: Key wavelengths screening using competitive adaptive reweighted sampling method for multivariate calibration. Anal. Chim. Acta **648**(1), 77–84 (2009)

Cloud-Based Video Monitoring System Applied in Control of Diseases and Pests in Orchards

Xue Xia, Yun Qiu[✉], Lin Hu, Jingchao Fan, Xiuming Guo,
and Guomin Zhou

Agricultural Information Institute of Chinese Academy of Agricultural Sciences,
Beijing 100081, China
qiuyun@caas.cn

Abstract. As the proposition of the 'Internet plus' concept and speedy progress of new media technology, traditional business have been increasingly shared in the development fruits of the informatization and the networking. Proceeding from the real plant protection demands, the construction of a cloud-based video monitoring system that surveillances diseases and pests in apple orchards has been discussed, aiming to solve the lack of timeliness and comprehensiveness in the control of diseases and pests in apple orchards. The system can not only monitor the growth state of all apple trees in orchards, but also detect apple trees' diseases and pests. The system featured a camera located under tree canopies that could precisely detect diseases and pests on the back of the leaves. On the cloud storage service side, plant protectors can get hold of the situation of apple trees' diseases and pests by using the software installed in a smartphone or a computer. The system provided pinpoint surveillance data and determining criterions for the control of diseases and pests of fruit trees, which had a great realistic meaning to the development of agricultural and rural informatization.

Keywords: Video cloud · Video monitoring · Diseases and pests · Apple orchards

1 Introduction

For a long time, traditional method for diseases and pests surveillance in orchards is going to the orchards for checking. This kind of monitoring method exist some drawbacks, such as the surveyed orchard is far from the plant protection station, uncertainly survey schedule and too much manpower input. On the other hand, during apple trees' growth period, some diseases and pests would occurred on the surface of the trees' leaves, but others would occurred on the back of the leaves. Orchard surveillance are typically monitored from top to bottom [1], which lack the function of capture the video of the back of leaves. Then, the state and severity of diseases and pests of the back of leaves cannot be detected automatically.

With the introduction of the 'Internet plus' concept and rapid development of new media technology, more and more traditional industries have been sharing the development fruits of the informatization and the networking. Liu et al. combined cloud

D. Li and Z. Li (Eds.): CCTA 2015, Part I, IFIP AICT 478, pp. 275–284, 2016.
DOI: 10.1007/978-3-319-48357-3_27

computing and environmental protection industry and discussed the application of cloud computing in the video monitoring of massive pollution sources [2]. In the field of food safety retracing, using the Hadoop file system, Du studied its distributed processing architecture, data storage method and operation procedure of data read and write, which provided a new solution for the application of cloud video information processing in food safety tracing [3]. According to the practical demand of agro-technology extension, Luan et al. discussed the cloud video service applied in visual agro-technology extension, and constructed a platform system of visual video service of agro-technology extension by using the spatial embedding technology of cloud video service [4, 5].

Cloud computing is a fresh service that traditional computer technology and networking technology intermingle. A new service pattern would be built by mobilized present information resources to satisfied different levels of users' needs of information service, which have characteristics of macro scale, virtualization and extendibility.

As an inseparable part of the application of Internet, the development of cloud computing has become more and more extensive. Cloud computing adopt software as a service (SaaS) as its type of service mode. This service mode is a wholly novel software application mode that is gradually formed along with the continuous development of the Internet technology. This is just a kind of extension to the host-client pattern [6]. Emergence of the cloud computing promoted video monitoring to further development, and gradually formed a brand new mode of cloud service–video surveillance as a service [7]. The mode and structure provided users a kind of video surveillance service, on which users just need concern about the demands of video preview and playback they wanted, and they do not care which devices run the service. This kind of fully virtualized method simplified its application steps, and enhanced its convenience. At present, the application of cloud-based video monitoring focus more on video meeting and Internet protocol television, but there is lack of application in the control of orchard's diseases and pests. In time surveillance to detect disease and pest is very important for fruit tree production, even directly influences the year-round fruit yield and quality. Yet, it is hard for traditionally manual operation to surveillance trees' diseases and pests in real time and Omnibearing. Therefore, a convenient monitoring solution is needed to deal with this issue.

The video monitoring of diseases and pests that combined agricultural protection with video technology, networking technology and cloud computing technology is the concrete reflection of 'Internet plus' thinking on agricultural protection surveillance. Proceeding from the needs of the practical plant protection, the construction of cloud-based video monitoring system that surveillance diseases and pests in apple orchards has been discussed. The system can not only monitor the overall growth state of apple trees in orchards, but also detect apple trees' diseases and pests. In particular, the system also can detect the diseases and pests hidden in the back of leaves. Surveillance diseases and pests by means of the spatial convenience of the cloud structure might save much time that plant protectors commuted between orchard in rural area and plant protection station in urban area, and provided more timely monitoring data more accurate distinguishing basis.

2   Overall Frame Structure of the System

The system of video monitoring based on cloud computing is called the "cloud-based video monitoring system" [8, 9]. The structure of cloud computing service mode allows cloud video monitoring nodes placed somewhere they want monitored. And then, the monitoring video stream would be accessed to the surveillance service center located on the cloud-side, and stored and managed uniformly [10, 11]. Users can access to the cloud-side via the Internet to check the real time monitoring video at any time.

Cloud-based video monitoring system of orchards' diseases and pests divide into four parts: acquisition side, gathering side, cloud storage service side and application side. Acquisition side is the equipment captured video data in orchards. It mainly includes cameras, transmission lines and control devices. Gathering side is the specialized digital video recorder which fuses the functions of Internet transmission, matrix display and encoding. Both of side cooperate with each other and fulfilled the tasks of video acquisition, data encoding and data transfer. Cloud storage service side is an important parts of the system that running data request from the acquisition side and the application side [12]. Besides, it also implements the functions of saving and managing a lot of monitoring video data of diseases and pests. Application side is

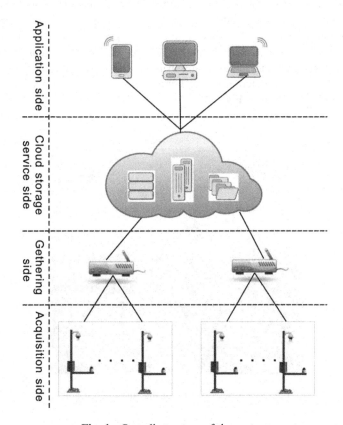

Fig. 1.   Overall structure of the system

belonging to the final client. Users issue a surveillance request to the cloud storage service side used by the desktop computer, notebook computer or smartphone to check the video of diseases and pests at a remote place. Sketch map of the overall system structure is shown in Fig. 1.

3   Fundamental Function Modules

3.1   The Module of Acquisition Side

The acquisition side mainly consists of high definition dome camera, hemispheric camera and some auxiliary equipment. According to the requirements of modern management style of fruit trees, in full fruit period, height of fruit trees often remain under about 3 m. And the height of surveillance pole need higher than trees' height in order to observe the entire orchard. So the height of surveillance pole designs with about 3.5 m. Surveillance pole adopts stainless steel material quality that not only ensured stiffness of bracket, but also prevents rain erosion and field weathering.

Cameras are fitted on the top and the middle of the surveillance pole. In order to observe the diseases and pests of the entire orchard, the top camera use a unibody waterproof dome camera that equipped with a 27-power binocular lens. Lower-middle part of the surveillance pole, under the canopy of the fruit trees, uses a hemispheric camera that equipped with a 10-power binocular lens. Camera stretched into the tree canopy by means of the transverse branch arm to observe the occurrence status of diseases and pests. Similarly, taking rainfall issue into consideration, the sealed rain-proof is installed on the external of the hemispheric camera. Physical picture of the surveillance pole is shown in Fig. 2.

Fig. 2. Physical picture of the surveillance pole

Power supply and control of the outdoor facilities use 4 cores copper wire, and divided the copper wire into two sets. One set is used as alternating current power supply. Anther set is used as signal line of PTZ control. Given that orchardists would regularly working in the orchard such as blossom and fruit thinning, fertilization and chemical spraying, electric shock are most likely to be occurred if 220 V alternating current directly accessed into orchard. Yet the output stability of operating voltage cannot be ensured if transmit electric power with low voltage. Thus, the wire would firstly pass through in 5 m high when routing and access to the 5 m height relay pole. And then, the wire would access to the surveillance pole after reduced voltage used by an electrical transformer, which can cut down the probability of risk. Video signal of the orchard transferred by coaxial-cable (Syv75-5, theoretical distance is 300 m). During the signal transmission, the signal strength would gradually attenuate. So the video signal-amplifier need appended to ensure the quality of the video transmission when the transmission distance is too long. Sketch map of wire arrangement in apple orchard is show in Fig. 3.

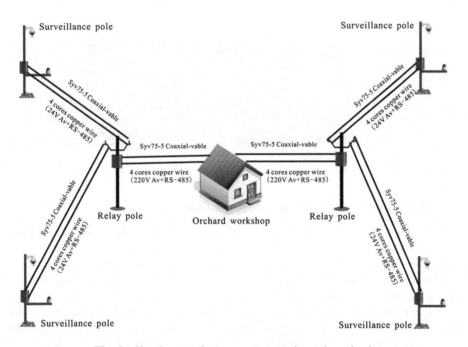

Fig. 3. Sketch map of wire arrangement in apple orchard

3.2   The Module of Gathering Side

The existing patterns of the orchard remote video monitoring most adopt wireless transmission. Although we do not have to consider much about wiring procedure, the steep drop of the video quality caused by the bottleneck of the wireless network bandwidth have limited the application of this technology in remote video monitoring, and it is hard to be practically promoted in remote video monitoring filed.

With the gradually optimizing of the base installation, electric supply and web network have been established in modern orchards. The gathering side of the orchard deployed a digital video recorder with the network transmission function. This side is use to converge the video signal from the surveillance pole, and output and display on the video matrix after converted the signal into image information. Meanwhile, by means of p2p network technology, the digitized video signal would access to the web via ADSL or optical fiber network and transferred to the server farm of the cloud. Users can penetrate the router and firewall through the p2p network technology to visit all of digital video recorder equipment in the cloud and capture the monitoring video of the tree's diseases and pests. In addition, the system would meet the downloading and accessing requirements of p2p pattern by lower the data storage granularity, which rendered sharing and interaction of video monitoring more immediate and simple, and avoided network congestion to some extent. The equipment drawing of the gathering side is show in Fig. 4.

Fig. 4. Equipment picture of the convergence-side

3.3 The Module of Cloud Storage Service Side

The cloud storage service side mainly consists of virtual machine, control servers and data storage servers. It can store the video data of diseases and pests from the digital video recorder equipment. The servers on the cloud can distinguish the video origin in the light of the particular ID heading code of the digital video recorder equipment, and saving the digitized video data into the databases of the server farms on the cloud side. Unlike the other platform of video monitoring, the cloud-based monitoring video storage server sides for orchards' diseases and pests not only need to completed the access and manage for the surveillance and browsing nodes by using the control servers, but also need to completed the scheduling and operating tasks from the acquisition side and gathering side. And finally achieved the functions in terms of the data save, manage and visit for the surveillance of diseases and pests. Structure map of the storage service cloud side is show in Fig. 5.

Compute nodes (physical node) is the major carrier of virtual machine. The compute nodes of the storage service cloud side contains a lot of host machines and physical servers that processed the virtual machines of monitoring and browsing of

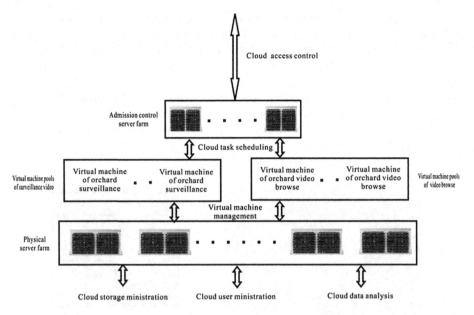

Fig. 5. Structure map of the storage server cloud side

orchard's diseases and pests video. Every virtual machine can execute multiple monitoring or browsing tasks in order to accommodate frequent operations of tasks access and tasks shift out. The storage service cloud that based on virtualization technology can not only unified manages and distributes the video source, but also shield the physical position between storage entities and its particular multi-institutional property. The major duty of the control servers is assign permissions to the user in line with the service they requests, and providing different services for the different users. Meanwhile, it can also complete the functions of network access and user authentication [14]. The underlying data center is the base of distributed storage of monitoring video in cloud structure environment. It consists of servers, disk array and network communication control facilities. In which the storage servers is the most part of the underlying data center, and it responsible for the storage and management of video stream that data volume and growth rate are rather large. The technology of sliced file segmentation storage is adopted on the storage servers. Making the best of the I/O ability of every memory cells and optimizing storage methods, the large video files are distributed stored in the physical storage facilities, which proposed to enhance the data rate of large video files between write and read, and lift the efficiency of storage and access of video data [15].

3.4   The Module of Application Side

Users on the application side can acquire the monitoring video data of the assigned orchard from the cloud server farm by committing video request, which achieve the distributed remote surveillance of orchards' diseases and pests. The software client installed in the users' computers or smartphones, and they can access to the cloud server farm by means of inputting the ID heading code of the digital video recorder equipment. Through the high definition transmission with low bit-rate video, users on the application side can not only capture the clear real-time video of the orchards' diseases and pests, but also preview and playback the video by means of the web storage function of cloud server farm so that users can get hold of the occurrence status of orchards' diseases and pests. The interface of video monitoring on application side is show in Fig. 6.

Fig. 6. The interface of video monitoring on application-side

4   Results and Analysis

Regard to the ordinary video monitoring, the sheer amount of image information is an evident characteristic, so the big enough memory space is required. At present, the domestic video monitoring recorder equipment on the market mostly equip with 1 TB hard disk. If store video followed 720p format that is the rudimentary HD format, the hourly data size can reached 4G ~ 8G per route. Even partial video compressing on the premise of remain video quality, the ordinary native video monitoring system required 17000G data size every month. Such a large video data is an insurmountable bottleneck for the native storage equipment's capacity, read/write performance and reliability in the mode of traditional video monitoring.

By use of clustered application, grid technology and distributed file system, the cloud-based video monitoring system of orchard's diseases and pests interconnect and facilitate collaboration between memory service devices, providing the functions of monitoring data storage and visit, which achieved a fully virtualized procedure of video monitoring and decreased the costs of system construction and application. The video monitoring cloud is fully transparent to users. They do not need regard the details of equipment's type, quantity and network structure, and just use an ordinary web connection to access to the cloud side. In general, the mode of cloud-based video monitoring has break through the barriers of performance and storage capacity in traditional video storage. It enable users share the video resources in the monitoring cloud of orchard's diseases and pests, and through the clearly monitoring images, the occurrence of diseases and pests in orchard can be efficiently and conveniently estimate.

5 Conclusion and Prospects

With the continuous development of the video monitoring technology, more and more surveillance system calls for the achievement of long-distance transmission of mass data. Yet the traditional surveillance pattern is no longer applicable to current monitoring needs. The cloud-based video surveillance of orchard's diseases and pests is based on clouds, pipes and sides. Clouds refer to video monitoring service clouds. Pipes are the routes of data transmission. Sides are the user clients. In this structure, cloud may not just a complete one. Pipes probably have much more. The number of sides is countless [16]. The video monitoring data of orchards' diseases and pests can intelligently shuttle and flow among clouds, pipes and sides in vertical and lateral styles.

On the unceasingly development and progress of the cloud technology, the applications of video monitoring technologies in terms of cloud storing, cloud unicasting and cloud searching are bound to become mature. Cloud-based video monitoring pattern has become a trend in future surveillance, and the trend is certain to develop towards high definition, webified, intelligent and stereoscopic, which can provide further professional solution programs of orchards' diseases and pests for the field of agriculture and plant protection.

Acknowledgment. This work was supported by National High Technology Research and Development Program 863 (2013AA102405), Agricultural Science and Technology Innovation Project of CAAS (CAAS-ASTIP-2015-AII-03).

References

1. Yang, M., Xu, Y.: Analysis of the application on cloud computing platform-based networking video technology. Telev. Eng. **3**, 10–12 (2012)
2. Liu, J., Zhang, X., Han, X.: Cloud computing-based application on the HD video monitoring of the massive pollution sources. China Secur. Prot. **10**, 32–35 (2014)
3. Du, A.: Application of Information Processing Technology for Cloud Video Monitoring in Food Traceability. Jinan University, Jinan (2014)
4. Luan, Y., Li, L., Ju, K., et al.: The cloud video services applied research of visualization agricultural science and technology. Agric. Outlook **8**(10), 38–40 (2012)
5. Luan, Y., Li, L., Ju, K., et al.: Farm-oriented requirements analysis of online video platform. Agric. Outlook **9**(11), 63–64 (2013)

6. Jia, S.: Application of cloud computing in TV news live broadcasting technology. Radio TV Broadcast Eng. **37**(11), 28–33 (2010)
7. Xiong, Y., Zhang, Y., Chen, X., et al.: Research of energy consumption optimization methods for cloud video surveillance system. J. Softw. **26**(3), 680–698 (2015)
8. Tsai, Y.: The cloud streaming service migration in cloud video storage system. In: Proceedings of the 27th IEEE Advanced Information Networking and Applications Workshops, pp. 672–677 (2013). doi:10.1109/WAINA.2013.146
9. Vetro, A., Christopoulos, C., Sun, H.: Video transcoding architectures and techniques: an overview. Signal Process. Mag. IEEE **20**(2), 18–29 (2003)
10. Xiong, Y., Wan, S., He, Y., et al.: Design and implementation of a prototype cloud video surveillance system. J. Adv. Comput. Intell. Intell. Inform. **18**(1), 40–47 (2014)
11. Wang, Q., Ren, L., Zhang, L.: Design and implementation of virtualization-based middleware for cloud simulation platform. In: 2011 4th International Conference on Computer Science and Information Technology, June 10–12, Chengdu, China (2011)
12. Xu, G.: Cloud-based video monitoring is running from theory to practice. China Public Secur. (16), 120–129 (2012)
13. Qin, F., Liu, J.: Key technologies in P2P media streaming. Acta Electronica Sinica **39**(4), 919–927 (2011)
14. Liu, K., Li, A., Dong, L.: Research on cloud data storage based on Hadoop and its implementation. Microcomput. Inf. **27**(7), 220–221 (2011)
15. Lei, Y.: Cloud technology and its application on video storage. China Public Secur. **20**, 178–187 (2013)
16. Li, J.: On the way of video cloud time. Sci. Technol. China's Mass Media (19), 22–23 (2012)

Research on Coordinated Development Between Animal Husbandry and Ecological Environment Protection in Australia

Yiming Zhu[(✉)] and Shasha Li

College of Economics and Management, China Agricultural University,
Beijing 100083, China
cauzhuyiming58@163.com, susanlss2008@sina.com

Abstract. Australia is one of the countries whose scientific level of using grassland resources and animal husbandry development level is very high and the way to coordinate Australia's animal husbandry and ecological environment protection is worth using for reference. This study focuses on facilities about the Australian government's controlling and protecting of ecological environment in the process of development of animal husbandry, and combines with the current problems in the process of China's animal husbandry development, in order to explore a suitable path for implementing the animal husbandry sustainable development in China.

Keywords: Australia · Animal husbandry · Ecological environment · Intensive management

1 Introduction

The animal husbandry in Australia is rather developed where sheep and cattle are the main livestock. In addition, the number of sheep in Australia is the top in the world, and Australia is known as "riding on the sheep's back country". In the early, the development model of Australia's animal husbandry tended to rough grazing. Since the early 19th century, European settlers' increasing investment in science and technology, introducing high quality forage grass and implementing policies and measures are conducive to the development of animal husbandry. Especially after the Second World War, the number of Australia's livestock breeding stock and livestock production increased significantly. In 1972, compared with 1950, the number of cattle and sheep breeding stock increased by 87 % and 44 % respectively and wool production grew by 89 % in comparison to the average output from 1947 to 1949. In addition, compared with 1955, the beef and egg production increased by 86 % and 95 % respectively [1]. During this period, there are mainly three points about the cause of the rapid development of animal husbandry in Australia. The first point is that during the post-war, world economy began to recover and the demand for animal products increased at home and abroad. Developing animal husbandry can not only meet the domestic demand, can increase income through export animal products. So animal husbandry industry had more obvious competitive advantage than other industry. The second

D. Li and Z. Li (Eds.): CCTA 2015, Part I, IFIP AICT 478, pp. 285–291, 2016.
DOI: 10.1007/978-3-319-48357-3_28

point is that the Australia government ordered the Italian prisoners of war to engage in livestock production, which increased the required labor input during the process of livestock production from 1941 to 1947. The third point is that combined with the domestic and international market demand, the Australia government guided the livestock farmers to adopt the latest Science and Technology and farming equipment, what is more, the government implemented financial support and policy inclination to development the animal husbandry. Considering the scarcity of resources and the protection of ecological environment and basing on the conditions of natural resources and markets all over the world, the Australian government continuously adjusted the animal husbandry development policy in order to realize the coordinated development of animal husbandry and ecological environment protection.

Since the reform and opening, with the aid of the push of market economy, government policy support and irregular factors, China's animal husbandry were integrated and were in the transition period from extensive to intensive little by little. Extensive development pattern makes natural environment deteriorating surrounding China's pastoral areas, which make the protection of natural resources face great challenge. This study summarizes the current development situation, the characteristics of the development of animal husbandry in Australia and the policy support and emphasis on the measures to protect the ecological environment in the process of sustainable development of animal husbandry. Analysis of Australian animal husbandry development can guide to seek a suitable way for China to coordinate animal husbandry development and ecological environment protection.

2 Current Development Situation of Graziery Industry in Australia

Australia's land area is about 7.68 million km^2, and the number of population is about 23.71 million people of which 80 % distribute in the eastern coastal areas. Most area of Australia is plain region, the climate is relatively dry and seasonal temperature difference is inconspicuous. The unique climate conditions in Australia forms a perennial natural pasture grazing. 55 % of the land area is used to develop animal husbandry whose output value accounted for about 80 % of the agricultural output in 2014. Animal husbandry in Australia occupies an irreplaceable position in the agricultural and even the entire national economy. In Australia, over average, everyone owned 5.00 sheep and 5.00 cowsin 2011. However, compared with the Australia, the Chinese person just has the number of sheep and cattle less than a quarter [2].

Australia's sheep and goat breeding stock is the largest, followed by chicken and beef. From 1993 to 2013, Australia mainly livestock breeding stock is on the decline on the whole. From Fig. 1, we can see that the sheep and goats breeding stock decline relatively obvious, down from 1.40 million in 1993 to 2013 in 0.71 million, dropping about 49.28 %. The number change of beef cattle and pig breeding stock is not obvious, of which the former increased but pig breeding stock has a downward trend.

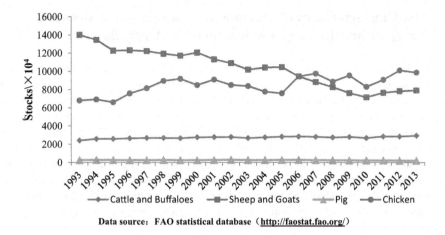

Data source: FAO statistical database (http://faostat.fao.org/)

Fig. 1. Main livestock number of Australia from 1993 to 2013

From 2006 to 2011, the number of beef, mutton and red meat production decline year by year (Table 1). Combined with Fig. 1, it can be seen that the Australian animal husbandry is in the transition to a new development model. There are two possible reasons: the first one is that animal products exports accounts for a large proportion of total output in Australian [3]. With the rapid development of animal husbandry in emerging countries, the international market for Australia's livestock reliance declines, resulting in the decrease of breeding stock. The second one is that with the all-round development of economy, the international market increases demand for some of the agricultural economic crops, such as soybean and rapeseed, so planting economic crops has more competitive advantage than raising livestock, which makes laborer to increase the crop planting area in parts of Australia animal husbandry area.

Table 1. Main livestock products of Australia from 2006–2010

Year	Beef (t)	Veal (t)	Mutton (t)	Lamb (t)	Pig (t)	Total red meat (t)	Chicken (a) (t)
2006–07	2 195 714	30 578	270 988	412 584	381 866	3 291 730	811 591
2007–08	2 105 706	26 417	243 119	428 388	374 409	3 178 038	797 280
2008–09	2 098 615	26 489	219 820	415 867	321 005	3 081 798	832 456
2009–10	2 056 514	52 262	161 774	412 537	331 261	3 014 347	834 409
2010–11	2 089 233	44 133	123 245	391 340	342 101	2 990 052	1 014 978

Data sources: Bureau of Statistics of Australia (2012).
Note: Excludes Northern Territory, Tasmania and the Australian Capital Territory.

According to above analysis, the Australian animal husbandry is transforming from a single animal husbandry to planting-culture combined animal husbandry develop-ment model gradually.

3 The Characteristics of Harmonious Development Between Graziery and Ecology Environment in Australia

Since the idiographic climate in Australia, natural pasture is the main food of animal composite feed. After more than 200 years of development, animal husbandry in Australia has completed transformation and upgrading and the ecological environment is protected simultaneously in the process of animal husbandry industry in Australia. In extensive development phase, the utilization of grassland resource in Australia also experienced overgrazing, pasture degradation and desertification process. In the early 20th century, the grassland in Australia has reached the ceiling of the bearing capacity and the deterioration of ecological environment has endangered the healthy development of animal husbandry. The stockholders took positive and effective cooperation to prohibit overgrazing behavior and improved ecological environment finally.

3.1 Seeding to Revive Ranch, Formulating Appropriate Grazing Capacity

According to soil conditions and climatic environment, planning different varieties of forage can not only balance the year's yield, but also helps to improve the quality of soil and increase the yield and to optimize the quality. Grassland yield of forage planting is five times as much as natural grassland [4]. Grassland yield and regeneration ability determine the grassland grazing capacity. The Australian government boosted grassland ecological construction and decided appropriate grazing capacity according to the production capacity of grassland and grassland recovery ability to prevent the grassland desertification caused by overgrazing grassland. Each family determines reasonable scale of breeding according to suitable grazing capacity to avoid the behavior of grassland overgrazing.

3.2 Rotation Grazing on an Area Basis and Grazing off Season

According to the different growth stages of herds, combining with the carrying capacity of grassland, the herdsman adjusts grazing stages scientifically and manages grassland pasture to guarantee the sustainable utilization of grassland resources. In addition, Australian government divided grazing area all over the country into four different types of the grassland animal husbandry according to the precipitation, temperature, and soil conditions such as low density livestock grazing district, natural grassland grazing area, mixed farming zone and high density grazing area [5]. In the same zone, livestock farmers divided family farm into several small areas of which 20 % of the pastoral areas of grazing to protect other pasture grass growing in order to maximize the grassland biomass.

3.3 Grazing in Accordance with the Law and Utilizing the Water Resource Rationally

The Australian government had strict rules on grassland construction, the development of environmental protection, water conservancy etc. The government would impose severe penalties on violators, which ensured the coordinated development between animal husbandry and ecological environment protection and cultivated the ecological environment protection consciousness of farmers and herdsmen at the same time. The weather of Australian outback is drought where the average annual rainfall is less than 200 mm. In order to protect the grassland yield, it is indispensable to exploit and utilize water resources. In Australia, livestock farmers built small water storage low dam, reservoir and other water engineering project generally. Storage of water resources ensures livestock drinking water and grassland irrigation.

4 Australian Animal Husbandry Development Model for China's Enlightenment

As people living standard rising, for Chinese people, demand for animal products is on the rise. What's more, the price of domestic beef and mutton market is higher, which causes part of the pastoral areas overgrazing phenomenon, resulting in bearing pressure increasing and grassland ecological environment destructed [6]. In recent years, in order to maintain sustainable use of grassland, the Chinese government is strengthening the construction of grassland ecology and intensifying the efforts on ecological protection, such as returning farmland to grassland project and ecological compensation mechanism. However, support on the investment of pastoral animal husbandry development still has a lot of space [7]. At the same time, livestock farmers list livestock production as the first goal at the current stage and lack the protection environment consciousness about grassland ecological [8]. Although there are some difference about the basis of existing on the animal husbandry between Australia and China, China can combine their own development period of animal husbandry and existing problems to explore a sustainable development way to coordinate animal husbandry development and ecological environment protection.

4.1 Planning and Constructing Regional and Special Pastoral Areas

According to the climatic conditions all over China and the annual average precipitation, China should divide and construct pastoral area and set up the appropriate grazing way to achieve balanced development in breeding according to the local territory characteristics to protect grassland resources and ecological environment eventually. Referring to ecological grassland division from Ren [9], we can divide Chinese grassland animal husbandry into the following areas, such as north desert scrub area, qinghai-tibetalpine region, onobrychisuiciaefolia, the northeast forest district, southwest karst mountain thickets grassland ecological economic zone and southeast evergreen broad leaved forest area. Planning and constructing regional and special pastoral areas is beneficial to promote the development of animal husbandry and ecological environment protection.

4.2 Planting Grass and Improving Pastures

To alleviate the prairie excessive load problem, Australia government promoted to cultivate grass artificially. This move can not only guarantee the supply of grass feed and improve the soil but accelerates the sustainable development. One of the methods to improve pasture construction is cultivating artificial grassland in the pastoral areas of China. In different area, herdsman should cultivate and plant suitable grass according to local climate and soil conditions. At the same time, the aid of soil testing and fertilizer technology on the shortage of trace elements in soilcan promote the growth of grass. In addition, the same pastoral areas should be divided into different farming area and the shepherd also should delimit the pasture use ratio to ensure the rest of the time grazing plot of grass growing.

4.3 Conducting Water Conservancy Facilities Construction

Water resources is the necessary means of production in the process of herbage growth, in pastoral areas with less precipitation, effective supply of water resources is a necessary condition for the sustainable development of animal husbandry. According to pasture area and grazing capacity, the farm family can select suitable water conservancy project, such as the construction of reservoirs, small reservoirs, deep well and so on. If the condition is appropriate, the government can also construct large water conservancy facilities.

4.4 Strengthening the Grassland Ecological Protection Ability and Passing Relevant Laws and Regulations for the Protection of Grassland Ecological Environment

The protection of the ecological environment is dependent on the government's support and guidance. The Chinese government should be on the basis of the existing ecological construction projects, strengthen ecological environmental protection, return grazing land to grassland continually, moderate grazing capacity and implement the effective governance of desertification and sandstorm. In recent years, the Chinese government issued some laws and regulations about ecological environment protection and the construction of grassland, which has obtained the good effect. These measures played a significant role in promoting transformation of animal husbandry and ecological breeding. However, under the comprehensive effect of the inherent mode of production and the external economic environment, ecological protection consciousness of farmers still need to be further improved. The Chinese government should draw lessons from the Australian government's relevant measures, such as strengthening the professional training of herdsmen, intensifying the communication and collaboration between different pastoral areas, etc.

5   Conclusions

This study argues that the animal husbandry of China should be based on the long-term view, pay attention to ecological environment protection in the process of animal husbandry and implement sustainable development strategy. Through summarizing the measures to coordinate the animal husbandry and ecological environment protection in Australian, combining with the current problems in the development of animal husbandry in China, this paper puts forward the planning and construction of regionalization and specialization, the construction of cultivated grassland in pastoral area and improvement of grassland, promoting water conservancy facilities construction, strengthening the protection of grassland ecology and introducing relevant laws and regulations for the protection of grassland ecological environment policy suggestions, meanwhile, should develop modern animal husbandry and improve the ecological environment to promote the sustainable development of animal husbandry in China ultimately.

Acknowledgment. Funds for this research was provided by the Modern Agricultural Industry Technology System of China (CARS-41-K26) and China's Livestock and Poultry Industry Research Project during the 13th Five-Year Plan.

References

1. Wu, Z.: Advanced animal husbandry of Australia. World Econ. **07**, 41–46 (1979)
2. Yan, X., Nan, Z., Tang, Z.: Introduction of animal husbandry in Australia and its implication for China. Pratacultural Sci. **03**, 482–487 (2012)
3. Mao, X.: The characteristic of animal husbandry economy in Australia. Guide of Sci-tech Mag. **05**, 37 (2003)
4. Zhang, L., Xin, G.: Experience in the development of grassland animal husbandry in Australia and New Zealand. World Agric. **04**, 22–24 (2008)
5. Chen, B.: Agriculture resources and their regional distribution in Australia. Chin. J. Agric. Resour. Reg. Plann. **04**, 55–58 (2006)
6. Liu, J.: Current status and main tasks of pratacultural development in China. Pratacultural Sci. **02**, 1–5 (2008)
7. Huang, T., Li, W., Zeng, Y.: Debate between grassland ecological protection and herdsmen's income. Pratacultural Sci. **09**, 1–4 (2010)
8. Ai, Y.: Animal husbandry in Australia. China Anim. Husb. Bull. **07**, 63–65 (2006)
9. Ren, J., Lin, H.: Assumed plan on grassland ecological reconstruction in the source region of Yangtse River, Yellow River and Lantsang River. Acta Pratacultural Sci. **02**, 1–8 (2005)

Determination of Lead (Pb) Content in Vetiver Grass Roots by Raman Spectroscopy

Yande Liu[✉], Yuxiang Zhang, Lixia Jiang, and Haiyang Wang

College of Mechanical Engineering, East China JiaoTong University,
Nanchang 330013, Jiangxi, People's Republic of China
jxliuyd@163.com

Abstract. In order to provide references for heavy metals diagnosis of Vetiver grass, Raman spectroscopy technology was employed, the partial least squares (PLS) quantitative analysis model of heavy metal lead of Vetiver grass root was established, different processing methods were used to optimize the Raman spectra of Vetiver grass roots, Successive projection algorithm (SPA) was applied to screen the bands of Raman spectrum and enhance the model's accuracy. The best Raman spectroscopy quantitative analysis model of lead content in Vetiver grass root were set up, 20 unknown samples were used to test the quality of optimized model. The prediction correlation coefficient and the root mean square error were 0.607 and 0.040 g/kg, respectively.

Keywords: Heavy metal · Vetiver grass · Raman spectroscopy

1 Introduction

As a kind of not easily biodegradable toxic substances, heavy metals are harmful to the vast majority of biology, especially the human. However, heavy metals can be gradually accumulated in plants and animals by the food chains [1]. Ingestion of heavy metals in agricultural products is one of the important factors that cause the accumulation of heavy metal in human body [2]. Therefore, the content of heavy metals in foods is an important index for food safety risk assessment. In order to reduce the harm caused by eating crops, which containing excessive heavy metals, seeking a kind of rapid and nondestructive method for detecting heavy metal content in plants is necessary.

However, the traditional methods for detecting heavy metal content in plants, including X-ray fluorescence spectroscopy (XRF) [3], inductively coupled argon (ICP) [4] and atomic absorption spectroscopy (AAS) [5] and so on are always combined with digestion and chemical analysis instrument, which are time-consuming and plant-damage [6]. Raman spectroscopy, which is a spectrum analysis technology with the advantages of fast, non-destructive, non-pretreatment, has the board application prospect in terms of detecting heavy metal contents in plants indirectly [7]. Although the pure heavy metals cannot be estimated directly by Raman spectroscopy, heavy metals in plants usually combined with the organic molecular groups that have Raman spectral information [8]. Therefore heavy metals in plants can be indirectly detected by using spectral technique basing on the chelation.

© IFIP International Federation for Information Processing 2016
Published by Springer International Publishing AG 2016. All Rights Reserved
D. Li and Z. Li (Eds.): CCTA 2015, Part I, IFIP AICT 478, pp. 292–299, 2016.
DOI: 10.1007/978-3-319-48357-3_29

Vetiver grass can live in the range of −22 °C to 55 °C temperature, which has strong vitality. It also can grow up in the environment of heavy metal pollution, shows the strong tolerance of heavy metals [9, 10]. Because of these characteristics, Vetiver grass becomes one of the ideal plants for soil and water conservation and phytoremediation of soil heavy metal pollution [11].

2 Materials and Methods

2.1 Samples and Reagents

The Vetiver grasses in this study were collected in the heavy metal pollution fields of Guixi city, China. The ultrapure water was produced by ultrapure water polishing system (TS-RO-10L/H, TAOSHI Water Equipment Engineering Co. Ltd., China), with 18.2 MΩ cm resistivity. All chemical reagents (HNO_3, $HClO_4$) in this study are of analytical purity grade, and purchased from Sinopharm chemical reagent Co. Ltd., China.

2.2 Samples Preparation

The vetiver grass were washed by tap water until removing the soil and dust on the surface of plants, and rinsed by ultrapure water twice in order to remove the interference of heavy metals in water. Then the roots of vetiver grass were cut by scissors for the samples for Raman spectroscopy detection.

2.3 Raman Spectral Measurement

All Raman spectra were measurement from 90 cm^{-1} to 3500 cm^{-1} with a confocal microprobe Raman spectrophotometer (SENTERRA, Bruker, Germany). Each root sample of Vetiver grass was put on the quartz glass slide, which has been put on the sample frame of spectrophotometer. Eyepiece of 10 times and objective lens of 50 times were employed respectively. In order to make the laser focus on the internal organizations of Vetiver grass roots, fine tune the sample frame until the clearest images of plant surface was observed.

At the same time, the impact to organizations of Vetiver grass roots by the quantity of heat of laser should be considered. Although the higher power laser can improve the spectrum information, it also can be more easily destroy the organizations, which would influence the outcome of Raman spectral measurement. Choose the optimal instrument experimental parameters through several tests to ensure that the plant in the laser focus will not lead to excessive heat structure of plant tissue damage: the excitation wavelength was 785 cm^{-1}, the Resolution was 0.5 cm^{-1}, the integration time was 5 s, the light spot diameter was 50*1000 microns, and the power of laser was 10 mw.

In order to reduce the spectral measurement error, each root sample of Vetiver grass was measured three times, so each spectrum of root sample of Vetiver grass was averaged from three parallel measurements.

2.4   Measurement of Quality Parameters

All root samples of Vetiver grass were dried to constant weight by drying oven (DHG-9101-19A, Shanghai Hong Electronic Technology Co. Ltd., China) with temperature of 75 to 80 centigrade after Raman spectral measurement, then ground into powder respectively with a pulverizer.

Each Vetiver grass root powder sample was weighed 0.2 g by electronic balance, soaked with 10 ml HNO_3 in the flask and sealed overnight, then digested by an electric hot plate (Mb, Beijing Kewei Yongxing Instrument Co. Ltd., China) until the large granular sediments were nearly disappeared. The flask was removed with the crucible tongs for cooling, 10 mL HNO_3 and 5 mL $HClO_4$ standard solution were added into the flask respectively after cooling, then heated by electric hot plate again until white smoke appeared in the flask. Removing the flask for cooling again, the sample solution in the flask was washed into 25 ml volumetric flask by deionized water, and then filtered into another 25 mLvolumetric flask. Finally, flame atomic absorption spectrometer (AA280FS, Varian, America) was used for measuring the total lead in solution of Vetiver grass roots.

3   Results and Discussion

3.1   Raman Spectral Features of Vetiver Grass Root

All Raman spectrums of Vetiver grass root samples were averaged to be a mean Raman spectrum, and shown in Fig. 1. Because of the low power of laser, the Raman signal of Vetiver grass roots is very weak, the Raman peak are not obviously. What's more, the weak Raman characteristic peaks were covered up by the interfering substance in the plants and the fluorescence effect should be another main reason for causing this phenomenon.

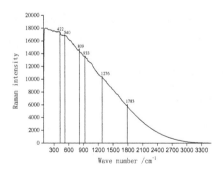

Fig. 1. Raman spectrum of Vetiver grass root

Different wave number Raman characteristic peaks represent the position of chemical bonds, and also reflect the corresponding molecular structure information. So the information of some molecules in Vetiver grass root, which have strong Raman effects, can be found according to the figure of the Raman spectrum peak position, and shown in Table 1. The result shows that the alkanes, hydrocarbons and ketone compounds in Vetiver grass root have stronger Raman information than the other compounds.

Table 1. The characteristic Raman bands of Vetiver grass roots

Frequency/cm^{-1}	Vibration mode	Reference compounds
422	Stretch	n-alkanes
540	CBr stretch	Alkanes
802	Ring breath	Cyclohexane
933	Ring vibration	Alkyl cyclobutane
1276	N=N=N stretch	CH_3N_3
1783	C=O stretch	Cyclobutanone

At the same time, in the range of 2000 to 3500 cm^{-1} of Raman spectrum of Vetiver grass root, the effective Raman characteristic information in Raman spectra of Vetiver grass root does not exist. Therefore, in order to reduce the interference from other useless information, excluding this region of Raman spectrum, $150 \sim 2020$ cm^{-1} spectral region are selected for establishing Raman mathematical analysis model of Vetiver grass root.

3.2   Reference Analysis

In order to facilitate the model's modeling and testing, 60 plants were randomly selected in 80 Vetiver grass samples as a calibration set, the remaining 20 plants as prediction sets. The statistical analyses including range, mean and standard deviation (S.D.) for lead values in Vetiver grass roots are performed in Table 2. It shows that the heavy metal lead values are within the wide range for the samples of Vetiver grass roots, which are beneficial to produce robust and stable calibration models and evaluate the models' predictive ability.

Table 2. Statistics of heavy metal Pb in Vetiver grass roots

Sample set	Minimum (g/kg)	Maximum (g/kg)	Mean (g/kg)	S.D.
Calibration (n = 60)	0.063	0.233	0.126	0.046
Validation (n = 20)	0.060	0.235	0.126	0.051
All (n = 80)	0.060	0.235	0.126	0.048

3.3   Quantitative Models of Vetiver Grass Root

Because of the weak Raman signal and the strong fluorescence intensity, a lot of Raman signal usually are covered up by the fluorescence background. Therefore, in order to establish a great Raman mathematical analysis model, enlarging the effective information in the Raman spectra and enhancing the signal-to-noise ratio of spectrum are indispensable. Different spectral preprocessing methods can effectively eliminate some interference information because of their different functions in spectrum optimization. In this study, six spectral preprocessing methods, smoothing, standard normal variate (SNV), multiplicative scatter correction (MSC), first derivative, second derivative and baseline correction, were employed to optimize the quality of models. The quantitative partial least squares (PLS) models of different preprocessed spectrums were established with Unscrambler 8.0 software package, the PLS model of raw spectrum was also established for the sake of contrastive analysis, and shown in Table 3.

Table 3. The results of calibration and cross-validation for the Pb content in the roots of Vetiver grass samples

Pre-processing methods	PCs	r_c	RMSEC (g/kg)	r_{cv}	RMSECV (g/kg)
Raw	10	0.851	0.024	0.626	0.037
Smoothing (3)	10	0.855	0.024	0.621	0.037
SNV	10	0.807	0.027	0.446	0.044
1st derivative (3)	6	0.832	0.025	0.648	0.035
2nd derivative (1)	**10**	**0.773**	**0.029**	**0.675**	**0.034**
MSC	10	0.838	0.025	0.452	0.044
Baseline correction	11	0.884	0.021	0.554	0.039

From the contrastive analysis of cross-validation correlation coefficients (r_{cv}) and cross-validation root mean square errors (RMSECV) in the Table 3, it can be found that two preprocessing methods (1st derivative, 2nd derivative) have the ability to optimize the spectrum information of Vetiver grass roots, with the higher R_{cv} and lower RMSECV than the raw spectrum. The 2nd derivative is the best preprocessing method for optimizing PLS model. However, the rest four preprocessing methods can not raise the quality of model, it should be reasonable. The main functions of derivative preprocessing method are distinguishing the overlapping peaks and improving the resolution of spectrum. It shows that the important interference factors of model established are overlapping peaks.

3.4   Identification of Optimum Wavelength

Each wave number of spectrum means the corresponding information of tested object. Some noises may have been collected by Raman spectrophotometer when the raman spectrums of Vetiver grass roots were collected. In order to reduce the negative influence of noises in the spectrum, eliminating the spectral ranges of noises and extracting the wavelengths of target component is necessary.

Successive projections algorithm (SPA) [12] is a forward selection method, it can promote the quality of model and ensure the realizability of model. Therefore, SPA has been widely praised in the multiple qualitative analysis of the spectrum of applications. With the help of Matlab R2010a software package, SPA was employed to solve the problem of wavelength selection after spectral preprocessing.

It can be seen from Fig. 2, the prediction root mean square error RMSE of model is changed along with the number of variables. In order to select the lowest prediction root mean square error, the number of selected variables of 8 was chose for establishing model. The corresponding PLS model parameters of Pb content in Vetiver grass roots was shown in Table 4.

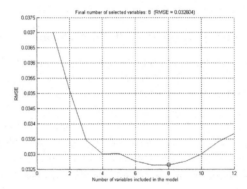

Fig. 2. Variable number screening of Pb model basing on SPA

Table 4. The results of Pb content in the roots of Vetiver grass by PLS with SPA

Variable selection method	PCs	r_c	RMSEC (g/kg)	r_{cv}	RMSECV (g/kg)
None	10	0.773	0.029	0.675	0.034
SPA	8	0.758	0.030	0.697	0.033

Table 4 shows that the effective variables of model was screened by SPA, the cross-validation correlation coefficients (r_{cv}) was rise from 0.675 to 0.697, the cross-validation root mean square errors (RMSECV) was decreased from 0.034 g/kg to 0.033 g/kg, and the principal component number (PCs) was changed to 8. Therefore, the algorithm of SPA can improve the model quality of detecting lead content in Vetiver grass roots, but the effect is not obvious.

In order to find the compounds chelated with heavy metal lead in Vetiver grass roots, the characteristics of Raman spectra wavelengths screened by SPA were listed in Table 5, it shows that a wide variety of compounds in Vetiver grass roots may be chelated with lead, including normal alkanes, bromides, chloroalkane, allenes, hydrazone and so on.

Table 5. The characteristic Raman bands of vetiver grass roots screened by SPA

Wave number screened by SPA/cm^{-1}	Vibration mode	Reference compounds
155.5	Stretch	n-alkanes
395		
560	CBr stretch/CCl stretch	Primary alkyl bromide/Tertiary chloroalkane
614.5	CCl stretch	Para chloro alkanes
1120.5	C=C=C stretch	Allenes
1314	CH$_2$ in-plane deformation	Trans two alkyl vinyl
1637	C=N stretch	Hydrazone (solid)
1964	C=C=C stretch	Allenes

3.5   Prediction of Quality Attributes with the Optimum Wavelength

In order to verify the feasibility of the Pb detecting models in Vetiver grass roots, external validation is further done with the 20 validation samples. The Raman scattering intensities corresponding wave numbers screened by SPA were intercepted for the variables of models, the atomic absorption values were served as measured values, and the optimized PLS models were used for prediction. Figure 3 shows the relationships of the predicted and the measured values of the validation samples, the correlation coefficients of prediction (rp) and root mean square errors of prediction (RMSEP) of Vetiver grass root are 0.607 and 0.040 g/kg, respectively.

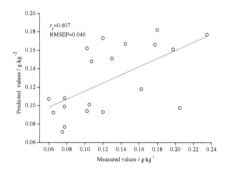

Fig. 3. External validation correlation of heavy metal Pb content in Vetiver grass roots

4   Conclusion

The calibration and validation statistics obtained in this study showed the potential of Raman spectroscopy to predict heavy metal Pb in the root of Vetiver grass. Although the accuracy of model with this method was lower than the routine analysis, Raman spectroscopy could be used as a fast and simple tool to diagnose the content of heavy metal Pb in Vetiver grass approximately. Overall, Raman spectroscopy can be used as a reference method for diagnosing the content of heavy metal Pb in Vetiver grass, but the accuracy needs to be improved.

Acknowledgements. This project is supported by the National Natural Science Foundation of China (61178036, 31160250), Financial support was provided by the National High Technology Research and Development Program of China (863program) (no. 2012aa101906), Ganpo excellence project 555 Talent Plan of Jiangxi Province (2011-64), Center of Photoelctric Detection Technology Engineering of Jiangxi Province (2012-155), Graduate Science and Technology Innovation Fund projects of Jiangxi Province (YC 2013-S157).

References

1. Devkota, B., Schmidt, G.H.: Accumulation of heavy metals in food plants and grasshoppers from the Taigetos Mountains, Greece. Agric. Ecosyst. Environ. **78**(1), 85–91 (2000)
2. Arora, M., Kiran, B., Rani, S., Rani, A., Kaur, B., Mittal, N.: Heavy metal accumulation in vegetables irrigated with water from different sources. Food Chem. **111**(4), 811–815 (2008)
3. Donner, E., Howard, D.L., Jonge, M.D.D., Paterson, D., Cheah, M.H., Naidu, R., Lombi, E.: X-ray absorption and micro X-ray fluorescence spectroscopy investigation of copper and zinc speciation in biosolids. Environ. Sci. Technol. **45**(17), 7249–7257 (2011)
4. Tormen, L., Torres, D.P., Dittert, I.M., Araújo, R.G., Frescura, V.L., Curtius, A.J.: Rapid assessment of metal contamination in commercial fruit juices by inductively coupled mass spectrometry after a simple dilution. J. Food Compos. Anal. **24**(1), 95–102 (2011)
5. Tufekci, M., Bulut, V.N., Elvan, H., Ozdes, D., Soylak, M., Duran, C.: Determination of Pb (II), Zn(II), Cd(II), and Co(II) ions by flame atomic absorption spectrometry in food and water samples after preconcentration by coprecipitation with Mo(VI)-diethyldithiocarba-mate. Environ. Monit. Assess. **185**(2), 1107–1115 (2013)
6. Zhen, X.S., Lu, A.H., Gao, X., Zhao, J., Zheng, D.S.: Contamination of heavy metals in soil present situation and method. Soil Environ. Sci. **11**(1), 79–84 (2002)
7. Torreggiani, A., Domènech, J., Atrian, S., Capdevila, M., Tinti, A.: Raman study of in vivo synthesized Zn(II)- metallothionein complexes: structural insight into metal clusters and protein folding. Biopolymers **89**(12), 1114–1124 (2008)
8. Tan, W.N., Li, Z.A., Zou, B.: Molecular mechanisms of plant tolerance to heavy metals. J. Plant Ecol. **30**(4), 703–712 (2006)
9. Pang, J., Chan, G.S.Y., Zhang, J., Liang, J., Wong, M.H.: Physiological aspects of vetiver grass for rehabilitation in abandoned metalliferous mine wastes. Chemosphere **52**(9), 1559–1570 (2003)
10. Dalton, P.A., Smith, R.J., Truong, P.N.V.: Vetiver grass hedges for erosion control on a cropped flood plain: hedge hydraulics. Agric. Water Manag. **31**(1–2), 91–104 (1996)
11. Datta, R., Das, P., Smith, S., Punamiya, P., Ramanathan, D.M., Reddy, R., Sarkar, D.: Phytoremediation potential of vetiver grass [chrysopogon zizanioides (L.)] for tetracycline. Int. J. Phytorem. **15**(4), 343–351 (2013)
12. Ghasemi-Varnamkhasti, M., Mohtasebi, S.S., Rodriguez-Mendez, M.L., Gomes, A.A., Araújo, M.C.U., Galvão, R.K.: Screening analysis of beer ageing using near infrared spectroscopy and the Successive Projections Algorithm for variable selection. Talanta **89**, 286–291 (2012)

Dynamic Analysis of Urban Landscape Patterns of Vegetation Coverage Based on Multi-temporal Landsat Dataset

Dong Liang, Ling Teng, Linsheng Huang, Xinhua Xie,
Yan Zuo, and Jingling Zhao[✉]

Key Laboratory of Intelligent Computing and Signal Processing,
Ministry of Education, Anhui University, Hefei 230601, China
dliang@ahu.edu.cn,
695805968,@qq.com, 872026319@qq.com,
linsheng0808@163.com, xiexinhuavip@163.com,
aling0123@163.com

Abstract. Dynamic monitoring of vegetation coverage changes, especially on a relatively large temporal scale, have important practical significance in urban planning and environmental protection. The objective of this study is to dynamically investigate the urban landscape patterns of vegetation coverage based on remote sensing techniques. Multi-temporal Landsat images of 1990, 2000 and 2013 were firstly used to produce three vegetation coverage maps of Hefei City, Anhui Province, China with five grades using the NDVI (Normalized Difference Vegetation Index) dimidiate pixel model. Subsequently, a total of eight landscape pattern indictors in FRAGSTATS 4.2 were selected to analyze the dynamic characteristics of area, quantity and density for the study area with different vegetation coverage grades. The results showed that (1) the dominant vegetation coverage of 1990, 2000 and 2013 were the high vegetation coverage, the moderate vegetation coverage and the moderate-to-high vegetation coverage, respectively. The acreage of non-vegetation coverage increased by 1.89 %, while the high vegetation coverage decreased by 10.48 % from 1990 to 2013; (2) the quantity and density of patches decreased by 33.42 % and 33.41 % during 1990–2013. Shannon's diversity index and Shannon's evenness index increased from 0.92 in 1990 to 0.97 in 2000, and then declined to 0.96 in 2013; and (3) the contagion index had an upward trend and conversely the aggregation index showed no significant changes, but both of them were close to 1 during 1990–2013. In comparison with natural influences, the primary driving forces causing the changes were ascribed to human factors including the rapid population growth and fast-growing urban areas.

Keywords: Landscape pattern · Vegetation coverage · Normalized difference vegetation index (NDVI) · Landsat · Remote sensing

D. Li and Z. Li (Eds.): CCTA 2015, Part I, IFIP AICT 478, pp. 300–316, 2016.
DOI: 10.1007/978-3-319-48357-3_30

1 Introduction

Vegetation is a general term that describes the phytocoenosium covering the land, which not only provides large amount of material resources (e.g., vegetation products and vegetation by-product), but also helps in maintaining soil and water, conserving water, fixing sand, purifying air, regulating climate and so on. Furthermore, it plays an important role in maintaining the ecological balance and promoting regional sustainable development [1–3]. With the continuous improvement of socio-economics, the built-up areas gradually expanded, and spatial regions are becoming more concentrated due to human activities, especially in urban areas. Consequently, quantitative monitoring of the spatio-temporal changes of urban vegetation cover has attracted more attention in urban planning and environmental protection [4, 5]. In traditional vegetation monitoring, statistical methods with lower updating frequency are usually used, but they are rather time-consuming, laborious, and costly, especially failing to monitor the changes of vegetation coverage on a large scale. Conversely, remote sensing has facilitated extraordinary advances in analyzing urban vegetation coverage changes by its real-time and wide coverage features, which has become a significant measurement for monitoring the urban ecological environment.

In recent years, different ecological problems have appeared in urban regions, and some landscape ecology approaches to combining remote sensing have been applied to explore the urban landscape pattern and its change process [6–9]. Tang et al. linked spatial pattern and biophysical parameters of urban vegetation by multi-temporal Landsat imagery [10]. Liu et al. monitored vegetation cover changes in Hefei based on normalized difference vegetation index (NDVI) dimidiate pixel model [11]. Deng et al. used MODIS-NDVI to monitor the vegetation coverage of Shangri La in Northwest Yunnan [12]. It can be found that in the previous studies, multi-temporal Landsat imagery including TM and ETM$^+$ were primarily used on a regional scale, while MODIS-NDVI products were mainly used on a national or continental scale. In addition, the methods used in the studies were primarily concentered on how to calculate the vegetation coverage precisely for monitoring the spatio-temporal vegetation changes. Conversely, the urban landscape patterns of vegetation coverage have not fully been investigated and explored. Specifically, two primary limitations can be found in the aforementioned studies. On the one hand, most of the studies are based on the changes in vegetation quantity and spatial distribution derived from remote sensing imagery. On the other hand, most remote sensing data are focused on Landsat TM/ETM$^+$ (before 2010) and didn't integrate the new data of Landsat8 launched in 2013.

In this study, Hefei, the capital city of Anhui Province, China were taken as the study area and three Landsat imageries (Landsat 5 TM images of 1990 and 2000 and Landsat 8 OLI (Operational Land Imager) image of 2013) were firstly used to calculate and classify the urban vegetation coverage grades based on the NDVI dimidiate pixel model. Subsequently, the landscape patterns of vegetation coverage were dynamically analyzed in FRAGSTATS 4.2 software. Finally, the primary driving forces were also investigated.

2 Data and Methodology

2.1 Study Area

Hefei, the capital city of Anhui Province, China is located between 30°56' N–32°32' N and 116°40' E–117°58' E, with an area of more than 11408.48 km². It has a subtropical monsoon climate, with an annual average temperature of 15.7 °C, an annual average precipitation of 1000 mm and an annual average sunshine of 2100 h. As of 2014, the city has a total resident population of 7.696 million. The general trend of topography is high in the north, southeast and southwest, and low in central and southern parts.

2.2 Data Source and Pre-processing

Landsat (name indicating Land + Satellite) imagery is available since 1972 from seven satellites in the Landsat series, especially the newly launched Landsat8 in 2013. The time-series remotely sensed imagery with various spectral and spatial resolutions in Landsat series provides good data sources for dynamic analysis of urban vegetation coverage. In this study, multi-temporal Landsat satellite images were acquired and used to dynamically investigate the landscape pattern changes of urban vegetation coverage. Specifically, the cloud-free Landsat5 TM data (path 121/row 38) of 1990 and 2000 and Landsat 8 OLI image of 2013 were acquired (Table 1 and Fig. 1), where the data of 2000 and 2013 was derived from the United States Geological Survey (USGS), and the data of 1990 was derived from the Global Land Cover Facility (GLCF). In addition, the vector data of Hefei was also used to mask the three Landsat images.

Table 1. Multi-temporal Landsat images used in the study

Year	Path/row	Acquisition time	Sensor	Resolution (m)
1990	121/38	1990.9.19	Landsat 5 TM	30
2000	121/38	2000.9.25	Landsat 5 TM	30
2013	121/38	2013.9.20	Landsat 8 OLI	30

The Landsat_5 TM_in 1990 derived from the GLCF is the Geocover data set which is a collection of high resolution satellite imagery provided in a standardized, orthorectified format. While the Landsat_5 TM in 2000 and Landsat_8 OLI in 2013 derived from the USGS are primary products and they were processed by referring to the data of 1990. The quadratic polynomial geometric correction model was used to geometrically correct the images of 2000 and 2013 within 0.5 pixels. Radiometric calibration and atmospheric correction were also performed. Finally, the vector file of Hefei was utilized to mask the three images.

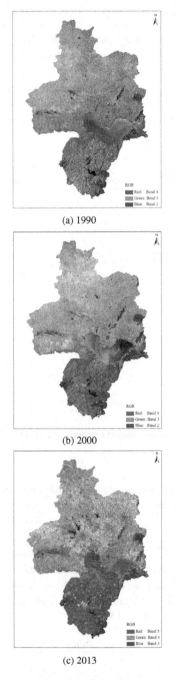

(a) 1990

(b) 2000

(c) 2013

Fig. 1. The false-color composite images of 1990, 2000 and 2013 of Hefei (Color figure online)

2.3   Method

As shown in Fig. 2, two primary steps would be required. Firstly, the NDVI images of three pre-processed images were respectively calculated according to the Eq. 1. Then, the appropriate values of $NDVI_{veg}$ and $NDVI_{soil}$ for three periods were selected from corresponding Landsat images, and the vegetation coverage were calculated in accordance with the Eq. 2. Finally, different landscape indices were calculated in FRAGSTATS 4.2 software for further analyzing the vegetation landscape patterns in the study area.

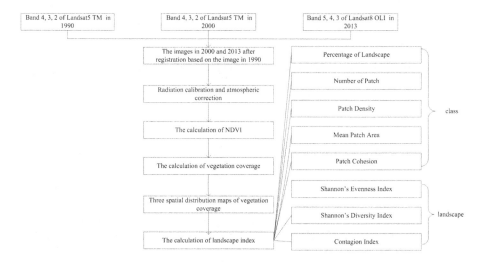

Fig. 2. Flow chart of calculating vegetation coverage and performing landscape pattern analysis

2.3.1   Calculation of NDVI

Vegetation index (VI) can show a strong absorption characteristics of plant chlorophyll in 0.69 μm. Consequently, the vegetation healthy status can be expressed by corresponding spectral information of red and near infrared bands by simple linear or non-linear combination. Multi-spectral remote sensing data have been extensively used to investigate the vegetation features based on different vegetation indices [13]. In comparison with other VIs, NDVI (Eq. 1) has been widely used in the monitoring of vegetation biomass by remote sensing techonology. It is not only related to the plant distribution density, but also reflects the plant growth and the spatial distribution density of plant growth factors [14, 15].

$$NDVI = \frac{\rho_{NIR} - \rho_{red}}{\rho_{NIR} + \rho_{red}} \tag{1}$$

Where, ρ_{NIR} and ρ_{red} represent the reflectance of the near infrared (NIR) band and the red band. For the Landsat 5 TM image, NIR corresponds to the Band 4 and red corresponds to the Band 3. For the Landsat 8 OLI image, they correspond to the Band 5 and the Band 4, respectively.

2.3.2   Calculation of Vegetation Coverage

Vegetation coverage refers to the ratio of the vegetation canopy area in vertical projection and the total soil area, namely planting soil ratio [16]. In our study, and the NDVI dimidiate pixel model was selected to estimate the vegetation coverage (Eq. 2), which is a simple and practical remote sensing estimation model by assuming that the earth's surface corresponding to the pixel is composed by the vegetation cover regions and non-vegetation regions [17, 18]. Similarly, the observed spectral information by remote sensing sensor is also composed by the linear weighted synthesis of the two components, and the weight of each factor is the proportion of the pixel area [15].

$$F_c = \frac{NDVI - NDVI_{soil}}{NDVI_{veg} - NDVI_{soil}} \tag{2}$$

Where, F_c represents the vegetation coverage, while $NDVI_{veg}$ and $NDVI_{soil}$ were the $NDVI$ values of pure vegetation and pure soil, respectively.

2.3.3   Selection and Calculation of Landscape Pattern Indices

Landscape indices are usually used in landscape ecology quantitative research methods, which can describe the landscape patterns and corresponding changes, and establish a link between the landscape pattern and changing process [19]. In this study, the landscape indices were calculated in FRAGSTATS 4.2 software which can provide more than 50 landscape indices at three levels (patches, landscape types and landscape). A total of eight classes can be divided according to the measurement aspects of landscape pattern, which are Area/Density/Edge Index, Shape Index, Core Area Index, Independent/Near Index, Contrast Index, Contagion/Discrete Degree Index, Connectivity Index and Diversity Index. Specifically, two aspects of landscape patterns were mainly carried out. One is the amount distribution of patches of different vegetation coverage grades to describe the structure and changes, which includes Number of Patch (NP), Patches Area (AREA), Aggregation Index (AI), Percentage of Landscape (PLAND), Shannon's Evenness Index (SHEI), Shannon's Diversity Index (SHDI), etc. The other is the spatial form and distribution characteristics of patches for describing spatial pattern and changes, which are measured by Contagion Index (CONTAG), Fractal Dimension Index (FRAC), Patch Cohesion Index (COHESION), etc. [20]. In our study, PLAND, NP, PD, AREA-MN, COHESION, CONTAG, SHDI and SHEI were selected by referring to [21] to describe the vegetation spatial pattern changes (Table 2).

Table 2. Landscape pattern indices and their ecological significance*

Index and formula	Level	Ecological significance
$PLAND = p_i$	Landscape class	Reflecting the abundance ratio of a certain type of patches in the landscape
$NP = n_i$	Landscape class	Reflecting the landscape heterogeneity and fragmentation
$PD = \frac{n_i}{A} * 10000 * 100$	Landscape class	Reflecting the amount of patches per unit area
$AREA - MN = \dfrac{\sum\limits_{j=1}^{n} x_{ij}}{n_i}$	Landscape class	Reflecting the landscape fragmentation
$COHESION = [1 - \dfrac{\sum\limits_{i=1}^{m}\sum\limits_{j=1}^{n} p_{ij}}{\sum\limits_{i=1}^{m}\sum\limits_{j=1}^{n} p_{ij}\sqrt{a_{ij}}}]$	Landscape class	Reflecting the natural status of connectivity of patch type
$CONTAG = [1 + \sum\limits_{i=1}^{m}\sum\limits_{j=1}^{n} \frac{p_{ij}\ln(p_{ij})}{2\ln(m)}]$	Landscape	Reflecting of the reunion degree and extending trend of different patch types
$SHDI = -\sum\limits_{i=1}^{m} (p_i.lnp_i)$	Landscape	Reflecting the amount of landscape factors and proportion of various landscape factors change
$SHEI = \dfrac{-\sum\limits_{i=1}^{m} (p_i.\ln(p_i))}{\ln m}$	Landscape	Reflecting the ratio of equilibrium degree of area proportion of different patch types as well as its maximum in the landscape

* Denotes that i represents the patch type; j represents the amount of patches; m represents the total number of all the patch types; n represents the number of a certain patch type; P_i represented the circumference of the ith patch type.

3 Results and Discussion

3.1 Classification Maps of Vegetation Coverage

Three vegetation coverage maps were produced according to the Eq. 2 (Fig. 3). Five grades were categorized: Grade 1: non-vegetation coverage ($0 \leq F_c < 20$); Grade 2: low vegetation coverage ($20 \leq F_c < 40$); Grade 3: moderate vegetation coverage ($40 \leq F_c < 60$); Grade 4: moderate-to-high vegetation coverage ($60 \leq F_c < 80$); and Grade 5: high vegetation coverage ($80 \leq F_c < 100$). It showed that the spatial distribution of vegetation coverage in Hefei showed an obvious characteristics, The peripheral vegetation coverage was higher and the central city was lower in 1990, 2000 and 2013. In 1990, the vegetation coverage of the study area was good, the high vegetation coverage was mainly in the south, southwest and northeast of, and the area of non-vegetation coverage was mainly concentrated in the central city. From 1990 to 2013, the vegetation coverage of the study area had been reduced obviously. The high vegetation coverage in southwest and northeast had been degraded into low and moderate-to-high vegetation coverage and the area of non-vegetation coverage in the central city showed an obvious trend of expanding. The regions with obvious decreasing vegetation coverage can be ascribed to active regional economic developments [13].

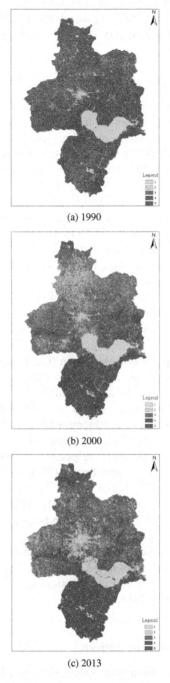

(a) 1990

(b) 2000

(c) 2013

Fig. 3. Three spatial distribution maps of five-grade vegetation coverage

On the other hand, the changed vegetation coverage regions can also reflect the development paths for the study area. In general, the urban sprawl has been transferred from central urban area to the periphery.

3.2   Analysis of the Number of Vegetation Landscape Structure Change

The Percentage of Landscape (PLAND), Number of Patches (NP), Shannon's Diversity Index (SHDI) and Shannon's Evenness Index (SHEI) were selected to analyze the changes of the number of vegetation landscape structure. As shown in Table 3, the percentage distribution (Fig. 4a) of different vegetation coverage of 1990, 2000 and 2013 could be obtained. In 1990, Grade 5 occupied the dominant position in the landscape, and followed by Grade 4, Grade 3, Grade 1 and Grade 2. In 2000, Grade 3 occupied the dominant position in the landscape, and the area ratio of Grade 5 was relatively high. In 2013, Grade 4 was the dominant type in the landscape. The area ratio of Grade 1 increased slightly from 12.35 % to 14.25 % during 1990–2013. The area ratio of Grade 2 increased from 8.25 % to 13.90 % during 1990–2000, and then declined to 10.89 % in 2013. The area ratio of Grade 3 increased from 15.54 % to 27.61 % during 1990–2000, and then declined to 21.18 % in 2013. On the whole, the area of Grade 2 and Grade 3 increased by 2.64 % and 5.64 % respectively during 1990–2013. The area of Grade 4 increase slightly from 28.36 % to 28.67 %. Conversely, the area of Grade 5 changed greatly and decreased from 35.50 % to 25.02 % during 1990–2013.

To describe the landscape heterogeneity and fragmentation, the number of patches was used. As shown in Table 3, the distribution of the number of patches for different vegetation coverage grades could be also obtained (Fig. 4b). It showed that the total number of patches in the study area reduced by 278555 during 1990–2013. Specifically, the number of patches of Grade 1 increased from 74867 to 74908 during 1990–2000, and then decreased to 47698 in 2013. From 1990 to 2013, that of Grade 2 and Grade 5 decreased by 42237 and 8497, respectively. That of Grade 3 and Grade 4 decreased from 259384 to 170252 and from 215976 to 183300 respectively during 1990–2000, and then decreased by 12546 and 66298. It was obvious that the number of patch of Grade 3 changed greatly during 1990-2013 (Fig. 4).

The increase of Shannon's Diversity Index indicated that various types of patches in the landscape showed the equalized distribution. Larger evenness index showed that the landscape patches in area had more uniform distribution. The results showed that SHDI and SHEI were relatively high in the study area. The Shannon's Diversity Index increased from 1.4785 to 1.5526 and the Shannon's Evenness Index increased from 0.9186 to 0.9647 during 1990–2013.

3.3   The Changes in Landscape Pattern

Patch Density Index, Mean Match Area Index, Cohesion Index and Contagion index were selected to analyze the landscape pattern changes in space of different vegetation coverage. As shown in Table 4, patch density distribution of different vegetation

Table 3. Spatial landscape pattern properties of vegetation coverage

Type	Percentage of landscape (%)			Patches number (a)			Shannon's Diversity Index			Shannon's Diversity Index		
	PLAND			NP			SHDI			SHEI		
	1990	2000	2013	1990	2000	2013	1990	2000	2013	1990	2000	2013
1	12.35	12.28	14.25	74867	74908	47698	**1.4785**	**1.5612**	**1.5526**	**0.9186**	**0.9700**	**0.9647**
2	8.25	13.90	10.89	189087	198783	146850						
3	15.54	27.61	21.18	259384	170252	157706						
4	28.36	25.75	28.67	215976	183300	117002						
5	35.50	20.46	25.02	94174	112682	85677						
LAND	100.00	100.00	100.00	833488	739925	554933	1.4785	1.5612	1.5526	0.9186	0.9700	0.9647

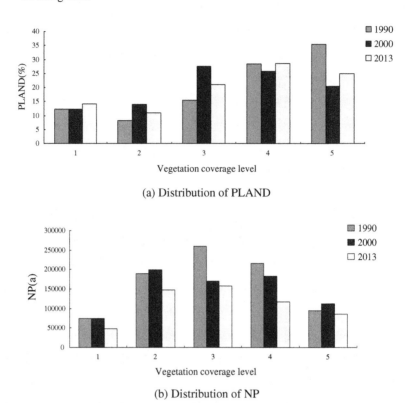

(a) Distribution of PLAND

(b) Distribution of NP

Fig. 4. Comparison of PLAND and NP for five vegetation coverage grades among three periods

Table 4. Spatial landscape pattern properties of vegetation coverage

Type	Patch density (%)			Mean patch area (hm²)			Patch Cohesion (%)			Contagion Index (%)		
	(PD)			(AREA-MN)			(COHESION)			(CONTAG)		
	1990	2000	2013	1990	2000	2013	1990	2000	2013	1990	2000	2013
1	6.55	6.55	4.17	1.89	1.87	3.41	98.90	98.89	99.52	**26.43**	**3.26**	**26.45**
2	16.55	17.39	12.85	0.50	0.80	0.85	76.73	98.83	88.67			
3	22.70	14.90	13.80	0.68	1.85	1.54	88.09	99.59	98.02			
4	18.90	16.04	10.24	1.50	1.61	2.80	98.96	98.65	99.24			
5	8.24	9.86	7.50	4.31	2.08	3.34	98.85	99.30	99.41			
LAND	72.93	64.74	48.56	1.37	1.54	2.06	98.29	99.29	98.97	26.43	23.26	26.45

coverage of three years could be obtained (Fig. 5a). From 1990 to 2013, the total patch density of landscape in the study area reduced from 72.93 % to 48.56 %. For urban areas, patch density decreased at different levels during 1990–2013. For Grade 1, patch density remained unchanged from 1990 to 2000, however it decreased from 6.55 % to

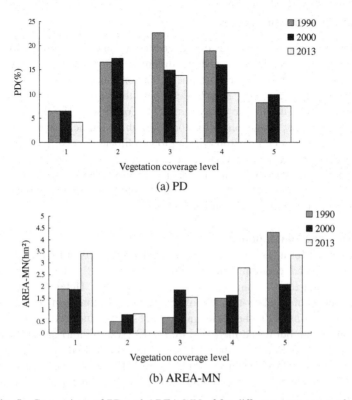

(a) PD

(b) AREA-MN

Fig. 5. Comparison of PD and AREA-MN of for different coverage grades

4.17 % during 1990–2013. For Grade 2 and Grade 5, the patch density increased firstly and then decreased, which showed that patches of two grades were getting less fragmented in the entire period. Similarly, Grade 3 and Grade 4 declined respectively during 1990–2000 and 2000–2013. The largest patch density zones were Grade 3, Grade 2 and Grade 3, respectively. The results could be explained that the lands of the three grades had become more fragmented. Conversely, the smallest patch density zones of three years were the Grade 1, which showed there were lower degree of fragmentation degree in these areas of Grade 1.

Concerning on the AREAR-MN (Table 3), comparison of the number of patches for different vegetation coverage grades could also be obtained (Fig. 5b). There was an increase for all the grades except the Grade 5e. For Grade 1, the mean patch area increased significantly by 1.52 hm^2, which indicated that the vegetation coverage was fragmentated in the study area during 1990–2013.

From 1990 to 2013, Contagion Index tended to increase and there was no significant changes for Patch Cohesion Index, but both of them were close to 1, which showed that the landscape overall trend gathered, and good natural connectivity. Grade 1, Grade 4 and Grade 5 in the three periods were close to 1, which indicated that the aggregation degree was higher. However, the Patch Cohesion Index of Grade 2 and Grade 3 were relatively lower, which indicated that those regions of three grades had lower degree of aggregation, but they increased for s in 2013.

3.4   Analysis of Driving Forces

Driving forces analysis is a way of understanding and accounting for vegetation coverage changes potentially caused by different factors such as regional climate change and human factors [22]. Environmental factors, such as slope, aspect, elevation, temperature, precipitation, *etc.*, usually determine the vegetation growth and agricultural development in the fragile ecosystem [23]. Conversely, Hefei city is the capital city and the influences of human factors paly a highly significant role in changing the vegetation coverage in comparison with regional climate change. Fast-growing urban areas have changed the natural vegetation coverage, and demographic and economic development are the primary driving factors for the growing metropolitan cities [24]. In this study, three driving forces were investigated including the rapid population growth, the fast-growing urban the higher urbanization level.

3.4.1   Economic Development

The urban sprawl for a fast-growing city is substantially driven by the urgent requirements of social and economic development. Consequently, the natural vegetation coverage is changed to adapt to the human activities. Hefei has experienced a remarkable period of rapid economic growth spanning in more than 20 years from 1978 to 2010 (Fig. 6). In 1990, the local gross domestic product (GDP) was just 58.19 (100 million) Yuan, but it had increased to 324.73 (100 million) of 2000 and 4672.91 (100 million) of 2013. Similarly, the GDP per capita also showed a rapid growth trend. It was 458, 7481 and 61555 Yuan in 1990, 2000, 2013, respectively. In more than 20 years, the GDP in Hefei had increased by 7,930 % and the GDP per capita have greatly

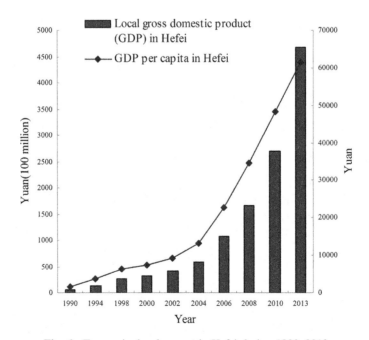

Fig. 6. Economic development in Hefei during 1990–2013

increased by 13340 %. It was inevitable that the natural vegetation coverage had been extremely change along with the improvement in economic development.

3.4.2 Growth of Population and Construction Areas

Since the 1990s, the total population have showed a sharp increase from 386.42 (10,000) in 1990 to 761.1 (10,000) in 2013 (Fig. 7a). According to the population statistics from Anhui Statistical Information Net (http://www.ahtjj.gov.cn/tjj/web/index.jsp), there were totally 5,702 million registered permanent residents in 2010 of Hefei City from the sixth national population census. A total of 1.235 million (27.65 %) was increased in

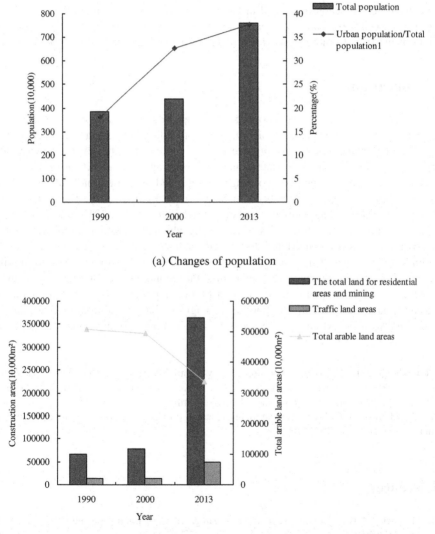

(a) Changes of population

(b) Area changes of housing construction, trafficland and arable land

Fig. 7. Changes of population and artificially changed areas in Hefei during 1990–2013

10 years and the average increase was 0.124 million (about 2.47 %) per year compared to the fifth population census in 2000. With the rapid urbanization and accelerated development in Hefei, the number of registered permanent residents have been increasing. Consequently, more and more houses and traffic facilities will be required along with the population growth. According to the Anhui Statistical Yearbook 2014, the construction areas and the traffic land areas have significantly increased from 1990 to 2013 (Fig. 7b). Specifically, the total land for residential areas and mining was just 66,477.64 (10,000) in 1990, but it reached 78,036.95 (10,000) in 2000 and 364040 (10,000) in 2013, respectively. Similarly, the traffic land area was just 13,593.15 (10,000) in 1990, but it reached 48,710 (10,000) in 2013. At the same time, arable land areas have kept a decreasing trend in Hefei (Fig. 7b). The total arable land areas was 507,868.42 (10,000), and it decreased to 494,464.80 (10,000) in 2000 and 336,025 (10,000) in 2013. In more than 20 years, the arable land increased by 33.84 %.

4 Conclusions

The landscape pattern of vegetation coverage can be dynamically explored based on multi-temporal remote sensing imagery. In general, the high vegetation coverage was significantly reduced, and the moderate-to-high degree had little changes, while low vegetation coverage and moderate vegetation coverage was significantly increased. Specifically, the high vegetation coverage in Hefei had decreased markedly from 1990 to 2000, but the situation had become better from 2000 to 2013. Furthermore, the degree of landscape fragmentation had decreased from 1990 to 2013, while landscape diversity index and evenness index were higher, which showed that the distribution of each patch type was balanced and distributed relatively evenly in area. To find out the driving forces causing the changes of vegetation coverage, three types of driving factors were primarily discussed, including the economic development, growth of population and construction areas, and institutional policies. Multi-temporal remote sensing images are useful to monitor ecological changes of regional vegetation. Additionally, it is extremely necessary to find out the driving forces causing the changes of vegetation coverage.

Acknowledgment. The project was supported by Anhui Provincial Natural Science Foundation (No. 1408085QF126), the Open Research Fund of Key Laboratory of Digital Earth Science, Institute of Remote Sensing and Digital Earth, Chinese Academy of Sciences (No. 2014LDE012), and the Leadership Introduction Project of Academy and Technology of Anhui University (No. 10117700024).

References

1. Morancho, A.B.: A hedonic valuation of urban green areas. Landscape and Urban Plann. **66**, 35–41 (2003)
2. Tang, J., Fang, C., Schwartz, S.S., et al.: Assessing spatiotemporal variations of greenness in the Baltimore-Washington corridor area. Landscape and Urban Plann. **105**, 296–306 (2012)

3. Gong, Z., Gong, H., Li, X., et al.: Ecological environment effect analysis of Wetland change in Beijing region using GIS and RS. In: Urban Remote Sensing Event, 2009 Joint, pp. 1–7. IEEE (2009)
4. Dale, V.H.: The relationship between land-use change and climate change. Ecol. Appl. **7**, 753–769 (1997)
5. Kalnay, E., Cai, M.: Impact of urbanization and land-use change on climate. Nature **423**, 528–531 (2003)
6. Tang, J.: Spatial analysis and temporal simulation of landscape patterns in two petroleum-oriented cities. IEEE J. Sel. Top. Appl. **2**, 54–60 (2009)
7. Griffith, J.A., Trettin, C.C., O'Neill, R.V.: A landscape ecology approach to assessing development impacts in the tropics: a geothermal energy example in Hawaii. Singap. J. Trop. Geogr. **23**(1), 1–22 (22) (2002)
8. Yin, L., Chen, H., Li, W.: Multi-scale analysis of landscape pattern of Hunan Province. In: 2014 3rd International Workshop on Earth Observation and Remote Sensing Applications (EORSA), pp. 309–313. IEEE (2014)
9. Peng, B., Hu, Y.: A study in stability of the regional land-use landscape pattern. In: 2011 International Conference on Remote Sensing, Environment and Transportation Engineering, pp. 1903–1906 (2011)
10. Tang, J., Tang, J.: Linking spatial pattern and biophysical parameters of urban vegetation by multitemporal landsat imagery. Geosci. Remote Sens. Lett. IEEE **10**(5), 1263–1267 (2013)
11. Liu, L., Yao, B.: Monitoring vegetation-cover changes based on NDVI dimidiate pixel model. Trans. Chin. Soc. Agric. Eng. **26**, 230–234 (2010)
12. Gong, Z., Gong, H., Li, X., et al.: Dynamic monitoring of vegetation coverage in Shangri-La county of northwest Yunnan based on MODIS-NDVI. In: International Conference on Geoinformatics, pp. 1–5. IEEE (2013)
13. Chen, S.: Study of Information Mechanism Remote Sensing, 1st edn., pp. 175–185. Guangxi Education Press, Guangxi (1998). (in Chinese)
14. Wang, M., Yang, W., Shi, P., et al.: Diagnosis of vegetation recovery in mountainous regions after the Wenchuan earthquake. IEEE J. Sel. Topics Appl. Earth Obs. Remote Sens. **7**, 3029–3037 (2014)
15. Wang, P., Li, X., Gong, J., et al.: Vegetation temperature condition index and its application for drought monitoring. In: Geoscience and Remote Sensing Symposium (2001)
16. Adams, J.E., Arkin, G.F.: A light interception method for measuring row crop ground cover. Soil Sci. Soc. Am. J. **4**, 789–792 (1977)
17. Leprieur, C., Verstraete, M.M., Pinty, B.: Evaluation of the performance of various vegetation indices to retrieve vegetation cover from AVHRR data. Remote Sens. Rev. **10**, 265–284 (1994)
18. Zribi, M., Le Hégarat-Mascle, S., Taconet, O., et al.: Derivation of wild vegetation cover density in semi-arid regions: ERS2/SAR evaluation. Int. J. Remote Sens. **24**(6), 1335–1352 (2003)
19. O'Neill, R.V., Krummel, J.R., Gardner, R.H., et al.: Indices of landscape pattern. Landscape Ecol. **1**(3), 153–162 (1988)
20. Zhen, X.: Landscape Spatial Analysis Technology and Application, 1st edn., pp. 93–147. Science Press, Beijing (2010). (in Chinese)
21. Jones, K.B., Wickham, J.D., O'Neill, R.V., et al.: Monitoring environmental quality at the landscape scale. Bioscience **47**, 513–519 (1997)
22. Zhou, H., Wang, J., Yue, Y., et al.: Research on spatial pattern of human-induced vegetation degradation and restoration: a case study of Shaanxi Province. Acta Ecologica Sinica **29**(9), 4847–4856 (2009)

23. Wang, G., Wang, Y., Jumpei, K.: Land-cover changes and its impacts on ecological variables in the headwaters area of the Yangtze River, China. Environ. Monit. Assess. **120** (1–3), 361–385 (2006)
24. Lin, G.C.S., Ho, S.P.S.: China's land resources and land-use change: insights from the 1996 land survey. Land Use Policy **20**, 87–107 (2003)

The Application of the OPTICS Algorithm in the Maize Precise Fertilization Decision-Making

Guowei Wang[1,2,3(✉)], Yu Chen[1,3], Jian Li[1,3], and Yunpeng Hao[1,3]

[1] School of Information Technology, Jilin Agricultural University,
Changchun 130118, China
41422306@qq.com, 123929697@qq.com,
2312852319@qq.com, 793097534@qq.com
[2] School of Biological and Agricultural Engineering,
Jilin University, Changchun 130022, China
[3] Jilin Province Research Center,
Changchun 130118, China

Abstract. With the development of computer science and information technology, data mining technology in the field of agriculture in recent years has become a hot research. Corn planting process, rational fertilization can effectively promote the growth of corn, however, no basis and targeted fertilization may cause shortage of low fertility soil fertilization, high fertility soil fertilization overdose. To solve this problem, In this paper, cluster analysis OPTICS algorithm based on density of soil classification, and press the nutrient balance method to calculate the level of soil fertility for each corresponding amount of fertilizer, farmers can be targeted based on fertilizer fertilization. In the town of Yushu City, Jilin Province by Gongpeng for application, compared with the traditional fertilization, fertilizer input savings of 20.5 %, maize yield of about 10 %, not only to meet the needs of farmers, but also achieve a reduction in fertilizer inputs, increase production purposes.

Keywords: Data mining · Precision fertilization · Fertility classification · OPTICS algorithm

1 Introduction

Precision fertilization is one of the core elements of precision agriculture [1]. The method using GPS positioning to obtain field information, including production monitoring, soil sampling and so on. The computer system through analysis of the data processing, decision of the management measures of agricultural land, the yield and soil status information is loaded with GPS devices, implemented on each operating unit due to the overall balance of the soil due to crop fertilization, thus greatly improving fertilization fertilizer use efficiency and economic benefits, reduce the adverse impact

© IFIP International Federation for Information Processing 2016
Published by Springer International Publishing AG 2016. All Rights Reserved
D. Li and Z. Li (Eds.): CCTA 2015, Part I, IFIP AICT 478, pp. 317–324, 2016.
DOI: 10.1007/978-3-319-48357-3_31

on the environment [2, 3]. However, the variable rate fertilization machine with GPS has not been used widely, so the method of dividing the operation unit is not conducive to the promotion of farmers.

Soil fertility is an important aspect of the constitution of land productivity, it depends on many factors and a variety of nutrients in the soil organic matter content, texture, thickness and soil tilth configuration, and these factors on soil fertility and sizes mutual restraint, is the ability to provide the soil for plant growth and coordination of nutrition and environmental conditions. The soil fertility grading, contribute to the establishment of scientific fertilization scheme, abandon the traditional concept of the more, the better. The soil nutrient status, crops need fertilizer characteristics determine the reasonable amount of fertilizer and nutrients mix, thereby reducing the production chain of agricultural input costs and the probability of the occurrence of pests and diseases, improve nutrient utilization. Reduce the background of cultivated land, control the point source pollution and non-point source pollution, and restore the ecological agriculture and sustainable development [4–7].

Currently, soil fertility studies, Luo Laijun [8] used fuzzy mathematics theory, to explore the reclamation area of farmland soil fertility grading method, which provides a reasonable reference for the target yield. Based on GIS Map info software, the establishment of land resources management information spatial database and attribute database. The soil nutrient content and distribution map of soil nutrient content and distribution map were made.

In summary, this paper adopts cluster analysis in optics algorithm based on density and on soil fertility grading, and according to the nutrient balance method to calculate the corresponding of soil fertility at each level of fertilization, provides a feasible scheme for large area promotion of precision fertilization.

2   OPTICS Algorithm

2.1   Basic Concepts of OPTICS Algorithm

The core idea of the density of clusters is a point of ε neighborhood neighbor points to measure the density of the point where the space [9]. If ε neighborhood neighbor exceeds a specified threshold MinPts, it is that the point is in a cluster, called the core point, or that the point is on the boundary of a cluster, called a boundary point [10, 11]. Here are some definitions:

Define 1. Core distance
The core distance object p is p as the core object of minimum. If p is not the core object, the core distance of the p is not meaningful. The core distance is the smallest radius of the neighborhood of the core points.

To explain the definition of core distance and reachable distance, we set up Minpts = 5, ε = 6 mm, if the core object, a field point p must have at least 5 points (including itself). In Fig. 1, the d point is the boundary point that satisfies the core object of p, so the core distance of p is c = 3 mm.

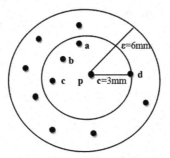

Fig. 1. Core distance

Define 2. Reachability-distance
Reachability-distance is the bigger one from the core of p and the Euclidean distance between n and p.

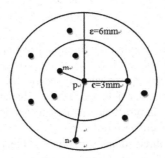

Fig. 2. Reachability-distance

In Minpts = 5, ε = 6 mm, according to n p on the definition of distance, distance of n to Euclidean distance p. Because the Euclidean distance of p to n is more than the core distance of 3 mm.

Define 3. Directly density-reachable
If the p is the core point, q in the - neighborhood of p, the p direct density of up to q.

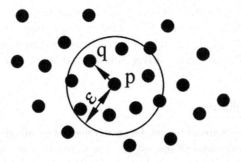

Fig. 3. Directly density-reachable

2.2 OPTICS Algorithm Process

The algorithm is as follows:

Step1: Create two queues, order queues and result queues. (Orderly queue is used to store the core object and its core direct access to an object, and arranged in ascending distance; Results the sequence of the output sequence of the sample points is stored

Step2: If all the points in the sample set D are processed, the algorithm ends. Otherwise, choose an untreated (is not in the queue) and as the core object of sample points, find the all density reachable sample points, such as the sample point does not exist in the queue, will it into the ordered queue, and are ordered by distances of up to

Step3: If the ordered queue is empty, the jump to the step 2, otherwise, the first sample point from the orderly queue to expand, and the sample points will be removed to the results queue

Step3.1: Judge whether the expansion point is the core object, if not, back to step 3, otherwise find the expansion point of all the direct density of up to point;

Step3.2: Judging whether the direct density of the sample point is already there is the result queue, is not processed, otherwise the next step;

Step3.3: If orderly queue already exists the directly density reachable point, if this new reachability distance less than old reachable from, sharp distance is used to replace old reachability distance, orderly queue reordering.

Step3.4: If the direct density of the sample points is not present in the ordered queue, then the insertion point is inserted, and the ordered queue is re ordered;

Step4: The algorithm is over and the ordered sample points of the output queue are ordered.

3 Core Precision Fertilization Decision

3.1 Data Acquisition

Gongpeng town in Yushu No. thirteen villages, land use GPS to obtain information, press 40 m * 40 m mesh distance. Each grid soil sampling, 117 soil samples were obtained, each soil sample for testing to give each sample soil organic matter, available phosphorus, available nitrogen, available potassium content data.

3.2 Data Standardization

Since the practical problems, different data generally has a different dimension, in order to make different dimensionless quantity can be compared, so the need for standardization of data, about to compress the data to the [0,1]. Data normalization formula is:

$$g = \frac{\max - \min}{G - \min} \tag{1}$$

3.3 Soil Fertility Grading

Using OPTICS algorithm to process the data, that is, soil fertility, soil fertility of the clustering, divided into different clusters. This paper divides the soil fertility into four clusters. Calculate the distance of the range of each cluster, and draw up the distance chart of the sample points according to the reachable distance. The output sequence is the horizontal coordinates, and the distance chart is the vertical coordinate. The reachable distance ordered graph is generated by the plot function in the MATLAB software.

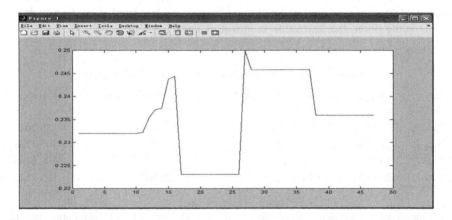

Fig. 4. Orderly figure of the reachability-distance

As shown in Fig. 4, it is clearly divided into four wave values, that is, the four clusters, we can according to the distance ordered map of the horizontal coordinates, determine the number of the four clusters contained in each of the soil number.

The point is, we can through the analysis of ordered graphs can be obtained directly when the parameter E (field radius) was 0.24, minpts (epsilon minimum points) for 10 of the clustering results. Can be seen from the chart, 0.24 the horizontal lines on the ordinate, data is obviously divided into three clusters, other not in the range of points is regarded as outlier.

According to the OPTICS algorithm, the soil fertility is effectively clustered into clusters. The average values of the four attributes of each cluster are respectively summation, and then according to the size of the classification, and the largest of the land, and so on.

According to the average values of the attributes of the four clusters, the results of the soil classification are shown in Table 1.

Table 1. Soil classification result

Soil level	Included sample points
Class A	a68 a102 a111 a75 a103 a92 a112 a105 a106 a96 a73 a72 a113 a95 a93 a108 a90 a79 a99 a109 a104 a85 a100 a81
Class B	a5 a83 a22 a60 a29 a15 a6 a97 a56 a37 a7 a17 a49 a8 a12 a26 a84 a59 a34 a16 a25 a23 a13 a24 a101 a38 a19 a116 a74 a45 a71 a94 a2 a18 a31 a3 a27 a47 a117 a20
Class C	a48 a44 a76 a11 a21 a30 a61 a4 a70 a33 a65 a50 a82 a14 a10 a36 a69 a51 a43
Class D	a41 a86 a78 a91 a28 a80 a115 a89 a88 a77 a39 a40 a42 a67 a55 a64 a46 a58 a35 a54 a110

3.4   Calculation of Fertilizer Application

Using nutrient balance method, we calculated the amount of soil nutrient, to meet the crop yield. Because the data in the nutrient balance method need to be the raw data of the sampling point, the data value is shown in Table 2.

Table 2. Average properties of different soil classification

Soil level	Organic mean	Available phosphorus mean	Available nitrogen mean	K mean
Class A	27.16404	11.24264	151.8825	82.20833
Class B	27.7042	15.93813	132.363	105.4524
Class C	30.28655	11.58181	128.2779	108.5789
Class D	26.6535	13.8507	129.7659	103.7143

Calculation of nutrient balance fertilization model by fertilizer application rate (**2**):

$$sf = \frac{cl \times xs - cd \times ys}{hl \times ly} \tag{2}$$

sf: Fertilization;cl:Corn yield target;
xs: Grain corn nutrient absorption amount per 100 kg;
cd: Soil nutrient determination;
ys: Soil available nutrient conversion factor;
hl: Fertilizer nutrient content;
ly: Fertilizer utilization season

Taking the mathematical model of fertilizer application rate in the elm city as an example (The content of P2O5 was 46 % in the application of diammonium phosphate):

$$\text{Phosphate fertilizer} = \frac{\text{Corn targer amount} * 0.07 - 0.03 * \text{Soil nutrient content} * \text{Soil available nutrient conversion factor}}{0.46 * \text{Fertilizer utilization rate}}$$

$$\text{Soil available nutrient conversion factor} = \begin{cases} \dfrac{1578.8 * \text{Soil nutrient content}^{-0.98}}{100} \\ \dfrac{1068 * \text{Soil nutrient content}^{-0.832}}{100} \\ \dfrac{732 * \text{Soil nutrient content}^{-0.749}}{100} \end{cases}$$

$$\text{Blank area yield} = \frac{0.3 * \text{Soil nutrient content} * \text{Soil available nutrient conversion factor}}{0.022}$$

$$\text{Fertilizer utilization rate} = \begin{cases} \dfrac{(43.4-0.024)*\text{Blank area yield}}{100} \\ \dfrac{(36.6-0.025)*\text{Blank area yield}}{100} \\ (4.6 - 0.035) * \text{Blank area yield} \end{cases}$$

Yield = 10000 kg/hm^2

According to the principle of nutrient balance, the Table 2 data and substitute it into formula (2), can be obtained as shown in Table 3 shows each level of soil fertility of specific fertilizer.

Table 3. Soil nutrient fertilizer rate

Soil level	N amount of fertilizer (kg/hm^2)	P amount of fertilizer (kg/hm^2)	K amount of fertilizer (kg/hm^2)
Class A	222.9694	195.0158	165
Class B	235.7228	166.1162	115
Class C	239.792	192.369	115
Class D	238.2537	177.1553	125

4 Conclusions

The application of OPTICS algorithm in precision fertilization has not been reported. The method in this paper, for example application in arch shed town of Yushu City in Jilin Province, the average fertilization amount was 332 kg/hm^2, compared with the traditional fertilization, and fertilizer 68 kg/hm^2; the average yield of 8313 kg/hm^2 was 813 kg/hm^2, and the yield was increased. Indeed achieve a reduction in fertilizer inputs, improve soil environment, increase production, income purposes.

References

1. Helong, Yu., Guifen, C., Chunguang, B.: Corn precision fertilization Database Modeling. Maize Sci. **16**(4), 184–188 (2008)
2. Guifen, C., Li, Ma., Hang, C.: Research status and development trend of precision fertilization technology. J. Jilin Agric. Univ. **35**(3), 253–259 (2013)
3. Xiaohua, Q., Yilong, Z., Ronggen, H.: Precision fertilization technology and application measures. Anhui Agric. Sci. Bull. **7**(2), 44–45 (2001)
4. Xu, J.Y., Wbster, R.: Optimal estimation of soil survey data by geostatistical method-semi variogram and block Kringing estimation of topsoil nitrogen of Zhangwu country. Acta Pedol Sci. **162**(4), 291–298 (1983)
5. Yost, R.S., Uehara, G., Fox, R.L.: Geostatistical analyst of soil chemical properties of large land areas iv. Semivariograms. Soil Sci. Soc. **46**, 1028–1032 (2000)
6. Mei, L., Xuelei, Z.: GIS-based evaluation of farm land soil fertility and its relationships with soil profile configuration pattern. J. Appl. Ecol. **22**(1), 129–136 (2011)
7. Yuhong, W., Xiaohong, T., Yanan, T.: Assessment of integrated soil fertility index based on principal components analysis. J. Ecol. **29**(1), 173–180 (2010)
8. Laijun, L.: Study on partition fertilization and soil testing and soil fertility of cultivated land reclamation area classification. Mod. Agric. Sci. Technol. **42**(2), 228–229 (2013)
9. Hongbo, Z.X., Shuo, B.: OPTICS-Plus for Text Clustering. Chin. J. Inf. **22**(1), 22–24 (2008)
10. Xiujie, W., Quanchao, Z., Haijun, L.: The value of clustering analysis and principal component analysis in the study of Anthropology. J. Anthropol. **29**(4), 35–37 (2007)
11. Shanjie, W.: Further thinking about the classification method. J. North China Inst. Sci. Technol. **5**(1), 108–110 (2008)

The Methodology of Monitoring Crops with Remote Sensing at the National Scale

Quan Wu[(⊠)], Li Sun, Yajuan He, Fei Wang, Danqiong Wang,
Weijie Jiao, Haijun Wang, and Xue Han

Remote Sensing Application Centre,
Chinese Academy of Agricultural Engineering, Beijing 100125, China
wuquan95@tom.com, sunli0618@163.com,
haijun076481@163.com, jiaoweijie0502@163.com,
hyjuan@gmail.com, wangfei@agri.gov.cn,
1226773217@qq.com, 346334593@qq.com

Abstract. Monitoring crops with Remote Sensing (RS) at the national scale is usually an operational work acted as a normal business for the government needs to the crop field conditions. The crop information is main content of agricultural condition. It mainly includes crop growth, crop areas and crop yields, which can be named 3 factors for crop monitoring with RS. Diversification is the general feature of crop monitoring with RS, which reflects in 3 parts of labor objects, labor materials and labor process. Monitoring the 3 factors with RS has similar process summarized as 3 periods which are data acquisition and transmission, model development and application, producing products. Monitoring crops with RS at the national scale needs to found an organizational and technical system, using the System Theory according to the 3 factors, the 3 parts and the 3 periods, mentioned above. The operational work of monitoring the 3 factors have a common goal, which is that the monitoring result is more accurate, the monitoring process is faster, more economic and more convenient. In China, Remote Sensing Application Centre (RSAC) has been working on monitoring the main crops as an operational task and a research project based on its system for several years. The monitoring methods to the 3 factors are presented in this paper along with the cases coming from the monitoring products produced by RSAC in 2014.

Keywords: Crop growth · Crop area · Crop yield · Agricultural condition · The system theory · Model · Operational work · National scale · RS

1   Introduction

The crop condition is an important project paid much attention by governments in the world because it is closely related with food security. Monitoring crops with RS at the national scale is usually an operational work acted as a normal business for the government needs to crop field conditions. Monitoring Agriculture with Remote Sensing (MARS), constituted by European Union Committee, is a project facing Europe in order to obtain the crop yield information [1]. In America, the prediction of all crop

© IFIP International Federation for Information Processing 2016
Published by Springer International Publishing AG 2016. All Rights Reserved
D. Li and Z. Li (Eds.): CCTA 2015, Part I, IFIP AICT 478, pp. 325–334, 2016.
DOI: 10.1007/978-3-319-48357-3_32

yields is acquired from crop areas and crop yields per unit. The crop area is mainly gotten by June Agricultural Survey (JAS) and Crop Data Layer (CDL) produced by NASS [2]. RSAC has been working on monitoring the main crops as an operational task and a research project based on its system for several years in China [3–5]. RSAC submits the monitoring results to Ministry of Agriculture at prescribed time according to the crop monitoring calendar every year [1]. Acted as main content of agricultural conditions, the crop condition mainly includes crop growth, crop area and crop yield, which can be named 3 factors for the crop monitoring with RS. The operating monitoring to the 3 factors is mainly provided by RSAC in China [6–8].

Monitoring crops with RS at the national scale needs to found an organizational and technical system. Diversification is the general feature of crop monitoring with RS, which reflects in 3 parts of labor objects, labor materials and labor process based on the complex system. At the same time, monitoring the 3 factors with RS has a similar process summarized as 3 periods which are data acquisition, model development and application, product production. The operational work of monitoring the 3 factors has a common goal, which is that the monitoring result is more accurate, the monitoring process is faster, more economic and more convenient.

2 Diversification of Crop Monitoring with RS

In the organizational and technical system for crop monitoring with RS, based on the operational task at national scale, presented in the crop area monitoring, crop growth monitoring and crop yield estimation, diversification is obviously reflected in the 3 parts of labor objects, labor materials and labor process.

2.1 On Labor Objects

In operational task for monitoring crop with RS, the objects of labor are of all kinds of data. The diversification in this part is reflected in multisource data. It includes different types of data, such as raster data, vector data and tabular data, etc. Raster data include RS data of various sensors, or of various resolution scales. The raster data also include data of DEM, DTM, etc. Vector data include these data which are basic geographic information, landuse data, meteorological station locations, etc. Tabular data is mainly Agricultural statistics, such as crop area, crop yield and meteorological data which is temperature, precipitation and daylight hour, etc. These multi types, multi formats and multi scales of data constitute the diversified feature of the crop monitoring with RS in the part of labor objects.

2.2 On Labor Materials

In the work of monitoring crop with RS, the labor materials mainly include the tools of production, such as various computers and all kinds of softwares. Acted as the physical platforms, the computers include graphics workstations and personal computers applied to different needs. Acted as non physical platforms, the software are just as important as the computers, which can be classified as RS software, such as ERDAS Imagine, ENVI, PCI, etc., GIS software, such as ArcGIS, Mapinfo, MapGIS, etc., statistical software, such as SAS, SPSS, STATA, etc. These multi platforms and multi applications to multisource data constitute the diversified feature of the crop monitoring with RS in the part of labor materials.

2.3 On Labor Process

Monitoring crops with RS at the national scale, it is necessary to set up an organizational and technical system. At a strategic point of view, it has various construction schemes for founding the system. Based on economical and convenient rules, a suitable scheme meeting the needs of the government can always be designed. Considering from the tactical point of view, three choices are unavoidable. The first is the choice of data. The second is of softwares The third is of methods. Multisource data provide the possibility for the selection of data. On the other hand, it also increases the difficulty of selection. The softwares used in the system do so. The methods of monitoring crops are many for different monitoring contents. Even if the monitoring contents are same, but there are also different methods to select. The multi levels and multi choices of methods constitute the diversification of labor process.

3 Elements and the System

As mentioned above, it is necessary to found a system for monitoring crops at the national scale. A system is a collection of organized objects [9, 10]. The organized objects can be called elements. The elements of the system can be summarized as the organizational and technical elements in two categories. There are designed relations between the technical elements and the organizational elements.

3.1 The Organizational Elements

The organizational elements are basic units to structure the perceived entity which may be or rely on a social organization. One of the organizational elements may be one person or a group consisted of some members. Obviously, the organizational elements are laborers, which the smallest unit is individual. Each element has a specific task to perform and has two-way connections with other related elements based on commands and requests. Each element has a clear position or level in the system. The Fig. 1 shown below presents that the system has one first level element consisted of one member, two second level elements respectively consisted of one member and the one third level element consisted of three members.

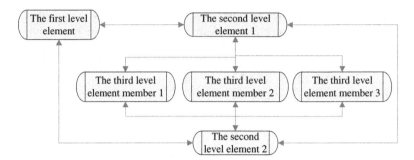

Fig. 1. The organizational elements in the system

3.2   The Technical Elements

Technical elements are labor materials and labor objects. It mainly include materialized production tools which are concrete and visible, such as computers, and the products of human spirit which are abstract and invisible, such as software. The mental technical elements mainly present as methods, software, programming languages, models and various data, etc., which must rely on some objects to exist. The objects are the materialized technical elements and the organizational elements. Obviously, only combining with the two kinds of elements, the mental technical elements can show vitality of life. The technical elements and organizational elements are not a one-to-one relationship. The technical elements are not divided into levels, not same with organizational elements. The technical elements are connected by the information flow which is one-way between two elements.

The organizational mechanism for managing organizational elements, which can be called the management method to members in the system, is a kind of special technical element which determines the structure of the system, the distribution of technical elements among the organizational elements, the direction and the speed of information flow. It also determines the dynamic behavior of the system [10]. The Fig. 2 shown below presents the mental technical elements and the direction of information flow.

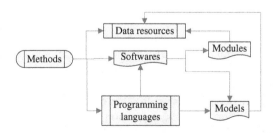

Fig. 2. The mental technical elements in the system

3.3   The System Structure and Function

Because the elements and the relationship among the elements are clear, therefore, the system structure is also clear. This system is not a black box [11]. Information can't exist without the material carrier, but it does not matter [12]. Information is transmitted in the channel [12]. In the system, the information is data, commands and requests. The channel is mainly network.

The system, composed of organizational and technical elements, is open and far away from the equilibrium state, not isolated. It operates by exchange of material and energy with environment at a non equilibrium. The non equilibrium is founded on a dissipative structure. Obviously, In order to reduce the system entropy, maintaining the dissipative structure is required to provide material and energy from the outside world. The system operation reflects the system function. Perfecting the system function means improving the quality of the system products and improving the production efficiency of the system. The system structure, operational mechanism and elements determine the system function. By optimizing the system structure, improving the operating mechanism and improving the quality of the elements, the system upgrade may be realized.

4   The Work Flow of the Operational System

The operational system of crop monitoring has a specific structure established based on the organizational and technical elements to perform specific tasks, which is called system functions. Whether it is to monitor crop growth, crop areas or crop yields, the monitoring process is always same. The work flow is from the collecting and processing of data to calculate data with models or modules, then to produce monitoring products. The entire process, from the collection of data to the product output, is always dependent on the overall scheme which can be integrated into the operational mechanism of the system and the system itself. The work flow of the operational system is briefly shown below in the Fig. 3.

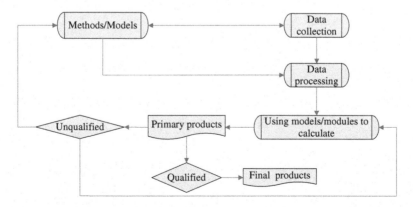

Fig. 3. The work flow of the operational system of crop monitoring

4.1   Monitoring the Crop Growth

The crop growth is states and trends of growing crops, which can be described by individual or the group characteristics [1]. Normalized difference vegetation index (NDVI) is a remote sensing index based on the spectral features of ground objects, which is related to leaf area index (LAI), vegetation cover, plant development and biomass. Based on this index, the monitoring models of crop growth can be established [1].

4.1.1   Introduction to the Crop Growth Model

There are many crop growth models based on NDVI [1]. At present, the yearly comparison model and the many year comparison model are used in operational work by RSAC. The yearly comparison model is to do subtraction operation and then to grade, which the value of this year' NDVI is acted as the minuend while the value of last year's NDVI is acted as the subtrahend. The model is presented below in formula 1. The difference is that the subtrahend is the average of the many years' NDVI in the many years comparison model. The model is shown below in formula 2. According to the difference of NDVI, the crop growth can be estimated. The grading standard of NDVI difference comes from China Meteorological Administration [1].

$$\Delta NDVI = NDVI_{this.year} - NDVI_{last.year} \qquad (1)$$

$$\Delta NDVI = NDVI_{this.year} - \overline{NDVI_{many.year}} \qquad (2)$$

4.1.2   A Case

In 2014, RSAC had adopted the above two models to estimate China's the crop growth of the late and single cropping rice mainly distributed on 19 provinces with Modis data. The Fig. 4(a), produced from the many years comparison model, is the map of the growth status of 5 month's the single cropping rice in northeast China. The Fig. 4(b),

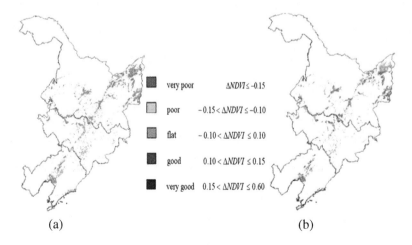

(a)   (b)

Fig. 4. The growth of single cropping rice of northeast China

produced from the yearly comparison model, is the map of growth status of 8 month's the single cropping rice in the same region.

4.2   Monitoring Crop Areas

In China, the operational prediction of crop areas is mainly provided by RSAC. The used RS data is mainly images of SPOT, Landsat-OLI, GF-1, etc. With this data, the model which is the Area Enlargement Estimation (AEE) is used to estimate the areas of main crops such as wheat, corn, cotton, soybean, rice, etc. The Stratified Sampling Method (SSM) is major content in the AEE [2–4].

4.2.1   Introduction to the AEE Model

The AEE model adopts the SSM to estimate crop areas. In the SSM, the cultivated land area derived from land use vector maps are used to stratify while the sampling unit is designed to the quadrangle frame of the relief map. The area estimation can be calculated by the follow equation.

$$\hat{Y} = \sum_{h=1}^{L} \sum_{j=1}^{N_h} \left(\frac{1}{n_h} \sum_{i=1}^{n_h} y_{hi} \right) \tag{3}$$

Where

$\hat{Y} =$ estimate value of total value of the population
$h = 1, 2, \ldots, L$
$n_h =$ the sample size of the h layer;
$y_{hi} =$ the sample value of the i unit of the h layer
$L =$ the amount of layers
$N_h =$ the population of the h layer

4.2.2   A Case

In 2014, RSAC had estimated China's the area of the late and single cropping rice mainly distributed on 19 provinces with the AEE model and the data of Landsat-OLI images, etc. The estimated result was compared with the one of last year. The annual change rate was obtained, which is −0 85 % shown below Table 1.

Table 1. The monitoring result of the area of the late and single rice in 2014

Region	Sample	Sampling ratio	Change rate	Confidence interval
19 provinces	841	0.0242	−0.0085	−0.0512–0.0361

The Fig. 5 is the map of the sample spatial distribution of the single cropping rice in northeast China, in 2014.

Fig. 5. The sample spatial distribution of single cropping rice of northeast China

4.3   Monitoring Crop Yields

The crop yield is closely related to LAI [1]. Based on NDVI designing models can invert RS data to LAI [1]. This research, called inversion algorithm, has been done by many persons in the world [1]. So, the relation between NDVI and crop yields can be set up by LAI [1]. Estimating crop yields with RS, from the experiment research to production application where RSAC is walking down, is possible.

4.3.1   Introduction to the Yield Estimation Model
The model to estimate crop yields includes two contents, which one is about crop yields and LAI. The other is about LAI and NDVI. The former is usually a linear model shown below in formula 4. The latter model is based on crop species, terrain and RS data to determine, shown below in formula 5.

$$Y_{yield} = aX_{LAI} + b \tag{4}$$

$$Y_{LAI} = f(X_{NDVI}) \tag{5}$$

4.3.2   A Case
In 2014, RSAC selected Zhaodong city, located in northeast China, to do an experiment of estimating the maize yield by RS. By collecting the maize LAI of 37 ground plots in the different growth periods of maize to analyze the relation between LAI and the measured output of maize, it is confirmed that the maize jointing stage is the key period for the model. So, the model of the relation between the maize yield and LAI is established, shown below formula 6. The maize NDVI of the same stage was calculated using multi-spectral data of GF-1. The relation between NDVI and LAI of maize

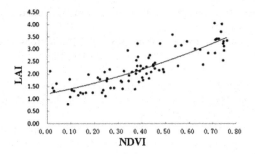

Fig. 6. The relationship between the LAI and the NDVI of the maize in jointing stage

of the jointing stage can be approximately explained by the Fig. 6 while the model is shown in formula 7. So, the maize yield can be estimated by the formula 6 and formula 7 with RS.

$$Y_{yield} = 5.7828X_{LAI} + 3.1674 \tag{6}$$

$$Y_{LAI} = 1.4913X_{NDVI}^2 + 1.8439X_{NDVI} + 1.2151 \tag{7}$$

5 Conclusions

Monitoring crops with RS at the national scale mainly includes monitoring crop growth, crop areas and crop yields, which can be named 3 factors. Diversification is the general feature of crop monitoring with RS reflected in 3 parts of labor objects, labor materials and labor process. The process of monitoring the 3 factors can be summarized as 3 periods of data acquisition and transmission, model development and application, producing products. The system composed of organizational and technical elements is necessary for monitoring crops with RS at the national scale, which is open and far away from the equilibrium state. It operates by exchange of material and energy with the environment at a non equilibrium founded on a dissipative structure. By optimizing the system structure, improving the operating mechanism and improving the quality of the elements, the system upgrade may be realized.

Acknowledgement. This paper is supported by the innovation team of crop monitoring by RS, authorized by Chinese Academy of Agricultural Engineering (CAAE), in 2015.

References

1. Yang, B.J.: Monitoring the agricultural condition using RS, pp. 19–29. China Agriculture Press, Beijing (2005). (in Chinese)
2. Wu, Q., Pei, Z.Y., Wang, F., Zhao, H., Guo, L., Sun, J.Y., Jia, L.J.: A sampling design for monitoring of the cultivated areas of main crops at national scale based 3S technologies in

China. In: Computer and Computing Technologies in Agriculture VI. Part II, IFIP 2013, AICT 393, pp. 10–19 (2013)

3. Wu, Q., Sun, L.: Sampling methods using RS and GPS in crops acreage monitoring at a national scale in China. In: Remote Sensing and Spatial Information Sciences, Beijing, Part B7, [WG VII/7]*, vol. XXXVII, pp. 1337–1342 (2008)

4. Wu, Q., Sun, L., Wang, F.: The applications of 3S in operational monitoring system of main crops acreage in China. In: Computer and Computing Technologies in Agriculture, vol. 24, pp. 319–324. TSI Press, USA (2010)

5. Wu, Q., Sun, L., Wang, F., Jia, S.R.: Theory of double sampling applied to main crops acreage monitoring at national scale based on 3S in China. In: Computer and Computing Technologies in Agriculture IV, Part III, PP. 198–211 (2011)

6. Wu, Q., Pei, Z.Y., Zhang, S.L., Wang, F., Wang, Q.F.: The methods for monitoring land-use change with RS at non-large scale, pp. 33–40. China Agriculture Press, Beijing (2010). (in Chinese)

7. Wu, Q., Pei, Z.Y., Guo, L., Liu, Y.C., Zhao, Z.Y.: A study of two methods for accuracy assessment to RS classification. In: 2012 First International Conference on Agro-Geoinformatics, pp. 1–5 (2012)

8. Wu, Q., Sun, L., Wang, F., Jia, S.R.: The quantificational evaluation of a sampling unit error derived from main crop area monitoring at national scale based 3S in China. Sens. Lett. **10**, 213–220 (2012)

9. Liu, S.F., Dang, Y.G., Fang, Z.G., Xie, N.M.: Grey system theory and application, pp. 122–130. Science Press, Beijing (2010). (in Chinese)

10. Zhong, Y.G., Jia, X.J., Li, X.: System dynamics, pp. 10–36. Science Press, Beijing (2012). (in Chinese)

11. Wan, B.W., Han, C.Z., Cai, Y.L.: Cybernetics: Concepts, Methods and Applications, pp. 4–16. Tsinghua University Press, Beijing (2014). (in Chinese)

12. Fu, Z.Y.: Information Theory, pp. 1–16. Publishing House of Electronics Industry, Beijing (2013). (in Chinese)

Is Time Series Smoothing Function Necessary for Crop Mapping? — Evidence from Spectral Angle Mapper After Empirical Analysis

Ailian Chen[✉], Hu Zhao, and Zhiyuan Pei

Chinese Academy of Agricultural Engineering,
Beijing 100125, People's Republic of China
cal-0601@163.com, zhhoo@126.com,
Peizhiyuan@hotmail.com

Abstract. Time series smoothing functions have been frequently applied to fit multi-temporal vegetation index for better extraction of plant seasonal/growing parameters. Questions are raised that whether the smoothing is necessary for crop mapping. Four time series smoothing functions, namely, HANTS, Savitzky-Golay (S-G), double logistics and asymmetric Gaussian, were used to smooth 23 MODIS 16-days composite NDVI images in one year. The effectiveness were compared through visual check, correlation coefficient R, root mean square error (RMSE), and local signal noise ratio (SNR). The best smoothing time series NDVI images, along with the original time series images, were then used to map corn and soybeans by spectral angle mapper (SAM) method and their mapping accuracies were compared. Comparison of smoothing results showed that S-G fitted data got the strongest correlation coefficient R, the lowest RMSE and lower local SNR. Comparison of mapping results further showed that time smoothing function does not improve the classification accuracy obviously with the same training sample and same temporal bands. The whole analysis indicates that it is the mapping method that matters more than time series smoothing function for classification precision.

Keywords: MODIS · NDVI · Crop phenology · TIMESAT · SAM

1 Introduction

High temporal resolution time series (TS) vegetation indices (VIs) have been well used in the research of global environment with respect to vegetation phenology. Large scale environmental parameters, such as global land use/cover change, could be derived with the knowledge of vegetation phenology. However, the high temporal resolution time series VIs often have coarse spatial resolution, which often make the time series VIs noisy and fluctuating due to the complex atmospheric condition within one coarse pixel (such as undetected cloud) and varying sun-sensor-surface geometries (Duggin 1985). Therefore, a number of noise-reduction functions, including maximum value composite (Huete et al. 2002), Fourier transform (Jentsch and Subba Rao 2014), wavelet transform (Sakamoto et al. 2010), Gaussian transform (Hird and McDermid 2009), etc., were proposed or utilized to solve this problem.

© IFIP International Federation for Information Processing 2016
Published by Springer International Publishing AG 2016. All Rights Reserved
D. Li and Z. Li (Eds.): CCTA 2015, Part I, IFIP AICT 478, pp. 335–347, 2016.
DOI: 10.1007/978-3-319-48357-3_33

As for crop mapping, some studies (Sakamoto et al. 2010; Brown et al. 2013; Zhang et al. 2014) have used TS fitting function before mapping, while many others have not (Chang et al. 2007; Wardlow and Egbert 2008; Gusso et al. 2014; de Souza et al. 2015). Some of those which have used time series smoothing functions often conducted more than one smoothing function to the time series images. For example, researchers used MODIS Level 3 VI products always conducted smoothing function upon the already composited VIs by maximum value composite (MVC) (Huete et al. 2002).

A scanty few studies have been conducted regarding the benefit of noise reduction for vegetation phenology (Hird and McDermid 2009), or regarding empirical information of noise of the smoothed images themselves, such as Atzberger and Eilers (2011), who have compared the effectiveness of four fitting methods. Fewer studies have investigated the necessity of fitting TS VIs for crop classification. Questions are raised that whether smoothing functions are necessary for crop type mapping. This goals of this paper are: (1) to study the effectiveness of four different time series fitting methods, and then with the best fitting results, (2) to compare the classification accuracy.

2 Materials and Methodology

2.1 Materials

MODIS 16 days composite NDVI of the year 2013 were used for time series analysis. The time series started at 2013001 as band 1, and ends at 2013353 as band 23, with an interval of 16 days. The value in band 1 represents the max NDVI from the date Jan. 1st to the date Jan. 16th, 2013, and so on. The MODIS NDVI 16 days composite NDVI named "MOD13Q1-Level 3 16-Day Vegetation Indices-250 m", is one of the MODIS 16-Day Tiled Land Products (collection 5). The data was downloaded from the U.S. National Aeronautics and Space Administration (NASA). The "MOD13Q1-Level 3 16-Day Vegetation Indices-250 m" included both NDVI and enhanced VI, EVI. The NDVI was extracted for research as it was a normalized value. MODIS data are distributed from NASA on a near real-time basis at no cost. It is available at two satellite platforms, Terra and Aqua, which were launched in December 1999 and May 2002, respectively. The spatial resolutions of the MODIS-250 m (MOD13Q1, MYD13Q1) on the sinusoidal projections are 231.7 m. The downloaded tiles were mosaicked, re-projected to Albers projection and resampled to 250 m.

Crop data layer (CDL) was used for training and accuracy check. CDL is a specific crop type classification data with resampled resolution of 30 m, and is produced and distributed in the name of NASS (national agricultural statistics service), USDA. The data is available through the web of USDA.

One U.S. state, the Ohio, was taken as case study to classify corn and soybeans from others, as in large area mapping, the large area was often divided into small regions. Corn and soybeans are the dominant crop in Ohio, where the yield of corn takes up 4 % of the whole states total, the yield of soybeans takes up 7 %. Other parts of the state are mainly forest and shrub lands. The spatial distribution of corn and soybeans from CDL 2013 was shown in Fig. 1.

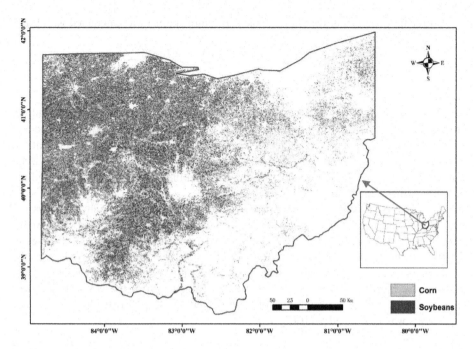

Fig. 1. Corn and soybeans distribution in 30 m resolution crop data layer (CDL) data in 2013

2.2 Time Series Smoothing Function and Result Check

Four popular (Hird and McDermid 2009; Zhang and Ren 2014) smoothing functions were conducted, including Harmonic analysis of TS (HANTS) (Zhou et al. 2015), S-G filtering function short for adaptive Savitzky-Golay filtering (Chen et al. 2004), asymmetric Gaussians (referred to as Gauss herein after) (Jönsson and Eklundh 2002) and double logistic functions (referred to as Logistics herein after) (Beck et al. 2006). The basic idea of HANTS is to calculate a Fourier series to model a TS of pixel-wise data, while meanwhile recognizing abnormal data value relative to the TS of Fourier model. HANTS deletes abnormal data value and substitutes it with the value given by the model. It is developed with the interactive data language (IDL). The other three smoothing algorithms are embedded in the free software TIMESAT (Jönsson and Eklundh 2004).

Visual and statistical check were conducted for the comparison of the four smoothing functions. Visual check was done first by the screenshot of the original images and all the smoothed images, and then by compared the time series NDVI profiles of two pixels, one of which was the corn pixel, and the other was soybeans.

Statistical check included the comparison of correlation coefficient R of one single band image (Band 14) and 20 sample points, local SNR and RMSE in Band 14, as Band 14 represents the NDVI value in late June and early July, when the NDVI corn and soybeans were near the max value (Wardlow and Egbert 2008).

Correlation coefficient R (whole image and sample point) was calculated with Eq. 1

$$R = \frac{\sum_{i=1}^{n}(xi - \bar{X})(yi - \bar{Y})}{\sqrt{\sum_{i=1}^{n}(xi - \bar{X})^2(yi - \bar{Y})^2)}} \tag{1}$$

Where, xi, yi represent the pixel value of the whole non-smoothed image and the smoothed image, and $\bar{X}$ and $\bar{Y}$ are mean value of the whole non-smoothed image and the smoothed image in Band 14, respectively. For sample points, xi, yi represent the pixel value of samples from the non-smoothed image and the smoothed image in Band 14, and $\bar{X}$ and $\bar{Y}$ are the mean value of samples from the non-smoothed image and the smoothed image in Band 14 respectively.

As global SNR requires a homogeneous underlying surface, which is always not easy to find, the local SNR was used, which calculated local SNR using the mode of several grids. The local SNR is also used for the determination of HJ-images (Zhu et al. 2010). Sixteen fishnet grid with edge of 10 km was used to calculate SNR, and the mode, mean, min and max values were given. Root mean square error (RMSE), local signal noise ratio SNR (Zhu et al. 2010) of 16 fishnet grid of the whole time series images were calculated through the following equations (Eq. 2 to Eq. 6).

RMSE was calculated with Eq. 2

$$RMSE = \sqrt{\frac{\sum_{i=1}^{N}(xi - yi)^2}{N}}, \tag{2}$$

Local SNR is calculated by Eq. 3.

$$SNR = \bar{X}/\sigma, \tag{3}$$

where, $\bar{X}$ is the mean value of the grid, and σ is the standard error, which are calculated by Eqs. 4 and 5.

$$\bar{X} = \frac{\sum_{i=1}^{N} xi}{N}, \tag{4}$$

$$\sigma = \sqrt{\frac{\sum_{i=1}^{N}(xi - \bar{X})^2}{N-1}}, \tag{5}$$

where, xi represents the pixel value of i in the analyzing grid, and N is the total pixel numbers in the grid.

For the calculated R, RMSE, and SNR, the criteria for effectiveness is higher R, lower RMSE and lower SNR.

2.3   Crop Mapping and Result Check

Spectral Angle Mapping (SAM) in the commercial software ENVI^{TM} was used for classification. SAM "uses an n dimension angle to identify pixels to reference spectra" (Kruse et al. 1993). The spectral angle is reversed by Eq. 6.

$$\cos\theta = \frac{XY}{|X||Y|} = \frac{\sum_{i=1}^{n} xiyi}{\sqrt{\sum_{i=1}^{n} xixi}\sqrt{\sum_{i=1}^{n} yiyi}} \quad \theta \in (0, \frac{\pi}{2}), \tag{6}$$

where, n is the total band number. It is 23 in this study, xi and yi are the data value (usually reflectance) of each time series images, and θ is the spectral angle which the mapper would be based on. The algorithm determines the spectral likeness between two spectrum profiles by figuring out the angle between the spectrum profiles in a space with dimensionality equaling to the number of bands. The SAM method is relatively less sensitive to illumination and albedo effects than other methods such as maximum likelihood.

End-member spectra (training data spectra) used in SAM mapping procedure was collected with the aid of CDL. The CDL was resampled to a resolution of 250*250 m, the same as the MODIS NDVI data, with each 250 * 250 m pixel assigned a value of percent of corn or soybeans. The pure corn and soybeans were extracted for random sampling. Ten percent of Random samples (points) of corn and soybeans were extracted from the pure corn or soybeans layer, respectively. One third of these random samples were used for training of SAM, which was done by performing a MOD function, going by like MOD(ID, 3) equaling to 0, and the left were used for accuracy check. The tolerance angle for SAM was set as 0.5 rad.

For the accuracy check of mapping result, three groups were compared. Overall, producer and user accuracy were all derived for comparison. The first group classified all the 23 original and S-G fitted time series images, the second group classified the 12 images from Band 7-18, and the third group classified 7 images including band 9, and Band12-17. Percent and pixel accuracy of corn and soybeans were given, as well as the overall accuracy. The mapping results with best overall accuracy was compared by ratio to the acreage estimated by CDL and acreage from national statistics released by NASS, USDA.

3   Results and Discussion

3.1   Time Series Fitting Results

3.1.1   Visual Check

The screen shots of one sub region of the original image and four smoothed images in Band 14 were shown in Fig. 2. The result of HANTS was different from others to the most extent, which has brought more noise to the original image.

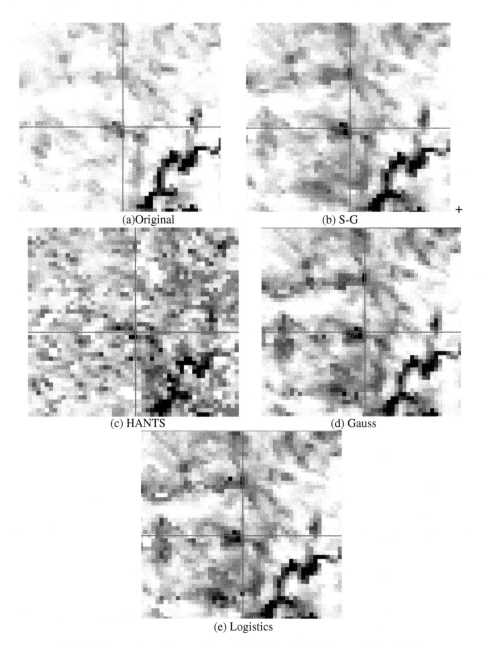

(a)Original   (b) S-G

(c) HANTS   (d) Gauss

(e) Logistics

Fig. 2. Screenshot of one random pixel in Band 14 before and after fitting

Time series NDVI profiles of one corn pixel and one soybeans pixel from the original images and four smoothed images were shown in Figs. 3 and 4, which indicated that HANTS tended to change the max value position, and obscured the start and end of season date. S-G has well kept the original curve.

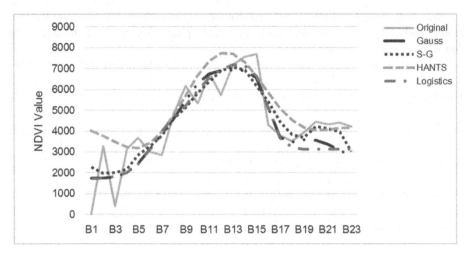

Fig. 3. NDVI time series profiles of corn pixel before and after smoothing

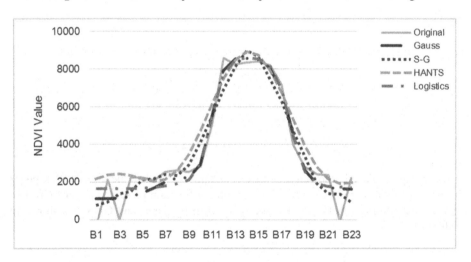

Fig. 4. NDVI time series profiles of soybeans pixel before and after smoothing

3.1.2 Correlation Coefficient, RMSE and SNR

Correlation coefficient R of the whole image and the 20 sample points in Band 14 between the smoothed and the non-smoothed images were given in Table 1.

Table 1. Correlation coefficient between each smoothed and the non-smoothed images

	Original whole image	Original 20 sample points
S-G	0.996	0.846
Logistics	0.995	0.751
Gauss	0.995	0.708
HANTS	0.995	0.826

The correlation coefficient R showed that the S-G smoothing functions got the best results both for the whole image and for sample points among the four smoothing functions conducted in this study.

Correlation analysis between smoothed data and the original data for the 16 fishnet grids also showed that S-G smoothing function presented higher median and mean correlation coefficient than Logistics, Gauss and HANTS as shown in Table 2:

Table 2. Correlation coefficient of each time series smoothing function of 16 fishnet grids

	S-G	Logistics	Gauss	HANTS
Median	0.83	0.81	0.81	0.77
Mean	0.82	0.81	0.81	0.74
Min	0.61	0.62	0.60	0.38
Max	0.92	0.94	0.96	0.88

S-G smoothing function presented the lowest RMSE than Logistics, Gauss and HANTS as shown in Table 3:

Table 3. RMSE of each time series smoothing function of 16 fishnet grid

	S-G	Logistics	Gauss	Hants
Median	962	1076	1025	1308
Mean	981	1046	1005	1303
Min	703	770	734	859
Max	1234	1271	1222	1583

Local SNR analysis showed that the original time series had the highest SNR either in mode, median or in mean value. S-G smoothing function presented lower SNR than HANTS smoothing function, but higher SNR than Logistics and Gauss, as shown in Table 4:

Table 4. Local SNR of the original and fitted time series images

Statics	Mode	Median	Mean
Original	19.586	16.306	14.858
S-G	15.383	12.496	11.686
Logistics	12.062	11.196	9.935
Gauss	12.232	11.716	10.225
HANTS	17.521	14.042	13.104

3.2   Classification Results

3.2.1   Visual Check

Corn and soybeans distribution mapped from time series Band 7 to Band 18 of the original images and S-G smoothed images were shown in Figs. 5 and 6. The coarse resolution of MODIS Vis has made the mapping results less specific compared to the CDL distribution of corn and soybeans as shown in Fig. 1. However, the trend of distribution were generally right where there was corn or soybeans. The red ellipse circle on the map showed slightly different details between the results from the original images and S-G smoothed images.

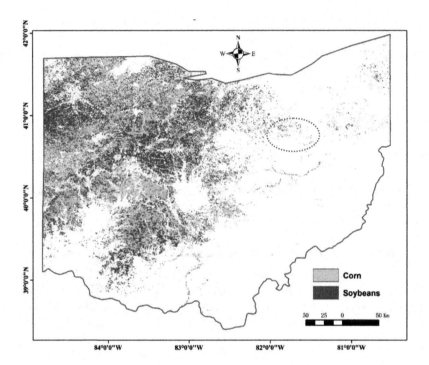

Fig. 5. Spectral angle mapper result from the original time series images (Color figure online)

3.2.2   Mapping Accuracy

Overall accuracy, producer and user accuracy of classification were compared in three groups, the first of which classified all 23 original and S-G fitted images, the second classified the 12 images from Band 7-18, and the third group classified 7 images including band 9, and Band 12-17. Percent and pixel accuracy of corn and soybeans were given, as well as the overall accuracy, as shown in Table 5:

The ratio of mapped acreage from MODIS time series NDVI images to the acreage of CDL and to the statistics reported by ASD released June 28, 2013, by NASS, USDA were given in Table 6.

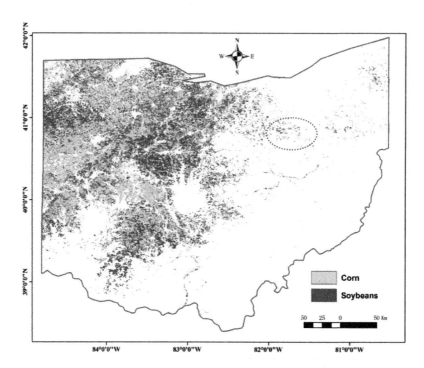

Fig. 6. Spectral angle mapper result from the S-G algorithm smoothed time series images (Color figure online)

Table 5. Overall, producer, and user accuracy of classification of three group

Group	Time series	Crop type	Prod. Acc.	User. Acc.	Overall Acc.
The whole 23 bands	Original	Corn	60.970	69.000	64.486 %
		Soy	68.530	60.480	
	S-G smoothing	Corn	50.630	66.220	59.667 %
		Soy	70.110	55.150	
Band 7 to Band 18	Original	Corn	65.250	68.910	65.613 %
		Soy	66.040	62.270	
	S-G smoothing	Corn	65.250	68.910	60.594 %
		Soy	66.040	62.270	
Band 9, Band 12 to band 17	Original	Corn	65.250	68.910	58.138 %
		Soy	66.040	62.270	
	S-G smoothing	Corn	48.210	64.130	58.902 %
		Soy	69.400	54.180	

Table 6. Mapped MODIS result acreage compared to pixel counted from CDL and statistical acreage released June 28, 2013, by NASS, USDA

Ratio	Corn	Soy
Original time series to CDL	95.273 %	74.977 %
S-G smoothing time series to CDL	96.463 %	73.690 %
PIF to statistic	89.477 %	72.415 %
S-G to statistic	90.595 %	71.173 %

The mapped acreage of corn was very near to the acreage of both CDL and state statistics released in June 28, 2013, by NASS, USDA.

4   Discussion

4.1   Time Series Fitting Effectiveness

Visual checking showed that HANTS altered the original data value massively, and thus has shown higher SNR than S-G, Gauss and Logistics. Visual checking by time series profiles showed that S-G, Gauss, and Logistics were similar in the growing season, but differed from each other in other part of the year. Generally, S-G smoothing function got the best results, especially with NDVI peak value neither obviously lagging nor postponing.

Statistical check showed that S-G smoothing algorithms had the highest correlation coefficient either for the whole image or for sample points. S-G smoothing algorithms also presented the lowest RMSE and relatively lower SNR. With these results, S-G was thought to be the best and stably effective smoothing function for crop mapping.

4.2   Crop Mapping Accuracy

With the same methods, the same input data bands, and the same training data, the overall accuracy of the original time series was generally better than the S-G smoothed time series, but the accuracy was not obviously high, no higher than 5 %. The ratio of mapped acreage to CDL pixel counted acreage was near 100 % for corn, and near 80 % for soybeans, and so was the ratio of mapped acreage to national statistical acreage. Besides, the detail in the distribution map also led to different conclusion compared to the accuracy report. Therefore, it was concluded here that the smoothing functions, the mapping method, the training data and the accuracy check method all had effect on the comparison. It is the goal the matters for the choosing of mapping method.

4.3   Limitations

Both time series fitting method and the selection of input bands have effect on the mapping result. This paper only investigated the result of spectral angle mapper mapping method. Other method may got different results as well.

5   Conclusion

This paper compared the effectiveness of four time series smoothing algorithms through a series of statistical indicators, and used the most effective smoothed result to conduct crop mapping. S-G smoothing was found to be the most effective fitting method. However, the most effective smoothed results did not provide the most accurate mapping results with the same method, the same training data and the same checking data. This findings imply that both mapping methods and the goals or purposes the user require matter more for a better result of crop mapping. It was not necessary to conduct a time series smoothing before a better mapping scheme was founded.

Acknowledgment. We express our deep gratitude to the free data and software providers: NASA for MODIS, USDA for CDL, GDSC (Geospatial Data Service Centre) for HANTS package, and Lars Eklundh for the software TIMESAT. The whole work was funded by the Civil Space Project in the 12th Five-Year Plan of China (2011–2015)

References

Atzberger, C., Eilers, P.H.: Evaluating the effectiveness of smoothing algorithms in the absence of ground reference measurements. Int. J. Remote Sens. **32**, 3689–3709 (2011)

Beck, P.S.A., Atzberger, C., Høgda, K.A., Johansen, B., Skidmore, A.K.: Improved monitoring of vegetation dynamics at very high latitudes: a new method using MODIS NDVI. Remote Sens. Environ. **100**, 321–334 (2006)

Brown, J.C., Kastens, J.H., Coutinho, A.C., de Castro Victoria, D., Bishop, C.R.: Classifying multiyear agricultural land use data from Mato Grosso using time-series MODIS vegetation index data. Remote Sens. Environ. **130**, 39–50 (2013)

Chang, J., Hansen, M.C., Pittman, K., Carroll, M., DiMiceli, C.: Corn and soybean mapping in the United States using MODIS time-series data sets. Agron. J. **99**, 1654–1664 (2007)

Chen, J., Jönsson, P., Tamura, M., Gu, Z., Matsushita, B., Eklundh, L.: A simple method for reconstructing a high-quality NDVI time-series data set based on the Savitzky-Golay filter. Remote Sens. Environ. **91**, 332–344 (2004)

de Souza, C.H.W., Mercante, E., Johann, J.A., Lamparelli, R.A.C., Uribe-Opazo, M.A.: Mapping and discrimination of soya bean and corn crops using spectro-temporal profiles of vegetation indices. Int. J. Remote Sens. **36**, 1809–1824 (2015)

Duggin, M.: Review article: factors limiting the discrimination and quantification of terrestrial features using remotely sensed radiance. Int. J. Remote Sens. **6**(1), 3–27 (1985)

Gusso, A., Arvor, D., Ducati, J.R., Veronez, M.R., Da Silveira, L.G.: Assessing the MODIS crop detection algorithm for soybean crop area mapping and expansion in the Mato Grosso State, Brazil. Sci. World J. **2014** (2014)

Hird, J.N., McDermid, G.J.: Noise reduction of NDVI time series: an empirical comparison of selected techniques. Remote Sens. Environ. **113**, 248–258 (2009)

Huete, A., Didan, K., Miura, T., Rodriguez, E.P., Gao, X., Ferreira, L.G.: Overview of the radiometric and biophysical performance of the MODIS vegetation indices. Remote Sens. Environ. **83**, 195–213 (2002)

Jönsson, P.K., Eklundh, L.: Seasonality extraction by function fitting to time series of satellite sensor data. IEEE Trans. Geosci. Remote Sens. **40**, 1824–1832 (2002)

Jönsson, P., Eklundh, L.: TIMESAT—a program for analyzing time-series of satellite sensor data. Comput. Geosci. **30**, 833–845 (2004)

Jentsch, C., Subba Rao, S.: A test for second order stationarity of a multivariate time series. J. Econometrics **185**(1), 124–161 (2015)

Kruse, F.A., Lefkoff, A.B., Boardman, J.W., Heidebrecht, K.B., Shapiro, A.T., Barloon, P.J., Goetz, A.F.H.: The spectral image processing system (SIPS)—interactive visualization and analysis of imaging spectrometer data. Remote Sens. Environ. **44**, 145–163 (1993)

Sakamoto, T., Wardlow, B.D., Gitelson, A.A., Verma, S.B., Suyker, A.E., Arkebauer, T.J.: A two-step filtering approach for detecting maize and soybean phenology with time-series MODIS data. Remote Sens. Environ. **114**, 2146–2159 (2010)

Wardlow, B.D., Egbert, S.L.: Large-area crop mapping using time-series MODIS 250 m NDVI data: an assessment for the US Central Great Plains. Remote Sens. Environ. **112**, 1096–1116 (2008)

Zhang, J., Feng, L., Yao, F.: Improved maize cultivated area estimation over a large scale combining MODIS–EVI time series data and crop phenological information. ISPRS J. Photogrammetry Remote Sens. **94**, 102–113 (2014)

Zhou, J., Jia, L., Menenti, M.: Reconstruction of global MODIS NDVI time series: performance of Harmonic ANalysis of Time Series (HANTS). Remote Sens. Environ. **163**, 217–228 (2015)

Zhang, H., Ren, Z.Y.: Comparison and application analysis of several NDVI time-series reconstruction methods. Scientia Agricultura Sinica **47**(15), 2998–3008 (2014). (in Chinese)

Zhu, B., Wang, X.H., Tang, L.L., Li, C.R.: Review on methods for SNR estimation of optical remote sensing imagery. Remote Sens. Technol. Appl. **25**(1), 303–309 (2010). (in Chinese)

Application Feasibility Analysis of Precision Agriculture in Equipment for Controlled Traffic Farming System: A Review

Caiyun Lu[1,2,3,4,5], Zhijun Meng[1,2,3,4(✉)], Xiu Wang[1,2,3,4],
Guangwei Wu[1,2,3,4], Nana Gao[1,2,3,4], and Jianjun Dong[1,2,3,4]

[1] Beijing Research Center of Intelligent Equipment for Agriculture,
Beijing Academy of Agriculture and Forestry Sciences,
Beijing 100097, China
{lucy,mengzj,wangx,wugw,gaonn,dongjj}@nercita.org.cn
[2] National Research Center of Intelligent Equipment for Agriculture,
Beijing 100097, China
[3] Beijing Key Laboratory of Intelligent Agricultural Equipment and Technology,
Beijing 100097, China
[4] Key Laboratory of Agri-Informatics,
Ministry of Agriculture, Beijing 100097, China
[5] Beijing Research Center for Information Technology in Agriculture,
Beijing 100097, China

Abstract. Equipment includes planter, field management equipment and harvester for controlled traffic farming system is elaborated in the paper. And the analysis results shows that main problems of equipment for controlled traffic farming system contains poor ability to navigate, uneven and bent wheel lanes. To solve the problem, the thought that precision agriculture is introduced into equipment for controlled traffic farming system is put forward. For this purpose, the key technology of precision agriculture is introduced, and the feasibility of combination of precision agriculture and controlled traffic farming system is discussed, which provides theoretical support for the development of equipment for controlled traffic farming system.

Keywords: Controlled traffic farming system · Equipment · Precision agriculture

1 Introduction

Controlled traffic farming system, as a new agricultural production system, combines the conservation tillage and controlled traffic technology. The principle includes that crop zone is independent of wheel lane, which means wheels of tractor and machines run only on permanent lanes, and crop is planted on other soil except wheel lanes. So the crop will not be compacted by wheels, which can keep better environment for crop growth. Controlled traffic farming system requires few tillage to keep the lanes, no tillage and straw cover to realize water saving, waterlogging, tillage reduction and

D. Li and Z. Li (Eds.): CCTA 2015, Part I, IFIP AICT 478, pp. 348–355, 2016.
DOI: 10.1007/978-3-319-48357-3_34

residue management before planting. Controlled traffic farming system was developed from Mexico, and has developed in many countries, which suffer drought and waterlogging.

Researches show that controlled traffic farming system has many effects [1, 2]. Wheel traffic can increase soil compaction which has negative influences on some critical soil properties, like saturated hydraulic conductivity, porosity and bulk density [3–6]. Bai et al. (2008) showed that 9 years controlled traffic farming system reduced bulk density 11.2 %, and increased porosity in 0-15 cm soil layer. In addition, the saturated hydraulic conductivity was greater under controlled traffic farming system [7]. Research by Chen et al. (2008) indicated that controlled traffic farming system prominently improved microbial biomass, soil organic matter and total N in 0–5 cm soil layer and the yield increased 10 % greater than conventional tillage [8].

The development of controlled traffic farming system bases on mechanization, for example, key operations such as ridge reforming and reshaping and planting can't be completed by manual. Extending controlled traffic farming system will be meaningless without mechanization, and mechanized controlled traffic farming system is the inevitable trend. The equipment for controlled traffic farming system includes planter, field management equipment and harvester etc. There must be any method to guide the tractor and agricultural equipment running on the wheel lane rather than crop zone due to the separated crop zone and wheel lane. The development of controlled traffic system is still in its infancy, and the corresponding equipment is not mature. The guidance is mainly completed by artificial marking, and the method only suits short field, and the ability of guidance is bad, and wastes time and energy.

Based on the positioning and navigation, precision agriculture manages each growth process and controls agriculture products, such as fertilizer, pesticide and seed, etc. to fulfill a full potential of soil and crop. Precision agriculture is comprised of 10 systems, including global positioning system (GPS), geographic information system (GIS), remote sensing (RS), field information collecting system, intelligent agricultural system, expert system, system integration, environmental monitoring system, network management system, and communication system. And GPS, GIS, RS and computer control system are the core of the precision agriculture [9–12]. The attribute data, which comes from the analysis of soil and crop information by GIS, and map data can make field management information system. On the basis, input is adjusted according to actual conditions on each unit to reduce waste, increase income and protect environment. GPS, one of the cores of the precision agriculture, is used for accurate positioning of agricultural machine. The application of precision agriculture in controlled traffic farming system can realize accurate orientation, which is conducive to the demonstration and application of controlled traffic.

Basis on the literature at home and abroad, the paper introduces equipment for controlled traffic farming system and the existing problems, and elaborates the key technology of precision agriculture. On the basis, application feasibility of precision agriculture on equipment for controlled traffic farming system is discussed.

2 Equipment for Controlled Traffic Farming System

The research of equipment for controlled traffic farming system focuses on machines, and aims to avoid blockage by mulch on the field. However, the research on how to ensure accurate wheel layout has rare research, which is more important to controlled traffic farming system [13].

2.1 Planter

(1) No-tillage maize and wheat seeder for controlled traffic farming system
   2BMDF-2/7 no-tillage maize and wheat seeder for controlled traffic farming system is used for wheat or maize planting in double cropping area. The planter can complete straw chopping, furrowing, fertilizing, planting and compacting in one time. Furrowing and fertilizing are completed by opener, and the mulch is chopped by high speed cutter. Combination of high speed cutter and double-disc opener is designed to avoid blockage by straw and weed. The cutter can put into sides of openers, and take away downed straw and weed to avoid blockage. The sharp tip of tine opener can cut off or hitch stubble, and the cutter chops the stubble hitched by tip to avoid blockage (Fig. 1).

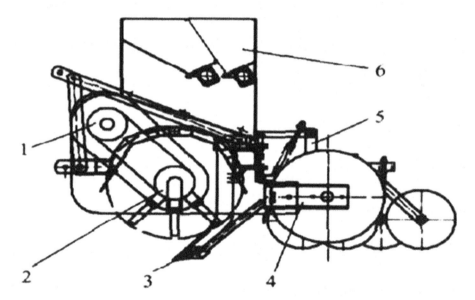

Fig. 1. 2BMDF-2/7 no-tillage maize and wheat seeder for controlled traffic farming system

(2) No-tillage fertilizing planter for controlled traffic farming system
   2BMFSG-3/6 no-tillage fertilizing planter for controlled traffic farming system can plant wheat and maize with the mating power 300HP and working speed 3 ~ 5 km/h. The planter works on sunken controlled traffic, which means permanent raised bed farming system. Installation position of land wheel can be adjusted according to the depth of furrow to satisfy the requirements of sowing

depth. The stubble and straw are cut by the rotary cutter. The diameter of land wheel is larger than that in flat field, and the installation position is lower, so that the furrow by opener satisfies the agronomy requirements when land wheel runs on the traffic lanes. The openers are distributed into two rows to avoid blockage [14] (Fig. 2).

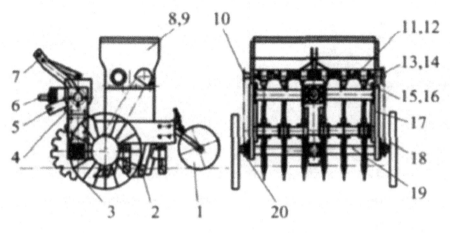

Fig. 2. 2BMFSG-3/6 no-tillage fertilizing planter under controlled traffic farming system

2.2 Field Management Equipment

Mating equipment of controlled traffic farming system is developed from planters to sprayers and other field management equipment with the wide application of controlled traffic farming technology in the world.

Figure 3 shows the sprayer for sunken controlled traffic farming system in Mexico. The spraying is finished during crop growth season to control weed and insects; furrow is used for guide, which improves the spraying with overlap and misses [2].

Fig. 3. Mexico sprayer for sunken controlled traffic farming system

2.3   Harvester

To improve efficiency of crop harvest, harvester for controlled traffic farming system is researched in the world, such as hanging harvester developed by China Agricultural University and two-row ridge harvester in Mexico. These two harvesters can not only ensure the harvest quality, but also keep ridge bed [2] (Fig. 4).

Fig. 4. Hanging harvester in China   **Fig. 5.** Two-row ridge harvester in Mexico

3   Key Technology of Precision Agriculture

Precision agriculture was put forward in the 1990s by soil scientists in the University of Minnesota, America. After years of development, precision agriculture has developed in many aspects, including variable rate fertilization, nutrient detection, and yield monitoring, etc. The key technology of precision agriculture mainly includes (Fig. 5):

3.1   Global Positional System

Global positioning system (GPS) refers to a satellite navigation system based on space, which can provide spatial and temporal information in all climatic situations, and anyplace where has a clear line of sight to at least four GPS satellites. This system can provide key capabilities for military, civilian, and commercial users all over the world. It was established and maintains by the United States government, and it is freely accessible to anyone who has a receiver. In 1973, the United States began the GPS project to overcome the restrictions of previous navigation systems, integrates advanced technologies from a few predecessors, and includes some classified engineering design studies in the 1960s. The system initially used 24 satellites. And it was completely operable in 1995 [15].

Global positional system has following functions in precision agriculture: measurement controlling, field information sampling positioning and navigation control [16]. The research of GPS mainly focuses on field information sampling and intelligent agricultural machinery system with GPS receiver. For example, GPS receiver is installed on rotary tiller, planter, field sampler, fertilizing sprayer and harvester, which can obtain accurate positioning information navigation monitoring, and get crop growth information and associated spatial position information [17, 18].

3.2 Geographic Information System

Geographic information system (GIS) is the core of precision agriculture, which is used for field data management, query of soil, natural condition, crop growth and yield etc., and agriculture theme maps drawing, as well as spatial data collecting and analyzing. In addition, GIS system can combine various data into new data sets, for example, GIS system deals with soil type, terrain, and mulch, and establishes the connection among them.

GIS system is software for dealing with spatial information for organizing, analyzing and showing all types of spatial information in the same area. Each kind of information can form one GIS layer, and information from different layers can form new layer by analysis. GIS system can combine various element maps into new map, and analyze the effect of various factors on crop. Take variable fertilizing for example; fertilizer rate is made in different site according to soil nutrition and yield, which guides the fertilizing [19].

3.3 Remote Sensing

Remote sensing (RS) refers to remote probing techniques. It uses sensors detecting electromagnetic waves, visible light and infrared rays from distant, and obtains detection and identification information by data analysis. Remote sensing technology is the key tool for large area obtaining field data rapidly in precision agriculture. It can support vast field spatial change information.

Remote sensing technology is an advanced probing techniques based on modern physics, space science, electronic computer technology, mathematics and earth scientific theory [20]. Various kinds of sensors were developed since 1900s, and both the probing ability and applying range have been significantly expanded.

Remote sensing mainly provides two types of crop and field spatial distribution information. One is basic information which has little change during crop growth season. The information mainly includes farm infrastructure, field contribution etc. And the other is dynamic change information of time and space, including yield, crop nutrition and insects, etc. [21].

4 Application Feasibility Discussion of Precision Agriculture in Equipment for Controlled Traffic Farming System

The research of equipment for controlled traffic farming system mainly focus on working performance with the assumption that wheel lanes exists and works well, and the tractor and equipment run on the wheel lanes. However, it is difficult to control the distance of two lanes constant, and the wheel lane is easy to bend during operation.

In China, when the equipment works on the soil, it is guided by the distance location method of pulling a rope, which means that the ropes are pulled on the rim of lanes, and a person stands in the middle of the two traffic lanes at the end of the rows, and the operator takes the person as the orientation. This method guides the equipment staying on the traffic lanes. But it needs many labors, and is very inconvenient to pull a rope, so it only fits to the small-scale fields. In addition, the size of the overlap and the potential reduction by using controlled traffic farming system depends to a large extent on the individual driver's experience and skills with the precise handling of equipment and tractor tools prior to the introduction of auto steering. To solve the problems, precision agriculture is introduced to controlled traffic farming system [22].

Precision agriculture is a new agricultural technique to realize accurate fertilizing, sowing, irrigation and harvesting etc. with the GPS, GIS and RS. Farmers try hard to improve efficiency by increasing field size and machinery. Precision agriculture adoption on this scale has been made possible due to the enormous decrease of GPS equipment cost in recent years. Yield mapping can be accessed by the GPS system, and farmers could see the yield variation in real time occurring on the farm. Precision agriculture helps farmers to balance management with land capability to improve profitability and protect environmental resources [23].

5 Conclusions

Traditional equipment for controlled traffic farming system including planter, field management equipment and harvester is elaborated and the analysis results shows that main problems contain poor ability to navigate, uneven and bent wheel lanes. To solve the problems, the thought that precision agriculture is introduced into equipment for controlled traffic farming system and key technology of precision agriculture is introduced. Precision agriculture is a new agricultural technique to realize accurate fertilizing, sowing, irrigation and harvesting etc. with the GPS, GIS and RS. Farmers try hard to improve efficiency by increasing field size and machinery, which promote precision agriculture adoption. In addition, precision agriculture can help farmers to balance management with land capability to improve profitability and protect environmental resources. From this perspective, the application of precision agriculture in equipment for controlled traffic farming system is feasible.

Acknowledgments. This work was financed by the Postdoctoral Science Foundation of Beijing Academy of Agriculture and Forestry Sciences of China (2014002) and 863 Program (2012AA101901).

References

1. He, J.: Study on permanent raised beds in irrigation areas of Northern China, China Agricultural University, Beijing (2007). (in Chinese with English abstract)
2. Jin, H., Zhiqi, Z., Qingjie, W.: Current status of permanent raised beds equipment. J. Agric. Mech. Res. **9**, 6–10 (2014). (in Chinese with English abstract)

3. Lamande, M., Hallaire, V., Curmi, P., Peres, G., Cluzeau, D.: Changes of pore morphology, infiltration and earthworm community in a loamy soil under different agricultural managements. Catena **54**, 637–649 (2003)

4. Pagliai, M., Vignozzi, N., Pellegrini, S.: Soil structure and the effect of management practices. Soil Tillage Res. **79**, 131–143 (2004)

5. Radford, B.J., Bridge, B.J., Davis, R.J., McGarry, D., Pillai, U.P., Rickman, J.F., Walsh, P.A., Yule, D.F.: Changes in the properties of a Vertisol and responses of wheat after compaction with harvester traffic. Soil Tillage Res. **61**, 157–166 (2000)

6. Green, T.R., Ahuja, L.R., Benjamin, J.G.: Advances and challenges in predicting agricultural management effects on soil hydraulic properties. Geoderma **116**, 3–27 (2003)

7. Yuhua, B., Chen, F., Hongwen, L., Hao, C., Jin, H., Qingjie, W., Tullberg, J.N., Gong, Y.: Traffic and tillage effects on wheat production on the Loess Plateau of China: 2. Soil physical properties. Aust. J. Soil Res. **46**, 652–658 (2008)

8. Hao, C., Yuhua, B., Qingjie, W., Chen, F., Hongwen, L., Tullberg, J.N., Murray, J.R., Gao, H., Gong, Y.: Traffic and tillage effects on wheat production on the Loess Plateau of China: 1. Crop yield and SOM. Aust. J. Soil Res. **46**, 645–651 (2008)

9. Yaqin, L., Feng, X.: The necessity analysis of the development of precision agriculture in China. J. Agric. Mech. Res. **6**, 4–6 (2006). (in Chinese with English abstract)

10. Maohua, W.: Development of precision agriculture and innovation of engineering technologies. Trans. CSAE **15**(1), 7–14 (1999). (in Chinese with English abstract)

11. Chunjiang, Z., Xuzhang, X., Xiu, W., Liping, C., Yuchun, P., Zhijun, M.: Advance and prospects of precision agriculture technology system. Trans. CSAE **19**(4), 7–12 (2003). (in Chinese with English abstract)

12. Pusheng, K., Gang, L., Jishuang, K.: On the precision agriculture technological system. Trans. CSAE **15**(3), 1–4 (1999). (in Chinese with English abstract)

13. Hao, C., Huang, H., Yali, Y., Hongwen, L.: Design of row-followed no-till wheat and maize planter under controlled traffic farming system. Trans. Chin. Soc. Agric. Mach. **40**(3), 72–76 (2009). (in Chinese with English abstract)

14. Tian, B., Han, S., Wu, J.: Design of 2BMFSG-3/6no-till fertilizing planter under controlled traffic farming system. Trans. Chin. Soc. Agric. Mach. **38**(6), 187–189, 198 (2007)

15. https://en.wikipedia.org/wiki/Global_Positioning_System

16. Pan, Y., Zhao, C.: Application of geographic information technologies in precision agriculture. Trans. CSAE **19**(4), 1–6 (2003)

17. Blackmore, B.S.: An information system for precision farming. In: Brighton Conference Pests and Diseases, pp. 18–21. British crop Protection Council, November, 1996

18. Wang, S., Liu, D.: Research on application of 3S in precision agriculture. J. Hotan Norm. Sch. **26**(5), 166–167 (2006). (in Chinese with English abstract)

19. Xu, S.X., Du, J.Q.: Application of geographic information system in precision agriculture. Mod. Agric. **2**, 33–34 (2003)

20. Hadjimitsis, D.G.: Advances in remote sensing and geo-information for the environment. Cent. Eur. J. Geosci. **6**(1), 1 (2014)

21. Zhou, S., Liang, Z., Yang, Y., Guo, Y.: Status and prospect of agricultural remote sensing. Trans. Chin. Soc. Agric. Mach. **46**(2), 247–260 (2015)

22. Branson, M.: Using conservation agriculture and precision agriculture to improve a farming system. In: Tow, P., Cooper, I., Partridge, I., Birch, C. (eds.) Rainfed Farming Systems, pp. 875–900. Springer, Heidelberg (2011)

23. Hans, G.J., Jacobsen, L.B., Soren, M.P., Elena, T.: Socioeconomic impact of widespread adoption of precision farming and controlled traffic systems in Demark. Precis. Agric. **13**, 661–677 (2012)

Accurate Inference of Rice Biomass Based on Support Vector Machine

Lingfeng Duan[1,2], Wanneng Yang[1,2,3], Guoxing Chen[4],
Lizhong Xiong[3], and Chenglong Huang[1,2(✉)]

[1] College of Engineering, Huazhong Agricultural University,
Wuhan 430070, People's Republic of China
{duanlingfeng,ywn,ehcl}@mail.hzau.edu.cn
[2] Agricultural Bioinformatics Key Laboratory of Hubei Province,
Huazhong Agricultural University, Wuhan 430070
People's Republic of China
[3] National Key Laboratory of Crop Genetic Improvement
and National Center of Plant Gene Research,
Huazhong Agricultural University, Wuhan 430070
People's Republic of China
lizhongx@mail.hzau.edu.cn
[4] MOA Key Laboratory of Crop Ecophysiology
and Farming System in the Middle Reaches of the Yangtze River,
Huazhong Agricultural University, Wuhan 430070, China
hchenguoxing@mail.hzau.edu.cn

Abstract. Biomass is an important phenotypic trait in plant growth analysis. In this study, we established and compared 8 models for measuring aboveground biomass of 402 rice varieties. Partial least squares (PLS) regression and all subsets regression (ASR) were carried out to determine the effective predictors. Then, 6 models were developed based on support vector regression (SVR). The kernel function used in this study was radial basis function (RBF). Three different optimization methods, Genetic Algorithm (GA) K-fold Cross Validation (K-CV), and Particle Swarm Optimization (PSO), were applied to optimize the penalty error C and RBF γ. We also compared SVR models with models based on PLS regression and ASR. The result showed the model in combination of ASR, GA optimization and SVR outperformed other models with coefficient of determination (R^2) of 0.85 for the 268 varieties in the training set and 0.79 for the 134 varieties in the testing set, respectively. This paper extends the application of SVR and intelligent algorithm in measurement of cereal biomass and has the potential of promoting the accuracy of biomass measurement for different varieties.

Keywords: Rice biomass · Support vector regression · Partial least squares · All subsets regression

© IFIP International Federation for Information Processing 2016
Published by Springer International Publishing AG 2016. All Rights Reserved
D. Li and Z. Li (Eds.): CCTA 2015, Part I, IFIP AICT 478, pp. 356–365, 2016.
DOI: 10.1007/978-3-319-48357-3_35

1   Introduction

Plant phenotyping is essential in the study of plant biology, plant functional genomic and plant breeding (Dhondt et al. 2013; Yang et al. 2013; Bolger et al. 2014). Yet plant phenotyping has become a new bottleneck in plant biology. In the recent 5 years, lots of efforts have been done on automatic phenotyping (Duan et al. 2011a, 2011b; Jiang et al. 2012; Huang et al. 2013). However, much work still needs to be done to fill the genotype-phenotype gap.

Biomass is an important phenotypic trait in functional plant biology and plant growth analysis (Honsdorf et al. 2014). Shoot dry weight (DW) is a popular measure of biomass in studying biomass of individual plants (Golzarian et al. 2011). In traditional measurement of DW, the shoot of the plant is cut off, oven-dried to constant weight and weighed by a balance. The low efficiency of the traditional method makes it almost impossible for investigation of a large population of plants. In addition, because the traditional measurement is destructive, continuous inspection of DW over time for an individual plant is infeasible.

Inference of biomass based on machine vision and image analysis allows for non-destructive, high-throughput and continuous measurement of a large quantity of samples. There are researches contributing to automatic measurement of plant biomass (Rajendran et al. 2009; Munns et al. 2010; Hairmansis et al. 2014). However, these researches were only satisfying for young plant (several weeks after sowing) of few varieties.

Based on the statistical learning theory, Support Vector Machine (SVM) is advantageous in robustness to high input space dimension and generalization capabilities (Vapnik, 1995). Support Vector Regression (SVR) is an extension of SVM for regression application and is especially useful in presence of outliers and non-linearities (Brereton and Lloyd 2010).

This study aims to establish a model for measuring aboveground biomass of different rice varieties based on SVR. To the best of our knowledge, no publication available use SVR for biomass measurement.

2   Materials and Methods

2.1   Plant Materials and Image Acquisition

402 rice plants (402 accessions with 1 replicate) were grown in the greenhouse. At late booting stage, all the plants were imaged with a rice automatic phenotyping platform (RAP) (Yang et al. 2014). A turntable rotated the plant and a charge-coupled device (CCD) camera (Stingray F-504C, Applied Vision Technologies, Germany) acquires images at 30° intervals. For each plant, 12 color images at different angles were taken. Simultaneously, a linear X-ray CT captured sinogram of the plant, from which section image were reconstructed and used to extract the tiller number (Yang et al. 2011). The images were saved in the computer for further processing. Next, the plants were harvested and manually measured for the shoot dry weight (DW).

2.2   Feature Extraction and Feature Selection

After image acquisition, the images were analyzed and 39 features, including tiller number, 8 texture features and 30 morphological features, were extracted for each plant. The 39 features included 33 features introduced in Yang et al. (Yang et al. 2014), differential boxing counting dimension (DBC), ratio of plant area to area of bounding rectangle (ABR), greenness area (A_G), yellowness area (A_Y), information fractal dimension (IFD), ratio of perimeter to area (PAR). The features were then used as the potential predictors for DW.

To determine the effective predictors, partial least squares (PLS) regression (Cho et al. 2007) and all subsets regression (ASR) were carried out (Montgomery et al. 2012). PLS regression was accomplished using Matlab 2012b. Prior to performing the PLS regression, the data were normalized so that the mean value and standard deviation of the data was zero and one, respectively. The leave-one-out cross-validation method was performed to determine the optimal number of PLS factors. ASR was done using SAS 9.3. The Cp criterion was used for selecting the best subset. The effective predictors were then used for model input.

2.3   Model Construction and Comparison

The 402 samples were randomly divided into two subsets at 2:1 ratio: 268 samples for training set and 134 samples for testing set. The training set and the testing set was applied for constructing model and evaluating the performance of the model, respectively.

6 models were developed based on support vector regression (SVR). The radial basis function (RBF) only needs to optimize one parameter (the value of γ) and was adopted as the kernel function in this study. Penalty error C and RBF γ were key to the performance of SVR (Brereton and Lloyd 2010). A larger C generates more significant misclassifications but meanwhile leads to a more complex boundary. And inappropriate RBF γ may lead to overfitting. In this study, three different optimization methods, K-fold Cross Validation (K-CV, in this study 5-CV), Genetic Algorithm (GA) (Storn and Price 1997) and Particle Swarm Optimization (PSO) (Clerc and Kennedy 2002), were applied and compared to optimize C and γ. The fitness function for GA and PSO was set as the mean squared error under 5-CV in this study. Libsvm, a popular SVM software package for Matlab designed by professor Lin Chih-Jen was used to accomplish SVR in this study.

In comparison with SVM models, we also built models based on PLS regression and multiple linear regression (MLR). In total, 8 models were developed and compared in this study (Table 1). Figure 1 shows the flowchart of the model construction.

For model comparison, coefficient of determination (R^2), mean absolute percentage error (MAPE, Eqs. 1-2) and standard deviation of the absolute percentage error (SAPE, Eq. 3) for training set and testing set were computed for each model.

Table 1. 8 models developed in this study

Models	Feature selection	C and γ optimization	Modelling
PLS-GA-SVR	PLS	GA	SVR
ASR-GA-SVR	ASR	GA	SVR
PLS-PSO-SVR	PLS	PSO	SVR
ASR-PSO-SVR	ASR	PSO	SVR
PLS-CV-SVR	PLS	K-CV	SVR
ASR-CV-SVR	ASR	K-CV	SVR
PLS	PLS		PLS regression
MLR	ASR		MLR

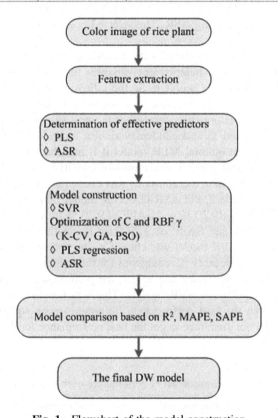

Fig. 1. Flowchart of the model construction

$$APE_i = \frac{|DW_{i.manual} - DW_{i.automatic}|}{DW_{i.manual}} \tag{1}$$

$$MAPE = \frac{1}{n}\sum_{i=1}^{n} APE_i \tag{2}$$

$$SAPE = \sqrt{\frac{1}{n-1} \sum_{i=1}^{n} (APE_i - MAPE)^2} \tag{3}$$

where $DW_{i.automatic}$ represents the dry weight measured automatically using the method described, $DW_{i.manual}$ represents the dry weight measured manually, and n represents the number of samples.

3   Results and Discussion

After PLS regression, 4 PLS factors were selected. And a subset with 18 features was selected as the best subset using ASR. The selected predictors were used as input for the models.

Table 2 illustrates the comparison of the 8 models. Note that when using the best subset by the Cp criterion as independent variables for MLR modelling, the model suffered from multi-collinearity problem. So the following strategy was used to select the feature subset for MLR: (1) the subset that has the maximum R^2 among all subsets with i $i = 1, 2, \cdots, 39$ features was deemed as the best subset with i features, (2) the 39 best subsets were used as independent variables to build MLR models and the model was chosen as the optimal MLR model if it had the largest number of independent variables and did not present multi-collinearity. Finally, a model with 3 independent variables (exclude the constant) was chosen as the optimal MLR model.

As seen from the Table 2, the ASR-GA-SVR model outperformed other models, with R^2 of 0.85, MAPE of 10.20 % and SAPE of 9.20 % for the training set and R^2 of 0.79, MAPE of 12.44 % and SAPE of 9.79 %for the testing set, respectively. Consequently, the ASR-GA-SVR model was chosen as the optimal DW model. The SVR models were generally noticeably advantageous for the training set compared with PLS and ASR model. However, for the testing set, the performance of the PLS and ASR model were comparative to the SVR models. This was because the optimal C and γ were chosen to obtain the best performance (minimum mean squared error) for the training set but could not guarantee to get the best performance for the testing set under the optimal C and γ.

Table 2. Comparison of performance of the 8 models

Method	Training set			Testing set		
	R^2	MAPE	SAPE	R^2	MAPE	SAPE
PLS-GA-SVR	0.82	11.90 %	9.23 %	0.79	12.62 %	10.14 %
ASR-GA-SVR	0.85	10.20 %	9.20 %	0.79	12.44 %	9.79 %
PLS-PSO-SVR	0.82	11.76 %	9.21 %	0.79	12.69 %	10.10 %
ASR-PSO-SVR	0.86	9.59 %	8.82 %	0.75	13.01 %	10.31 %
PLS-CV-SVR	0.83	11.22 %	9.07 %	0.78	12.78 %	9.73 %
ASR-CV-SVR	0.86	10.03 %	9.04 %	0.77	12.69 %	9.93 %
PLS	0.81	12.12 %	9.28 %	0.79	12.71 %	10.41 %
ASR	0.80	12.74 %	10.18 %	0.77	13.25 %	10.64 %

Figures 2, 3 and 4 show the performance of the final DW model (ASR-GA-SVR model), the PLS model and ASR model, respectively.

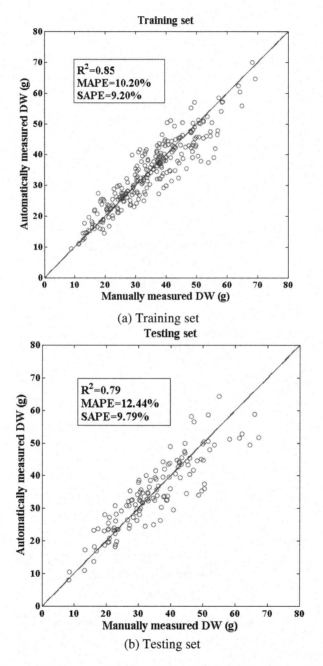

(a) Training set

(b) Testing set

Fig. 2. Performance of the final DW model (ASR-GA-SVR model)

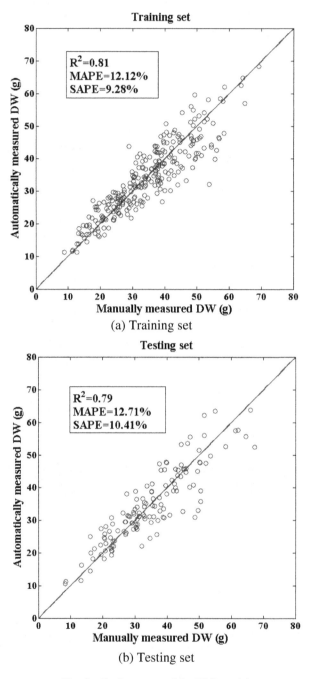

(a) Training set

(b) Testing set

Fig. 3. Performance of the PLS model

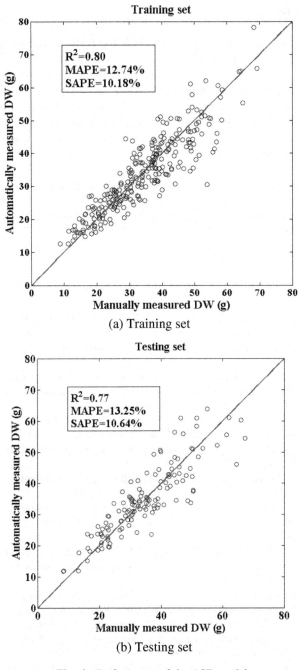

(a) Training set

(b) Testing set

Fig. 4. Performance of the ASR model

4 Conclusions

This paper presents 8 models based on SVR, PLS and ASR for measuring aboveground biomass of different rice varieties. The result showed the ASR-GA-SVR model outperformed other models with R^2 of 0.85, MAPE of 10.20 % and SAPE of 9.20 % for the training set and R^2 of 0.79, MAPE of 12.44 % and SAPE of 9.79 %for the testing set, respectively. The study extends the application of SVR and intelligent algorithm in the measurement of plant biomass. The method has the potential to promote the accuracy of biomass measurement for different varieties and thus contributes to automatic plant phenotyping.

Acknowledgment. This work was supported by the Fundamental Research Funds for the Central Universities (2662015QC006, 2662015QC016, 2013PY034, 2662014BQ036), the National High Technology Research and Development Program of China (2013AA102403), the National Natural Science Foundation of China (30921091, 31200274).

References

Bolger, M., Weisshaar, B., Scholz, U., Stein, N., Usadel, B., Mayer, K.: Plant genome sequencing - applications for crop improvement. Plant Biotechnol. J. **8**(1), 31–37 (2014)

Brereton, R., Lloyd, G.: Support vector machines for classification and regression. Analyst **135** (2), 230–267 (2010)

Cho, M., Skidmore, A., Corsi, F., van Wieren, S., Sobhan, I.: Estimation of green grass/herb biomass from airborne hyperspectral imagery using spectral indices and partial least squares regression. Int. J. Appl. Earth Obs. Geoinf. **9**, 414–424 (2007)

Clerc, M., Kennedy, J.: The particle swarm - explosion, stability, and convergence in a multidimensional complex space. IEEE Trans. Evol. Comput. **6**(1), 58–73 (2002)

Dhondt, S., Wuyts, N., Inzé, D.: Cell to whole-plant phenotyping: the best is yet to come. Trends Plant Sci. **18**(8), 428–439 (2013)

Duan, L., Yang, W., Huang, C., Liu, Q.: A novel machine-vision-based facility for the automatic evaluation of yield-related traits in rice. Plant Methods **7**, 44 (2011a)

Duan, L., Yang, W., Bi, K., Chen, S., Luo, Q., Liu, Q.: Fast discrimination and counting of filled/unfilled rice spikelets based on two modal imaging. Comput. Electron. Agric. **75**(1), 196–203 (2011b)

Golzarian, M., Frick, R., Rajendran, K., Berger, B., Roy, S., Tester, M., Lun, D.: Accurate inference of shoot biomass from high-throughput images of cereal plants. Plant Methods **7**, 11 (2011)

Hairmansis, A., Berger, B., Tester, M., Roy, S.: Image-based phenotyping for non-destructive screening of different salinity tolerance traits in rice. Rice **7**, 16 (2014)

Honsdorf, N., March, T., Berger, B., Tester, M., Pillen, K.: High-throughput phenotyping to detect drought tolerance QTL in wild barley introgression lines. PLoS ONE **9**(5), e97047 (2014). doi:10.1371/journal.pone.0097047

Huang, C., Yang, W., Duan, L., Jiang, N., Chen, G., Xiong, L., Liu, Q.: Rice panicle length measuring system based on dual-camera imaging. Comput. Electron. Agric. **98**, 158–165 (2013)

Jiang, N., Yang, W., Duan, L., Xu, X., Huang, C., Liu, Q.: Acceleration of CT reconstruction for wheat tiller inspection based on adaptive minimum enclosing rectangle. Comput. Electron. Agric. **85**, 123–133 (2012)

Montgomery, D., Peck, E., Vining, G.: Introduction to Linear Regression Analysis. Wiley, Hoboken (2012)

Munns, R., James, R., Sirault, X., Furbank, R., Jones, H.: New phenotyping methods for screening wheat and barley for beneficial responses to water deficit. J. Exp. Bot. **61**, 3499–3507 (2010)

Rajendran, K., Tester, M., Roy, S.: Quantifying the three main components of salinity tolerance in cereals. Plant Cell Environ. **32**(3), 237–249 (2009)

Storn, R., Price, K.: Differential evolution – a simple and efficient heuristic for global optimization over continuous spaces. J. Global Optim. **11**(4), 341–359 (1997)

Vapnik, V.: The Nature of Statistical Learning Theory. Springer, New York (1995)

Yang, W., Xu, X., Duan, L., Luo, Q., Chen, S., Zeng, S., Liu, Q.: High-throughput measurement of rice tillers using a conveyor equipped with X-ray computed tomography. Rev. Sci. Instrum. **82**(2), 025102–025109 (2011)

Yang, W., Duan, L., Chen, G., Xiong, L., Liu, Q.: Plant phenomics and high-throughput phenotyping: accelerating rice functional genomics using multidisciplinary technologies. Curr. Opin. Plant Biol. **16**, 180–187 (2013)

Yang, W., Guo, Z., Huang, C., Duan, L., Chen, G., Jiang, N., Fang, W., Feng, H., Xie, W., Lian, X., Wang, G., Luo, Q., Zhang, Q., Liu, Q., Xiong, L.: Combining high-throughput phenotyping and genome-wide association studies to reveal natural genetic variation in rice. Nat. Commun. **5**, 5087 (2014)

Brazil Soybean Area Estimation Based on Average Samples Change Rate of Two Years and Official Statistics of a Year Before

Kejian Shen[(⊠)], Weifang Li, Zhiyuan Pei, Fei Wang,
Xiaoqian Zhang, Guannan Sun, Jiong You, Quan Wu,
and Yuechen Liu

Chinese Academy of Agricultural Engineering, Innovation Team of Crop
Monitoring by Remote Sensing, Beijing Chaoyang District Maizidian Street
No. 41, Beijing 100125, China
ashenkejian@126.com, guannan_sun@126.com,
wflil988@aliyun.com, {peizhiyuan,wangfei,
zhangxiaoqian,liuyuechen}@agri.gov.cn,

Abstract. Comprehensive, reliable and timely information of Brazil's soybean area is necessary for China to make decisions on agricultural related problems. Spatial sampling method which combined remote sensing and sampling survey is widely used. Due to limitations of width and revisit cycle of medium resolution satellite, This study designed a typical investigation method about Brazil soybean area based on average samples change rate of two years and official statistics of a year before, typical samples were selected to survey, sampling frame was constructed on soybean planting state, the sampling unit was designed as 40 km × 40 km, the sampling proportion was 2 %, average samples change rate of two years were 2013 and 2014. Estimated area was compared with Brazil official harvested area in 2014 (published on 2015 April by Brazilian Institute of Geography and Statistics), the relative error is 2.37 %.

Keywords: Brazil soybean area estimation · Change rate of two years · Official statistics of a year before · Sampling

1 Introduction

Chinese soybean planting area has decreased year by year, soybean self-supply ability had been to about 20 % in 2013, mainly imports country are United States, Brazil, and Argentina [1]. Comprehensive, reliable and timely information of Brazil's soybean area is necessary for China to make decisions on agricultural related problems. Compared with the traditional survey method, Remote sensing survey has advantage of large coverage, low cost and less investigation time [2]. Spatial sampling method which combined remote sensing and sampling survey is widely used in the investigation of large scale crop area estimation [3]. The survey accuracy is mainly effect by Population, sampling proportion, sample distribution. Population can be defined by

© IFIP International Federation for Information Processing 2016
Published by Springer International Publishing AG 2016. All Rights Reserved
D. Li and Z. Li (Eds.): CCTA 2015, Part I, IFIP AICT 478, pp. 366–374, 2016.
DOI: 10.1007/978-3-319-48357-3_36

historical cultivated land [4] or administrative divisions with historical statistical data [5]. Sampling proportion is the bigger the better with the premise of meeting the minimum sampling proportion, but the actual survey should consider the accuracy requirements, cost and time. The ideal sample distribution is random distribution, but it is subject to the satellite width limitation and revisit cycle.

With the rapid development of remote sensing technology, medium resolution satellite (10 m–30 m) are gradually meet the sampling survey requirements and even full coverage. But in the soybean growth period, it is difficult to get full coverage image with cloud free. Considering the weather, satellite width limitation and revisit cycle, using landsat7/8 as a data source, this study designed a typical investigation method about Brazil soybean area estimation based on average samples change rate of two years and official statistics of a year before.

2   Study Area and Data Source

2.1   Study Area

Brazil is located in the west by 35 to 74°, 5° north latitude to 35° south latitude. Brazil's total area is about 8514900 square kilometers, which is about 46 % of the South America total area. The terrain of Brazil is divided into two parts, one part is plateau of Brazil with altitude of 500 m above, located in the south of Brazil, the other part is plains with elevation of 200 m below, mainly distributed in the Amazon River Basin in the north and the west. Throughout the terrain is divided into the

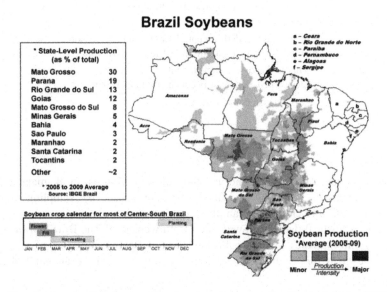

Fig. 1. The sketch map of study area (http://www.usda.gov/oce/weather/pubs/Other/MWCACP/Graphs/Brazil/BrzSoyProd_0509.pdf)

Amazon plain, Paraguay basin, Brazil and the Guyana plateau, the Amazon plain area accounting for about 1/3. Most of the area of Brazil belongs to the tropical climate, parts of the South belongs to the subtropical climate. Annual average temperature of The Amazon plain is 25 ~ 28°, the annual average temperature of south is 16 to 19°.

Soybean mainly distribute in central Brazil and southern Brazil (Fig. 1), Due to the tropical climate and long growing season, the crop production cycles are much more complicated. Below is a month-by-month (Table 1) account of what to expect during the growing season.

Table 1. Brazil soybean month-by-month crop cycle

September	• Early soybean begins in Mato Grosso and central Brazil.
October	• Soybean planting in full swing in southern Brazil.
November	• Early November is main planting period.
December	• Finish planting, early-planted soybeans flowering and setting pods. • Begin spraying to control soybean rust.
January	• Soybeans flowering and setting pods. • Some very early soybeans in central Brazil may be harvested this month. • Continue spraying to control soybean rust.
February	• Main pod filling month. • Early soybeans being harvested. • Soybean rust control now focused on later maturing soybeans.
March	• Main soybean harvesting month. • Critical time for soybean rust to affect late maturing soybeans.
April	• Finish soybean harvest.
May	• Rains have ended in central Brazil and dry season has started. • Scattered rains continue to fall in southern Brazil.
June-July-August	• This is the dry season in central Brazil. • Occasional rains can occur in southern Brazil.

From: http://www.soybeansandcorn.com/Brazil-Crop-Cycles

2.2   Data

Landsat Multi-spectral image: Landsat7 and landsat8 Multi-spectral image listed as Table 2 were used. Landsat7 and landsat8 were subset by sampling frame of 40 km × 40 km, only cloud free samples were selected.

Soybean statistical data are downloaded from website of The Brazilian Institute of Geography and Statistics[1], which publishes harvests figures consisting of area, output and average yield for 35 different crops of previous year in the annual 1–4 month.

---

[1] http://www.ibge.gov.br/english/estatistica/indicadores/agropecuaria/lspa/default_publ_completa.shtm.

Table 2. Landsat multi-spectral image

Path/row	Number of samples	Data	Acquired time	Purpose
221/78	16	Lansat7	10-jan-13	2013 soybean extract
		Lansat8	21-jan-14	2014 soybean extract
221/79	6	Lansat7	10-jan-13	2013 soybean extract
		Lansat8	21-jan-14	2014 soybean extract
222/79	14	Lansat7	17-jan-13	2013 soybean extract
		Lansat8	28-jan-14	2014 soybean extract
222/81	9	Lansat7	06-mar-13	2013 soybean extract
		Lansat8	28-jan-13	2014 soybean extract
223/77	11	Lansat7	24-jan-13	2013 soybean extract
		Lansat8	19-jan-14	2014 soybean extract
224/68	8	Lansat7	30-dec-12	2013 soybean extract
		Lansat7	16-feb-13	
		Lansat7	13-apr-13	
		Lansat8	02-jan-14	2014 soybean extract
		Lansat8	27-feb-14	
		Lansat8	08-apr-14	
224/69	5	Lansat7	30-dec-12	2013 soybean extract
		Lansat7	16-feb-13	
		Lansat7	05-apr-13	
		Lansat8	02-jan-14	2014 soybean extract
		Lansat8	27-feb-14	
		Lansat8	08-apr-14	
225/75	14	Lansat7	21-dec-12	2013 soybean extract
		Lansat7	07-feb-13	
		Lansat7	20-apr-13	
		Lansat8	30-nov-13	2014 soybean extract
		Lansat8	02-feb-14	
		Lansat8	07-apr-14	

Soybean official harvested area in 2013 is 27736 thousand hectares, which is used for are estimation. Soybean official harvested area in 2014 is 30241 thousand hectares, which is used for accuracy assessment.

3 Methodology: Sampling Design

The flow of this experiment (Fig. 2) includes: (1) Construction of sampling frame; (2) Determine the Sampling proportion and distribution of samples, (3) Soybean extraction by unsupervised classification and visual interpretation; (4) The average change rate of samples between 2013 and 2014; (5) Area estimation; (6) Accuracy assessment.

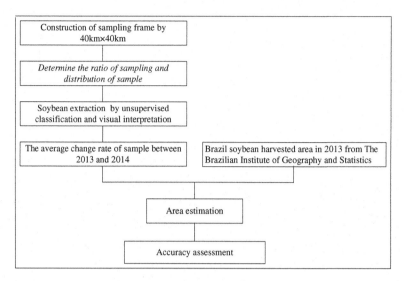

Fig. 2. Flow chart of experiment

3.1   Construction of Sampling Frame

The sampling frame covers 17 soybean planting states (from 2013 Brazil official statistics), which can be seen from Fig. 3, the sampling unit was designed as 40 km × 40 km. The population is 4215, which is shown on Fig. 3.

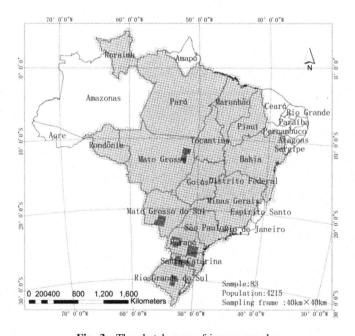

Fig. 3. The sketch map of image samples

3.2   Determine the Sampling Proportion and Distribution of Samples

We considered relevant research to determine the sampling proportion. The rate of sampling of Monitoring Agriculture with Remote Sensing (MARS) of European Union is about 1 % (60sites × 40 km × 40 km/10160000 km^2) [6]; The rate of paper about paddy rice area estimation using a stratified sampling method with remote sensing in China is 1.3 % [7]. In order to improve the estimation accuracy, the rate of sampling is increased to 2 %. The number of samples is 83. Selected samples could be seen from Fig. 3.

Distribution of samples should consider several factors. Firstly, the samples should cover the major and minor soybean planting areas; Secondly, in the soybean growth season, the images can effectively extract the spatial distribution of soybean. Thirdly, sample is cloud free. The samples are shown in Fig. 3.

3.3   Soybean Extraction by Unsupervised Classification and Visual Interpretation

Each sample has soybean classification results of 2013 and 2014, samples were classified by unsupervised classification. The classification results were corrected by visual interpretation of ArcGIS software. Statistics of samples classification results are shown as Table 3.

3.4   The Average Change Rate of Samples Between 2013 and 2014

$$Change\_rate = (Area2014_{sample\_i} - Area2013_{sample\_i})/Area2013_{sample\_i} \quad (1)$$

where change_rate represents change rate of samples between 2013 and 2014, sample_i represents sample number, i from 1 to 83. The average change rate of 83 samples between 2013 and 2014 is 6.45 %, which is shown on Table 3.

3.5   Area Estimation

$$\hat{P} = Area2013_{official} \times (1 + Average\ change\_rate) \quad (2)$$

where $\hat{P}$ represents estimated area; $Area2013_{official}$ represents Brazil soybean statistical data of 2013, which are downloaded from website of The Brazilian Institute of Geography and Statistics.

3.6   Accuracy Assessment

Sampling results was appraised by relative error r, which is defined as follow:

Table 3. Statistics of samples classification result

Sample	Path/row	Area13	Area14	Rate of change	Sample	Path/row	Area13	Area14	Rate of change
1	221/78	226	195	−14.00 %	43	222/81	375	408	8.75 %
2	221/78	66	60	−9.19 %	44	222/81	321	350	8.84 %
3	221/78	138	130	−5.59 %	45	222/81	15	55	273.35 %
4	221/78	162	159	−2.21 %	46	223/77	422	424	0.46 %
5	221/78	516	515	−0.33 %	47	223/77	621	646	3.95 %
6	221/78	281	289	2.96 %	48	223/77	92	96	4.48 %
7	221/78	67	69	3.56 %	49	223/77	210	226	7.64 %
8	221/78	382	398	4.09 %	50	223/77	574	626	9.08 %
9	221/78	190	201	6.09 %	51	223/77	437	479	9.41 %
10	221/78	325	349	7.21 %	52	223/77	517	644	24.65 %
11	221/78	223	239	7.38 %	53	223/77	488	624	27.99 %
12	221/78	404	435	7.48 %	54	223/77	687	942	37.15 %
13	221/78	343	371	8.04 %	55	223/77	82	157	91.33 %
14	221/78	356	386	8.45 %	56	223/77	15	34	120.32 %
15	221/78	241	265	9.98 %	57	224/68	264	188	−28.87 %
16	221/78	231	262	13.40 %	58	224/68	34	26	−24.84 %
17	221/79	23	9	−61.04 %	59	224/68	420	379	−9.69 %
18	221/79	94	53	−43.82 %	60	224/68	566	538	−5.05 %
19	221/79	263	162	−38.50 %	61	224/68	515	497	−3.59 %
20	221/79	454	328	−27.79 %	62	224/68	353	366	3.89 %
21	221/79	80	59	−26.40 %	63	224/68	58	61	5.55 %
22	221/79	303	332	9.46 %	64	224/68	55	66	18.95 %
23	221/79	78	47	−39.41 %	65	224/69	795	736	−7.39 %
24	221/79	146	120	−18.30 %	66	224/69	485	470	−3.07 %
25	221/79	35	35	0.32 %	67	224/69	1295	1300	0.32 %
26	221/79	725	734	1.24 %	68	224/69	288	309	7.32 %
27	221/79	326	332	2.11 %	69	224/69	144	164	14.49 %
28	221/79	12	13	8.66 %	70	225/75	513	372	−27.43 %
29	221/79	293	318	8.69 %	71	225/75	513	429	−16.41 %
30	221/79	526	573	8.92 %	72	225/75	675	637	−5.61 %
31	221/79	801	878	9.70 %	73	225/75	8	8	−3.38 %
32	221/79	253	281	10.96 %	74	225/75	798	779	−2.41 %
33	221/79	289	335	15.90 %	75	225/75	282	278	−1.28 %
34	221/79	594	695	17.15 %	76	225/75	401	405	1.07 %
35	221/79	78	93	19.55 %	77	225/75	23	23	2.61 %
36	221/79	227	314	38.14 %	78	225/75	388	399	2.81 %
37	222/81	352	348	−1.39 %	79	225/75	376	388	3.23 %
38	222/81	101	103	1.61 %	80	225/75	117	123	4.94 %
39	222/81	186	193	3.78 %	81	225/75	585	614	5.05 %
40	222/81	290	304	4.89 %	82	225/75	201	212	5.29 %
41	222/81	260	275	5.46 %	83	225/75	195	208	6.76 %
42	222/81	193	207	7.56 %	Average change rate of samples				6.45 %

$$r = 100 \times (P - \hat{P})/P \qquad (3)$$

where r represents relative error, $\hat{P}$ represents estimated area, P represents true area, Brazil soybean harvested area in 2014 are used.

4 Results

Soybean official harvested area in 2013 is 27736 thousand hectares, average change rate between 2013 and 2014 is 6.45 %, Estimated soybean harvested area in 2014 is:

$\hat{P} = Area2013_{official} \times (1 + Average\ change\_rate) = 27736 \times (1 + 6.45\ \%)$
$= 29525$ thousand hectares

Soybean official harvested area in 2014 is 30241 thousand hectares, relative error is:

$$r = 100 \times (P - \hat{P})/P = 100\ \% \times (30241 - 29525)/30241 = 2.37\ \%.$$

5 Discussion and Conclusion

Discussion: in the previous study of stratified sampling, stratified variable often from Modis data or land use/cover data or statistical data, the location of sample is determined by stratified variable. However, the number of sample with determined location is hard to be satisfied with cloud free image. A question then worth asking is How to maximize the use of available images?

This study designed a typical investigation method about Brazil soybean area based on average samples change rate of two years and official statistics of a year before, typical samples were selected to survey, sampling frame was constructed on soybean planting state, the sampling unit was designed as 40 km × 40 km, the sampling proportion was 2 %, average samples change rate of two years were 2013 and 2014. Estimated area was compared with Brazil official harvested area in 2014 (published on 2015 April by Brazilian Institute of Geography and Statistics), the relative error is 2.37 %.

Acknowledgment. Funds for this research was provided by National Natural Science Foundation of China (No. 41301506) and Civil Space Project in the 12th Five-Year Plan of China (2011–2015).

References

1. Liu, Z.: Some thoughts concerning development strategy for soybean industry in China. Soybean Sci. **32**(3), 283–285 (2013)
2. Ahmed, R., Sajjad, H.: Crop acreage estimation of Boro Paddy using remote sensing and GIS techniques: a case from Nagaon district, Assam, India. Adv. Appl. Agric. Sci. **3**(3), 16–25 (2015)
3. Huajun, T., Wu, W., Yang, P., et al.: Recent progresses in monitoring crop spatial patterns by using remote sensing technologies. Sci. Agric. Sin. **43**(14), 2879–2888 (2010)
4. Wu, B.-F., Li, Q.-Z.: Crop acreage estimation using two individual sampling frameworks with stratification. J. Remote Sens. **8**(6), 551–569 (2005)
5. Chen, Z., Liu, H., Zhou, Q., et al.: Sampling and scaling scheme for monitoring the change of winter wheat acreage in China. Trans. Chin. Soc. Agric. Eng. **16**(5), 126–129 (2000)
6. Gallego, F.J., Delincé, J., Carfagna, E.: Two-stage area frame sampling on square segments for farm surveys. Surv. Methodol. **20**(2), 107–115 (1994)
7. Jiao, X., Yang, B., Pei, Z., et al.: Paddy rice area estimation using a stratified sampling method with remote sensing in China. Trans. Chin. Soc. Agric. Eng. **22**(5), 105–110 (2006)

Meta-Synthetic Methodology: A New Way to Study Agricultural Rumor Intervention

Ruya Tian[1,2], Lei Wu[1,2], Yijun Liu[3], and Xuefu Zhang[1,2(✉)]

[1] Agricultural Information Institute,
Chinese Academy of Agricultural Sciences (CAAS),
Beijing 100081, People's Republic of China
{tianruya,girlrable}@126.com, zhangxuefu@caas.cn
[2] Key Laboratory of Agri-Information Service Technology,
Ministry of Agriculture, Beijing 100081, People's Republic of China
[3] Institute of Policy and Management, Chinese Academy of Sciences,
Beijing 100190, People's Republic of China
yijunliu@casipm.ac.cn

Abstract. Intervention of online public opinion, especially agricultural rumors, has increasingly become an important issue that is related to the online society safety and reflecting the managers' social management capabilities. In this paper, meta-synthetic methodology was used to investigate the intervention mechanism of online public opinion system. General characteristics of online public opinion system were characterized qualitatively, and an opinion super-network model was built to analyze the intervention mechanism quantitatively. We put forward a rumor intervention mechanism including intervention timing, intervention mode, and intervention intensity. Based on system modeling and simulation, an agricultural rumor case was used to verify the practical effects of online agricultural rumor intervention. The results show that rumor intervention based on meta-synthetic methodology produces good results. This investigation will help to provide a reliable basis for the development of more sophisticated and effective rumor intervention strategies.

Keywords: Meta-synthetic methodology · Online public opinion system · Agricultural rumor intervention · Intervention mechanism

1 Introduction

With the popularity of the web, BBS and Twitter on Internet have increasingly become the most important channel for people to publish, exchange and access to information. It is now an important platform for Internet users to leave comments on emergencies, to express attitudes and to vent their emotions. Furthermore, it even becomes an important force in promoting the development of the incidents. Monitoring and management of agricultural rumors are increasingly becoming an important aspect of the national interests and a matter of national security, which make the emergency management of the online rumors become a hot spot [1–4]. At present, researches on online public opinion intervention in China focused more on the qualitative ways, which takes case analysis as the main method. While scientific, systematic, and quantitative research is rare [5].

© IFIP International Federation for Information Processing 2016
Published by Springer International Publishing AG 2016. All Rights Reserved
D. Li and Z. Li (Eds.): CCTA 2015, Part I, IFIP AICT 478, pp. 375–389, 2016.
DOI: 10.1007/978-3-319-48357-3_37

System is the general form of the existence of things. One of the founders of general systems theory, Bertalanffy, defined the system as "A synthesis of all related factors", which is an organic integrity with a specific function combined of several interactional and interdependent components [6]. General systems theory requires people to see things from a global standpoint, to comprehensively analyze the relationships between the elements in the system, elements and their system, system and its environment, this system and other systems. By doing so, internal relations and laws of a system can be grasped, as well as the way to control and transform it effectively. Meta-synthetic methodology proposed by the Chinese famous scholar Xue-Sen Qian, illustrates a comprehensive approach of qualitative and quantitative methods [7]. It is a creative sublimation to systems theory, providing a new way of thinking and methods to solve the problem of open complex giant system [8].

This paper investigated the intervention mechanism of online public opinion from a meta-synthetic methodological point of view. Online public opinion can be seen as a system. It has general constituent elements. Each element in the system is on a particular position with specific functions. Links and constraints between elements of each other constitute an indivisible whole. On the basis of elaborating its composition, structure, and function qualitatively, this paper quantified the intervention mechanism for online public opinion to provide a new research perspective for online public opinion intervention.

The rest of this paper is organized as follows. Section 2 describes online public opinion system, including its composition, structure, and function. In Sect. 3, we built an Opinion SuperNetwork according to the system characteristics to explore the online public opinion, and then discussed the intervention mechanism based on this supernetwork. In Sect. 4, we applied the methods described in Sect. 3 to a certain food safety event, and the results are presented and compared. Section 5 concludes, and suggests possible future research approaches.

2   Online Public Opinion System

Online public opinion is a system of particular composition, structure, and function. Intervention and guide for online public opinion can be highly effective or even gets twice the result with half the effort by fully considering the inherent properties and operation mechanism of the online public opinion system. Online public opinion system is the sum of beliefs, attitudes, opinions and emotions, etc. It is public opinion subject published on the Internet about certain social events under certain psychologies at a certain time and space. It can impact on the development of the events.

Online public opinion system consists of opinion subjects, opinion objects, and system dynamics, as is shown in Fig. 1. Public opinion system exists in a particular system environment, and is divided from its environment by the system boundary. There are certain inputs and outputs between the public opinion system and its environment. Opinion subjects of public opinion system includes the opinion agents discussing the events and the parties of the events. For instance, Netizens are opinion subjects in public opinion system. The opinion object is the sum of beliefs, attitudes, opinions and emotions, etc., which are published by public opinion subjects on the

Internet, such as netizens' posts. System dynamic is the psychology of public opinion subjects, which is the internal driving force for subjects to express different objects. For example, hatred answers for the deep psychological reason of certain netizens who publish rumors.

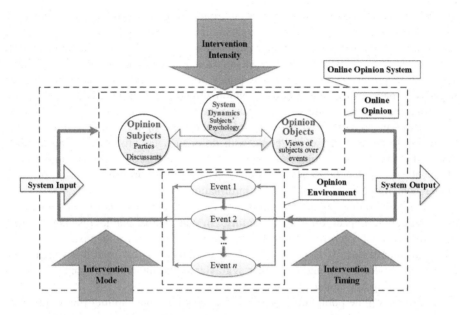

Fig. 1. Online public opinion system and its intervention mechanism

Public opinion system environment is the event evolvement, which is the input of the system, and also the external driving force of the development of the system. System boundary is the certain space and time, as well as specific event. Effect of opinion system on the event is the system output. There are two fundamentally different types of system output. One is the direct result caused by the system. The other type is the conclusion with more universal significances. As in "Guo Mei-mei rumor event", the direct result of the online public opinion system is Guo Mei-mei clarified that there is no relationship between her and the Chinese Red Cross, and Red Cross disclosed its money flows. While universal conclusions are we need to reflect on the values of the post-90s generation, to explore methods resolving the government confidence crisis, and to ponder the role of the media.

Relation mode between various elements of the opinion system is the system's structure. First, opinion subjects interact one another. For instance, opinion of the opinion leaders influences the others' opinion [9, 10]. Second, there are temporal and spatial dependencies between opinion objects. Some posts published earlier on time result in preconception impacts on later posts. And information nonidentity of different forums caused the isolation between the opinion objects in space. Thirdly, there is a relationship of action and reaction between opinion subject and object. Subjects express different viewpoints under different system dynamics, and different viewpoints in turn affect the psychology of the subjects. Last but not least, system dynamic is the connector, mediator, and promoter associating system subjects and objects. It is the internal driving force of system development and evolvement. Interactions between this subject and that subject, this object and that object, as well as between subject and object are driven by the system dynamic.

Online public opinion impacts on its environment, i.e. the evolvement of incidents. This is the output of the public opinion system. It can reflect the function of the system. Discussions in online public opinion system have positive effects on the development of the event. It is a collection and a reflection of public opinion. Online public opinion system is a promoter for government policy to be transparent and open. But the negative impacts of the online public opinion should not be overlooked.

First, the online public opinion can sometimes become an amplifier of false information and rumors, leading to misinformation and opinion anomie. Internet is a place mixed of good and evil. People with different ages, professions, and interests gathered in this discourse space. Anyone can feel free to post or retweet the information of social events after he or she registered an ID as a forum user or had a personal twitter, which opens the door for false information and rumors.

Second, the online public opinion can be the fuse of public discontent sometimes, with negligible impact on social stability. In the online public discourse space, multi-dimensional interacts among Internet users include not only text, but also an emotional interaction. It is noteworthy that, discontent of netizens often start from criticizing a social event, and be further aggravated with the criticism going deeper. It sometimes bring bad impacts on social stability.

Third, the online public opinion can sometimes become a splitter digesting social cohesion, which leads to social crisis of confidence. With the increase of online rumor, social confidence is facing a deeper crisis. In recent years, the social crisis of confidence problems caused by the spread of online rumor is worth more increasingly of concern.

3 Intervention Mechanism of Online Public Opinion System

Qualitative descriptions for components, structure, and function of the online public opinion system provide a theoretical basis for quantitative calculation of system mechanism. Online public opinion intervention can be implemented based on the characteristics of the system. Regulating the composition and structure of the system changes function of the system. This will make public opinion interacting with the event positively and achieve the goal of rumor intervention.

3.1 Online Public Opinion System Modeling

A supernetwork is defined as a network that exists above and beyond existing networks [11]. In recent years, it is applied in many aspects such as knowledge networks [12], logistics (supply chain) [13], traffic [14], ecology [15], and so on. According to the compositional and structural characteristics of online public opinion system, a super-network model including Social Subnetwork, Environment Subnetwork, Psychological

Subnetwork, and Viewpoint Subnetwork can be created (Fig. 2). The model in this study can not only depict the relationships between the elements in the public opinion formation process, but better reveal the dynamics and evolution mechanisms of the system [16, 17].

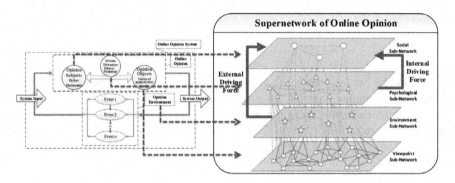

Fig. 2. Online public opinion system modeling

3.1.1   Social Subnetwork

Take the nodes of the network for the agents issued public opinion, and the network edges for the reply relationships between the agents. By doing so, we form the "Social Subnetwork of Opinion SuperNetwork (G_A)". Opinion agents who issued positive viewpoints are positive agents in the public opinion system. The one who issued negative viewpoints are negative agents.

3.1.2   Environment Subnetwork

Take the nodes of the network for the environments under which public opinion issued, and the network edges for relationships between issued time of pieces of information. By doing so, we form the "Environment Subnetwork of Opinion SuperNetwork (G_E)". A negative environment is the one with rumor spreading, while positive environment refutes rumor spreading.

3.1.3   Psychological Subnetwork

Take the nodes of the network for agents' psychology who issued public opinion, and the network edges for their transformation relationships. By doing so, we create the "Psychological Subnetwork of Opinion SuperNetwork (G_P)". The agents with positive psychology are reluctant to believe the rumors, and the ones with negative psychology are inclined to believe rumors.

3.1.4   Viewpoint Subnetwork

Take the nodes of the network for viewpoint keywords issued by public opinion agents, and network edges for various keywords' affiliation relationships. By doing so, we create the "Viewpoint Subnetwork of Opinion SuperNetwork (G_K)". The viewpoint with rumor spreading is negative, while the one with rumor clarifying is positive.

3.1.5   Opinion SuperNetwork

Take opinion agents, opinion environment, agent's psychology and viewpoints as nodes, the relationships in and between the Subnetworks for superedges (SE) to create "Opinion SuperNetwork (OSN)".

$$OSN = \left(G_A, G_E, G_P, G_K, SE_{a-e-p-k}\right) \tag{1}$$

This model can be used to describe the overall features and the general rules of online public opinion system. Based on this model, we explore the quantitative intervention mechanisms, including how to make the best timing to intervene, select feasible intervention mode, as well as master the appropriate intervention intensity (Fig. 1).

3.2   Intervention Timing

Currently, the intervention of online agricultural rumor is implemented after the event had become opinion hotspot. Government departments are involved in the interference with netizens' urgent requirement, missed the opportunity to change the negative momentum of the public opinion. This often caused netizens' emotion siltation and conditions that public opinion is detrimental to the development of the event. Due to time rush, a lot of intervention agents often have too little expertise to manage the crisis. Researchers have proved that it is better to manage public opinion crisis with responding to the crisis within "golden four hours" [18].

It is an ideal choice to implement intervention according to the characteristics of the online public opinion system, before system output, i.e. before the system caused impact or significant influence to the event. So quantitative prediction of system output is particularly important. Superlink (i.e. superedges) prediction referred to the prediction of existent yet unknown superlinks and future superlinks in a supernetwork structure [19]. Superlink prediction between opinion agents can be conducted by method shown in Fig. 3 [20].

Through superlink prediction, we can know superedges that may appear in the next moment in opinion supernetwork. The predicted superedges contain opinion subjects, information environment, opinion psychology, and opinion viewpoints. This portrays how the subjects issued their point of view under an external driven force of environment and an internal driven force of psychology. And these viewpoints will in turn affect the development of events, becoming an output of the system. So we can predict the intervention timing of the system in accordance with opinion superlink prediction results. Suppose a supernetwork model of a public opinion system is as shown in

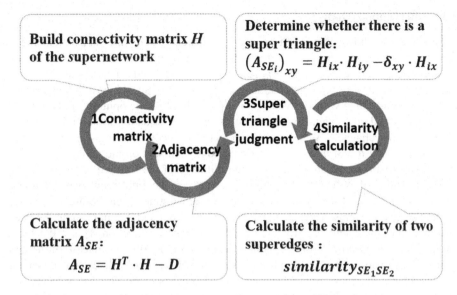

Fig. 3. Flow diagram of superlink prediction

Fig. 4. In the figure, red hollow graphics are those with negative attribute. Green solid graphics are those with positive attribute. The red dashed line represents a negative predicted superedge, and the green double-dashed line represents a positive predicted superedge. Intervention for the system is required when the predicted superedges in negative is more than positive ones.

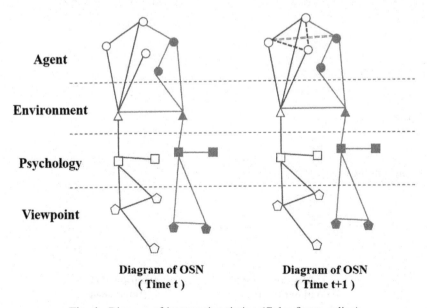

Fig. 4. Diagram of intervention timing (Color figure online)

3.3   Intervention Mode

Experience has shown that when serious emergencies occur, the mainstream media should proactive and timely report the truth of emergencies objectively, with the support and guidance of the government and the propaganda department. This may inhibit the spread of rumors, and maintain social stability. Speaking of the mainstream media in the first time will resolve irrational pressure of public opinion. In the Internet era, it is fast and convenient for people accessing to information through the Internet, SMS and other channels. When a serious emergency occurs, the message will spread like wildfire, and it will have a strong psychological impact on the public in the short term. Mainstream media should occupy the initiative of the incident reporting.

Intervention of online public opinion system can be implemented by changing system input. Input of the system is the opinion environment, that is, different development stages of events. Thus changing opinion environment, i.e. publishing the latest information of the event, can promote the event to positive direction. This will achieve the intervention purpose of online public opinion system. After the release of the new positive information, the psychology of the opinion subjects will change under the impacts of the external driving force of opinion environment. Opinion subjects will update their opinion objects under the internal driving force of opinion psychology, namely the views expressed by opinion objects further varies with changes in their psychology. In Fig. 5, the bigger solid green triangle represents the newly added opinion environment. The green double-dashed line represents intervention of public opinion environment on the various elements of the system. It should be emphasized that environmental impact will be different between the opinion subjects in the center position of reply relationship in the system (core agent) and the ones in the edge position (leaf agent). So how to control the intervention intensity to get best effect under particular cost becomes the key point of the rumor intervention study.

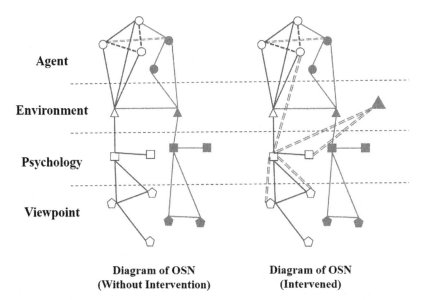

Fig. 5. Diagram of intervention mode (Color figure online)

3.4   Intervention Intensity

Psychological state of public opinion subject is the dynamic of system evolution. Each subject will produce specific psychology in a particular environment, and published their viewpoints on social events under this psychological impact. Dynamic of system evolution is adjusted after subject's psychology changes, i.e. condition basis of viewpoints expressing is changed. Therefore, for public opinion system, intervention intensity must be strong enough to change the subject's psychology.

After the release of positive information to the public opinion system for intervention, the overall evolution rules of the model are: changes of the environment cause the changes of the subject's psychology; changes of the subject's psychology cause the changes of subject's attribute; changes of subject's attribute affect the attribute of newly published viewpoint. Specific changes of each Subnetwork in the model are as follows:

3.4.1   Evolution of System Environment
The new environment of agent A_i is E'. Its property depends on environment information E and the neighbors of agent A_i (i.e. other agents have replying relationships with A_i), namely,

$$E\prime = \frac{\partial_E \times Num + \sum \partial_j \times 2}{Num} \tag{2}$$

Num is the neighbors' number of agent A_i. $\partial_E = +1$, it is the property of a new environmental information. A_j are neighbors of agent A_i, ∂_j are their points of view. If a viewpoint is a positive one, $\partial_j = +1$; otherwise $\partial_j = -1, j = 1, 2, \cdots, Num$, and $j \neq i$.

3.4.2   Evolution of System Dynamics
Opinion subject's psychology changes when the environment changes. If the subject is exposed to a changed environment, we may calculate the new environment of the agent in according to formula (2). And then we calculate the psychology transition probability under the new positive environment. The probability that the positive psychology of an agent's transform to a negative one is

$$P(+ \rightarrow -) = \frac{e^{(-E'/C)}}{e^{(E'/C)} + e^{(-E'/C)}} \tag{3}$$

And the probability that the negative psychology of an agent's transform to a positive one is

$$P(- \rightarrow +) = \frac{e^{(E'/C)}}{e^{(E'/C)} + e^{(-E'/C)}} \tag{4}$$

C reflects the concern degree on an event by the public.

3.4.3   Evolution of Opinion Subjects

The opinion subject property is judged according to the attribute of its psychology in accordance with the following rules:

a. ∃ subject psychology is negative, the subject is negative.
b. ∀ subject psychology is positive, the subject is positive.

3.4.4   Evolution of Opinion Objects

Objects, i.e. viewpoints of the subject changes when the subject's property changed.

a. If the subject property changes from a negative one to a positive one, the negative standpoints of the subject change to the majority of the positive standpoints of the neighbors'.
b. If the subject property changes from a positive one to a negative one, the positive standpoints of the subject change to the majority of the negative standpoints of the neighbors'.

4   Case Study

4.1   Experiment Design

Matlab programming is used in this study to seek the variation of the proportion of positive agents over time under the release of positive information. We picked an agricultural rumor event as a study case. The model of the case contains four Sub-networks, including 108 nodes of agents in G_A, 3 nodes of environmental information in G_E, 5 nodes of psychological statuses in G_P, and 28 nodes of viewpoints in G_K. The relationships of nodes in and between different Subnetworks form the superedges of the Opinion SuperNetwork. This simulation experiment aims at analyzing how the proportion of positive agents change with time under the opinion intervention implement, and exploring the efficiency of rumor intervention based on the meta-synthetic methodology. Here we proposed three different intervention strategies, and compared the efficiency of various strategies. The experiment procedure is as follows:

Strategy 1. The released positive information affects all subjects at the same time.

Step 1: Initialize the property values of all nodes and the superedges of the Opinion SuperNetwork.
Step 2: Calculate the positive agents' proportion. When the ratio is <50 %, continue to next step, otherwise end.
Step 3: Simulate the release of a positive information. Add a positive environment node to simulate intervention implement.
Step 4: Choose a negative agent node randomly to update the node's property and other nodes' in the opinion supernetworks according to the evolution rules described in Sect. 3.4.
Step 5: Return to Step 2 and cycle.

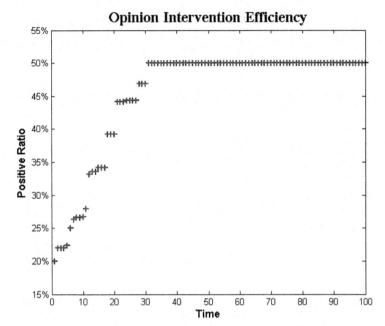

(a) Strategy 1. Both core and leaf agents at the same time

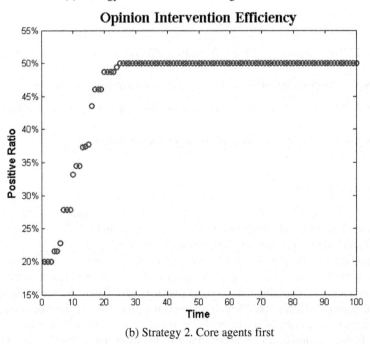

(b) Strategy 2. Core agents first

Fig. 6. Rumor intervention efficiency of information release

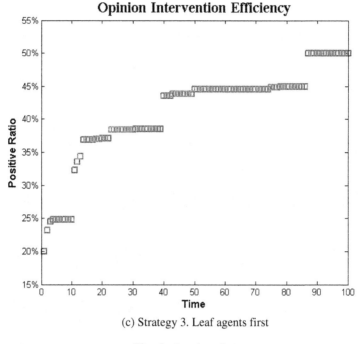

(c) Strategy 3. Leaf agents first

Fig. 6. (continued)

Strategy 2. The released positive information affects core subjects at the first time.

In *Step 4*, choose core subjects of negative property first.

Strategy 3. The released positive information affects leaf subjects at the first time.

In *Step 4*, choose leaf subjects of negative property first.

4.2   Result Analysis

Implement all the above three intervention strategies. Conduct 1000 simulation experiments independently under each of the strategies, and take the average value of the 1000 simulation experiments as the result (Fig. 6).

As can be seen from Fig. 6, after the release of positive information, the proportion of positive public opinion gradually increased under all the three intervention strategies. It indicates that establishing supernetwork model based on the meta-synthetic methodology to conduct rumor interventions can achieve the desired outcomes.

In addition, efficiency of the three different intervention strategies is in comparison (Fig. 7). In the beginning, the strategy that positive information first affects the of leaf public opinion agents has the highest efficiency, followed by the strategy that both kinds of opinion agents being affected at the same time. The strategy that information impacts core public opinion agents first is the least efficient. That may because the leaf agent kind is superior in numbers to core ones. Soon, the advantage of affecting core agents first shows up. By the 15th time step, effect of affecting core agents first surpasses the other two strategies. After the system is steady, strategy of affecting core agents first has a dominant effect, followed by the strategy that both kinds of opinion agents being affected at the same time.

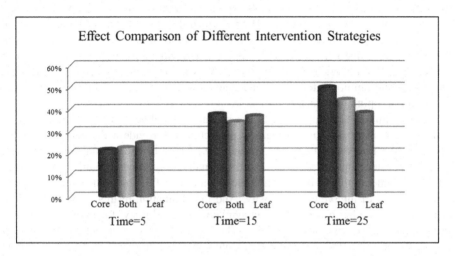

Fig. 7. Effect comparison of different intervention strategies

It is commonly known that publishing positive information, and making it affect the core public opinion agents first is the best strategy for rumor intervention. This intervention mechanism research based on a meta-synthetic methodology approach has provided a reliable quantitative basis for evidence. However, it can be seen from the efficiency comparison of the three kinds of strategies that the efficiency of affecting all the public opinion agents together is only slightly inferior to the one of affecting core public opinion agents first. While in practice the letter takes a much higher cost. So in the real agricultural rumor intervention, take the strategy of affecting all public opinion agents at the same time can achieve a higher effect under a lower cost.

Simulation results show that analyzing rumor intervention mechanism from the perspective of meta-synthetic methodology to develop intervention strategies can not only quantify intervention strategies, but also draw influence mechanisms and laws of intervention strategies by changes and interlocking mechanisms of each element in the online public opinion system. This will help to provide a reliable basis for the development of more sophisticated and effective intervention strategies.

5   Conclusions

Meta-synthetic methodology provides a new perspective to study online agricultural rumor intervention. Online public opinion is seen as an organic integrity with specific functions composed of several elements of interaction and interdependence. This helps us to make a comprehensive analysis of the relationships among elements, elements and the system, as well as the system and its environment. Thus we can grasp its internal relationship and law, and achieve the purpose of controlling and transforming the system effectively.

Our investigation made two important contributions to the investigation of online agricultural rumor intervention. First, based on the meta-synthetic methodology, intervention mechanism of online public opinion system was investigated under a way of "from qualitative to quantitative". We described the general characteristics of online public opinion system qualitatively, including its composition, structure, and function. Then we built a supernetwork model which includes Social Subnetwork, Environment Subnetwork, Psychological Subnetwork and Viewpoint Subnetwork, to quantitatively research public opinion intervention mechanism to get the general intervention methods of public opinion system.

Second, this paper contributes to the research of online agricultural rumor intervention mechanism. We used superlink prediction to provide a theoretical basis for predicting the intervention timing. The release of new environmental information to impact the system input is used as the specific intervention mode. Changing the system dynamic by intervening in the psychological condition of the opinion agents was taken as the measure of intervention intensity. And the simulation results show that build a supernetwork model based on meta-synthetic methodology to implement rumor intervention can achieve desired effects. The analysis of the intervention strategies in this paper provides the government a comprehensive measure standard of online agricultural rumor events, which allows them to get a much deeper understanding to the rumor spreading influence elements, and make more strategic decisions intervening in rumor spreading.

Our study provides a useful consultation to agricultural rumor intervention theoretical studies from a meta-synthetic methodology perspective. And in future studies, more work may be done to improve our research. This article only analyzed the system changes caused by the environmental Subnetwork. The next step, we will continue to investigate the system changes caused by the other three Subnetworks, and also explore the corresponding intervention strategies.

Acknowledgements. Funds for this research was provided by the National Social Science Foundation of China (15CTQ030), the National Natural Science Foundation of China (91324009), the National Sci-Tech Pillar Program during the 12th Five-year Plan Period (2011BAH10B06), and the Science and Technology Innovation Projects of the Chinese Academy of Agricultural Sciences (CAAS).

References

1. Liu, Y.J., Li, Q.Q., Tian, R.Y., Ma, N.: Formation and application of public opinion based on supernetwork analysis. Bull. Chin. Acad. Sci. **28**(5), 560–568 (2012)
2. Li, Q.Q., Liu, Y.J.: Dynamical model of public opinion and its application based on supernetwork. Bull. Chin. Acad. Sci. **28**(5), 569–577 (2012)

3. Tian, R.Y., Liu, Y.J.: Intervention of public opinion and its application based on supernetwork analysis. Bull. Chin. Acad. Sci. **28**(5), 578–585 (2012)
4. Ma, N., Liu, Y.J.: Recognition of online opinion leaders based on supernetwork analysis. Bull. Chin. Acad. Sci. **28**(5), 586–594 (2012)
5. Tian, R.Y., Chen, Z.L.: Advances in Intervention of Online Public Opinion. Social Physics Series, vol. 4. Science Press, Beijing (2013)
6. Von Bertalanffy, L.: General System Theory: Foundations, Development, Applications. George Braziller, New York (1968)
7. Qian, X.S.: A new discipline of science: the study of open complex giant system and its methodology. Urban Stud. **12**(5), 0001 (2005)
8. Gu, J.F.: Practice on Wuli-Shili-Renli system approach. Chin. J. Manag. **8**(3), 317–322 (2011). 355
9. Katz, E., Lazarsfeld, P.F.: Personal Influence: The Part Played by People in the Flow of Mass Communications. Transaction Publishers, New Brunswick (1970)
10. Tian, R.Y., Liu, Y.J., Niu, W.Y.: Leader-guiding model of online opinion supernetwork. Chin. J. Manag. Sci. **22**(10), 136–141 (2014)
11. Nagurney, A.: On the relationship between supply chain and transportation network equilibria: a supernetwork equivalence with computations. Transp. Res. Part E **42**(4), 293–316 (2006)
12. Xi, Y.J., Dang, Y.Z., Liao, K.J.: Knowledge supernetwork model and its application in organizational knowledge systems. J. Manag. Sci. China **12**(3), 12–21 (2009)
13. Estrada, E., Rodríguez-Velázquez, J.A.: Subgraph centrality and clustering in complex hyper-networks. Phys. A **364**, 581–594 (2006)
14. Wakolbinger, T., Nagurney, A.: Dynamic supernetworks for the integration of social networks and supply chains with electronic commerce: modeling and analysis of buyer-seller relationships with computations. Netnomics **6**(2), 153–185 (2004)
15. Nagurney, A., Dong, J.: Management of knowledge intensive systems as supernetworks: modeling, analysis, computations, and applications. Math. Comput. Model. **42**(3–4), 397–417 (2005)
16. Tian, R.Y., Liu, Y.J.: Isolation, insertion, and reconstruction: three strategies to intervene in rumor spread based on supernetwork model. Decis. Support Syst. **67**, 121–130 (2014)
17. Tian, R.Y., Zhang, X.F., Liu, Y.J.: SSIC model: a multi-layer model for intervention of online rumors spreading. Phys. A **427**, 181–191 (2015)
18. Liu, Y.J., Ma, N., Wang, H.B.: Study on the guidance of online public opinion for innovation social management. Bull. Chin. Acad. Sci. **27**(1), 9–16 (2012)
19. Liu, Y.J., Li, Q.Q., Tang, X.Y., Ma, N., Tian, R.Y.: Superedge prediction. Manag. Rev. **24** (12), 137–145 (2012)
20. Liu, Y.J., Li, Q.Q., Tang, X.Y., Ma, N., Tian, R.Y.: Superedge prediction: what opinions will be mined based on an opinion supernetwork model? Decis. Support Syst. **64**, 118–129 (2014)

Rapid Identification of Rice Varieties by Grain Shape and Yield-Related Features Combined with Multi-class SVM

Chenglong Huang[1,2], Lingbo Liu[3], Wanneng Yang[1,2,4], Lizhong Xiong[4], and Lingfeng Duan[1,2(✉)]

[1] College of Engineering, Huazhong Agricultural University,
Wuhan 430070, People's Republic of China
hcl@mail.hzau.edu.cn, ywn@mail.hzau.edu.cn,
duanlingfeng@mail.hzau.edu.cn
[2] Agricultural Bioinformatics Key Laboratory of Hubei Province,
Huazhong Agricultural University, Wuhan 430070, People's Republic of China
[3] Britton Chance Center for Biomedical Photonics,
Wuhan National Laboratory for Optoelectronics-Huazhong University of Science
and Technology, 1037 Luoyu Rd., Wuhan 430074, People's Republic of China
firbo007@gmail.com
[4] National Key Laboratory of Crop Genetic Improvement
and National Center of Plant Gene Research, Huazhong Agricultural University,
Wuhan 430070, People's Republic of China
lizhongx@mail.hzau.edu.cn

Abstract. Rice is the major food of approximately half world population and thousands of rice varieties are planted in the world. The identification of rice varieties is of great significance, especially to the breeders. In this study, a feasible method for rapid identification of rice varieties was developed. For each rice variety, rice grains per plant were imaged and analyzed to acquire grain shape features and a weighing device was used to obtain the yield-related parameters. Then, a Support Vector Machine (SVM) classifier was employed to discriminate the rice varieties by these features. The average accuracy for the grain traits extraction is 98.41 %, and the average accuracy for the SVM classifier is 79.74 % by using cross validation. The results demonstrated that this method could yield an accurate identification of rice varieties and could be integrated into new knowledge in developing computer vision systems used in automated rice-evaluated system.

Keywords: Computer vision · Rice varieties identification · Grain shape · Rice yield · Multi-class SVM

1 Introduction

Rice is one of the most significant cereals in the world, especially for china (Zhu et al. 2011). Thousands of rice varieties could be produced daily by modern breeding technique (Bagge and Lubberstedt, 2008). And a large number of rice

© IFIP International Federation for Information Processing 2016
Published by Springer International Publishing AG 2016. All Rights Reserved
D. Li and Z. Li (Eds.): CCTA 2015, Part I, IFIP AICT 478, pp. 390–398, 2016.
DOI: 10.1007/978-3-319-48357-3_38

germplasm-resources need to be exploited by breeders for the rice improvement (Xing and Zhang, 2010). However, the characterization for the various rice varieties are technically challenging due to the slight difference (Tanabata et al. 2012). Rice variety is also regarded as one of the most important factors related to cooking and processing quality, which was resulted by the variations in size, shape, and constitution (Zhang, 2007). Therefore rice variety identification is of great significance.

Since the identification of rice varieties is so important for rice-related research. A lot of work had been reported about it. Namaporn Attaviroj tried to identify the rough and pure rice varieties using fourier-transform NIR (Attaviroj et al., 2011). Liu Hongyun had tried to indentify rice varieties by tolerance and sensitivity to copper (Liu et al. 2007). Liu Feng tried to identify rice vinegar variety using visible and near infrared spectroscopy (Liu et al. 2011). However, the above study only focused on a few of special rice varieties, and the identification for the massive ordinary rice varieties were still an urgent problem.

Machine vision was a practical technology and had recently been widely applied in the agriculture. Dual-camera rice panicle length measuring system was proposed by Dr. Huang (Huang et al. 2013). A machine-vision-facility was developed for rice traits evaluation (Duan et al. 2011). A hyperspectral imaging system was designed for biomass prediction (Feng et al. 2013). Yang et al. applied x-ray computed tomography for rice tiller measurement (Yang et al. 2011). Duan et al. had counted filled/unfilled spikelets using Bi-modal imaging (Duan et al. 2011). Support vector machine (SVM), first proposed in 1995 by Cortes and Vapkin, has a lot of advantages, such as nonlinear, small-sample, and high dimensional pattern recognition and can be easily extended to other machine learning problems. However, since it is originally used for binary classification (Cortes and Vapnik 1995; Vapnik, 1999), it requires extra algorithm support to meet practical needs.

This research aimed to propose a feasible method for rapid identification of rice varieties. In this study, the features of grain shape and yield-related traits were extracted by image analysis. And the specific Muti-SVM classifier was developed to discriminate the rice variety.

2 Materials and Methods

The Rice varieties used in this study are selected from the Chinese core-germplasm resources. 79 rice varieties were tested and each variety had four samples. Three quarter of the rice samples were taken as training set, meanwhile the other were testing set. The rice grains were threshed from the panicles manually. And the filled spikelets were selected out by wind separator.

The technical method for rice variety identification is described as Fig. 1. Firstly, the rice grain were imaged and analyzed for shape and weight parameters. Then features of the training set were applied to build the SVM model. With the SVM model, the testing set was applied to evaluate the rice variety identification accuracy.

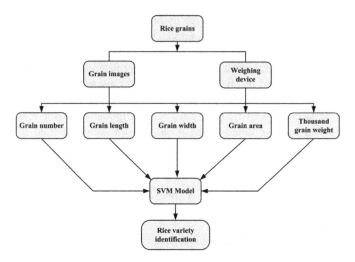

Fig. 1. The technical method for rice variety identification

2.1   Rice Feature Extraction

The features of each rice sample were obtained as shown in Fig. 2. The rice grains per sample were spread on the scanner manually (Fig. 2a). And the image was acquired and transferred to the computer. Then grain image was analyzed for grain number (GN), grain width (GW), grain length (GL), grain area (GA) (Figs. 2b and c). And the grain weight was obtained by the electro-weighing device (Fig. 2d).

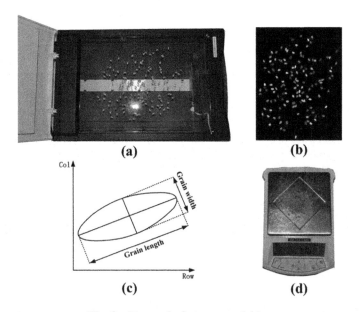

Fig. 2. Rice grain features acquisition

2.2   Multi-SVM Classifier

Since we focus on the classification of rice varieties, we should know the biological classification criteria of rice. Generally, different type of rice will have different genotypes, which usually leads to different phenotypes. Obviously, one of the important problems is how to divide the training rice samples into different subsets by binary tree based SVM-BTA algorithms. When it comes to plants classification, we wish the two subsets will have at least one totally different genotype. Knowing the fact that disparate varieties all have different genotypes, one possible way is using the K-Means clustering method, which will resign each data to the nearest cluster repeatedly, just like combining analogous genes. The problem is to determine the evaluation function.

In order to reduce the algorithm running time of the partitioning process, it's necessary to improve the KMeans clustering. Since we always want to divide the input set into two clusters, a pretreatment of the data using average threshold algorithm will work. In the next section, we introduce one partition function for evaluation, and then propose MBT-SVM based on the K-Means clustering with optimal partition function.

2.2.1   Partition Function

Suppose the problem's center $c_{problem}$ as the all data mean value in the i.th column of the input of a non-leaf node. The following partition function can be adopted to split the node (Huang et al. 2013):

$$PF(I_1 \cup I_2) = \sum_{j=1}^{2} \sum_{i=1}^{l} \frac{d(c_i, c_{problem})}{\sigma_i} \tag{1}$$

Where $d(c_i, c_{problem})$ is the Euclidean distance and σ_i is the variance of column i.

The larger PF is, the better it works. So we have also determined our termination criterion, just to traversal all the possible combinations to find the largest PF, or to reassign the object one by one until the value PF won't become larger in a whole round. An initial partition is needed to reduce the algorithm running time. The judging function using average threshold algorithm was described as followed:

$$J(x) = \begin{cases} 1, & \sum_{i=1}^{v} x_i \geq \frac{1}{l} \sum_{j=1}^{l} \sum_{i=1}^{v} x_{ij} \\ -1, & else \end{cases} \tag{2}$$

Where v is the number of features. And x_{li} is the i.th feature of the j.th sample in the training set.

The judging function compares the average value of all the features of a sample with the average value obtained by the whole training set. Since different features of the rice will have different magnitude, we have to standardize the input data. The standardization includes data integrity check, linear unification and repacking.

2.3   Kernel Function

The efficiency and accuracy of SVM is determined by the kernel type and parameters, as well as the parameter c. To determine the best type of kernel function, we can try the three basic kernel functions and pick the one with the best accuracy according to cross-valid. In general, the Gaussian kernel with a single parameter γ is a good choice. The c and γ is usually calculated by a grid searching method, in which the cross validation is applied, then we will pick the one with the highest accuracy, such as $c = \{2^{-5}, 2^{-4}, 2^{-3}, \ldots 2^{8}\}$; $\gamma = \{2^{-10}, 2^{-9}, 2^{-8}, \ldots 2^{3}\}$. The final model, will then training set was applied by the chosen type of kernel function and with the optimized parameters Duan et al. 2011. As is shown in Fig. 3, an inappropriate combination of kernel type and parameters will cause under-fitting or over-fitting problems.

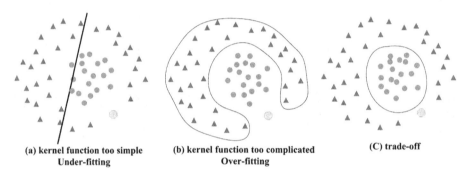

(a) kernel function too simple	(b) kernel function too complicated	(C) trade-off
Under-fitting	Over-fitting	

Fig. 3. Classification models generated by different kernel type and parameters

3   Results and Discussion

3.1   Grain Traits Extraction Accuracy

Totally, 79 copys of rice grains were measured automatically and manually, and the parameters of the GN, GL, GW, GA, grain weight were all obtained for each copy. The measurement accuracy for each traits were shown in Fig. 4, the MAPE were calculated according tho the Eq. 3.

$$MAPE = \frac{1}{n} \sum_{i=1}^{n} \frac{|x_{ai} - x_{mi}|}{x_{mi}} \qquad (3)$$

The measurement results showed that the average MAPE for grain number measurements was 1.33 %; the average MAPE for grain length measurement was 1.25 %; the average MAPE for grain width measurement was 2.20 %. The results demonstrated that the automatical measurements performed a good relationship with manual measurements.

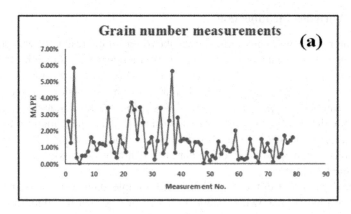

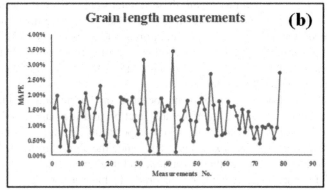

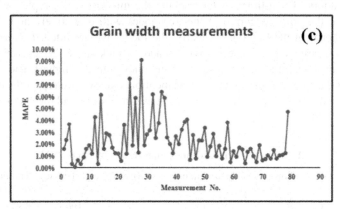

Fig. 4. System measurements accuracy evaluation. (a) Grain number, (b) Grain length, (c) Grain width

3.2   Rice Varieties Identification

The rice feature measurements were shown above. From the results, it was proved that the difference between outer-varieties and inner-varieties were both slight. In the SVM identification, for each tree node, an RBF-kernel SVM was first adopted after partitioning the input data by a clustering method and the linear function was the last one to try. A grid search based on cross-validation was employed for parameter optimization. The result for the whole experiment is shown in Table 1. The training time includes partitioning, tree construction and SVM training. It represented the CPU running time (milli second), for a quad core computer, the real-time approximately equals to a quarter of the CPU time.

The standard data from UCI and the rice sample data were all tested and the classifying results were shown in Table 1. It was proved that the algorithm had high classification accuracy when processing these standard data. Since the testing samples are randomly selected, every kind of data set was tested only three times and the worst result was recorded in order to avoid anthropogenic interference (like continually running the algorithm until it gets a good output). From the data associated to rice in Table 1, it was seen that building a classification binary tree for a set of data with many classes needed a relatively long time.

Rice-s79-1 is a basic pre-experiment focusing on verification of the algorithm compatibility. There was not a linear kernel function and the number of attributions was fixed to 13. Rice-s79-2 was a grid-search experiment aiming at finding a suitable number of attributions for the next experiment. Since the algorithm needed to traverse all the possible combinations of the attribution, and there were four kernel functions to be examined for each combination, it's naturally to have a very large training time (nearly ten hours of real-time). After tracking the misclassified samples, we find that about half of the errors occurred in the tree nodes with a relatively high cross-valid accuracy other than the low ones. Clearly, the model is over-fitted. It is necessary to improve the optimization function. We needed to pick the SVM classifier with an appropriate cross-valid accuracy instead of the ones with the highest. Rice-s79-3 was the result of the formal experiment. As was mentioned above, we repeat the experiment three times and record the result with the worst accuracy rate. And the average accuracy was about 79.74 %.

Table 1. The class identification results by Muti-SVM

Data	Class	Dimensions	Size	Training set	Testing set	Accuracy	Training time (CPU time ms)
iris	3	4	150	120	30	100 %	0.875
wine	3	13	178	133	45	97.78 %	28.03
seeds	3	7	210	100	110	96.67 %	161.03
vehicle	4	18	846	746	100	86 %	213.07
glass	6	9	214	166	48	72.91 %	321.04
rice s79-1	79	13	316	237	79	64.56 %	1226.6
rice s79-2	79	13	316	237	79	25.32 %	124457
rice s79-3	79	13	316	237	79	79.74 %	55108

4   Conclusions

In this paper, a support vector machine working in the multi-space-mapped mode (MBT) was proposed for a rice multi-class classification task. The result showed that this study performed high accuracy for the grain traits extraction and also proved a good performance for the rice varieties classification. In future work, we will further analyze the data of tree nodes from experiments to develop more effective algorithms. The range of the parameters will change according to the size of the input training set which will greatly reduce the computation time, which will reduce the time complexity of the algorithm. Therefore we could apply this method for larger rice sample sets for varieties recognition. The results also demonstrated that this method would provide new knowledge for automated rice-vision-evaluated system.

Acknowledgment. This work was supported by the Fundamental Research Funds for the Central Universities (2662014BQ036, 2662015QC006, and 2662015QC016), the Natural Science Foundation of Hubei Province (2015CFB529), the National High Technology Research and Development Program of China (2013AA102403), the National Natural Science Foundation of China (30921091, 31200274).

References

Bagge, M., Lubberstedt, T.: Functional markers in wheat: technical and economic aspects. Mol. Breed. **22**(3), 319–328 (2008)

Zhu, J., Zhou, Y., Liu, Y., Wang, Z., Tang, Z., Yi, C., Tang, S., Gu, M., Liang, G.: Fine mapping of a major QTL controlling panicle number in rice. Mol. Breed. **27**, 171–180 (2011)

Xing, Y., Zhang, Q.: Genetic and molecular bases of rice yield. Annu. Rev. Plant Biol. **61**, 11.1–11.22 (2010)

Tanabata, T., Shibaya, T., Hori, K., Ebana, K., Yano, M.: SmartGrain: high-throughput phenotyping software for measuring seed shape through image analysis. Plant Physiol. **160**, 1871–1880 (2012)

Liu, H., Zhang, H., Wang, G., Shen, Z.: Identification of rice varieties with high tolerance or sensitive to copper. J. Plant Nutr. **31**(1), 121–136 (2007)

Liu, F., Yusuf, B., Zhong, J., et al.: Variety identification of rice vinegars using visible and near infrared spectroscopy and multivariate calibrations. Int. J. Food Prop. **14**(6), 1264–1276 (2011)

Zhang, Q.-F.: Strategies for developing green super rice. PNAS **104**(42), 16402–16409 (2007)

Attaviroj, N., Kasemsumran, S., Noomhorm, A.: Rapid variety identification of pure rough rice by fourier-transform near-infrared spectroscopy. Cereal Chem. **88**(5), 490–496 (2011)

Huang, C., et al.: Rice panicle length measuring system based on dual-camera imaging. Comput. Electron. Agric. **98**, 158–165 (2013)

Duan, L., et al.: A novel machine-vision-based facility for the automatic evaluation of yield-related traits in rice. Plant Meth. **7**, 44 (2011a)

Feng, H., et al.: A hyperspectral imaging system for an accurate prediction of the above-ground biomass of individual rice plants. Rev. Sci. Instrum. **84**, 095107 (2013)

Duan, L., et al.: Fast discrimination and counting of filled/unfilled rice spikelets based on bio-modal imaging. Comput. Electron. Agric. **75**, 196–203 (2011b)

Yang, W., Xu, X., Duan, L., Luo, Q., Chen, S., Zeng, S., Liu, Q.: High-throughput measurement of rice tillers using a conveyor equipped with x-ray computed tomography. Rev. Sci. Instrum. **82**, 025102-1–025102-7 (2011)

Cortes, C., Vapnik, V.: Support-vector networks. Mach. Learn. **20**, 273–297 (1995)

Vapnik, V.: An overview of statistical learning theory. IEEE Trans. Neural Netw. **10**(5), 988–999 (1999)

Kumar, M., Gopal, M.: Reduced one-against-all method for multiclass SVM classification. Exp. Syst. Appl. **38**, 14238–14248 (2011)

Hsu, C., Chang, C., Lin, C.: A Practical Guide to Support Vector Classification. Department of Computer Science and Information Engineering. National Taiwan University (2003)

The Acquisition of Kiwifruit Feature Point Coordinates Based on the Spatial Coordinates of Image

Bin Wang, Zixiao Chen, Jianmin Gao, Longsheng Fu,
Baofeng Su, and Yongjie Cui[✉]

College of Mechanical and Electronic Engineering,
Northwest A&F University, Yangling 712100, Shaanxi, China
{wangbinwork, cuiyongjie}@nwsuaf.edu.cn

Abstract. How to obtain the spatial coordinates of kiwi fruit has been one of the key techniques for kiwi fruit harvesting robot. In this paper, the writer proposes a unique way to obtain the spatial coordinates of the features of kiwi fruit from the bottom of the target fruit based on the growth characteristics and scaffolding cultivation pattern characteristics of kiwi fruit, plus the help of Microsoft camera and Kinect sensor. Also included in this paper is the coordinate conversion between the images come from Microsoft camera and the images of the Kinect sensor, which is followed by an analysis of the precision of the spatial coordinates of Kiwi fruit captured by the Microsoft camera and Kinect sensor. The process is like this: first, capture images of the target fruit from the bottom of the fruit with Microsoft camera, and then extract coordinates of the target fruits' feature points to determine the corresponding target fruit feature point coordinates in the Kinect sensor; second, analyze the correspondence between the Microsoft camera image coordinate system and the Kinect sensor image coordinate system so as to establish a mathematical model for the image coordinate conversion; finally, capture target feature points' spatial coordinates with Kinect sensor and conduct tests. The results show that the precision of coordinate conversion mode and Kiwifruit spatial coordinates can meet the requirements of the harvesting robots.

Keywords: Kiwi fruit · Harvesting robot · Image coordinate · Mathematical model · Kinect sensor

1 Introduction

The acreage and production of China's kiwi fruit rank first in the world. However, at present, the kiwi fruit is mainly harvested manually, which is highly labor-intensive. With the progress of urbanization and industrialization, more and more young and middle aged people are attracted to work in cities. As a result, the loss of labor force in agriculture is becoming serious, which in turn raises the cost of agricultural production and lowers market competitiveness of our agricultural products. Therefore, the development of kiwi fruit picking robot is of great significance to the development of China's kiwi fruit industry.

© IFIP International Federation for Information Processing 2016
Published by Springer International Publishing AG 2016. All Rights Reserved
D. Li and Z. Li (Eds.): CCTA 2015, Part I, IFIP AICT 478, pp. 399–411, 2016.
DOI: 10.1007/978-3-319-48357-3_39

The key techniques of Kiwi picking robot involve three parts: fruit identification, location and nondestructive picking. The widespread adoption of standardized scaffolding pattern in kiwi fruit production makes robot picking fruit feasible. However, there are still several factors that hinders the development of kiwi fruit picking robot. Firstly, kiwi plants grow in clusters, each of which is usually composed of 3–5 fruits, and the fruits usually grow too close to one another and even overlap. Moreover, foliage sheltering and similar color between fruits and the background make the harvesting robot difficult to perform precise fruit identification and separation as well as feature extraction of fruits. Secondly, kiwi fruit positioning and spatial coordinate acquisition are also problems to be solved for the development of the harvesting robot. The existing fruit and vegetable harvesting robot positioning system is low in precision, time-consuming, complex in structure and high in cost. So it is imperative to develop a new, efficient positioning system.

Among the existing fruit and vegetable harvesting robots at home and abroad, some can harvest fruits whose colors are greatly different from the background colors, such as strawberry picking robot [1], tomato picking robot [2–4], citrus harvesting robots [5, 6], and some can harvest the target fruits whose colors are similar to the background colors, such as cucumber picking robots [7]. Such robots usually adopt near-infrared spectroscopy or laser technology for detection. In terms of detection and identification of kiwi fruits, Zhan et al. [8] used Adaboot algorithm. Ding et al. [9] used RB color component method to separate kiwi fruits. These two methods can only separate the regions with fruits from the ones without fruits, but they failed to identify the fruit individually. Cui et al. [10] used 0.9R-G color features in the fruit image segmentation, but in a complex background environment, this method involves a large amount of calculation and time. In terms of kiwi fruit positioning and coordinate acquisition, some other methods are used, such as monocular vision, binocular vision, multi-purpose vision, hyper-spectrum, laser etc. However, these methods have drawbacks, such as complex computation, low accuracy, high cost, poor reliability and so on. Meanwhile, the conversion among pixel coordinates, spatial coordinates and mechanical arms is still a problem to be solved.

In on-site investigation, it is found that the places below the fruits are spacious with less sheltering and the background is simple, so the writer proposes that the fruits be identified, positioned, and picked from the bottom parts of the plants. The principle is like this: determine the sequence of fruit identification, feature point extraction and fruit picking by using elliptic Hough conversion; acquire the feature point coordinates of the images with Kinect sensors made by Microsoft Company; obtain the image coordinates of feature points by Kinect sensor referring the foreign research results for Microsoft Kinect sensor in robot navigation [11, 12] and feature recognition [13–15]; finally, conduct coordinates conversion between the camera and sensors and construct mathematical model of the coordinates conversion to obtain the 3D coordinates of the feature points.

2   Information Perception

2.1   Feature Extraction and Image Acquisition

The kiwi fruit pictures were taken in October 2014 during harvest time at Kiwi Experimental Station of Northwest Agriculture and Forestry University and the breed is "Hayward". Camera used is Microsoft Life Camera studio with COMS sensor and auto-focus. Image pixel acquisition is of 640 × 360, jpg format. Each picture containing 2–5 fruits was taken from the bottom with a distance of 20 cm to the fruits. Image acquisition mainly comes from the side and bottom, as shown in Fig. 1.

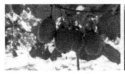 

(a) Picture shotfrom side   (b)Picture shoot frombottom

Fig. 1.   Two different shooting directions

In Fig. 1, due to the greater scene depth, and complicated background, it can be seen that the images of fruits taken from the side contain not only the leaves of the near-byplant branches, but also distant non-target fruits, and serious mutual occlusion between the target fruits. All of these affect the accuracy of target fruit segmentation and recognition. In contrast, the picture shot from the bottom has less mutual occlusion between fruits and no interference of other distant non-target fruits, which is favorable for extracting target fruits. Due to the mutual occlusion between fruits, fruit identification can only be performed from outside to inside, picking one by one, which results in low harvest efficiency. In contrast, the shadow area between fruits is less in image shot from the bottom, which makes it possible that all the fruits can be identified at a time. As a result, the picking sequence can be determined and efficiency improved.

In order to improve fruit recognition success rate under complex background, Cui et al. [16], researchers of Northwest Agriculture and Forestry University, presented a comprehensive method to identify fruits and extract fruit features according to kiwi fruit characteristics and color features, and elliptical Hough conversion. This method can minimize the impact of different complex background and illumination on the identification and the extraction of fruit features. The specific steps are shown in Fig. 2.

2.2   Picking Order

Figure 3 is the pixel coordinates of the feature points when each fruit is identified. X-axis is within the range of 0–360, Y axis 0–640.

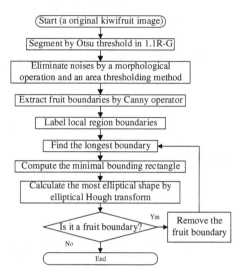

Fig. 2. Flow chart of fruit recognition

Fig. 3. Coordinates of target fruit

The picking sequence is determined according to the values of the feature point coordinates. In Fig. 3, 1, 2, 3, 4, 5 represent picking sequence, which is determined according to the Y coordinate values of the feature points from small to big, thus the picking arm of the robot reaches minimum stroke, and maximum efficiency in the whole picking process.

3   Coordinate Conversion

The principle of Kiwi fruit extraction is shown in Fig. 4. On the left is a front view and on the right is atop view, where 1 stands for kiwi fruits, 2 for Microsoft camera, and 3 for Kinect sensors. The camera is used to identify fruits, extract features and determine the picking sequence. Kinect sensors are used to obtain the spatial coordinates of the feature points of kiwi fruits. The intersection between the optical center and the outer

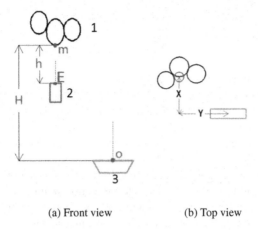

(a) Front view   (b) Top view

Fig. 4. The Schematic of fruit space coordinates acquisition

surface of the infrared camera is used as the origin of the coordinates; the intersection between Microsoft camera lens surface and the optical center is used as the coordinate origin of its coordinate system as is shown in Fig. 4, point 'O' and point 'E'.

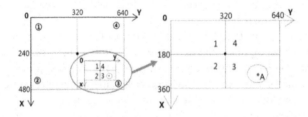

(a) Microsoft camera image coordinate system (b) Kinect sensor image coordinate system

Fig. 5. Relationship between two coordinate systems

As is shown in Fig. 4, the horizontal distance between the feature point and the Microsoft camera is 'h'; the distance between the feature point and Kinect sensor is 'H', then the distance between the Microsoft camera and Kinect sensor is 'H-h'. When the distance between the Microsoft camera and Kinect sensor remains unchanged, let the spatial coordinates between the camera and Kinect sensor be (X, Y, Z), then we get Fig. 5.

The coordinate system diagram of Microsoft camera image is shown in Fig. 5(a) with a pixel area of 640 × 360. The value of the center coordinates is (320,180), which is also the projection position of the camera's optical center in the image. Letthe pixel coordinates of the feature point A recognized by the camera at the image be (x, y), so the value of the relative pixel coordinates in the image is $(\Delta x, \Delta y)$. Then:

$$\Delta x = x - 320 \tag{1}$$

$$\Delta y = y - 180 \tag{2}$$

The relationship between Δx and Δy is positive and negative, and the image can be divided into our regions, that is, '1', '2', '3', '4'. When $\Delta x < 0$ and $\Delta y < 0$, it corresponds to region '1', representing that the feature point is on the upper left of the projection point. When $\Delta x < 0$ and $\Delta y > 0$, it corresponds to region '2', showing that the feature point is on the lower left of the projection point. When $\Delta x > 0$ and $\Delta y > 0$, it corresponds to region '3', meaning that the feature point is at the right bottom of the projection point; When $\Delta x > 0$ and $\Delta y < 0$, it corresponds to region '4', indicating that the feature point is on the upper right of the projection point.

When the distance between the camera and the feature point is h, letone of the image pixels be a (mm), then the three 3D coordinates X_w, Y_w, Z_w of the feature points recognized relative to the origin are respectively as follows:

$$X_w = a\Delta x \tag{3}$$

$$Y_w = a\Delta y \tag{4}$$

$$Z_w = h \tag{5}$$

In addition, the positive or negative values of X_w and Y_w determines the regions where the feature points distribute.

The Kinect sensor is installed below the Microsoft camera with a fairly great distance between them, so that images acquired by the Kinect sensor include the shooting area of the Microsoft camera. As is shown in Fig. 5 (b), 'XOY' is the Kinect image screen coordinates, and 'xoy' is the Microsoft camera image screen coordinates. When the pixel range of the Kinect sensor in capturing image is 640×480, the projection of the optical center of the infrared video camera in the image is the pixel coordinates (320×240) of the center point of image projected.

When the spatial position of the feature points recognized by the Microsoft camera remains unchanged, and let the coordinates of feature point A in the screen coordinate system of the Kinect sensor be x', y', then its relative pixel coordinates to the center point in the image are:

$$\Delta x' = x' - 320 \tag{6}$$

$$\Delta y' = y' - 240 \tag{7}$$

Supposing one pixel of the image on plane H of the Kinect sensor represents the actual length b (mm), the 3D coordinates X_k, Y_k, Z_k of feature point A relative to the origin of the Kinect sensor are respectively:

$$X_k = b\Delta x' \qquad (8)$$

$$Y_k = b\Delta y' \qquad (9)$$

$$Z_k = H \qquad (10)$$

Similarly, the positive and negative of X_k, Y_k correspond to the locations of four images ① ② ③ ④ and the actual locations of information. Since the spatial position of the Microsoft camera and Kinect sensors remain unchanged, pixel coordinates of feature point A in the imaged captured by Kinect sensor can be derived from the pixel coordinates of the feature point 'A' in the camera image shot by the Microsoft camera. That is

$$a(x - 320) + X = b(x' - 320) \qquad (11)$$

$$a(y - 180) + Y = b(y' - 240) \qquad (12)$$

From Eqs. (11) and (12), we can get the following formulas:

$$x' = \frac{a}{b}(x - 320) + \frac{X}{b} + 320 \qquad (13)$$

$$y' = \frac{a}{b}(y - 180) + \frac{Y}{b} + 240 \qquad (14)$$

When the values of distance H and h remain unchanged, the values of X, Y, a and b would be constant, so would be the values of $\frac{a}{b}, \frac{X}{b}, \frac{Y}{b}$. When Microsoft camera recognizes and extracts pixel coordinates(x, y) of the feature point of kiwi fruits, the corresponding pixel on the Kinect sensor screen is (x', y') in theory. However, in practice, image acquired by the obtained by RGB video camera on the Kinect sensor is just opposite along the left-right direction. That is to say, it is reversed along the direction of axis X. In this case, the value of the coordinates (x, y) of the point corresponding to the coordinates (x'', y'') is $(640 - X', y')$.

4 Obtaining Spatial Coordinates

As far as the existing fruit and vegetable harvesting robots at home and abroad are concerned, a variety of methods are adopted for target positioning and coordinates extraction, such as monocular vision, binocular vision, multi-purpose visualization, close infrared spectroscopy, laser scanners etc. but each of them has some problems to be solved.

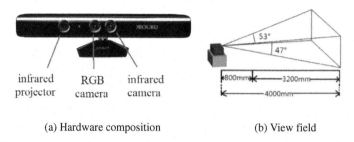

infrared projector	RGB camera	infrared camera	

(a) Hardware composition   (b) View field

Fig. 6. Hardware components and field of view of the Kinect sensor

In this study, Microsoft's Kinect sensor is used to obtain the spatial coordinates of the feature points of the fruits, and the development platform is Kinect for Windows SDK. The sensing device is shown in Fig. 6(a). It mainly consists of three parts. They are, from left to right, an infrared projector, a RGB camera, and an infrared camera. The function of the infrared projector is to project near-infrared spectrum actively. As is known, when the infrared spectrum is projected ontothe objects with rough surfaces or ground glass, there would be distorted spectrum, which in turn would will generate random points of reflected light (also called speckles). The speckles are then read by the infrared camera. The infrared camera is used to analyze the close infrared spectrum and to create depth images of the objects within our vision. RGB camera is used to shoot colored images within our vision. The measurement range is shown in Fig. 6(b). The range centers on the infrared camera with upper angle 43° and lower angle 43°, 400–4000 mm away in front of the video camera. The precision of the depth images captured within this area can reach millimeter.

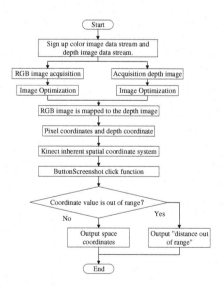

Fig. 7. Flow chart of spatial coordinates acquisition

The flow chart of Kiwi fruit spatial coordinates acquisition is shown in Fig. 7. First, register color image data and depth image data flow to get respective color image and depth image, and to map RGB image onto the depth image. Then load the mapped image into the Kinect built-in spatial coordinates system, flowed by loading Map Depth Point to Skeleton Point function. Afterwards, judge whether the coordinate values are within the range with higher accuracy. If the values are within the range, then output the spatial coordinates with the infrared camera as its origin; if not, the output distance is beyond range.

5 Test and Analysis

5.1 Test Method

In order to verify the conversion relationship between feature point coordinates acquired by Microsoft camera and the coordinates of the Kinect sensor, as well as the accuracy of the mathematical model, verification test was carried out in the laboratory. Firstly, graph paper was used to calibrate the length represented by one pixel (e.g. numerical values of 'a' and 'b') in the images captured by Microsoft camera and Kinect sensor in fixed positions. Then the feature point images and 3D coordinates were acquired with the Microsoft camera and the Kinect sensor. The specific steps are as follows:

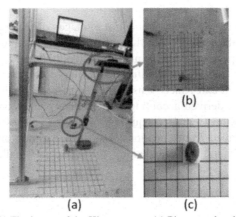

(a) Testing apparatus (b) The image of the Kinect sensor (c) Pictures taken by Microsoft camera.

Fig. 8. Test equipment and images acquired

(1) In order to facilitate verification testing, the whole coordinate conversion system was inverted: Kiwi fruit was placed at the bottom, and Microsoft camera was placed at the upper part with the Kinect sensor on top. The testing platform constructed is shown in Fig. 8(a). Kinect sensor and Microsoft camera are fixed onto the bracket on the same level with desktop supporting Kiwi fruit. In this test, in order to facilitate verification, the kiwi fruit surface was randomly marked with a point as the recognition feature point. Figure 8(b) is an image acquired by the sensor and Fig. 8(c) is the image captured by Microsoft camera.

(2) In this test, the vertical distance from the Microsoft camera to the feature point is 200 mm, and the vertical distance from the Kinect sensor to the feature point is 928 mm. Graph paper was used to calibrate the actual length between the image plane and the place 200 mm away from the Microsoft camera as well as the actual length of one pixel 928 mm away from the Kinect sensor (i.e. the values of a and b). The way of calibration is shown in Fig. 9, where Microsoft camera is fastened to the height gauge parallel to the coordinate plane. Kinect sensor calibration method is same with Microsoft camera calibration method.

Fig. 9. Calibration correspondence between the pixel value and the actual distance

(3) Microsoft camera was used to obtain the images of feature points and the pixel coordinates of the feature point. Kinect sensor was used to obtain the images of feature points, pixel coordinates and spatial coordinates.
(4) Equations (13), (14) were verified with the pixel coordinates acquired by Microsoft camera and Kinect sensor and the values of a, b. Errors in coordinate conversion were derived according to the actual pixel coordinates of the feature points while Kinect sensor acquiring images, D-values between the images coordinates derived from equations, D-values between pixel coordinates from Kinect sensor image and calculation, and the actual length represented by each D-value.

5.2   Result and Analysis

Through calibration, it is found that the actual length represented by one pixel 200 mm away from at Microsoft camera is 0.445 mm, that is to say, a = 0.445 m and that the actual length represented by one pixel 928 mm away from the Kinect sensor is 1.32 mm, that is to say, b = 1.778. In this experiment, we got the pixel coordinates of 24 groups of feature points in different positions on the images captured by Microsoft camera and the images captured by Kinect sensor. In addition, with the help of Kinect sensor, we got the spatial coordinates of the feature points at such positions.

By plugging into formulas (13), (14) the pixel coordinates of point 1 and point 24 on the images acquired by Microsoft camera and the symmetrical image of the image acquired by Kinect sensor, we have the following results:

$$\frac{a}{b} = 0.25$$

$$\frac{X}{b} = 32.00$$

$$\frac{Y}{b} = 186.374$$

The value of a/b coincides with the previous calibration. By plugging into formula (13), (14) the values of a/b, X/b, Y/b and the coordinate values of the rest 22 points on the images acquired by the Microsoft camera, we can work out the coordinates of these 22 points on the symmetrical images acquired of the Kinect sensor image.

Table 1. The coordinates of points in different coordinate systems

	(x, y)	(x', y')	(x'', y'')	The coordinates after the formula extrapolated	The spatial coordinates of the feature point coordinates obtained by Kinect (x, y, z)
1	591,42	218,393	422,393	420,391	−214, −247,928
2	594,154	218,423	422,423	421,420	−216, −301,928
3	597,265	218,451	422,451	421,448	−213, −349,928
4	597,324	218,467	422,467	421,462	−214, −375,926
5	481,42	246,393	394,393	392,391	−166, −249,928
6	482,151	246,421	394,421	393,419	−161, −294,928
7	486,265	248,451	392,451	394,448	−158, −349,928
8	484,322	248,465	392,465	393,462	−162, −375,928
9	369,41	276,391	364,391	364,392	−112, −244,928
10	372,153	276,421	364,421	365,420	−112, −299,928
11	373,264	276,449	364,449	365,447	−112, −348,928
12	373,319	276,464	364,464	365,461	−112, −371,926
13	260,38	304,385	336,385	337,391	−62, −242,928
14	260,153	304,419	336,419	337,420	−62, −294,928
15	261,266	304,449	336,449	337,448	−63, −344,928
16	259,321	306,463	334,463	337,462	−62, −369,928
17	144,38	332,389	308,389	308,391	−10, −244,928
18	145,152	334,419	306,419	308,419	−11, −294,928
19	144,266	334,447	306,447	308,448	−8, −343,928
20	144,326	334,463	306,463	208,463	−10, −370,926
21	26,39	362,388	278,388	278,391	41, −240,928
22	25,153	364,417	276,417	278,420	44, −289,928
23	26,268	364,447	276,447	278,448	42, −345,928
24	26,323	364,461	276,461	278,462	40, −364,927

Figure 10 is a diagram drawn with MATLAB to express the calculated coordinate values and the actual coordinate values.

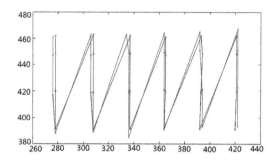

Fig. 10. The calculated and actual coordinate distribution curves

In the diagram, the red curves represent the actual distribution of points, and the blue ones represent the distribution of calculated points. It can be seen that the two curves coincide with each other. Table 1 indicates that the error between the calculated point coordinates and the actual coordinates is less than 3 pixels or 5 mm, thus formulas (13), (14) can accurately reflect the correspondence of a same point on the image acquired by Microsoft camera and on the image acquired by the Kinect sensor.

6 Conclusions

(1) In light of the kiwi fruit growth characteristics, an automatic identification method is studied, including how to acquire fruit image from the bottom and an integrated application of fruit shape and color features in recognition.
(2) Considering the drawbacks of the existing fruit and vegetable harvest robots, a fresh method based on Kinect sensor is proposed to acquire the coordinates of the target kiwi fruits.
(3) This paper discusses the coordinate conversion between Microsoft camera and Kinect sensor and the mathematical model constructed can perform accurate coordinates conversion.

Acknowledgment. This research was supported by grants from Natural Science Foundation of China (No. 61175099) and Sci-Tech Co-Innovation engineering plan projects of Shaanxi Province (2015KTCQ02-12).

References

1. Zhang, K., Yang, L., Wang, L., et al.: Design and experiment of elevated substrate culture strawberry picking robot. Trans. Chin. Soc. Agric. Mach. **43**(9), 165–172 (2012)

2. Monta, M., Kondo, N., Shibano, Y.: Agricultural robot in grape production system. In: IEEE International Conference on Robotics and Automation, pp. 2504–2509 (1995)
3. Arima, S., Kondo, N.: Cucumber harvesting robot and plant training system. J. Robot. Mechatron. **11**(3), 208–212 (1999)
4. Kondo, N., Monta, M., Fujiura, T.: Fruit harvesting robot in Japan. Adv. Space Res. **18**(1–2), 181–184 (1996)
5. Cai, J., Zhou, X., Wang, F., et al.: Obstacle identification of citrus harvesting robot. Trans. Chin. Soc. Agric. Mach. **40**(11), 171–175 (2009)
6. Lu, W., Song, A., Cai, J., et al.: Structural design and kinematics algorithm research for orange harvesting robot. J. Southeast Univ. (Nat. Sci. Ed.) **41**(1), 95–100 (2011)
7. van Henten, E.J., Hemming, J., van Tuijl, B.A.J., et al.: An autonomous robot for harvesting cucumbers in greenhouses. Auton. Robots **13**(3), 241–258 (2002)
8. Zhan, W., He, D., Shi, S., et al.: Recognition of kiwifruit in field based on Adaboost algorithm. Trans. Chin. Soc. Agric. Eng. **23**, 140–146 (2013)
9. Ding, Y., Geng, N., Zhou, Q.: Reacher on the object extraction of kiwifruit based on images. Microcomput. Inf. **25**(18), 294–295 (2009)
10. Cui, Y., Su, S., Lv, Z., et al.: A method for separation of kiwifruit adjacent fruits based on Hough transformation. J. Agric. Mechanization Res. **34**(12), 166–169 (2012)
11. Zainuddin, N.A., Mustafah, Y.M., Shawgi, Y.A.M., Rashid, N.K.A.M.: Autonomous navigation of mobile robot using Kinect sensor. In: 2014 International Conference on Computer and Communication Engineering (ICCCE), pp. 28–31. IEEE (2014)
12. Ruiz, E., Acuña, R., Certad, N., Terrones, A., Cabrera, M.E.: Development of a control platform for the mobile robot Roomba using ROS and a Kinect sensor. In: Robotics Symposium and Competition (LARS/LARC), 2013 Latin American, pp. 55–60. IEEE (2013)
13. Clark, M., Feldpausch, D., Tewolde, G.S.: Microsoft Kinect sensor for real-time color tracking robot. In: 2014 IEEE International Conference on Electro/Information Technology (EIT), pp. 416–421. IEEE (2014)
14. Dutta, T.: Evaluation of the Kinect™ sensor for 3-D kinematic measurement in the workplace. Appl. Ergon. **43**, 645–649 (2012)
15. Sgorbissa, A., Verda, D.: Structure-based object representation and classification in mobile robotics through a microsoft Kinect. Robot. Auton. Syst. **61**, 1665–1679 (2013)
16. Cui, Y., Su, S., Wang, X., et al.: Recognition and feature extraction of kiwifruit in natural environment based on machine vision. Trans. Chin. Soc. Agric. Mach. **44**(5), 247–252 (2013)

The Soil Nutrient Spatial Interpolation Algorithm Based on KNN and IDW

Xin Xu[1,3], Hua Yu[1,3], Guang Zheng[1,3],
Hao Zhang[1,2,3], and Lei Xi[1,2,3(✉)]

[1] College of Information and Management Science,
Henan Agricultural University, Zhengzhou 450002, China
hnaustu@126.com
[2] Henan grain crops Collaborative Innovation Center, Zhengzhou 450002, China
[3] Ministry of Agriculture Agricultural Information Technology Science
Observation Station, Huanghuaihai District, Zhengzhou 450002, China

Abstract. For breaking the limitation of the GIS platform and realizing the soil nutrients spatial interpolation algorithm for any points in the monitoring area to transplant to the mobile platforms, this paper established the spatial index of the soil nutrient sampling points utilizing the K-D Tree as the space splitting algorithm of the soil nutrient sampling points. On this basis, the K nearest neighbor search of the soil nutrient sampling points was also implemented employing KNN algorithm. Finally, the soil nutrient spatial interpolation was realized based on KNN and IDW algorithm. Meanwhile, the accuracy of the algorithm and the influence to the different soil nutrient elements affected by the K value in KNN algorithm were also verified. The results show that the soil nutrient spatial interpolation algorithm was viable to predict the element contents of soil nutrient during the running time was less than 3 s. To reach the best accuracy, the values of the proximal point K for predicting the PH, organic matter, rapid available phosphorus and rapid available potassium should be set as 85, 15, the largest sample space and 65 respectively. The optimal average absolute error of the pH, organic matter, rapid available phosphorus and rapid available potassium was 0.0405, 0.3870, 0.0015 respectively.

Keywords: IDW · KNN · KD-Tree · Soil nutrient · Spatial interpolation

1 Introduction

Due to the problems of the obtaining the sampling points data of the soil testing and formulated fertilization such as difficult sampling [1], heavy workload [2], high cost [3] and so on, the data points during the fertilizing decision stage in the process of testing soil for formulated fertilization was not enough to achieve the complete coverage for the field parcel in each region. Therefore, the sampling point soil nutrient data of the currently unknown spatial points were obtained on the basis of interpolation operation of the existing sampling point data frequently in real application. The utilizing of spatial interpolation technology was commonly used method. In the method, the statistical methods were applied to the nutrient data of some smaller density soil sample

© IFIP International Federation for Information Processing 2016
Published by Springer International Publishing AG 2016. All Rights Reserved
D. Li and Z. Li (Eds.): CCTA 2015, Part I, IFIP AICT 478, pp. 412–424, 2016.
DOI: 10.1007/978-3-319-48357-3_40

points to conduct interpolation operation to the data from the points that were not sampled to form the more dense point data distribution or the area data of different areal unit. The soil nutrient values of the points which were not sampled were predicted scientifically. Then, the precise fertilization could be implemented to the whole area.

Following the rapid development of the mobile intelligent terminal and the distributed computing technology, the computing and data processing of more and more applications was transferred to client-side [4]. Due to the frequent disconnection [5] and low reliability [6] of mobile GIS, most of space interpolation for mobile side had to rely on the service side in the using. It made the usage scenarios of the mobile-oriented space interpolation limited. Thus, making full advantage of modern computer technology could provide more configurable, extensible and customizable distributed applications and made traditional soil testing and formulated fertilization recommendation system running in embedded devices such as high performance and low cost mobile intelligent terminal possible [7, 8]. However, most of current soil testing and formulated fertilization system [9–11] and soil nutrient spatial interpolation technology [12–14] were based on the GIS platform and hard to be transplanted to mobile platform. Hence, to realize the soil nutrient inquiry to any land parcel and the fertilization information recommendation by the users through intelligent mobile terminals and the personalized recommendation of the soil testing and fertilizer recommendation technology by fully integrating mobile technology and soil testing formula technology, the researching of a type of soil nutrient spatial interpolation technology which was divorced from the limitation of the GIS platform was significant for the application and popularization of soil testing formula fertilization technology.

2 Materials and Methods

2.1 Source of Material

The data in this paper came from the 238 soil nutrient sampling points in 10 thousand mu precision agriculture model production field of Changge city, Henan province in 2014. The sampling point data was consist of sampling number, numbering of land parcel, soil texture, preceding crop, PH values, organic matter, rapidly available phosphorus, rapidly available potassium, latitude and longitude information.

2.2 Research Method

2.2.1 Inverse Distance Weighted (IDW) Interpolation

At present, there were many kinds of methods for the soil nutrient interpolation. In which methods, the Inverse Distance Weighted interpolation (IDW) and Kriging interpolation [15–17] were considered to be the most widely used two common interpolation methods. The Kriging interpolation was often influenced by many factors in actual application [18, 19]. Considering the space complexity of Kriging interpolation algorithm, the Inverse Distance Weighted interpolation was adopted in this study to design the interpolation algorithm for the soil nutrient data.

The inverse distance weighted interpolation was based on the theory of "like similarity principle". Through the calculating of the inverse distance weighted average of the discrete points of nearby area, the value in the cell was also calculated. According to the principle that the point with closer distance had the greater weighting values, the value of the estimate point was fitted employing the linear weighting of some neighboring points around.

This algorithm was a simple and effective data interpolation method with relatively fast computing speed. However, it also had the obvious flaws. For example, the method only considered the space distance between the estimation points and the interpolation points. Moreover, the weight calculation used in the algorithm lacked the exact physical basis [20].

2.2.2   K-Nearest Neighbor (KNN) Algorithm and Spatial Index of Soil Nutrient Sampling Points Based on K-D Tree

The soil nutrient content of the sampling point was not only affected by the change of the sample point distance, but also was affected by many other factors such as soil texture variation. Therefore, the K-Nearest Neighbor algorithm was also introduced in the process of the adjacent points selecting and searching during the process of the interpolation to the soil nutrient using inverse distance weighted interpolation to improve the IDW algorithm.

K-Nearest Neighbor (KNN) sorting algorithm was proposed by Cover and Hart in 1967 [20]. It was a simple algorithm based on analogy and had become one of the most mature machine learning algorithms in theory at present. The idea of the algorithm was: if most of the k most similar (i.e., the adjacent) samples of a sample in the characteristic space belonged to a category, then the sample also belonged to this category. If the query point and the positive integer K were given, the k closest data to the query point were also found from the data set.

The realization of the KNN algorithm was based on the establishment of the spatial index for data. The commonly used algorithms for establishment of a spatial index in GIS interpolation algorithm has K-D tree, K-D-B tree, BSP tree, R tree series, quadtree, grid and many other spatial index [21]. K-D (K dimension search binary tree) was a main memory data structure to generalize binary tree to multidimensional data and a type of binary tree in K dimension space. It was also a type of dynamic index structure for the segmentation of data space and suited for the space point target indexing [22]. Therefore, this paper constructed the spatial index algorithm of the soil nutrient utilizing the K-D tree.

3   Design and Simulation of Algorithm

3.1   Workflow Design

The idea of the soil nutrient spatial interpolation algorithm based on KNN and IDW proposed in this paper was: Constructing the KD-Tree firstly to establish the index space of the soil nutrient data sampling points. After that, the sampling point data set was further screened. The specific method was: using the given soil nutrient training data set, the tree structure data index was constructed adopting the space partitioning tree KD tree to conduct the hierarchical division for the search space of soil sampling point. After that, the fast matching was conducted to the new input sampling point data to find the nearest neighboring K sampling points of the sampling point in the training data set. The K sampling point data would be used as the spatial interpolation data. The algorithm process was shown in Fig. 1.

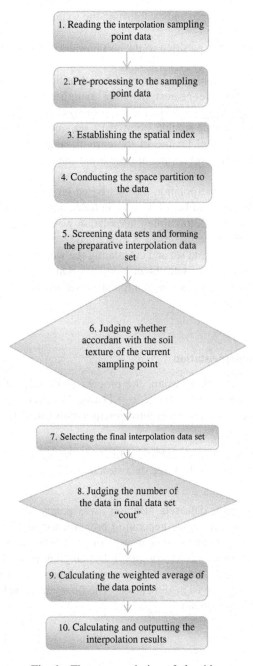

Fig. 1. The process design of algorithm

(1) Reading the current sampling point data set and ready for operation;
(2) Conducting the data pre-processing to the sampling point data and removing critical value such as the outliers and the value beyond administrative boundary and so on to reduce the measuring error of the interpolation results;
(3) Constructing the KD tree and establishing the spatial index of the sampling point data;
(4) Conducting the space partition to the soil nutrient sampling point data using the KD tree;
(5) Conducting the screening to current data set utilizing the K-Nearest Neighbor algorithm to seek the nearest neighbor sampling point data of the current interpolation sampling point data to form the preparative interpolation data set;
(6) Traversing the current sample point data set, judging the data whether was accordant with the soil texture of the current sampling point. If the data was inconsonant with the soil texture of the current sampling point, the data would be discarded.
(7) Selecting the final interpolation data set based on the step 6;
(8) Judging whether the number of the data in final data set "cout" was less than 2, if "cout" was less than 2, this time interpolation was fail and returned the "false" information. Otherwise, the process continued.
(9) Looping through the data set and calculating the weighted average of the current unknown sample points with the current data points.
(10) Calculating and outputting the interpolation results based on inverse distance weighted interpolation.

3.2   Algorithm Implementation

Based on the above algorithm design, the process of the algorithm implementation consisted of the following three steps: 1. KD tree Constructing, 2. K neighboring searching, 3. the inverse distance weighted interpolation calculation.

1. KD Tree Constructing

The KD tree data structure was established firstly in the algorithm. After that, the KD tree was constructed based on KD tree data structure. The steps of the algorithm were as follows:

(1) If the sampling point data sets "Dataset" was empty, the empty KD tree was returned. Otherwise, the node generation program would be executed.
(2) Determine the maximum variance dimension the "Split" domain. The maximum variances in each space dimension of all the sampling point data which prepared for interpolation were counted. The maximal dimension in their variance was elected as the space division dimension. The quantitative value was used as the value of the "Split" domain.

(3) The root Node "Node" was confirmed. The data of the selected space partition dimension was sorted. The sampling point data located in the middle was used as the data of the "Node" node.

(4) The left subspace "leftNodeData" and the right subspace "rightNodeData" of the "Node" were determined. The data of the sampling points whose space dimension was less than the "Node" node was used as the data of the left subspace of the "Node" node. The data of the sampling points whose space dimension was larger than the "Node" node was used as the data of the right subspace of the "Node" node.

(5) The above process was recursive to conduct the space division partition for the "leftNodeData" and the "rightNodeData" until the subspace had not data.

2. K-Nearest Neighbor Searching

The KD tree nearest neighbor search query algorithm was adopted in this study to construct the KD tree "Kd". Adopting the target point "target" as the input, employing the obtained K nearest neighboring "nearest" sets of the "target" as the output, the steps of the K search algorithm establishment were as follows:

(1) The leaf nodes containing the target point "target" were found from the KD tree. From the root node, the KD tree was recursively searched downward. If the coordinates of the current dimension of the target point "target" was less than the coordinate of the split points, the target point "target" was moved to the left child node space. Otherwise, it was moved to the right child node.

(2) The leaf node searched in step 1 was used as the current nearest point "currnearest".

(3) The other possible nearest points were backtracked recursively. If there was a node closer the target point "target" than the current nearest point "currnearest", the value of the current "currnearest" was updated.

(4) When the backtracking backed to the root node, the search ended. The last nearest point "currnearest" was the nearest neighboring point of the "target".

(5) The current nearest neighboring point was deposited in the "nearest" set and removed from the tree of Kd.

(6) The above process was repeated until the number of set elements in the "nearest" was equal to K. Stopped the search and returned back to the "nearest" set.

3. Inverse Distance Weighted Interpolation

Based on the KD tree generating algorithm, the spatial index tree of the current sample sampling points was constructed. On the basis of this, the K nearest-neighbor sampling points were searched out combining k-nearest neighbor algorithm to form the nearest-neighbor set "nearest". Employing the nearest-neighbor set "nearest" and the target point "target" as the input, the sampling points data of the current sampling points was conducted the spatial interpolation processing adopting the inverse distance weighted interpolation. The algorithm steps were as follows:

(1) Calculating the distance between the target point "target" and the sampling points in the nearest-neighbor set "nearest". Traversal the "nearest" set and calculating the distance from the "target" to each nearest-neighbor sampling point.

(2) Calculating the weight of each point. The weight of each sampling point in the nearest-neighbor set "nearest" was calculated to be used as the weight during the interpolation.
(3) Calculating the results of the weighted interpolation. Traversal the "nearest" neighboring set. The results was worked out based on the weight of each point and returned.

4   Design and Analysis of Experiments

Adopting the data from 238 soil nutrient sampling points in 10 thousand mu precision agriculture model production field of Changge city, Henan province in 2014 as the data source, the ARM CPU architecture as mobile hardware environment(1.6 GHz, 4 core), the above soil nutrient interpolation calculation method was realized based on the Android 4.4 mobile platform. As shown in Fig. 2, the pH, organic material, rapid available phosphorus and rapid available potassium in the sample data were conducted the spatial interpolation processing and effect analysis during the algorithm implementation efficiency was less than 3 s.

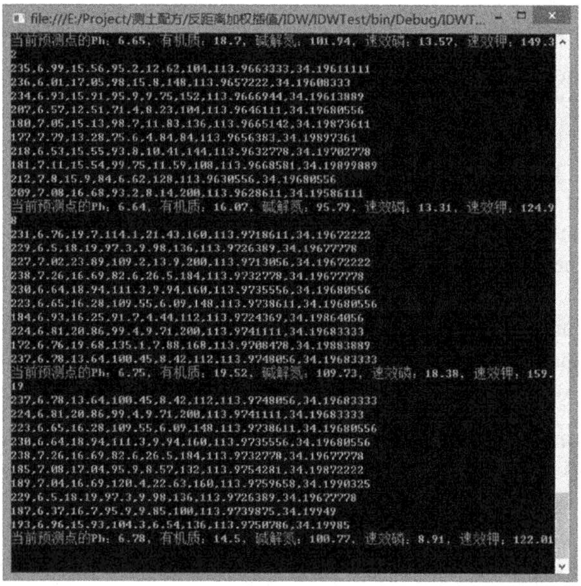

Fig. 2. Algorithm running results

The Fig. 2 showed the results page after interpolation to the 10 nearest-neighbor sampling point data. By the program, the nutrient content of the current point was predicted through the interpolation to the pH, organic matter, rapid available phosphorus and rapid available potassium of the current point and contrasted with the actual value.

The effect analysis of algorithm adopting the cross validation method: a small number of samples of the sampling data of nutrients were screened and reserved to be not involved in the interpolation. These reserved samples were conducted comparative evaluation with the interpolation results. That meant that the model established by the most of the sample was used to conduct the small sample prediction. Meanwhile, the error of prediction was also obtained. The mean absolute error, average relative error and root-mean-square error was used as the standard of measuring the interpolation precision [14, 23].

In this experiment, the randomly selected 50 sample points from the total samples were used as the data waiting for interpolation and need to be predicted. The rest of the sample points were used as the interpolation data. After many times of experiment, the influence of the different number of the nearest-neighbor to the result data of the sampling point was tested under the same soil texture. The interpolation results analysis were shown in the tables below (Tables 1, 2, 3 and 4):

Table 1. Results statistic of pH values

Number of adjacent points	Absolute error			Relative error			Root-mean-square error
	Maximum value	Minimum value	Mean value	Maximum value	Minimum value	Mean value	
5	1.2	−1.75	0.0365	0.203725	0.004342	0.0688	0.4681
10	1.17	−1.72	0.0405	0.200233	0.004342	0.06689	0.4569
15	1.15	−1.7	0.0390	0.197905	0.004342	0.065512	0.4484
20	1.14	−1.7	0.0405	0.197905	0.004292	0.065741	0.4478
25	1.13	−1.7	0.0395	0.197905	0.002894	0.06531	0.4448
30	1.12	−1.7	0.0390	0.197905	0.002894	0.065089	0.4431
35	1.12	−1.7	0.0390	0.197905	0.002894	0.065234	0.4427
40	1.12	−1.7	0.0385	0.197905	0.002894	0.065159	0.4422
45	1.12	−1.7	0.0380	0.197905	0.002894	0.06495	0.4415
50	1.11	−1.69	0.0395	0.19674	0.002861	0.064609	0.4388
55	1.11	−1.69	0.0385	0.19674	0.002861	0.064465	0.4383
60	1.11	−1.69	0.0390	0.19674	0.002861	0.064537	0.4386
65	1.11	−1.69	0.0385	0.19674	0.002861	0.064317	0.4380
70	1.11	−1.69	0.0385	0.19674	0.002861	0.064317	0.4380
75	1.11	−1.68	0.0400	0.195576	0.002861	0.064119	0.4366
80	1.1	−1.68	0.0405	0.195576	0.002861	0.064052	0.4362
85	1.1	−1.68	0.0405	0.195576	0.002861	0.064052	0.4360
90	1.1	−1.68	0.0425	0.195576	0.002861	0.064205	0.4363
95	1.1	−1.68	0.0435	0.195576	0.001431	0.064063	0.4364
100	1.1	−1.68	0.0440	0.195576	0.001431	0.063994	0.4365

Table 2. Results statistic of organic material

Number of adjacent points	Absolute error			Relative error			Root-mean-square error
	Maximum value	Minimum value	Mean value	Maximum value	Minimum value	Mean value	
5	11.76	−5.08	0.4455	1.4067	0.0089	0.1507	2.5395
10	11.17	−5.21	0.4070	1.3361	0.0013	0.1415	2.4262
15	11	−5.09	0.3870	1.3158	0.0009	0.1417	2.4147
20	10.96	−4.98	0.4205	1.3110	0.0012	0.1389	2.4227
25	10.92	−4.97	0.4415	1.3062	0.0029	0.1381	2.4306
30	10.9	−4.96	0.4500	1.3038	0.0053	0.1380	2.4306
35	10.89	−5	0.4565	1.3026	0.0059	0.1390	2.4506
40	10.88	−5	0.4540	1.3014	0.0076	0.1389	2.4506
45	10.87	−5	0.4525	1.3002	0.0082	0.1390	2.4489
50	10.86	−5.02	0.4490	1.2990	0.0094	0.1395	2.4512
55	10.88	−5.06	0.4440	1.3014	0.0099	0.1405	2.4623
60	10.88	−5.09	0.4460	1.3014	0.0099	0.1409	2.4648
65	10.89	−5.13	0.4495	1.3026	0.0099	0.1413	2.4676
70	10.88	−5.15	0.4635	1.3014	0.0099	0.1416	2.4678
75	10.89	−5.19	0.4640	1.3026	0.0099	0.1425	2.4734
80	10.9	−5.21	0.4720	1.3038	0.0099	0.1429	2.4785
85	10.91	−5.22	0.4780	1.3050	0.0094	0.1436	2.4840
90	10.91	−5.21	0.4830	1.3050	0.0099	0.1442	2.4881
95	10.91	−5.21	0.4835	1.3050	0.0103	0.1446	2.4909
100	10.92	−5.2	0.4790	1.3062	0.0108	0.1448	2.4939

Because the algorithms was designed based on mobile computing, it could not only guarantee the accuracy but also provide high efficiency. Through many times of experiment contrasting, when the execution efficiency was less than 3s, the variation situation of the soil nutrients interpolation error in the algorithm was examined in 100 neighboring set. The comparing results as follows:

(1) The prediction error of Ph decreased gradually with the increase of the neighbor point firstly. The prediction error reach minimum when the number of the neighbor point increased to 85. The mean absolute error, average relative error and root-mean-square error was 0.0405, 0.064052 and 0.4360 respectively. Then, the error increased gradually again.

(2) The prediction error of organic material decreased gradually with the increase of the neighbor point firstly. The prediction error reach minimum when the number of the neighbor point increased to 15. The mean absolute error, average relative error and root-mean-square error was 0.3870, 0.1417 and 2.4147 respectively. Then, the error increased gradually again.

(3) The prediction error of rapid available phosphorus decreased gradually with the increase of the neighbor point.

Table 3. Results statistic of rapidly available phosphorus

Number of adjacent points	Absolute error			Relative error			Root-mean-square error
	Maximum value	Minimum value	Mean value	Maximum value	Minimum value	Mean value	
5	8.16	−18.28	−1.7905	1.1209	0.0019	0.3219	6.9795
10	8.15	−17.75	−1.7320	1.1195	0.0102	0.3178	6.6646
15	8.06	−17.84	−1.7665	1.1071	0.0015	0.3166	6.5797
20	8.01	−17.76	−1.7455	1.1003	0.0029	0.3162	6.5122
25	7.99	−17.76	−1.7300	1.0975	0.0031	0.3169	6.4578
30	7.97	−17.79	−1.7420	1.0948	0.0051	0.3171	6.4259
35	7.95	−17.8	−1.7550	1.0920	0.0072	0.3177	6.4065
40	7.94	−17.77	−1.7735	1.0907	0.0072	0.3170	6.3900
45	7.93	−17.74	−1.7910	1.0893	0.0093	0.3179	6.3834
50	7.92	−17.7	−1.7915	1.0879	0.0103	0.3185	6.3733
55	7.92	−17.68	−1.7970	1.0879	0.0113	0.3181	6.3628
60	7.9	−17.68	−1.7950	1.0852	0.0144	0.3186	6.3545
65	7.9	−17.68	−1.7955	1.0852	0.0175	0.3184	6.3513
70	7.9	−17.68	−1.8010	1.0852	0.0175	0.3182	6.3490
75	7.9	−17.67	−1.8060	1.0852	0.0175	0.3179	6.3439
80	7.9	−17.67	−1.8145	1.0852	0.0163	0.3180	6.3433
85	7.89	−17.66	−1.8160	1.0838	0.0116	0.3181	6.3410
90	7.89	−17.67	−1.8160	1.0838	0.0088	0.3184	6.3395
95	7.89	−17.67	−1.8205	1.0838	0.0075	0.3185	6.3384
100	7.89	−17.67	−1.8285	1.0838	0.0034	0.3183	6.3368

Table 4. Results statistic of rapidly available potassium

Number of adjacent points	Absolute error			Relative error			Root-mean-square error
	Maximum value	Minimum value	Mean value	Maximum value	Minimum value	Mean value	
5	57.71	−47.5	0.0200	0.4809	0.0110	0.1964	24.2538
10	57.16	−48.44	−0.4010	0.5057	0.0180	0.1958	24.3808
15	56.74	−49.41	−0.2720	0.5072	0.0124	0.1942	24.3740
20	56.52	−50.03	−0.3100	0.4966	0.0090	0.1923	24.2862
25	56.25	−50.09	−0.2755	0.4954	0.0084	0.1908	24.1878
30	55.95	−50.21	−0.2020	0.4999	0.0037	0.1899	24.1507
35	55.73	−50.29	−0.0700	0.5020	0.0015	0.1897	24.1325
40	55.57	−50.35	−0.0160	0.5028	0.0012	0.1893	24.0965
45	55.41	−50.49	−0.0160	0.5035	0.0011	0.1887	24.0536
50	55.28	−50.56	0.0645	0.5024	0.0023	0.1891	24.0341
55	55.15	−50.65	0.0175	0.5003	0.0019	0.1888	23.9893
60	55.03	−50.69	0.0065	0.4993	0.0023	0.1888	23.9824
65	55	−50.74	0.0015	0.4980	0.0031	0.1885	23.9818

(*continued*)

Table 4. (*continued*)

Number of adjacent points	Absolute error			Relative error			Root-mean-square error
	Maximum value	Minimum value	Mean value	Maximum value	Minimum value	Mean value	
70	54.95	−50.84	−0.0190	0.4979	0.0035	0.1885	23.9995
75	54.88	−50.94	−0.0210	0.4967	0.0042	0.1886	24.0169
80	54.84	−50.97	−0.0220	0.4962	0.0043	0.1885	24.0312
85	54.79	−51.04	0.0215	0.4958	0.0044	0.1885	24.0372
90	54.75	−51.04	0.0115	0.4952	0.0044	0.1882	24.0579
95	54.71	−51.1	−0.0100	0.4942	0.0042	0.1882	24.0725
100	54.68	−51.18	−0.0450	0.4933	0.0040	0.1880	24.0808

(4) The prediction error of rapid available potassium decreased gradually with the increase of the neighbor point. The prediction error reach minimum when the number of the neighbor point increased to 65. The mean absolute error, average relative error and root-mean-square error was 0.0015, 0.1885 and 23.9818 respectively. Then, the error increased gradually again.

5   Conclusions

The K-D Tree was used as the space division algorithm for the soil nutrient sampling point in this study. The spatial index of the soil nutrient sampling point was also established. On this basis, the K-nearest neighbor search of the soil nutrient sampling point was realized using KNN algorithm. Furthermore, combined with KNN and IDW algorithm, the soil nutrients spatial interpolation algorithm was implemented and breaked away from the limitation of GIS platform.

(1) The experiment showed that the soil nutrient spatial interpolation algorithm based on KNN and IDW was effective and feasible for predicting the content values of the Ph, organic matter, rapid available phosphorus and rapidly available potassium in soil.

(2) This paper tested the influence to different soil nutrient elements affected by the K value in KNN algorithm. The result showed that the neighboring points for the prediction of Ph, organic matter, rapid available phosphorus and rapid available potassium of respectively take 85, 15, maximal sample space, 65 could achieve the best accuracy. Of which, the optimal average absolute errors of Ph, organic matter, rapidly available potassium were 0.0405, 0.3870, and 0.0015 respectively.

(3) Through the analysis of the error size, it was found that the interpolation precision to pH and organic matter of the algorithm is higher than the interpolation precision of the rapid available phosphorus and rapid available potassium. The error of the rapid available phosphorus and rapid available potassium was larger. This was largely relevant with their large spatial variability [24].

Following the rapid development of the mobile intelligent terminal and the distributed computing technology, more and more applications transferred the computing and the data processing to thin client. The soil nutrient spatial interpolation algorithm based on KNN and IDW enabled the traditional soil testing and formulated fertilization recommended system to run on embedded devices such as high performance and low cost mobile intelligent terminal. The advantages of the mobile technology and the soil testing formula technology could thus fully integrate to achieve the positioning and query to the soil nutrient and recommended fertilization information of any plot for the users through the intelligent mobile platform in the fields. It had an important significance for the accurate utilizing of the soil testing and formulated fertilization technology and the resolving of the problem of "the last kilometer" for the scientific fertilization information promotion.

Acknowledgment. Funds for this research was provided by the Henan Province Scientific and Technological Achievements Transformation Plan (132201110025), the National 12th Five-year Science and Technology Support Plan (2014BAD10B06), the Colleges and Universities in Henan Province Key Scientific Research Project (15A520019).

References

1. Jing-san, W., Gui-cheng, C.: The existing problems and countermeasures of unit soil samples collecting in soil testing and formulated fertilization. Mod. Agric. Sci. Technol. **1**, 274 (2010)
2. Yue, T., Xiao-xu, Z., Ce, W., et al.: The existing problems and countermeasures in soil testing and formulated fertilization project. Mod. Agric. Sci. Technol. **13**, 262 (2014)
3. Donghai, F.: The existing problems and countermeasures in soil testing and formulated fertilization. Mod. Agric. Sci. Technol. **18**, 298–299 (2011)
4. Yue, W., Wei-ning, Y., Heng, W., et al.: Multi-modal interaction in handheld mobile computing. J. Softw. **16**(1), 29–30 (2005)
5. Yisheng, Y., Fenggen, H.: Processing model of mobile data in disconnecting computing environments. Comput. Eng. Des. **27**(14), 2686–2687 (2006)
6. Tao, Y., Lizhong, Y., Zheng, W.: The development and application of GIS under mobile computing environment. Bull. Surv. Mapp. (2), 40–41(2002)
7. Li, Z., Yin-lei, T.: PDA-based mobile query system for testing soil for wheat formulated fertilization. J. Henan Agric. Univ. **44**(3), 341–342 (2010)
8. Zhang Guo-feng, H.E., Li-yuan, H.S., et al.: Research and development of recommended fertilization system based on mobile GIS. J. Huazhong Agric. Univ. **30**(4), 484–486 (2011)
9. Xin, X., Hao, Z., Lei, X., et al.: Decision-making system for wheat precision fertilization based on WebGIS. Trans. Chin. Soc. Agric. Eng. **27**(14), 95–97 (2011). (in Chinese with English abstract)
10. Hai-yan, J., Jin-hui, M., Xiao-ming., X.: Support system for wheat production management based on service-oriented architecture and WebGIS. Trans. Chin. Soc. Agric. Eng. **28**(8), 160–165 (2012)
11. Zhao Qing-song, X., Tao, JH.-y., et al.: Cropping system design service system based on SOA and WebGIS. Comput. Appl. Chem. **31**(6), 757–759 (2014)

12. Qiuan, Z., Wanchang, Z., Junhui, Y.: The spatial interpolations in GIS. J. Jiangxi Norm. Univ. (Nat. Sci.) **28**(2), 184–186 (2004)
13. Xiu, W., Xiao-ke, M., Zhi-jun, M., et al.: Effect on soil nutrition interpolation result of different interpolation styles. Chinese. J. Soil Sci. **36**(6), 827–829 (2005)
14. Juan, Z.: Researched and Developed a Soil Testing and Formulated Fertilization Application System Based on PDA-GIS. Anhui Agricultural University (2012)
15. Li-Hua, X.: Comparisons Among Different Prediction Method on the Soil Nutrient-Taking Wangjiagou Small Watershed of Three Gorges Reservoir Area as Researching Zone. Southwest University (2012)
16. Zeng-bing, L., Geng-xing, Z., Qian-qian, Z.: Comparision of spatial interpolation methods for soil nutrients in cultivated land fertility evaluation. Chin. Agric. Sci. Bull. **28**(20), 231–236 (2012)
17. Guo-dong, J., Yan-cong, L., Wen-jie, N.: Comparison between inverse distance weighting method and kriging. J. Changchun Univ. Technol. **24**(3), 54–56 (2003)
18. Guang, C., Li-yuan, H., Xiang-wen, Z.: Comparison of spatial interpolation technique of soil nutrient and reasonable sampling density. Chin. J. Soil Sci. **39**(5), 1008–1011 (2008)
19. Minl, D.: Study on Soil Property with GA-RBF-Neural-Network-Based Spatial Interpolation Method. Sichuan Agricultural University (2009)
20. Cover, T.M., Hart, P.E.: Nearest neighbor pattern classification. IEEE Trans. Inf. Theor. **13**(1), 21–27 (1967)
21. Zebao, Z.: Research of Spatial Indexes Optimization and Implementation. Harbin Engineering University (2005)
22. Chaode, Y., Xuesheng, Z.: The review of spatial indexes in GIS. Geogr. Geo – Inf. Sci. **20**(4), 23–26 (2004)
23. Xiao-hua, S., lian-an, Y., Lei, Z.: Comparison of spatial interpolation methods for soil available kalium. J. Soll Water Conserv. **20**(2), 70
24. Ke, W., Zhangquan, S., Bailey, J.S., et al.: Spatital variants and sampling strategies of soil properties for precision agriculture. Trans. Chin. Soc. Agric. Eng. **17**(2), 34–35 (2001)

Segmentation of Cotton Leaves Based on Improved Watershed Algorithm

Chong Niu[1,2,3], Han Li[2,3(✉)], Yuguang Niu[1], Zengchan Zhou[4],
Yunlong Bu[4], and Wengang Zheng[2,3]

[1] College of Information Engineering, Taiyuan University of Technology,
Taiyuan 030024, China
niuchong0503@163.com, ygniu@sina.com
[2] Beijing Research Center for Information Technology in Agriculture,
Beijing 100097, China
{lih, zhengwg}@nercita.org.cn
[3] National Engineering Research Center for Information Technology
in Agriculture, Beijing Academy of Agriculture and Forestry Sciences,
Beijing 100097, China
[4] Beijing Kingpeng International Hi-Tech Corporation, Beijing 100094, China
zengchan@sina.com, buyunlong2001@163.com

Abstract. Crop leaf segmentation was one important research content in agricultural machine vision applications. In order to study and solve the segmentation problem of occlusive leaves, an improved watershed algorithm was proposed in this paper. Firstly, the color threshold component $(G-R)/(G+R)$ was used to extract the green component of the cotton leaf image and remove the shadow and invalid background. Then the lifting wavelet algorithm and Canny operator were applied to extract the edge of the pre-processed image to extract cotton leaf region and enhance the leaf edge. Finally, the image of the leaf was labeled with morphological methods to improve the traditional watershed algorithm. By comparing the cotton leaf area segmented using the proposed algorithm with the manually extracted cotton leaf area, successful rates for all the images were higher than 97 %. The results not only demonstrated the effectiveness of the algorithm, but also laid the foundation for the construction of cotton growth monitoring system.

Keywords: Machine vision · Image segmentation · Lifting wavelet · Watershed algorithm

1 Introduction

Crop growth information is the basis of precise management of crop production, which plays a decisive role in the management of growth, quality and yield of crops. Leaf information is a direct reflection of crop growth status, and it is a research focus to extract leaf region effectively from crop image in the present research [1, 2]. Cotton is strategic materials relating to the national economy and people's livelihood, and leaf is also an important organ for photosynthesis of cotton. The size of leaf area has a direct effect on the yield of cotton in a certain extent. Therefore, establishing a convenient and

© IFIP International Federation for Information Processing 2016
Published by Springer International Publishing AG 2016. All Rights Reserved
D. Li and Z. Li (Eds.): CCTA 2015, Part I, IFIP AICT 478, pp. 425–436, 2016.
DOI: 10.1007/978-3-319-48357-3_41

accurate method for obtaining leaf area is of positive significance to guide cotton production time activity and develop high yield, high quality and high efficiency cultivation technique measures [3, 4]. In the process of agricultural automation, machine vision technology has become an indispensable part. This technology has been used in many fields of agricultural automation to mine the data from crop images, such as crop water stress [5, 6] and detection of crop diseases [7], etc. In recent years, many researchers have developed leaf separation algorithms to separate leaves from the crop images by machine vision [8, 9], and these algorithms have been applied to crop identification, weed control, and some other fields. However, there are different degrees of over segmentation and under segmentation in the extraction of leaves with shadow and overlapping. Therefore, in order to meet the requirements of the practical application, it is needed to develop an effective algorithm to extract the area of overlapping and shaded leaves. The targets of this paper were to extract the single cotton leaf quickly and accurately in cotton images obtained in natural light condition, and to compute the leaf area size to provide data reference for crop growth monitoring [10].

The extraction of cotton leaf area based on machine vision technology will generally use the image segmentation. The basic idea of the threshold method is to calculate one or more gray thresholds based on the gray level of the image, and the gray value of each pixel in the image is compared with the threshold value, the pixels are classified according to the comparison result [11–13]. The difficulty of this algorithm is to calculate the optimal segmentation threshold value and the segmentation effect is uneasy to grasp. In image segmentation, image edge information is often used. The edge is a collection of two different regions of the boundary line of the image, which is a reflection of the discontinuity of the local feature of the image and the change of image features, such as gray, color, texture, and so on [14–16]. However, the traditional edge detection is not ideal for the detection of overlapping edges. Lifting wavelet transform not only has good time-frequency local characteristic and multi-resolution analysis characteristic and avoids the loss of information due to the limitation of the computer accuracy when dealing with the image. Canny operator can detect the real weak edge, and it is the optimal edge detection operator. Researchers have used the combination of them to get a good edge detection effect [17], but they did not have a further image segmentation.

Watershed algorithm [18] is a morphological segmentation algorithm, which is based on the gradient of the image. The weak edge can be disposed effectively with this method, but the noise in the image can cause over segmentation of the watershed algorithm. So the traditional watershed segmentation algorithm is not fit for direct application. Therefore, many researchers have proposed the improvement methods for the traditional watershed algorithm [9, 19].

Major difficulty of the extraction of single leaf area in natural light is to handle the cotton leaf images with occlusion and shadows. A method based on lifted wavelet and improved watershed method was presented to extract cotton single leaf in this paper.

2   Experiments and Methods

2.1   Experimental Material

In laboratory conditions, images were collected in Beijing Academy of Agricultural and Forestry Sciences. In order to simulate the natural conditions of the day, the cotton plants were placed in the laboratory by the window to receive natural light irradiation, the laboratory temperature was 21 °C, humidity was 39 %. A CMOS camera (Nikon Inc., J1, Japan) with a resolution of 3872 × 2592 was used to get the RGB image, which was saved as a JPG format. According to the number of cotton leaves, cotton images were divided into two groups, images with 2 leaves and images with 4 leaves. According to whether there was occlusion or shadow between cotton leaves in the cotton image, the cotton images was further divided into four kinds. In this paper, Matlab (R2010b) and photoshop 7 software were applied to process the image. In order to improve the processing speed, the cotton RGB image was reduced to one fifth of the original image size, which was 510 × 775.

2.2   Experimental Methods

2.2.1   Overall Description of the Algorithm

Firstly, the green area of cotton in an RGB image was extracted based on color threshold and filtered the region that were not interested in the image, including most of the shadows, background and cotton stalk, in this way the cotton leaf area could be get. Then lifting wavelet and Canny operator for respectively accomplished the image gray enhancement and edge detection so that image edge could be effectively and accurately extracted from occlusive leaves, based on the operations above, watershed algorithm could achieve better segmentation effect. Finally, in order to solve the over segmentation caused by directly using the watershed transform, the leaf area image was converted to HSV space and morphological marking of the foreground and background was processed in the image to improve the watershed algorithm to reach a better effect of segmentation.

2.2.2   Extraction of Leaf Area in the Image Based on the Color Threshold, Lifting Wavelet Transform and Canny Edge Detection Operator

Color threshold $(G-R)/(G+R)$ applied to highlight the green component in the cotton image, the threshold value of the component was automatically calculated by the maximum difference method. Wavelet transform could enhance the gray image, but in practical applications, due to the limitation of computer calculation precision, the image would produce information loss after wavelet transform, lifting wavelet overcome the shortcomings. Canny operator was the most effective edge detection operator of edge function. After applying the lifting wavelet and Canny operator, the gray level of the image could be effectively enhanced. Finally, the leaf region was extracted from the enhanced edge, which allowed us to use the watershed algorithm obtain a good segmentation effect.

2.2.3   Cotton Image Segmentation Based on Improved Watershed Algorithm

2.2.3.1 Principle of Watershed Algorithm

The ideological source of the watershed algorithm [18] is in topography, the topography image is seen as a natural landscape covered by water, the gray value of each pixel in the image indicates the altitude of the point, each of its local minimum and its influence area is called catchment basin, two adjacent catchment basin boundary is the watershed. The main process of the algorithm is to find the catchment basin and the algorithm is described as follows.

$$T[n] = \{(s,t)|f(s,t) < n\} \tag{2.1}$$

where, $f(x,y)$ is image, $T[n]$ represents a collection of coordinate $[s,t]$, the points in the collection are located below the level of the $f(x,y) = n$.

$$C_n(M_i) = C(M_i) \cap T(n) \tag{2.2}$$

where, $C_n(M_i)$ can be seen as the two value image given by formula 2.2, $\{M_1, M_2, \ldots \ldots M_n\}$ is the coordinate of the local minimum point of $f(x,y)$, set $C[M_i]$ as the coordinate of a point, this point is located in the catchment basin associated with the local minimum in the M_i.

$$C[n] = \bigcup_{i=1}^{n} C_n(M_i) \tag{2.3}$$

where, $C[n]$ is the collection of the catchment basin of the flooded part at the stage n.

$$C[\max + 1] = \bigcup_{i=1}^{n} C_n(M_i) \tag{2.4}$$

where, $C[\max + 1]$ is the collection of all the catchment basin, min and max represents the minimum value and the maximum value of $f(x,y)$, as the water level in whole number is increasing from $n = \min + 1$ to $n = \max + 1$, the image of the terrain will diffuse through the water.

From analyses above, the result is that every connected component in $C[n-1]$ is a connected component of $T[n]$.

Set $C[\min + 1] = T[\min + 1]$ in search of the dividing line, then process the recursive call to obtain $C[n]$ according to the $C[n-1]$.

The watershed algorithm can achieve better results when the targets is connected each other, however, the watershed algorithm is highly sensitive to the change of the image, and the image contains noise and other factors often lead to over segmentation, so the contour of the hope is not covered by a large number of irrelevant contours. The direct application of the watershed segmentation algorithm is not good, so it is necessary to do the pre-processing of the input image.

2.2.3.2 Image Pre-processing Before Using Watershed Algorithm

Morphological marker and watershed algorithm are combined to segment the cotton leaf and get the accurate segmentation results. Foreground markers are the dot pixels that are attached to each object and they will create a maximum of each object and find out the target region by these maximums, laying the foundation for the correct segmentation of the image. Background markers are pixels that are not belonging to any object. The method of threshold is used to distinguish the background and the target region to segment images accurately.

The gray scale dilation of image f with structure element b is called $f \oplus b$ and described as follows.

$$(f \oplus b)(x,y) = \max\{f(x - x', y - y') + b(x', y') | (x', y') \in D_b\} \qquad (2.5)$$

where D_b is the definition domain of b, $f(x, y)$ is $-\infty$ out of the definition domain of f.

The gray scale erosion of image f with structure element b is called $f \Theta b$ and described as follows.

$$(f \Theta b)(x,y) = \min\{f(x + x', y + y') - b(x', y') | (x', y') \in D_b\} \qquad (2.6)$$

where D_b is the definition domain of b, $f(x, y)$ is $+\infty$ out of the definition domain of f.

The morphological opening operation of image f with structure element b is called $f \bullet b$ and described as follows.

$$f \bullet b = (f \Theta b) \oplus b \qquad (2.7)$$

Morphological opening operation completely removes the object region which didn't contain structural elements, smooths the object contour, disconnects the narrow connection and removes the tiny protruding part.

The morphological closing operation of image f with structure element b is called $f \bullet b$ and described as follows.

$$f \bullet b = (f \oplus b) \Theta b \qquad (2.8)$$

Morphological closing operation will connect the narrow the gap to form slender curved mouth, and filling the hole with smaller structure elements.

After morphological opening and closing, the image is reconstructed by morphological reconstruction to modify the image and find a maximum value of the unit within each object.

After the operation above, leaf region of gray image becomes smooth, then the revised local extreme value and the image labeled were combined with the watershed algorithm to find the catchment basin and watershed ridge line to realize the algorithm of image segmentation correctly.

2.2.3.3 Image Processing Based on Improved Watershed Algorithm

For the sake of objectively evaluating the effect of the method, the effect of automatic segmentation and manual segmentation was compared. After the use of watershed algorithm on the image of cotton leaf, region labeling method was applied for every cotton leaf, then the area of each leaf was found by using the region property. Photoshop 7 software was applied to extract a single leaf and obtain the pixel area. By comparing the two area of the corresponding single leaf, the accuracy of our experimental algorithm could be verified.

3 Results and Discussion

3.1 Single Sample Experimental Results Exhibition

To demonstrate the performance of key steps of the proposed algorithm, an image taken for a four-leaf cotton with partial occlusion was processed as an example. The color threshold of the original image was combined with the lifting wavelet and Canny operator to extract the region of the leaf, the effect was shown in Fig. 1, Fig. 1a is original image, Fig. 1b extracts the leaf region based on the color threshold, lifting wavelet and Canny operator.

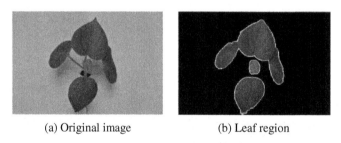

(a) Original image   (b) Leaf region

Fig. 1. Extraction of leaf area based on color threshold, lifting wavelet and Canny operator (Color figure online)

After the extraction of leaf area and its conversion to the HSV color space, the HSV image was transformed to gray scale image. The horizontal and vertical direction of the filtering of the image were carried out by using Sobel edge operator to obtain the modulus. Then the traditional watershed algorithm was applied to edge image. Figure 2a: leaf image is converted to HSV color space. Figure 2b is gradient mode value of HSV image, Fig. 2c is the result of the direct use of the watershed method based on the gradient mode image. It could be seen that direct application of the watershed algorithm would lead to over segmentation phenomenon.

In order to solve the over segmentation problem, morphological marker for images was labeled before using watershed algorithm. Firstly, foreground markers must be connected to the foreground object. Morphological techniques based on the open and close reconstruction were applied to clean up the image. After opening operation,

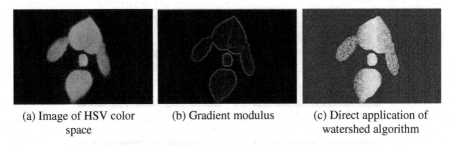

| (a) Image of HSV color space | (b) Gradient modulus | (c) Direct application of watershed algorithm |

Fig. 2. Direct application of watershed segmentation algorithm (Color figure online)

followed by closing operation, the darker spots and stem markers could be removed, treatment effect is shown in Fig. 3, Fig. 3a, b are effect of open operation and reconstructed operation; Fig. 3c, d is the image of the closed operation and reconstruction based on the open reconstruction image. Opening operation could filter out the thrusting which was smaller than structure elements, cut thin lap and played a separate role. Closing operation could fill the gap smaller than the structural elements, lapped short intervals and played a connect role. The reconstruction operation was to clean the image to find flat maximum in each object.

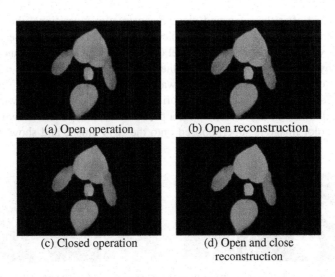

| (a) Open operation | (b) Open reconstruction |
| (c) Closed operation | (d) Open and close reconstruction |

Fig. 3. Morphological markers foreground image

Then, the local maximum of the image was calculated based on the open and close reconstruction to obtain better foreground markers. Processing results are shown in Fig. 4. Figure 4a is the region extreme value based on opening and closing operation, which is added to the gray image in Fig. 4b. Figure 4c modifies local maximum values and removes the flecks. Figure 4d applies the threshold method to distinguish the foreground and background.

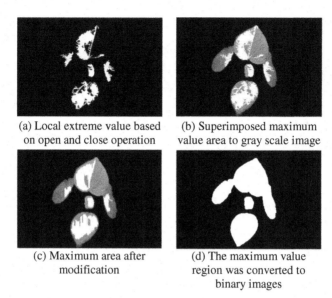

(a) Local extreme value based   (b) Superimposed maximum
   on open and close operation   value area to gray scale image

(c) Maximum area after   (d) The maximum value
      modification   region was converted to
                                   binary images

Fig. 4. Better foreground markers through regional extreme value

In the image after the open and close reconstruction operations, the background pixels were marked as black area. The calculation value of two Euclidean matrix of binary image was calculated. For each pixel of the binary image, the distance transform is specified by the distance of pixel and the nearest BW nonzero pixel of the binary image. Then of the watershed ridge line was found. Finally, the image segmentation was implemented based on watershed algorithm, and the results are shown in Fig. 5a. Then the marker matrix is converted into a pseudo color image and shown in Fig. 5b. The pseudo color image is superimposed onto the HSV image and shown in Fig. 5c.

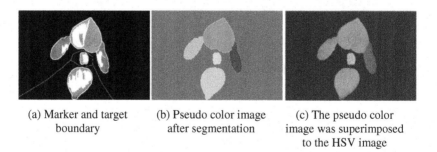

(a) Marker and target   (b) Pseudo color image   (c) The pseudo color
      boundary   after segmentation   image was superimposed
                                                             to the HSV image

Fig. 5. Image segmentation of watershed algorithm and converted it into pseudo-color image (Color figure online)

4 Overall Experimental Results

In order to illustrate the test results, the image of 6 cotton images were numbered in 6 conditions, as shown in Fig. 6. I: 2 leaf of cotton image without shadow; II: 2 leaf of cotton image with shadow; III: 4 leaf of cotton image without overlap and shadow; IV: 4 leaf of cotton image without occlusion and with shadow. V: 4 leaf of cotton image with occlusion and without shadow; VI: 4 leaf of cotton image with occlusion and shadow. The algorithm of this paper was applied to process these images and the segmentation effect are shown in Fig. 6. It is obvious that the 6 images had been

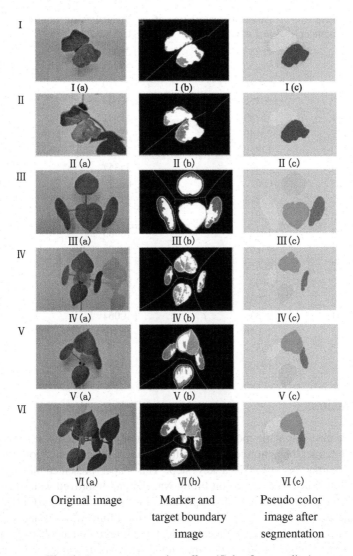

Fig. 6. Image segmentation effect (Color figure online)

correctly segmented. Figure 6a represents the original image, Fig. 6b represents the markers and the target boundary, Fig. 6c represents the pseudo color marker image after segmentation.

Table 1 compares automatic segmentation with manual segmentation results. The extracted single cotton leaf area was numbered in a clockwise direction. It could be seen that the algorithm in this paper could successfully segment cotton leaves under various conditions and the average correct rate of leaf segmentation was higher than 98 %.

Table 1. Leaf area results of cotton image segmentation

Cotton image	Leaf number	Automatic calculation of leaf area	Manual calculation of leaf area	Correct rate
I	1	25283	25006	98.89 %
	2	28474	27753	97.3 %
II	1	23886	23439	98.09 %
	2	30158	30044	99.62 %
III	1	15783	15952	98.94 %
	2	25149	25251	99.6 %
	3	17850	17999	99.17 %
	4	35463	35357	99.6 %
IV	1	6961	6912	99.29 %
	2	23371	23204	99.28 %
	3	7481	7426	99.26 %
	4	19127	18983	99.24 %
V	1	9389	9502	98.81 %
	2	26652	26945	98.91 %
	3	8103	8057	99.43 %
	4	19049	19101	99.73 %
VI	1	9933	10064	98.7 %
	2	26046	26217	99.35 %
	3	6557	6625	98.96 %
	4	14406	14557	98.96 %

5   Conclusions

In order to realize automatic monitoring of the cotton growth status, it is necessary to segment and extract the cotton leaf area. In this process, the segmentation of the cotton leaves with partial occlusion is a key step. This paper presented an algorithm to segment cotton leaf images based on lifting wavelet and improved watershed method. Firstly, the image was segmented by applying color threshold $(G-R)/(G+R)$. Then the region of the leaf and the edge was extracted by the lifting wavelet and Canny edge detection. Through using morphological method, the foreground and background of the image were marked to improve the traditional watershed segmentation algorithm.

The leaf extraction accuracy was greatly prompted and the image was segmented successfully. Experimental results on six cases of cotton images showed that the algorithm in this paper was able to successfully deal with the presence of shadow and occlusion in the cotton images, and the single leaf was successfully segmented. By comparing with the results of manually obtained leaf area, the correct segmentation rate of single leaf area was higher than 98 %, which had met the requirements for the single leaf area extraction in practical application. This approach could be used as a preliminary step to build a monitoring system and monitor the growth status of other natural objects, such as wheat, corn, or any other crops or vegetable.

Acknowledgment. This study was supported by Beijing Academy of Agriculture and Forestry postdoctoral scientific research funds (2014003), and The Beijing municipal science and technology plan (D151100003715002).

References

1. Ye, M., Cao, Z., Yu, Z., et al.: Crop feature extraction from images with probabilistic superpixel Markov random field. Comput. Electr. Agric. **114**, 247–260 (2015)
2. Pachidis, T.P., Sarafis, I.T., Lygouras, I.N.: Real time feature extraction and standard cutting models fitting in grape leaves. Comput. Electr. Agric. **74**, 293–304 (2010)
3. Trooien, T.P., Heermann, D.F.: Measurement and simulation of potato leaf area using image processing. I. Model development. Inf. Electr. Technol. Div. ASAE **35**(5), 1709–1712 (1992)
4. Chien, C.F., Lin, T.T.: Non-destructive growth measurement of selected vegetable seedings using orthogonal images. Inf. Electr. Technol. Div. ASABE **48**(5), 1953–1961 (2005)
5. Leinonen, I., Jones, H.G.: Combining thermal and visible imagery for estimating canopy temperature and identifying plant stress. J. Exp. Bot. **55**, 1423–1431 (2004)
6. Meron, M., Sprintsin, M., Tsipris, J., Alchanatis, V.: Foliage temperature extraction from thermal imagery for crop water stress determination. Precision Agric. **14**, 467–477 (2013)
7. Phadikaer, S., Sil, J., Das, A.K.: Rice diseases classification using feature selection and rule generation techniques. Comput. Electr. Agric. **90**, 76–85 (2013)
8. Neto, J.C., Meyer, G.E., Jones, D.D.: Individual leaf extractions from young canopy images using Gustafson-Kessel clustering and a genetic algorithm. Comput. Electr. Agric. **51**, 66–85 (2006)
9. Lee, W.S., Slaughter, D.C.: Recognition of partially occluded plant leaves using a modified watershed algorithm. Inf. Electr. Technol. Div. ASAE **47**(4), 1269–1280 (2004)
10. Peper, P.J.: McPherson: evaluation of four methods for estimating leaf area of isolated trees. Urban Fischer Verlag **2**, 019–029 (2003)
11. Thorp, K.R., Dierig, D.A.: Color image segmentation approach to monitor flowering in lesquerella. Ind. Crops Prod. **34**, 1150–1159 (2011)
12. Xia, C., Lee, J.-M., Li, Y., et al.: Plant leaf detection using modified active shape models. Biosyst. Eng. **116**, 23–25 (2013)
13. Kavitha, A.R., Chellamuthu, C.: Implementation of gray-level clustering algorithm for image segmentation. Procedia Comput. Sci. **2**, 314–320 (2010)
14. Yang, G., Fengchang, X.: Research and analysis of image edge detection algorithm based on the MATLAB. Procedia Eng. **15**, 1313–1318 (2011)

15. Asano, T., Katoh, N., Tokuyama, T.: A unified scheme for detecting fundamental curves in binary edge images. Comput. Geom. **18**, 73–93 (2001)
16. Yuan-Hui, Yu., Chang, C.-C.: A new edge detection approach based on image context analysis. Image Vis. Comput. **24**, 1090–1102 (2006)
17. Han, L., Wang, K., Haoyi, B.: Cotton leaf image edge detection using mean-shift algorithm and lifting wavelet transform. Trans. CSAE **26**(Supp. 1), 182–186 (2010). (in Chinese with English abstract)
18. Meyer, F.: Topographic distance and watershed lines. Sig. Process. **38**, 113–125 (1994)
19. Orbert, C.L., Bengtsson, E.W., Nordin, B.G.: Watershed segmentation of binary images using distance transformations. Nonlinear Image Process. IV **1902**, 159–170 (1993)

Research on Knowledge Base Construction of Agricultural Ontology Based on HNC Theory

Hao Xinning, Xie Nengfu, Sunwei, Zhong Xiaochun[(✉)],
and Zhang Xuefu

Agricultural Information Institute of Chinese Academy of Agricultural Sciences,
Beijing 10081, China
zhongxiaochun@caas.cn

Abstract. With the development of agricultural research and production, agricultural information and data processing technology have become increasingly demanding. In recent years, organizing and expressing semantic knowledge based on the form of ontology has become the focus of research in artificial intelligence field. And the construction of ontology-based knowledge base is the basic condition. HNC theory has great advantages in processing of Chinese character. However, complexity of the system has restricted its development. This paper summarized the existing research results, took rice production (cultivation) and processing as an example, proposed an idea and methods for the knowledge base construction of agricultural ontology based on HNC theory.

Keywords: Agriculture ontology · HNC theory · Knowledge base

1 Introduction

The rapid development of modern agriculture puts forward new and higher requirements for research on agricultural information. How to make a better use of modern information technology to innovate ideas and methods in the field of agricultural information science, and how to continuously promote research to enhance the performance of agricultural information have become an inescapable mission for research scientists in this area.

The rapid development of Internet, cloud computing, big data and other information technologies, and the continuous improvement of a variety of advanced mathematical theory spawn more IT research methods that are more convenient. Meanwhile, with the continuous advances in information science and technology, data management and information management have been difficult to fulfill modern demand for high-quality information. As a result, knowledge management has become one of the latest ways of information management. The main objective of knowledge management is to build domain knowledge in the form of knowledge networks, knowledge base and so on. This can reveal the relationship between semantic knowledge. Thus it is possible to form a more complete, proven and accurate knowledge platform and to support knowledge service, effective management and use of knowledge on the basis [1].

D. Li and Z. Li (Eds.): CCTA 2015, Part I, IFIP AICT 478, pp. 437–445, 2016.
DOI: 10.1007/978-3-319-48357-3_42

Nowadays, the mainstream technology of classification, retrieval, processing and filtering of agriculture information is still traditional key technology. However, this literal information retrieval system can handle only the literal meaning without carrying out more in-depth analysis, which leads to the lack of semantic understanding. Although it can make effective semantic analysis in artificial way to filter, process, classify and analyze information, the processing capacity and efficiency are insufficient to deal with the vast amounts of data generated daily. Therefore, only the thematic retrieval way, through the establishment of ontology knowledge base to describe relationship between domain concepts in an accurate, clear and standard way and use abundant examples to fully describe the conceptual system of domain knowledge, can effectively improve the matching degree of retrieving objects and then make the best use of agricultural information resources in a fast way to achieve the knowledge management of agricultural information.

2 Basic Technical Concepts

2.1 HNC Theory

Hierarchical Network of Concept (HNC) is a theoretical system for the entire natural language understanding and processing, which is founded by Professor Huang Zengyang, Institute of Acoustics. The theory attributes semantic processing to conceptual representation. It uses primitive symbols and their combinations that highlight the relevance between concepts to represent conceptual semantic connotation. It represents the semantic content explicitly in natural language by mapping the natural language symbol system to HNC concept notation system. HNC concept notation system can describe lexical semantic, concept type and composition of the statement. The system can complete statement analysis of natural language through the concept analysis of statement and words [2]. HNC provides the basis for the computer to grasp the semantics by using HNC1, HNC2, HNC3 and HNC4 to digitize the word, sentence, sentence group and chapter. It contains not only the word level of knowledge, but also statement and context level of knowledge. The key point is to use HNC theory to process natural language and establish the mapping relationship between natural language and the concept space [3, 4].

2.2 Ontology

Ontology is originated from philosophy area and it is a relative concept of Epistemology. Epistemology focuses on subjective perception, while ontology researches objective existence. So far, academic circles have not yet formed a unanimous conclusion on the definition of ontology.

Ontology uses its accurate, accepted concept definition to achieve specificity and semantic disambiguation of conceptual and terminology, and also completes the standardization and uniformity of reality concept definition. So it can guarantee the consistency between Human-Computer Interaction and machines and maximum the degree of realization of semantic disambiguation. The relationships between concepts, properties, instances and other components can achieve logical reasoning of knowledge concept as well as the description of knowledge conceptual system in the field. The relationships can effectively improve knowledge dissemination, sharing and retrieval efficiency in the aspects of semantics and pragmatics. It can also provide knowledge services under the big Data era background.

2.3   Ontology-Based Knowledge Base

As a special form of knowledge, ontology aims to describe the facts and common vocabulary that are long-standing and unanimously approved by the knowledge workers. Knowledge base is aimed at describing the particular state of things that are related to facts and vocabulary, as well as the cognition state of knowledge workers. From another perspective, ontology refers to information that is unrelated to the particular state of things, while knowledge base refers to information that is related. From the point of structure to analyze, ontology provides a set of terms and concepts to describe a field, while knowledge base use these terms to express the fact in this area.

In natural language processing, semantic analysis requires the support of huge knowledge base. Since the 1980 s, a number of semantic knowledge base have been developed in China and abroad, such as the famous WordNet, HowNet, Beijing University of CCD, etc. All of these can be called ontology-based knowledge base from the perspective of knowledge representation.

Ontology mainly contains ontology concept layer and ontology instance level. Ontology Concept layer is a definition on the concept within a specific range and the relationship between them, which mainly aims to reveal the rich relationships between the concepts. Ontology instance level refers to instances corresponding to a concept. It is a concrete expression of certain concept. Numerous and full instances will greatly enhance the capacity and application value of ontology knowledge base. Ontology knowledge base can illustrate domain concepts and their relationships in an accurate, clear and standard way and fully describe the conceptual system of knowledge in the field. So it is good at semantic description, logical reasoning, and revealing the hidden relationship between the concepts. As a result, it can effectively solve the unhandled problems of information retrieval and information sharing [5].

Ontology-based knowledge base can not only reveal the framework of the domain knowledge, but also help domain ontology function in concept unification, standardization, knowledge retrieval, knowledge sharing and other applications. Unlike the traditional information retrieval techniques, Ontology-based knowledge base will accurately locate the required knowledge; profoundly reveal the meaning of semantic information. And with the continued expansion of the boundaries of the domain ontology, the boundary between some areas becomes fuzzy and interdisciplinary emerges. As a result, the scope of knowledge domain that ontology knowledge base covers will continue to extend to make it play a greater role. Paying more attention to information semantic meaning analysis will be the future trend of information retrieval and information world development of association data and Semantic Web.

Different from the symbolic representation of HNC theory, ontology-based knowledge base can clearly express hierarchy and relationship between concepts to facilitate people's understanding and application. Meanwhile, ontology using formal description language can also be directly applied to natural language processing. If the concept expression in HNC theory can be represented by the general form of ontology, the accuracy of the human-computer interaction can be effectively improved [6].

3   Knowledge Base Construction of Agricultural Ontology Based on HNC Theory

3.1   Ontology Construction Method

Ontology construction method mainly starts with the domain concepts and their relationship. With more suitable and mature ontology construction methods to guide the building process, we can effectively guarantee the consistency of domain ontology construction and also make ontology construction standardized and modularized. Duties, tasks and requirements in each part of the building process are clearly defined. There are no uniform, generally accepted principles of ontology construction [7–9]. In the ontology community, five principles proposed by Gruber are the most widespread:

(1) Clarity: Ontology provides concepts involved with clear, authoritative, accurate and standardized description. The definition of concept should be combined with specific areas and professional background. It should be objective, independent, authoritative and described formally with logical axiom.
(2) Consistency: Logic rules of the ontology should be strict and rigorous. The definition of concept inferred by logical axiom should be correct.
(3) Extensibility: The design of ontology should go with the changes and development in the field. And it can be adjusted and extended continuously.
(4) Smallest coding error: Conceptualization should specify the level of knowledge, not varying according to different symbolic coding.
(5) Minimum ontology commitment: The establishment of the ontology needs to satisfy specific knowledge sharing needs. If the coverage of the ontology is too large, it tends to lead to low specificity and ambiguous concepts, which will result in losing the characteristics of ontology itself.

Currently, the mainstreams of ontology construction method are as follows.

(1) TOVE: It is a relatively new method of ontology, which regards demand problems and Completeness Theorem as the considerations of ontology construction. However, it lacks documented process description and specific steps description of ontology construction.
(2) Skeletal Methodology: The method provides a framework and guidelines for construction in each stage, requires a documented process and gives the steps of ontology assessment. Therefore, it is full of reference value. However, it lacks specific methods and techniques, and only provides guidelines for the development of corporate ontology.
(3) METHONTOLOGY: The method provides a description of the steps of ontology assessment and is suitable for the development of large ontology programs. It first proposes the concept of "writing specification" and also details the ontology construction tools, concept sources and concept extraction method. This promotes

the standardization and normalization of ontology construction. However, it doesn't provide any ontology assessment method, so it can't evaluate the quality of ontology construction.

(4) Cyclic Acquisition Process: The main contribution of it is a new method using a cyclic structure for ontology acquisition. However, it does not provide details of guidance and technical explanation for specific method of ontology acquisition. Therefore, it is difficult to put the method into practice.

(5) "Seven Steps" by Noy and McGuinness: Because of its comprehensiveness and systematisms, it is favored by many people, including us. Combined with Protege software, the main steps are as follows:

1) Determine the field and scope of ontology;
2) Consider reusing existing ontology;
3) List important terms in ontology;
4) Define the class and its hierarchy;
5) Define the property of the class;
6) Define the restrictions of properties;
7) Create an instance.

3.2 Knowledge Base Construction of Agricultural Ontology Based on HNC Theory

Compare with traditional English words, agricultural vocabulary is significantly professional, it is necessary to establish a new vocabulary base. Taking rice as an example based on HNC theory, this paper establishes an agricultural ontology knowledge base and realizes the establishment, management and update of knowledge base by Protege software. Since the terms in the HNC Theory based agricultural ontology knowledge base are clearly defined, accurately expressed and unified in the conceptual level, it will not return duplicate or irrelevant results when retrieving. The architecture is shown in Fig. 1.

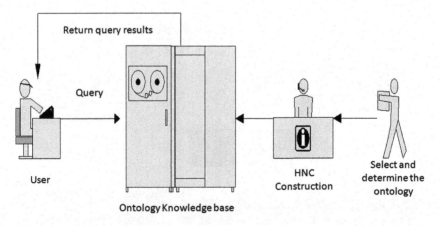

Fig. 1. The structure of ontology knowledge base

HNC theory digitizes words, sentences, sentence groups and discourses, which provides a basis for computer to grasp the semantics. The key point is to establish a mapping relationship between natural language and concept space by HNC codes. The main techniques include Concept Matrix, Link expression, composite structure and so on. The building model is shown in Fig. 2.

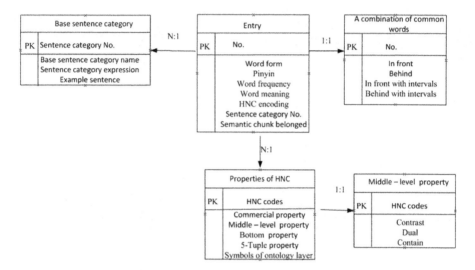

Fig. 2. HNC based ontology knowledge base model

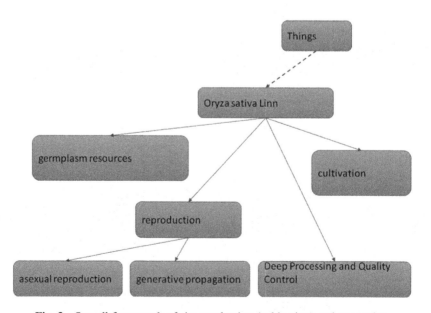

Fig. 3. Overall framework of rice production (cultivation) and processing

The ontology knowledge base shows the semantic description of words in six fields: word form, concept category, HNC symbol, semantic category belonged, relevant semantic category and synonymous appellation. The features are as follows. First, the words included are different from common words when considering the word form. The words may contain the form of non-Chinese characters, such as numbers, letters, etc. Second, for the same thing, the corresponding appellation may be more than one. Third, it focuses on the deep semantic relationship between words.

3.3   A Case Study–Rice Production (Cultivation) and Processing

The construction of agricultural ontology knowledge base is a huge project. This research applied rice field as an example to briefly introduce the construction of ontology knowledge base. The overall framework is shown in Fig. 3.

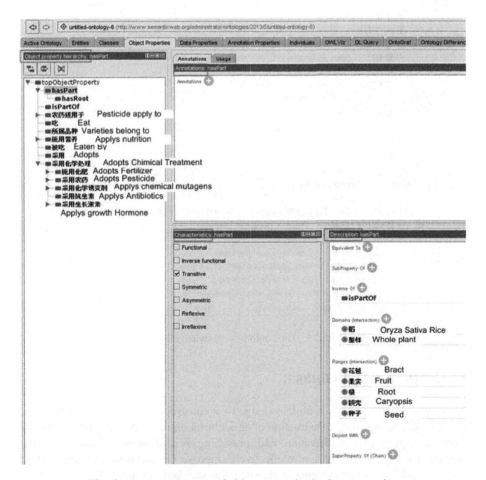

Fig. 4. An example page of object properties in data processing

The structure of ontology base contains 4 second-level directories and 10 third-level directories, Fig. 4 is shown an example page of object properties in data processing, this experiment cumulatively includes more than 700 rice ontology terms, 108 object type attributes and 23 data type attributes to organize rice ontology knowledge base. In every single object properties, there are 4 different parts, every single part was to make a connection to an object property.

The final structure proposed rice production (cultivation) and processing ontology, including field cultivation techniques, rice harvesting, storage, processing and product, rice production machinery, rice varieties. The final result is shown in Fig. 5.

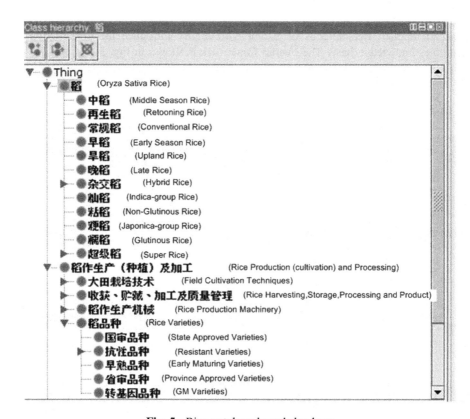

Fig. 5. Rice ontology knowledge base

4   Conclusion and Outlook

By combining HNC theory with knowledge base construction, this paper proposed the vision of knowledge base construction of agricultural ontology based on HNC Theory. Taking rice production (cultivation) and processing as an example, it initially realizes intelligent query of information to effectively improve the precision and recall. However, for lack of unified systematic classification standards in the field of agricultural production, future research should pay more attention to the systematicity and rationality of classification. The design of HNC codes is relatively simple and its relevance is not sufficient. This remains to be applied in further study.

Acknowledgment. Funds for this research was provided by the CAAS Agricultural Science and Technology Innovation Program 2015

References

1. Yigang, Z.: A review of dynamic knowledge management under semantic web. J. Modern Inf. **9**, 166–170 (2014)
2. Zengyang, H.: HNC theory summary. J. Chin. Inf. Process **11**(4), 12–21 (1997)
3. Zengyang, H.: HNC (Hierarchical Network Concepts) Theory. Qinghua University Press, Beijing (1998)
4. Shanshan, L., Wang, Y.: Retrieval of periodical literature knowledge element based on hierarchical network of concepts theory. J. Intell. **32**(9), 190–194 (2013)
5. Shouxue, Z., Bo, Y.: Construction method of knowledge base based on ontology. Comput. Era **4**, 45–46 (2014)
6. Yang, J.: Construction and application of agricultural word knowledge base of HCN. In: Beijing: The Second International Seminar on Computer and Computing Technologies in Agriculture and The Second China Development Forum of Rural Information, October 2008
7. Chuanjiang, M.: The research of HNC sentence category knowledge. Beijing: Institute of Acoustics of Chinese Academy of Sciences Ph.D thesis (2001)
8. Yaohong, J.: The language understanding technology and its application of Hierarchical Network Concepts (HNC). J. Yunan Normal University (Philosophy and Social Sciences) **42**(4), 19–23 (2010)
9. Chuan, W., Shangwang, L.: Study on construction of ontology knowledge base for wheat-weed. College of Computer and Information Engineering, Henan Normal University **42**(6), 138–142 (2014)

Method and System of Maize Hybridized Combination Based on Inbred SSR and Field Test

Zhe Liu, Zhenhong Zhang, Shaolong Fu, Xiaodong Zhang,
Dehai Zhu, and Shaoming Li[(⊠)]

College of Information and Electrical Engineering,
China Agricultural University, Beijing 100083, China
{liuz, zhangxd, zhudehai}@cau.edu.cn,
987627194@qq.com, 812238567@qq.com, lishaoming@sina.com

Abstract. Molecular breeding is considered an important way to improve the breeding efficiency. But due to the deletion of data, method and instrument, molecular design breeding is basically at the concept stage, without operational technology process and breeding practice. On the basis of breeding data from Beijing Kings Nower Seed S&T CO, LTD, this paper explored a set of methods for maize design breeding based on molecular detection and phenotypic testing information. Firstly, the parents of the combinations were obtained from the inbred lines with high homozygosity through SSR detecting and good comprehensive traits through field testing; secondly, the heterosis rate of the parents was got by calculating both the genetic differences and phenotypic differences of the parents according to the SSR detection results and field testing results respectively to express the special combining ability of their own; finally, in this paper it constructed the hybridization group model by using the comprehensive characters, special combining ability, orthogonal anti value, calculated the comprehensive characters, advantages, disadvantages of the hybridizations, and screened hybrid combinations to the next round field breeding, it also developed a software system of hybrid combination to support the technology route. Applying the software, 37 hybrid combinations resistance to Ralstonia solanacearum were got based on 179 inbred lines with molecular and phenotypic data. Thus, the method and software preliminary provides technical support for our country to carry out and perfect the molecular design breeding.

Keywords: Maize · Breeding · Heterosis · DNA fingerprint

1 Introduction

Breeding output is a small probability event, only through large-scale investment can we convert the small probability event into an inevitable event, making the originally accident breeding output become predictable and designable. Due to the lack of proper methods and tools support, domestic breeding team cannot manage and control the large-scale breeding materials and data, and it is difficult to accumulate the breeding rules of selections and combination, which leads to it popular that combination with experience

© IFIP International Federation for Information Processing 2016
Published by Springer International Publishing AG 2016. All Rights Reserved
D. Li and Z. Li (Eds.): CCTA 2015, Part I, IFIP AICT 478, pp. 446–458, 2016.
DOI: 10.1007/978-3-319-48357-3_43

and there are a lot of uncertainty with the maize breeding output in China [1]. Molecular breeding is considered an important way to improve the breeding efficiency. But due to the deletion of data, method and instrument, molecular design breeding is basically at the concept stage, without operational technology process and breeding practice. Based on this, this paper explored a set of design breeding method which included from the integrated management of maize varieties' molecular marker data and phenotypic data of field testing to maize inbred lines comprehensive evaluation, auxiliary screening. Besides, the design breeding method was also based on the heterosis analysis of inbred lines with molecular markers and phenotypic differences and the simulation distribution of maize hybrids and performance prediction. The paper also developed a software system of hybrid combination to support the technology route, providing methods and tools support to improve China's maize breeding scale and information management level.

2   Materials and Methods

2.1   Research Technical Route

In this paper, it calculated the parent value, yield and disease resistance characteristics, superiority and defect index for each variety to evaluate the screening inbred lines and calculate the phenotypic distance after making the inbred lines' phenotypic testing data be standardized; Similarly, it calculated the effect locis, heterozygosity index for each variety to evaluate the screening inbred lines and calculated the genetic distance based on the inbred lines' SSR molecular marker data [3–7]; combined with phenotypic distance and genetic distance to establish the indices of selected parents' heterosis rate; combined the seed performance indices of selected parents such as flowering and height difference coordination to establish the indices of double cross and reciprocal value; finally, composited the comprehensive value of the parents, the hybrid advantage, reciprocal value, the yield and resistance characteristics to establish hybridization group model and predict the comprehensive value, superiority, defect degree, the yield and resistance characteristics of the hybrid combinations, then screened the hybrid combination meeting the demands. (The technical route was shown in Fig. 1).

2.2   The Experimental Data

In this paper, the inbred lines' field identification data and DNA fingerprint data were obtained from the Beijing Kings Nower Seed S&T CO., LTD. Then it chose the data of 2009 in JunXian, Gongzhuling, Tieling, Dandong, Shunyi, Zhumadian six pilots totally of 482 inbred lines of field identification materials, putting them into the database (Table 1), of which 179 copies of materials were analyzed by SSR molecular marker detection (Table 2).

SSR molecular markers are mainly used for identification of new germplasm and inbred lines heterotic group in maize breeding [2, 9, 10]. Molecular marker is a direct reflection of genetic polymorphism on DNA level, the earliest molecular markers developed were RFLP and then RAPD, AFLP, ISSR, SSR, which were based on PCR, SNP were developed in recent years based on single base mutation. Simple sequence

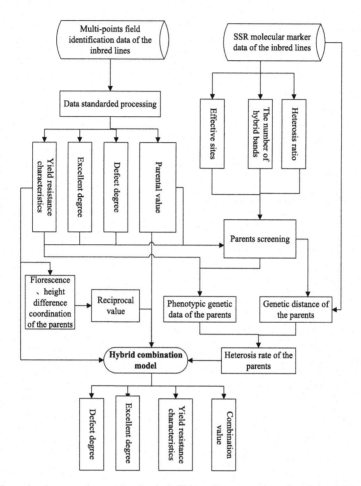

Fig. 1. The technology roadmap of maize hybrid breeding group based on inbred lines SSR and field test data

Table 1. The field identification standard data of the 482 inbred lines (the recorded indices of the inbred lines)

Pilot name	Cell number	Sowing date	Emergence date	Total number of emergence
Gongzhuling	CZ10001	2010/5/7	2010/5/21	43
Gongzhuling	CZ10002	2010/5/7	2010/5/21	37

Table 2. The primers of SSR molecular detection to the 179 inbred lines

Code of inbred lines	Primer 1	Primer2	Primer3	Primer4	Primer5
A01	–	–	a	–	–
B01	A	b	a	a	a

repeat (SSR) markers have been widely used because of its advantages, such as simple, fast, high repeatability, high polymorphism and codominant markers. Xue et al. research showed that the differences between SSR and SNP two kind of markers in data integrity, ability to distinguish among species and site stability was samall. A in another study it showed that both RAPD and SSR two kind of markers were suitable for the study of genetic diversity on maize germplasm, but SSR was more desirable.

Detection technology of SSR molecules is an acceptable cost of gene sequencing technology [11, 13, 15, 17] for domestic seed companies; therefore, Kings Nower Seed S&T CO., LTD. began to establish detection department of molecular based on SSR from 2008, which provided data support to this study. In this paper, the SSR molecular marker data and the phenotypic test data were applied to the estimation of genetic differences between the hybrid parents, and inspected the comprehensive development value of hybrid combination from these aspects such as genetic differences, combining ability, major defects, the advantage performances, the results were applied to hybrid combinations.

2.3 Standardization and Evaluation of Inbred Lines Field Identification Data

The standardization of inbred lines field identification data refers to the transform process of field data to relative value using the pilot average as the reference frame. The inbred lines usually use the absolute value or relative value of data to describe in different environment. The absolute value of data generally are obtained through direct observation to the indices, the dimension used is the independent reference system which has nothing to do with the specific observations. If the recorded data do not mark the data unit, the absolute value of data will be meaningless. The relative value of data take the data recorded in the general order or associated with the overall statistics as reference. Since rich in information and easy to undersatand, the relative value is the main evaluating method in the inbred lines evaluation. Zhe [8] proposed 5 kinds of relative value transform methods of variety test data such as the 3 s and 2 s conversion value taking group mean as the reference value, standard of value, the class NDVI value taking the sample mean as the reference value and so on, and used the "standard deviation ratio" to analysis the transform effect of the 5 methods, the analysis results indicated that the 3 s and 2 s conversion value taking the group mean as reference were more close to the genetic characteristics of the varieties. This paper adopted the method to transform the inbred lines identification data of which the type was numerical to relative value.

When the inbred lines identification data were normalized, it calculated the production, seeding, earning, economic value, disease resistance, insect resistance, lodging resistance, growth period and plant panicle characteristics and the comprehensive value, and used them to calculate the phenotypic distance of inbred lines selected.

The comprehensive value of inbred lines is a comprehensive study to the inbred lines field observation value, which is calculated by the following formula:

$$HD = \frac{\sum v_k * w_k}{\sum w_k} \qquad (1)$$

In the formula (1), v_k is the standard value of the kth field observation index value, w_k is the weight of the kth field observation index value.

To evaluate the advantages and disadvantages of the inbred lines and hybrid combinations, this paper designed the defect degree and excellence degree two indices on the basis of the existing observed phenotypic traits.

Defect degree and excellent degree are the statistics of inbred lines (or hybrid combinations) whose field observation indices perform "very good", "good", "bad", "poor". The statistics are the reached degree of the field identification index, followed by "very good", "good", "bad", "poor" to calculate the weighted average income in a ratio 3:2:2:3 of attached weights. The "very good" standard means that the value of the hybrid combination in a certain index is more than the average value of all combinations in the index pulsing 2times the standard deviation, then we regard the combination performing "very good" in the index. While if the hybrid combination in a certain index is below the average value of all combinations in the index subtracting 2times the standard deviation, it will be regarded performing "poor" in the index. Among them, the standard deviation multiple of which "very good", "good", "bad", "poor" can be set by software. Then the inbred lines were screened by using the difference, defect, inbred lines output resistance characteristics and comprehensive value.

2.4 Treatment and Evaluation of Inbred SSR Data

The effective sites, hybrid bands and heterozygous ratio for each inbred line can be calculated based on the inbred lines SSR molecular marker data. The number of the effective sites refers to the n loci of the selected SSR data, of which having the value is effective site; Generally the measured SSR data of inbred lines are homozygous, which means each site should be the same value, but actually it may not be homozygous; Heterozygous brand refers to how many different values in the n sites, and it will be the number of the heterozygous brand; Heterozygous ratio refers to the number of the heterozygous brand/n. These three indicators are used to judge the pure degree of the inbred lines, screen the inbred lines which have higher pure degree, more effective sites into hybrid combination, and then calculate the genetic distance of the inbred lines selected.

2.5 Calculation of Parent Heterosis Rate

The additive gene effect is the main factor of genetic effects of quantitative characters in maize, the molecular marker data and phenotypic value of the field test data are approximate value of the gene additive effect, both of them contain some information of additive effects, at the same time the information may be complement to each. If combine the information of these two parts to predict the comprehensive heterosis of the single cross hybrid F1 generation, it is expected to get higher choosing efficiency. Specifically to say, to show the phenotypic differences of hybrid parents we use phenotypic data to calculate the phenotypic distances, to show the genotype differences we use the DNA fingerprint data to calculate the molecular marker genetic distances, both of them do not completely overlap, which describe the genetic differences of parents from different angles. This paper used a parent heterosis (MD) to generalize the genetic differences of the parents, which was written as linear equations of the phe-notypic genetic distance and SSR molecular marker genetic distance:

$$MD = r1 * PD + r2 * GD \tag{2}$$

Among them, $r1 + r2 = 1$, PD refers to the phenotypic difference of the parents, GD refers to the genetic difference of the parents, R1 and R2 refers to the weight of phenotypic distance and genetic distance.

2.5.1   Algorithm of Molecular Marker Genetic Distance

Genetic distance is the index that measures the size of some characteristics of the comprehensive genetic differences among varieties. More than one traits are required in the breeding target, in order to be able to more fully reflect the genetic differences between the parental varieties, multiple traits need to be comprehensively considered, thus will extend the concept of genetic distance. Calculate the multidimensional geo-metric distance consisted of multiple traits according to the method of multivariate statistics analysis, called genetic distance. There are mainly in two ways to calculate the DNA fingerprint genetic distance:

$$GD = 1 - \frac{m}{m+n} \tag{3}$$

M denotes the number of same bands between the two varieties, n donates the number of different bands between the two varieties.

$$GD = 1 - \frac{2N_{ij}}{N_i + N_j} \tag{4}$$

N_{ij} denotes the number of same bands between the two varieties (lines), N_i and N_j respectively denotes the band number of their own in i and j varieties (lines). The differences between two kinds of molecular markers in genetic distance calculation are as flowing:

(1) When there are invalid detection sites at least in one of the both detection sides, the calculated results of formula (4) are more conservative than formula (3), which means the method 2 tends to judge the two sides involved in the calculation have a genetic difference. For example, for the DNA fingerprint data shown in Table 2, the invalid sites do not participate in calculation, the GD value is 0 according to formula (3), meaning that the genetic basis of inbred line "A01" and "B01" are very close, almost the same; the GD value is 0. 67 according to formula (4), apparently it is more conservative using formula (4) to calculate.

(2) Formula (4) is suitable for the parents' fingerprint data that are not using the same primer combinations to detect. If the fingerprint data are obtained using different primer combinations to detect, you cannot determine the different bands of varieties (lines) to be detect (i.e. n in formula (3)).

Considering the two points above, this paper used formula (4) to calculate the molecular marker genetic distance.

2.5.2 Calculation of the Phenotypic Distance

It makes principal component analysis to quantitative traits in the conventional phenotypic genetic distance calculation, after which uses the Euclidean distance to calculate the comprehensive index. The essence of the principal component analysis is based on the multivariate statistical method, extracting a few comprehensive indices as the distance analysis dimensions, and uses the index contribution rate to distribute weights of the comprehensive indices. Its purpose is to reduce the dimension so as to simplify the problem, and the weight distribution is just the result of the calculation. The disadvantages are:

(1) It will spend a lot of time to calculate multiple inbred lines, the more the data of inbred lines, the bigger the calculation scale, which is not suitable for real-time calculation in software.
(2) It only considers the genetic differences of quantitative traits, ignoring the genetic differences of varieties (lines) in quality traits.
(3) There is no clear meaning of the comprehensive index extracted by principle component analysis.

In order to calculate the phenotypic genetic distance of both quantitative traits and quality traits, and be easy to soft program, combining with the data standardization processing described in this article, we adopted the following calculation:

$$PD_{ij}^2 = \frac{\sum (V_{ik} - V_{jk})^2 * w_k}{n * \sum w_k} \tag{5}$$

Among them, PD_{ij} is the phenotypic distance of inbred lines i and j, V_{ik} and V_{jk} respectively denote the standard value of inbred lines i and j in index k, w_k is the weight of kth index, n is the effective index number. In this formula, the quality traits related to breeding after numerical processing are also used to express the phenotypic differences among inbred lines. Because the index weight can be configured flexibly, the data do not need to go through the orthogonal transformation and calculate the principal component values, thus the cost of the system decreases, and the calculation goes fast.

2.6   Calculation of Reciprocal Value

Reciprocal value mainly considers the special traits' combining degree of the parents, the parents in these characters are not the closer the better but with certain differences can play a greater role for hybrid seed production or maximizing the heterosis. These special traits are such as plant height-ear position difference of the parents, Anthesis-silking interval (ASI). This paper calculated the orthogonal value and reciprocal value respectively, choosing the larger of them as the exchange value of crossbreds. Orthogonal value (reciprocal value) is calculated as flowing:

$$V_{ij} = HD + r_1 * (t_i - t_j) + r_2 * (h_i - h_j) \tag{6}$$

Among them, HD is the comprehensive value of the inbred lines, V_{ij} is the orthogonal of the parents both i and j, t_i is the ith parent's standard value at the anthesis stage, t_j is the jth parent's standard value at the silking stage, h_i is the plant height of the ith parent, h_j is the plant height of the jth parent, h_j is the ear position height of the jth parent, r_1 and r_2 respectively denote the important weight of spinning-powder interval and plant height-ear position difference. The calculation of reciprocal value is as flowing:

$$SV = Max(V_{ij}, V_{ji}) \tag{7}$$

2.7 Hybrid Group Model and Evaluation and Screening of the Combination

Hybrid combination value (HV) is a comprehensive index which is used to describe the development value of the hybrid combination, this paper used the dear in heterosis rate (MD), the average of the parents' comprehensive value ($\bar{H}\bar{D}$) and the reciprocal value (SV) to denote the value of hybrid combination. The average of the parents' comprehensive value ($\bar{H}\bar{D}$) is calculated by (the male comprehensive value + the female comprehensive value)/2, which is HD. So the hybrid combination value is calculated as following, where r is the weight.

$$HV = MD * HD + r * SV \tag{8}$$

3 Realization of the System and Case Analysis

3.1 The Screening Results of the SSR Fingerprint Information

Screening can be carried out only after put the SSR fingerprint into database and complete the eigenvalue calculation: the example in this paper screened according to the heterozygous ratio >=0.1 or the number of the effective site <=18 or the number of mixed band >=2. According to the screening conditions 94 inbred lines were removed (The screening condition interface is shown in (Fig. 2):

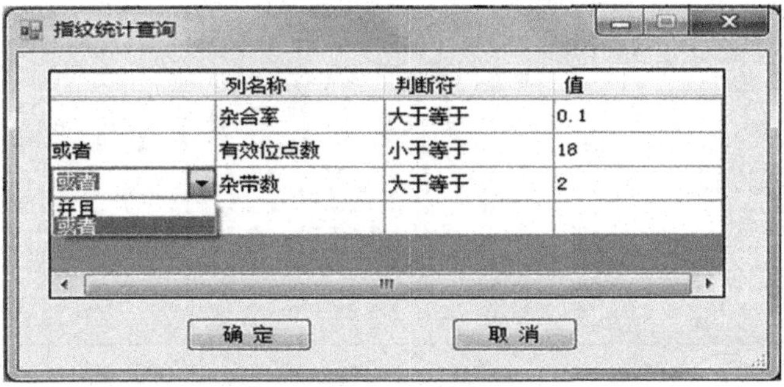

Fig. 2. The SSR fingerprint screening conditions of the inbred lines

3.2   The Screening Results of the Parental Phenotypic Traits

The screening of the parental phenotypic traits consists of three steps:

(1) According to the indices of defect degree and excellent degree in the "very good", "good", "bad", "poor" 4 categories of computing standard, it calculated the category each inbred lines belonged to for each of the phenotypic traits, and then calculated the defect degree and excellent degree of each inbred lines, and sorted them, then screened the inbred lines before 50 % to enter the next step.

(2) This paper aimed to prepare the hybrid combination resistance to Ralstonia solanacearum, this step according to the performance of each inbred line' resistance to Ralstonia solanacearum, we screened the inbred lines which were "very good" to enter the next step.

(3) According to the formula 1 which calculated the comprehensive evaluation index, we got the comprehensive value of each inbred line and sorted them, then screened the top 50 % to enter the final hybrid group.

According to the SSR molecular marker data and phenotypic testing data, 78 inbred lines were got for combination after a quality of indices screening. Then these 78 inbred lines would be combined each other.

3.3   The Results of Hybridization Group and Combination Screening

78 inbred lines were got according to the inbred lines SSR data and phenotypic data screening. For each hybrid combination in this paper, it calculated the heterosis rates using formula 2, and calculated the orthogonal value or reciprocal value using formula 6, and calculated the comprehensive value using formula 8, then sorted the hybrid combinations by the comprehensive value (the sorting interface was shown in Fig. 3 below).

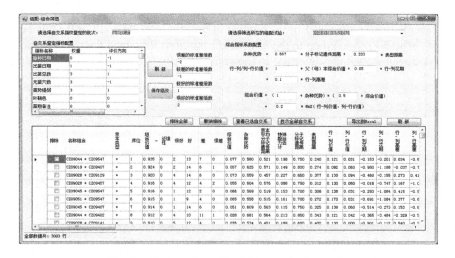

Fig. 3. The interface of hybrid combination order

Due to the field identification data of the inbred lines only contained 2009 one year's data in the case, the number of test points was a little less and its reliability was less higher than DNA fingerprint data, so we gave the molecular marker genetic distance a bigger weight. On the reciprocal cross (row-column, column-row) value of the parameter setting, this paper thought that the elevation difference was more important than spin-powder interval, thus gave the row-column flowering, column-row height difference respectively the weight of 0. 05 and 0. 1. According to the hybrid combinations screening conditions, finally we got 37 hybrid combinations. (partial results were shown in Table 3).

Table 3. The results of hybrid combination

Name combination	Order	Combination value	Very good	good	bad	poor	Comprehensive value	heterosis	Average molecular marker genetic distance of the parents	special combining ability	Molecular marker genetic distance	Phenotypic distance	Row-column value	Column-row value	Row-column flowering	Column-row flowering	Row-column height difference	Column-row height difference
A11&A12	1	0.94	2	13	7	0	0.077	0.58	0.52	0.20	0.75	0.24	0.12	0.03	-0.15	-0.20	0.03	-0.61
B11&B12	3	0.92	4	14	6	0	0.073	0.56	0.46	0.23	0.65	0.38	0.13	0.09	-0.46	-0.15	0.27	0.41
C11&C12	7	0.91	1	14	6	0	0.051	0.61	0.56	0.12	0.75	0.33	0.14	0.06	-0.51	-0.27	0.15	-0.85

(In this table, the results are ranked assending by combination value. Table 3 is corresponding to Fig. 3, which is the result calculated in the software system.
"Name combination" refers to the hybrid combination who are their parents, such as "A11&A12" is the parents of the first hybrid combination, because the maize varieties need to keep secret, it just uses A11、B11 to substitute in the paper."order" refers to the hybrid combinations' rank."combination value" describes the development value of the hybrid combination, which is calculated by formula 8 in the paper. "very good, good, bad, poor" refer to the defect degree and excellent degree of the combinations. "comprehensive value" refers to the comprehensive advantages or distanvages of the combinations. "heterosis" refers to the additive gene effect, the bigger the better. "average molecular marker genetic distance of the parents" refers to the parent heterosis(MD). "special combining ability" refers to the productivity and other characters of the combinations. "Molecular marker genetic distance" measures the size of some characteristics of the comprehensive genetic differences among varieties, calculating by formula 4. "phenotypic distance" refers to the calculation of the quantitative traits. "row-column value, column-row value" refers to reciprocal value. "row-column flowering, column-row flowering, row-column height difference, column-row height difference" refer to the seed performance indices, aiming to establish the indices of double cross and reciprocal value.)

4   Discussions

In this paper, a kind of hybrid combination technology and method on maize was proposed, in which the main idea was:

Based on the SSR molecular detection results and field test results of current or previous years, first, the paper screened the inbred lines with high homozygote through

SSR detection results, then it screened the inbred lines with good comprehensive traits as the parents of the combination; then it calculated the genetic differences of both of the parents each other according to the SSR detection results, similarly, it calculated the phenotypic differences of both of the parents each other according to field testing results, thus we could calculate the heterosis rate of the parents to express their special combining ability; in the last according to the comprehensive characters, special combining ability, orthogonal anti value, the paper constructed the hybridization group model, and calculated the comprehensive characters, advantages, disadvantages of the hybridizations which the parents combined with each other and screened hybrid combinations to the next round field breeding.

With the combination technology methods applied, this paper simulated combined 37 hybridizations resistance to Ralstonia solanacearum, but the distribution group only based on one year's performance data of the parents. Whether the combination had good anti-bacterial blight resistance and comprehensive performance, it also needed hybrid combinations to prove with multiple years and points testing.

In this paper, the software system was developed suitable to the combination technology method, which supported the management of a quality of SSR molecular marker data, field testing data together with the calculation of massive indices, formulas and models, but the efficiency still needed to be improved.

It was also a molecular design breeding practice carried by the Beijing Kings Nower Seed S&T CO., LTD. There were some key technological differences here compared with traditional group technological process:

First, it did not consider the selection process of inbred lines, but directly determined the parents' selection, simulated distribution group through the molecular and phenotypic characters of the inbred lines. Second, in this paper it did not use the genetic background of inbred lines, genealogical relationship and other information, especially when calculating hybrid heterosis rate it was only based on genetic distance and phenotypic distance without analyzing group relationship of each inbred line. Thus, it was likely to emerge the phenomenon that the parents had a high heterosis rate but belonged to the same group, which did not match the actual combining ability. Third, the phenotypic information of the inbred lines were obtained through multiple years and points field identification, but at present the domestic breeding institutions have not regarded the inbred lines multi-point identification as a conventional breeding work, mainly inferred the possible performance of the parents through measuring the combination process. Fourth, there was a good effect on classification using inbred lines SSR data [12, 14, 16], but because this method did not do division group to inbred lines, so it was difficult to increase measurement-matching through selecting test species, which may have an effect on the simulated group. It was the focus of the method in next step.

5 Conclusions

In this paper, it made some new attempts which were as follows:

Firstly, in the method, it first attempted to calculate using SSR fingerprint data together with phenotypic data, in this way more information about the inbred lines were considered to get good maize hybrids.

Secondly, it attempted to get the phenotypic information of the inbred lines through multiple years and points field identification, but at present the domestic breeding institutions mainly inferred the possible performance of the parents through measuring the combination process.

Thirdly, the paper explored a set of design breeding method which included from the integrated management of maize varieties' molecular marker data and phenotypic data of field testing to maize inbred lines comprehensive evaluation, auxiliary screening. Besides, the design breeding method was also based on the heterosis analysis of inbred lines with molecular markers and phenotypic differences and the simulation distribution of maize hybrids and performance prediction. It also developed a software system of hybrid combination to support the technology route, providing methods and tools support to improve China's maize breeding scale and information management level. Applying the software, 37 hybrid combinations resistance to Ralstonia solanacearum were got with 179 inbred lines' molecular and phenotypic data. The method and software preliminary provides technological support for our country to carry out and perfect molecular design breeding.

Acknowledgment. This work is supported by the Chinese Universities Scientific Fund (Method and Software of hybridized Combination for Maize, 2013XJ022), and the National Science-technology Support Plan Projects (Research and Demonstration of North China Corn Commercialized Breeding Technique, 2014BAD01B01) and Key Laboratory of Agricultural Information Acquisition Technology, Ministry of Agriculture.

References

1. Jingrui, D., Lizhu, E.: Science and technological innovation of maize breeding in China. J. Maize Sci. **18**(1), 1–5 (2010)
2. Xiangtuo, L., Jianchang, M., Quanming, W.: Molecular markers and maize breeding. J. Maize Sci. **12**(1), 26–29 (2004)
3. Jiuran, Z., Fengge, W., Hongmei, Y., et al.: Progress of construction of Chinese maize varieties standard DNA fingerprint database. Crops **2**, 1–6 (2015)
4. Yanfang, Z.: Research on cultivar identification and DNA fingerprinting of crops based on molecular markers. Zhejiang university (2013)
5. Fengge, W., TianHongli, Z.J., et al.: Genetic diversity analysis of 328 maize varieties (hybridized combinations) using SSR markers. Scientia Agricultura Sinica **47**(5), 856–864 (2014)
6. Xue, L., Hongli, T., Fengge, W., et al.: Comparison of SSR and SNP markers in maize varieties genuineness identification. Mol. Plant Breed. **12**(5), 1000–1004 (2014)
7. Zheng, Z., Lingyan, L., Gaohong, W., et al.: The application of molecular markers in maize breeding. Curr. Biotechnol. **5**(4), 259–264 (2015)
8. Zhe, L., Jianyu, Y., Shaoming, L., et al.: Optimal method of transforming observables into relative values for multi-environment trials in maize. Trans. CSAE **27**(7), 205–209 (2011). (in Chinese with English abstract)
9. Changsheng, L., Muhammad, I., et al.: Genetic diversity analysis of maize verities based on SSR markers. Res. J. Biotechnol. **9**(6), 48–51 (2014)

10. Song, X.F., Song, T.M., Dai, J.R., et al.: QTL mapping of kernel oil concentration with high-oil maize by SSR markers. Maydica **49**(1), 41–48 (2004)

11. Pegic, I., Ajmone Marsan, P., Morgante, M., et al.: Comparative analysis of genetic similarity among maize inbred lines detected by RFLPs, RAPDs, SSRs, and AFLPs. Theor. Appl. Genet. **97**(8), 1248–1255 (1998)

12. Senior, M.L., Murphy, J.P., Goodman, M.M., et al.: Utility of SSRs for determining genetic similarities an relationships in maize using an agarose gel system. Crop Sci. **38**(4), 1088–1998 (1998)

13. Enoki, H., Sato, H., Koinuma, K.: SSR analysis of genetic diversity among maize inbred lines adapted to cold regins of Japan. Theor. Appl. Genet. **104**(8), 1270–1277 (2002)

14. Reif, J.C., Melchinger, A.E., Xia, X.C., et al.: Use of SSRs for establishing heterotic groups in subtropical maize. Theor. Appl. Genet. **107**(5), 947–957 (2003)

15. Xia, X.C., Reif, J.C., Hoisington, D.A., et al.: Genetic diversity among CIMMYT maize inbred lines investigated with SSR markers. Crop Sci. **44**(6), 2230–2237 (2004)

16. Wende, A., Shimelis, et al.: Genetic interrelationships among medium to late maturing tropical maize inbred lines using selected SSR markers. Euphytica **191**(2), 269–277 (2013)

17. Kumar, A., Rakshit, A., Mangilipelli, N.K., et al.: Genetic diversity of maize genotypes on the basis of morpho-physiological and simple sequence repeat (SSR) markers. Afr. J. Biotechnol. **11**(99), 16468–16477 (2012)

Biomass-Based Leaf Curvilinear Model for Rapeseed (*Brassica napus* L.)

Wenyu Zhang, Weixin Zhang, Daokuo Ge, Hongxin Cao[✉],
Yan Liu, Kunya Fu, Chunhuan Feng, Weitao Chen, and Chuwei Song

Institute of Agricultural Economics and Information/Engineering
Research Center for Digital Agriculture,
Jiangsu Academy of Agricultural Sciences, Nanjing 210014, China
research@wwery.cn, nkyzwx@126.com, gedakuo@163.com,
caohongxin@hotmail.com, liuyan0203@aliyun.com,
921186907@qq.com, 1286234727@qq.com,
1303079141@qq.com, 923903764@qq.com

Abstract. Leaf is one of the most important photosynthetic organs of rapeseed (*Brassica napus* L.). To quantify relationships between the leaf curve and the corresponding leaf biomass for rapeseed on main stem, this paper presents a biomass-based leaf curvilinear model for rapeseed. Various model variables, including leaf length, bowstring length, tangential angle, and bowstring angle, were parameterized based on data derived from the field experiments with varieties, fertilizer, and transplanting densities during 2011 to 2012, and 2012 to 2013 growing seasons. And then we analysed the biological significance of curvilinear equation for straight leaves, constructed the straight leaf probabilistic model on main stem, quantified the relationship between leaf curvature and the corresponding leaf biomass, and constructed the leaf curvilinear model based on the assumption and verification of the curvilinear equation form for curving leaf. The probability of straight leaf can be quantified with piecewise function according to the different trend in the normalized leaf ranks ((0, 0.4], and (0.4, 1]). The leaf curvature decreased with the increasing of leaf biomass, and can be described with reciprocal function. The curve of straight leaf and the curving leaf can be simulated by linear equation and the quadratic function, respectively. Our models were validated with the independent dataset from the field experiment, and the results indicated that the model could effectively predict the straight leaf probability and leaf curvature, which would be useful for linking the rapeseed growth model with the rapeseed morphological model, and set the stage for the development of functional-structural rapeseed models.

Keywords: Rapeseed (*Brassica napus* L.) · Biomass · Leaf curve · Functional-structural plant models (FSPMs)

1 Introduction

Rapeseed is the world's important oil crops [1] with harvest area of 25.3 to 30.9 million ha and total yield of $46.5 \sim 72.5$ million tons during $2004 \sim 2013$ [2]. At the same time, it is the main oilseed crop in China [3], whose harvest area is about $5.6 \sim 7.5$ million ha,

D. Li and Z. Li (Eds.): CCTA 2015, Part I, IFIP AICT 478, pp. 459–472, 2016.
DOI: 10.1007/978-3-319-48357-3_44

and the total yield is about $10.6 \sim 14.4$ million tons [2] in general. Also, it is one of the main raw material of biodiesel [4]. Therefore, it is very important for ensure food and ecological security that promote the development of rapeseed production.

Light distribution characteristics in crop canopies directly affect the light energy utilization efficiency for photosynthesis, dry matter accumulation, and yield formation. All most all the growth models predicted the crop canopy light distribution through the Beer's law [5–8], in that the two key factors for light distribution simulation process, the extinction coefficient and the layered leaf area index, are closely related with the leaf curving characteristics [9]. Therefore, quantitatively modeling of the leaf curve could provide a mechanistic way for precisely simulating the crop canopy structure, light distribution, and photosynthesis, and lay a foundation for the predicting of light energy utilization efficiency and yield formation.

At present, there are many studies on leaf curve modeling. In the study of mathematical characterization of maize canopies [10], leaf curve was described as a general quadratic equation expressed by the initial leaf angle, the coordinates of the leaf tip and the leaf's maximum height. The general quadratic equation was also used to simulate the leaf curve of maize [11–13], rice [14], and other crops by many researchers. Leaf curve was also fitted into a quadratic function for rice [15, 16] and winter wheat [17], or a Gaussian function for spring barley [18] and rice [19]. Furthermore, Espana et al. [20] decomposed leaf curvature into two parts, the ascending part was described as a parabolic curve, and the descending part, when existing, was characterized by a portion of an ellipse, and then applied the leaf curvature model to maize canopy 3D architecture and reflectance simulation. Watanabe et al. [21] found that leaf curves could be fitted using Hermite functions though analyzing three angles related to the basal, mid, and tip of leaf. Shi et al. [22] characterized the rice leaf curve by a second order differential equation, including the synthetic effect of leaf blade length, width, specific leaf weight, initial leaf angle, and the deformation coefficient on leaf space shape, using force analyzing on rice leaf. Zheng et al. [23] obtain leaf midrib coordinate points by cubic spatial B-spline interpolation, and characterized the leaf curves as the connecting line of these points.

It is difficult to measure leaf curve because of the intricate leaf shape for rapeseed. Therefore, the objectives of this research were to develop straight leaf probabilistic model, straight leaf curvilinear model, and biomass-based leaf curvilinear model by linking leaf morphological parameters with the corresponding leaf biomass, to validate the hypothesis that the curvilinear function for curving leaves could be fitted as quadratic function, and to provide a reference for linking morphological parameters with corresponding organ biomass, and for the establishment of the FSPMs.

2   Materials and Methods

2.1   Materials

We used 2 rapeseed cultivars, they are: "Ningyou 18" (V1, conventional), and "Ningza 19" (V2, hybrid), breed by Institute of Economic Crops Research, Jiangsu Academy of Agricultural Sciences.

2.2 Methods

2.2.1 Experimental Conditions and Design

In order to determine the parameters and verify the models, three experiments were conducted involving different varieties, transplanting densities, and fertilizer during the 2011–2012, and 2012–2013 growing seasons at the experimental farm of our Academy (32.03°N, 118.87°E). The soil type is a hydragric anthrosol (organic carbon, 31.4 g kg^{-1}; total nitrogen, 2.03 g kg^{-1}; available phosphorus, 20.3 mg kg^{-1}; available potassium, 139.0 mg kg^{-1}; and pH 7.31).

Exp. 1, variety and the fertilizer experiment (2011–2012): The Experiment was deployed in split block design with three replications. Two fertilizer levels (N0 = no fertilizer; N2 = 180 kg ha^{-1}) were the whole-plot treatments while two cultivars (V1 and V2) constituted the sub-plots. The plots arranged random with 0.4 m row spacing, 0.17–0.20 m plant spacing in 7.0 × 5.7 m area. Fertilizer contained 12 kg P_2O_5 ha^{-1},18 kg K_2O ha^{-1}, and 15 kg boron ha^{-1}.

Exp. 2, variety experiment (2012–2013): The experiment was deployed in randomized complete block design with 2 varieties (V1 and V2) and 3 replications. Nitrogen fertilizer included 90 kg N ha^{-1}(N1), and the transplanting density was 1.2 × 10^5 plant ha^{-1}(D2).

Exp. 3, variety, fertilizer and transplanting density experiment (2012–2013): The Experiment was deployed in split block design with three replications. Three fertilizer levels (N0 = no fertilizer; N1 = 90 kg ha^{-1}; N2 = 180 kg ha^{-1}) were the whole-plot treatments while variety (V1) and three transplanting densities (D1 = 6 × 10^4 plant ha^{-1};   D2 = 1.2 × 10^5 plant ha^{-1};   D3 = 1.8 × 10^5 plant ha^{-1})   constituted   the sub-plots.

The plots of Exp. 2 and Exp. 3 arranged random with 0.42 m row spacing in 3.99 by 3.5 m area, and the plant spacing was calculated by row spacing and transplanting density. Fertilizer contained 90 kg P_2O_5 ha^{-1} and 90 kg K_2O ha^{-1} for N1 plots, and 180 kg P_2O_5 ha^{-1} and 180 kg K_2O ha^{-1} for N2 plots, and 15 kgboron ha^{-1} was used as foliage spray for both N1 and N2 plots after bolting.

2.2.2 Measurements

The leaf rank on the main stem of 50 randomly selected seedlings for each plot were marked using a red number stamp before transplanting. Leaf morphological parameters including leaf length (the distance between leaf basal and leaf tip in the straight state, including the leaf blade and petiole, if it exists. LL, for short), leaf tangential angle (the angle between the tangential direction of leaf basal and the main stem. TA, for short), leaf bowstring angle (the angle between the straight line from leaf basal to leaf tip in natural state and the main stem. BA, for short), and leaf bowstring length (the distance between leaf basal and leaf tip in natural state. LBL, for short) were measured using straightedge and protractor directly (Fig. 1).

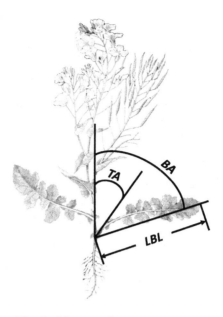

Fig. 1. Diagram of *TA*, *BA* and *LBL*

2.2.3 Data Analysis

We wrote a C# Program to solve the approximate solution of leaf curve equation parameters through a step of 10^{-5} cm. Leaf rank data were normalized to (0, 1] interval, in order to eliminate the apparent differences between treatments and replications. The data from Exp. 1 was used for model development, data from Exp. 2, and Exp. 3 were used for validation.

2.2.4 Model Validation

We validated the models developed in this paper by calculating the correlation (r), the root mean square error (*RMSE*), the average absolute difference (d_a), and the ratio of d_a to the average observation (d_{ap}) [24], and 1:1 line of simulated and observed properties. Some statistical indices were defined as follows:

$$RMSE = \sqrt{\frac{\sum_{i=1}^{n}(O_i - S_i)^2}{n}}$$

$$d_a = \frac{1}{n}\sum_{i=1}^{n}(O_i - S_i)$$

$$d_{ap}(\%) = |d_a|/\bar{O} \times 100$$

where i is sample number, n is total number of measurements, $n - 1 = n$ when $n \geq 30$, S_i is simulated value, and O_i is observed value.

3   Results

3.1   Model Description

3.1.1   Probability and Curvilinear Model for Straight Leaves

In order to represent the extension state of leaf better, we set the leaf in the Cartesian coordinate system with the leaf basal point as the origin and growth direction of main stem as y-axis, regardless of leaf distorting. According to observations in the Exp. 1, some rapeseed leaves could be considered as straight leaves with small difference between leaf length and leaf bowstring length, as well as leaf tangential angle and leaf bowstring angle. Therefore, we treated the leaves as straight leaves if difference between the leaf length and the leaf bowstring length was less than 1 cm or difference between the leaf tangential angle and the leaf bowstring angle is less than 10°. The curvilinear equation ($f(x)$) of straight leaves could be expressed as a linear function which passes through the origin and with the cotangent value of the leaf tangential angle as the slope, and the function could be described by Eq. (1).

$$f(x) = \cot(TA) \cdot x \tag{1}$$

where TA can be simulated by our previous model [25].

The data in the Exp. 1 showed that changes in the probability of different treatments for straight leaves by the normalized leaf ranks was close to quadratic curve with significant r ($r = 0.725$, $P < 0.01$, $n = 16$, $r_{(14,\ 0.01)} = 0.623$, Table 1) in the interval $(0, 0.4]$ (Fig. 2a), and logarithmic curve with significant r ($r = 0.925$, $P < 0.001$, $n = 33$, $r_{(31,\ 0.001)} = 0.547$, Table 1) in the interval $(0.4, 1]$ (Fig. 2b). So that the probability of straight leaves by the normalized leaf ranks (P_{LS}) could be expressed as a piecewise function as Eq. 2.

$$P_{LS} = \begin{cases} A_1 \cdot NLRs^2 + B_1 \cdot NLRs + C_1 & NLRs \in (0, 0.4] \\ A_2 \cdot \ln(NLRs) + B_2 & NLRs \in (0.4, 1] \end{cases} \tag{2}$$

where $NLRs$ is normalized leaf rank; A_1, A_2, B_1, B_2, and C_1 are model parameters whose values and testing data shown in Table 1.

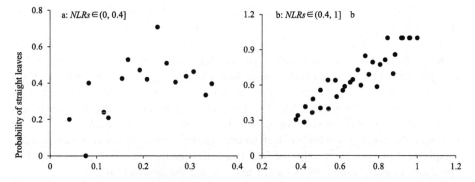

Fig. 2. Changes in the probability of straight leaves with the normalized leaf ranks

Table 1. Significance test of the straight leaf probabilistic model and its parameters

Interval of normalized leaf ranks	Function	n	r	Sig. for r	Sig. for F	Parameter symbolic	Unstandardized coefficients	t
(0, 0.4]	$A_1 \cdot NLRs^2 + B_1 \cdot NLRs + C_1$	16	0.725**	$r_{(14,\ 0.01)} = 0.623$	0.008	A_1	−10.519	−2.709*
						B_1	1.572	3.199**
						C_1	0.139	−0.800
(0.4, 1]	$A_2 \cdot \ln(NLRs) + B_2$	33	0.925***	$r_{(31,\ 0.001)} = 0.547$	0.000	A_2	0.716	13.575***
						B_2	0.965	36.223***

***, **, and * denote significance at $P < 0.001$, $P < 0.01$, and $P < 0.05$, respectively. The same as below.

3.1.2 Curvilinear Model for Curving Leaves

3.1.2.1 Assumed Functions and There Biological Significance for Leaf Curve

According to the leaves curving characteristics observed in experiments, and the research on curvilinear models for other crops, we supposed that curvilinear function of rapeseed leaf ($f(x)$) could be fitted by functions such as quadratic function ($ax^2 + bx$), quartic function ($ax^4 + bx^3 + cx^2 + dx$), sine function ($a\sin(bx)$), Hoerl-like function (axb^x), $(a + bx^c)/(d + x^c)$, and $(a + bx)/(1 + cx + dx^2)$, whose diagram were similar to leaf curves in a interval. On the basis of geometric meaning of derivative, $\cot(TA)$ can be interpreted as $f'(0)$ (value of leaf curvilinear equation's derived function at origin of coordinates), thus, biological significance of leaf curvilinear equation parameters are showed in Table 2.

Table 2. The various functions of leaf curve, their derivative function and the derivative value at the origin, and the biological significance of the parameters

$f(x)$	$f'(x)$	$f'(0)$	Biological significance of the parameters
$ax^2 + bx$	$2ax + b$	b	$-a$: curvature; $b = \cot(TA)$
$ax^4 + bx^3 + cx^2 + dx$	$4ax^3 + 3bx^2 + 2cx + d$	d	$d = \cot(TA)$
$a\sin(bx)$	$ab\cos(bx)$	ab	leaf curve peak: $(\pi/2b, a)$; $ab = \cot(TA)$
axb^x	$ab^x\log(b)x + ab^x$	a	$a = \cot(TA)$; $-b$: curvature
$(a + bx^c)/(d + x^c)$	$\frac{bcx^{c-1}}{x^c+d} - \frac{cx^{c-1}(bx^c + a)}{(x^c + d)^2}$	0	N/A
$(a + bx)/$ $(1 + cx + dx^2)$	$\frac{b}{dx^2 + cx + 1} - \frac{(bx+a)(2dx+c)}{(dx^2 + cx + 1)^2}$	b $-ac$	$b - ac = \cot(TA)$

From Table 2, we saw that only quadratic function, sine function, and Hoerl-like function had specific biological significance for all the parameters: for quadratic function $ax^2 + bx$, $-a$ expresses the leaf curvature, and b expresses the cotangent value of leaf tangential angle; for sine function $a\sin(bx)$, a expresses the ordinate value of leaf curve peak point, $\pi/2b$ abscissa value of expresses the leaf curve peak point, and ab expresses the cotangent value of leaf tangential angle; for Hoerl-like function axb^x, a expresses the cotangent value of leaf tangential angle, and $-b$ expresses the leaf curvature. So that we addressed them and validated the assumptions.

3.1.2.2 Solution of the Leaf Curvilinear Equation

As shown in Fig. 3, the leaf tip could be expressed as $(\sin(BA) \cdot LBL, \cos(BA) \cdot LBL)$ by solving $\triangle OL_t y_t$. We substituted coordinates of origin and L_t into curvilinear function for curving leaves to get equation set as Eq. 3.

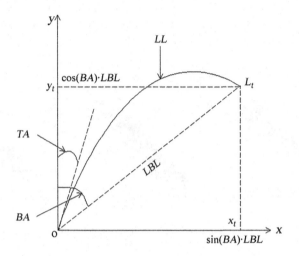

Fig. 3. The geometrical properties of leaf curve

$$\begin{cases} f(\sin(BA) \cdot LBL) = \cos(BA) \cdot LBL \\ f'(0) = \cot(TA) \end{cases} \tag{3}$$

The equation sets and their solutions corresponding to the three curvilinear functions for curving leaves we supposed above were shown in Table 3.

Table 3. Three curvilinear functions for curving leaves and the corresponding equation sets and their solutions

$f(x)$	Equation set	Solution
$ax^2 + bx$	$\begin{cases} f(\sin(BA) \cdot LBL) = a \cdot \sin(BA)^2 \cdot LBL^2 + b \cdot LBL \cdot \sin(BA) = \cos(BA) \cdot LBL \\ f'(0) = b = \cot(TA) \end{cases}$	$a = -\frac{\sin(BA) \cdot \cot(TA) - \cos(BA)}{LBL \cdot \sin(BA)^2}$ $b = \cot(TA)$
$a\sin(bx)$	$\begin{cases} f(\sin(BA) \cdot LBL) = a \cdot \sin(b \cdot LBL \cdot \sin(BA)) = \cos(BA) \cdot LBL \\ f'(0) = a \cdot b = \cot(TA) \end{cases}$	$a = \frac{LBL \cdot \cos(BA)}{\sin(b \cdot LBL \cdot \sin(BA))}$ $a \cdot b = \cot(TA)$
axb^x	$\begin{cases} f(\sin(BA) \cdot LBL) = a \cdot LBL \cdot \sin(BA) \cdot b^{LBL \cdot \sin(BA)} = \cos(BA) \cdot LBL \\ f'(0) = a = \cot(TA) \end{cases}$	$a = \cot(TA)$ $b = \left(\frac{\cos(BA)}{\sin(\sin(BA) \cdot \sin(TA))}\right)^{\frac{1}{\sin(BA) \cdot LBL}}$

As shown in Table 3, quadratic function, and Hoerl-like function could be solved directly, but there was no analytical solution for sine function, and we wrote a C# program to calculate the approximate solution with a step of 10^{-5} cm.

3.1.2.3 Validation for Leaf Curvilinear Functions

We used the observed value of tangential angle, bowstring angle, and bowstring length for solving equation set, except leaf length. Meanwhile, the leaf length also could be calculated as the arc length between leaf basal and leaf tip by the formula of arc length. Therefore, we can validate the leaf curvilinear functions through comparing the observed leaf length with the calculated arc length. The three leaf curvilinear functions were validated by 1:1 line of observed and calculated leaf length (Fig. 4), and the statistical parameters were shown in Table 4.

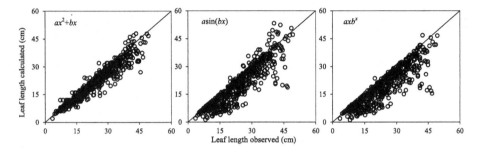

Fig. 4. Comparison of the observed and the stimulated leaf length on main stem in 2011–2012

Table 4. Comparison of statistical parameters of the observed and the stimulated leaf blade length on main stem in 2011–2012

$f(x)$	Statistic parameters of simulation and observation					
	n	d_a(cm)	d_{ap}(%)	$RMSE$(cm)	r	Sig.
$ax^2 + bx$	509	2.521	10.426	3.974	0.956[***]	$r_{(507,0.001)} = 0.145$
$a\sin(bx)$	509	3.730	15.427	6.083	0.904[***]	$r_{(507,0.001)} = 0.145$
axb^x	509	4.693	19.410	6.749	0.889[***]	$r_{(507,0.001)} = 0.145$

The results showed that the best fitted one (with the highest r and lowest d_a,d_{ap}, and $RMSE$, Table 4) with specific biological significance (Table 2) for leaf curve was quadratic function, so that the curvilinear model for curving leaves could be described as a quadratic function as Eq. (4).

$$f(x) = -Lc_i \cdot x^2 + \cot(TA) \cdot x \qquad (4)$$

where, Lc_i is the leaf curvature on the ith day after emergence.

3.1.2.4 Biomass-Based Leaf Curvature Simulation

The data in the experiments showed that quadratic coefficients of leaf curvilinear equations increased with the increase of corresponding leaf dry weight from leaf fully expanded until senescence. It means that leaf curvature decreased with the corresponding leaf biomass (Fig. 5), and could by fitted by reciprocal function with significant r (V1: $r = 0.974$, $P < 0.001$, $n = 34$, $r_{(32,\ 0.001)} = 0.539$; V2: $r = 0.637$, $P < 0.001$, $n = 51$, $r_{(49,\ 0.001)} = 0.447$. Table 5).

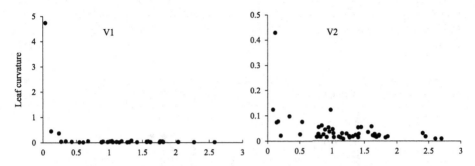

Fig. 5. Changes in the leaf curvature with the leaf dry weight in 2011–2012

Table 5. Significance test of the leaf curvature model and its parameter

Cultivar	n	r	Sig. for r	Sig. for F	Parameter symbolic	Unstandardized coefficients	t
V1	34	0.974^{***}	$r_{(32,\ 0.001)} = 0.539$	0.000	Lc_a	−0.165	-24.127^{***}
					Lc_b	0.163	4.625^{***}
V2	51	0.637^{***}	$r_{(49,\ 0.001)} = 0.447$	0.000	Lc_a	−0.018	-5.789^{***}
					Lc_b	−0.015	−1.826

$$Lc_i = \frac{-Lc_a}{DWLB_i} - Lc_b \tag{5}$$

where $DWLB_i$ is the dry weight on the ith day after emergence; Lc_a and Lc_b are model parameters whose values and testing data shown in Table 5.

3.2 Validation

The models developed above were validated were validated with the independent datasets from Exp. 2 and Exp. 3, and the results showed that the correlation (r) of simulation and observation probability for straight leaves and curvature for curving leaves all had significant level at $P < 0.001$, and that the average absolute difference (d_a), the ratio of d_a to the average observation (d_{ap}), and the root mean square error ($RMSE$) all were smaller (Table 6). Figures 6 and 7 indicated that the observed and simulated probability for straight leaves and curvature for curving leaves were all close to the 1:1 line.

Table 6. Comparison of statistical parameters of simulation and observation in the probability for straight leaves and curvature for curving leaves on main stem in 2012–2013

Models	Cultivar	Statistic parameters of simulation and observation					
		n	d_a	$d_{ap}(\%)$	RMSE	r	Sig.
Probability for straight leaves	V1, V2	52	0.001	0.245	0.191	0.762***	$r_{(50,0.001)} = 0.443$
Curvature for curving leaves	V1	93	0.007	−9.196	0.060	0.648***	$r_{(91,0.001)} = 0.336$
	V2	99	−0.005	−7.000	0.042	0.541***	$r_{(97,0.001)} = 0.326$

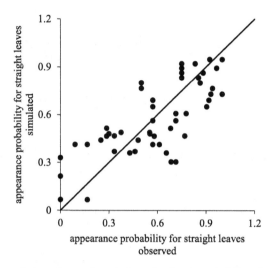

Fig. 6. Comparison of the observed and the stimulated probability for straight leaves in 2012–2013

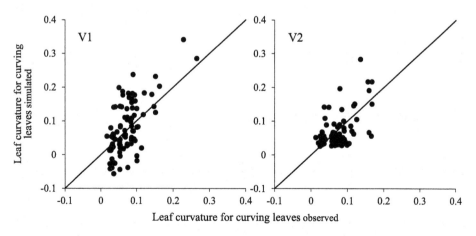

Fig. 7. Comparison of the observed and the stimulated leaf curvature for curving leaves in 2012–2013

4   Discussion

4.1   Establishment of Biomass-Based Leaf Curvilinear Model will Helps to Develop FSRM

Functional-structural plant model with better mechanistic embodies the interaction between plant morphogenesis and cultivars and environmental factors by integrated the functions of growth model with the structures of morphological model [26]. So far, there were lot of research on modeling the leaf curve [10–21] which aimed at 3D reconstruction of crop canopy with a high precision. They well explained the effects of structures on functions by combining technical instrumentalities like geometrical ray trace [27, 28], but could not change the structures as a response of the changed functions. Shi et al. [22] linked leaf curve with corresponding dry weight through force analyzing on rice leaf, but the leaf curvilinear equation which was limited by a form of second order differential equation in this research was difficult to use. Groer et al. [29] established a dynamic 3D model of rapeseed using the modelling language XL [30] and made the morphological model response of different nitrogen levels using a sinks and sources system like GREENLAB and the LEAFC3-N model [31] for photosynthesis at different N-regimes. But the time from sowing to the rosette stage and the conditions except nitrogen was neglected in this model. Jullien et al. [32] constructed a FSPM by characterizing the interactions between architecture and source-sink relationships in winter oilseed rape using the GREENLAB model. However, it mainly considered the relationship between biomass and leaf area [33], and the description of blade shape was relatively simple. Cao et al. [34] and Zhang et al. [25] established models to meticulous predict leaf morphological parameters like leaf length, width, and angles, apart from leaf curve. We developed biomass-based leaf curvilinear model for rapeseed by linking the leaf curvature with the corresponding leaf biomass, which could realize combination of structures with functions, explained effects of environmental conditions on leaf morphogenesis, and set the stage for the development of functional-structural rapeseed models.

4.2   The Research Provided a Mechanistic and Universal Method for Leaf Curve Modeling

As to facilitate observation and simulation, we set the leaf into a Cartesian coordinate system with the leaf basal point as the origin and growth direction of main stem as y-axis. To make the model more precisely, two cases of curving characteristic (straight or curving) were analyzed. In order to determining the form of leaf curvilinear equation to be difficult to measure directly, we compared various functions based on their biological significance, and calculated and observed leaf length. For eliminating the apparent differences between treatments and replications, we normalized the leaf rank into the interval of (0, 1]. All these practices made the model to have better mechanistic and universal.

4.3   Models Developed in This Paper Needs to Be Improved

The biomass-based rapeseed leaf curvilinear model developed in this paper for the stage from leaf fully expanded until senescence, but the processes of leaf extension, senescence, and distorting is neglected. Thus, it needs to be studied further.

5   Conclusions

This paper presents a biomass-based leaf curvilinear model for rapeseed designed to explain effects of cultivars and environmental conditions on leaf curve. Various model variables, including leaf biomass, length, and angles were parameterized based on datasets derived from the experiments with rapeseed cv. Ningyou 18, and Ningza 19.

With the help of our descriptive model, it will be easy to fulfill calculation of leaf curve via biomass, canopy structure via leaf curve, light distribution and the photosynthesis via canopy structure, and biomass via photosynthesis. It should be possible to connection morphological model with physiological model via biomass, and to development the FSRM.

A similar method could also be applied in other crops like maize and rice, and it could help to the regulation and selection for ideal plant type in the future.

Acknowledgments. This work is supported by the grants from the National Natural Science Foundation of China (31171455, 31201127, and 31471415), the National High-Tech Research and Development Program of China (2013AA102305-1), the Jiangsu Agriculture Science and Technology Innovation Fund of China [CX(12)5060], and Jiangsu Academy of Agricultural Sciences Basic Scientific Research Work Special Fund [ZX(15)2008, ZX(15)4001].

References

1. Cao, H., Zhang, C., Li, G., et al.: Researches of optimum shoot and ramification number dynamic models for rapeseed (*Brassica napus* L.). In: World Automation Congress (WAC), pp. 129–135. IEEE (2010)
2. FAO Statistics Division. FAOSTAT [DB/OL] (2015). http://faostat.fao.org/
3. Cao, H., Zhang, C., Li, G., et al.: Researches of optimum leaf area index dynamic models for rape (*Brassica napus* L.). In: Zhao, C., Li, D. (eds.) Computer and Computing Technologies in Agriculture II, Volume 3. IFIP AICT, vol. 295, pp. 1585–1594. Springer, New York (2009)
4. Hama, S., Kondo, A.: Enzymatic biodiesel production: an overview of potential feed stocks and process development. Bioresour. Technol. **135**(2), 386–395 (2013)
5. Kropff, M.J., Laar, H.H.V., Matthews, R.B.: ORYZA1: an ecophysiological model for irrigated rice production. In: SARP Research Proceedings, p. 1994 (1994)
6. Boote, K.J., Pickering, N.B.: Modeling photosynthesis of row crop canopies. HortScience **29**(12), 1423–1434 (1994)
7. Gao, L., Jin, Z., Zheng, G., et al.: Wheat cultivational simulation-optimization-decision making system (WCSODS). Jiangsu J. Agric. Sci. **16**(02), 65–72 (2000). (in Chinese)

8. Tang, L., Zhu, Y., Hannaway, D., et al.: RiceGrow: a rice growth and productivity model. NJAS - Wageningen J. Life Sci. **57**(1), 83–92 (2009)
9. Wang, W., Li, Z., Su, H.: Comparison of leaf angle distribution functions: effects on extinction coefficient and fraction of sunlit foliage. Agric. For. Meteorol. **143**(1–2), 106–122 (2007)
10. Stewart, D.W., Dwyer, L.M.: Mathematical characterization of maize canopies. Agric. For. Meteorol. **66**(3–4), 247–265 (1993)
11. Guo, Y., Li, B.: Mathematical description and three-dimensional reconstruction of maize canopy. Chin. J. Appl. Ecol. **10**(01), 39–41 (1999). (in Chinese)
12. Zhao, C., Zheng, W., Guo, X., et al.: The computer simulation of maize leaf. J. Biomath. **19** (04), 493–496 (2004). (in Chinese)
13. Zheng, W., Guo, X., Zhao, C., et al.: Geometry modeling of maize leaf canopy. Trans. CSAE **20**(01), 152–154 (2004). (in Chinese)
14. Liu, H., Wu, B., Zhang, H., et al.: Research on rice leaf geometric model and its visualization. Comput. Eng. **35**(23), 263–264+268 (2009). (in Chinese)
15. Yang, H., Luo, W., He, H., et al.: 3D morphology modeling and computer simulation of rice main stem. Acta Agriculturae Universitis Jiangxiensis **30**(06), 1153–1156+1160 (2008). (in Chinese)
16. Luo, W., Yang, H., Deng, S., et al.: A study based on multi-segment and proportion geometrical model of rice leaf and 3D morphology modeling. Acta Agriculturae Universitis Jiangxiensis **31**(05), 970–974 (2009). (in Chinese)
17. Chen, G.: Studies on Simulation and Visualization of Morphogenesis in Wheat. Nanjing Agricultural University, Nanjing (2004). (in Chinese)
18. Dornbusch, T., Wernecke, P., Diepenbrock, W.: Description and visualization of graminaceous plants with an organ-based 3D architectural model, exemplified for spring barley (Hordeumvulgare L.). Vis. Comput. **23**(8), 569–581 (2007)
19. Zhang, Y., Tang, L., Liu, X., et al.: Dynamic simulation of leaf curve in rice based on Gaussian function. Sci. Agric. Sin. **46**(01), 215–224 (2013). (in Chinese)
20. España, M.L., Baret, F., Aries, F., et al.: Modeling maize canopy 3D architecture: application to reflectance simulation. Ecol. Model. **122**(1–2), 25–43 (1999)
21. Watanabe, T., Hanan, J.S., Room, P.M., et al.: Rice morphogenesis and plant architecture: measurement, specification and the reconstruction of structural development by 3D architectural modelling. Ann. Bot. **95**(7), 1131–1143 (2005)
22. Shi, C., Zhu, Y., Cao, W.: A quantitative analysis on leaf curvature characteristics in rice. Acta Agron. Sin. **32**(05), 656–660 (2006). (in Chinese)
23. Zheng, B., Shi, L., Ma, Y., et al.: Three-dimensional digitization in situ of rice canopies and virtual stratified-clipping method. Sci. Agric. Sin. **42**(04), 1181–1189 (2009). (in Chinese)
24. Cao, H., Hanan, J.S., Liu, Y., et al.: Comparison of crop model validation methods. J. Integr. Agric. **11**(8), 1274–1285 (2012)
25. Zhang, W., Cao, H., Zhu, Y., et al.: Morphological structure model of leaf space based on biomass at pre-overwintering stage in rapeseed (*Brassica napus* L.) plant. Acta Agron. Sin. **41**(02), 318–328 (2015). (in Chinese)
26. Cao, H., Zhao, S., Ge, D., et al.: Discussion on development of crop models. Sci. Agric. Sin. **44**, 3520–3528 (2011). (in Chinese)
27. Chelle, M., Saint-Jean, S.: Taking into account the 3D canopy structure to study the physical environment of plants: the Monte Carlo solution. In: 4th International Workshop on Functional-Structural Plant Models, pp. 176–180 (2004)
28. Ma, Y., Wen, M., Li, B., et al.: Efficient model for computing the distribution of direct solar radiation in maize canopy at organ level. Trans. CSAE **23**(10), 151–155 (2007). (in Chinese)

29. Groer, C., Kniemeyer, O., Hemmerling, R., et al.: A dynamic 3D model of rape (Brassica napus L.) computing yield components under variable nitrogen fertilization regimes. In: 5th International Workshop on Functional-Structural Plant Models, pp. 4.1–4.3, November 2007

30. Kniemeyer, O.: Rule-based modelling with the XL/GroIMP software. In: Proceedings of 6th GWAL, pp. 56–65, 14–16 April 2004

31. Müller, J., Wernecke, P., Diepenbrock, W.: LEAFC3-N: a nitrogen-sensitive extension of the CO2 and H2O gas exchange model LEAFC3 parameterised and tested for winter wheat (Triticumaestivum L.). Ecol. Model. **183**, 183–210 (2005)

32. Jullien, A., Mathieu, A., Allirand, J.-M., et al.: Characterization of the interactions between architecture and source-sink relationships in winter oilseed rape (*Brassica napus*) using the GreenLab model. Ann. Bot. **107**(5), 765–779 (2011)

33. Jullien, A., Allirand, J.-M., Mathieu, A., et al.: Variations in leaf mass per area according to N nutrition, plant age, and leaf position reflect ontogenetic plasticity in winter oilseed rape (Brassica napus L.). Field Crops Res. **114**(2), 188–197 (2009)

34. Cao, H., Zhang, W., Zhang, W., et al.: Biomass-based rapeseed (*Brassica napus* L.) leaf geometric parameter model. In: Proceedings of the 7th International Conference on Functional-Structural Plant Models, p. 26, 9–14 June 2013

Exploring the Effect Rules
of Paddy Drying on a Deep Fixed-Bed

Danyang Wang[1,2(✉)], Chenghua Li[3(✉)], Benhua Zhang[1],
and Ling Tong[1]

[1] College of Engineering, Shenyang Agricultural University,
Shenyang 110866, China
[2] Institute of Agricultural Product Processing,
Chinese Academy of Agricultural Engineering, Beijing 100125, China
[3] College of Mechanical Engineering, Shenyang Ligong University,
Shenyang 100168, China

Abstract. In this paper, a series of paddy drying experiments were conducted on a deep fixed-bed and we investigated the effect rules of five influencing parameters on drying time of paddy. By using the quadratic orthogonal rotation combination design, the nonlinear function between the drying time and the five influencing parameters are built up. Then a detailed study of the qualitative and quantitative effect rules of each influencing factor is elaborated. The results of this paper conclude the rules of how drying parameters influencing the dry time of paddy. And also, it reveals how to reduce paddy drying time and improve the productivity, which have significance in practical productions.

Keywords: Drying time · Paddy · Deep fixed-bed drying · Drying parameters

1 Introduction

Drying process is very important to post-harvest paddy rice, and also is the weakest link in modernization technology for grain storage. It costs relatively more energy in paddy rice production, and has a complicated exchange process of heat and moisture. Due to intricate interaction of many influence factors, the system of paddy rice drying process is of multifactor and multicriteria. Therefore, analyzing the effect rules of the drying process to obtain low consumption and quality drying parameters is a hard and important work. Drying time is a direct measure for drying energy consumption and drying efficiency, and it is a important influence factor of drying quality. GB1350-1999 stipulates that the standard moisture content of paddy rice purchasing is 14.5 % (w.b), and the sale total would be reduce 0.75 % if moisture content of paddy rice is per more than 5 %. Besides the end moisture content of paddy rice drying, drying energy and drying quality can be controlled through controlling the drying time [1, 12]. However, empirical drying time data in the past is relatively single, and is hard to the requirements of regional planting for paddy rice. Furthermore, the data is more difficult to adapt to many kinds of new drying equipment and many new drying technology. So, this paper take drying time as evaluation index, and the laws of drying parameters effecting on the drying time is discussed, which would Provide the basis to improve the production efficiency and economic indicators assessment.

© IFIP International Federation for Information Processing 2016
Published by Springer International Publishing AG 2016. All Rights Reserved
D. Li and Z. Li (Eds.): CCTA 2015, Part I, IFIP AICT 478, pp. 473–484, 2016.
DOI: 10.1007/978-3-319-48357-3_45

2 Experimental Materials and Methods

2.1 Materials

The experimental paddy cultivar is 'Liaojing-294' mainly planted in Liaoning Province. This cultivar has a yield of 650 kg per acre and its planting area accounts about 10 % of the total paddy planting area in Liaoning Province. In the 48 h before the experiment, the moisture content of the paddy to be tested is firstly adjusted to the required value. To keep the balance of the water content and make the paddy's temperature consistent with the ambient temperature, the paddy are sealed in double plastic bags and then these bags are turned over in each 3 or 4 h. The original moisture content of the paddy is measured as 12 % ∼ 14 % (wet-basis) via the oven drying method.

2.2 Method

The experiment is carried out on a deep fixed-bed drying test-bed in the Drying Laboratory of College of Engineering, Shenyang Agricultural University, where the ambient temperature and relative humidity are 15 ∼ 23°C and 54 % ∼ 72 %, respectively. The 5 influencing parameters are original paddy moisture content (W), paddy layer thickness (h), hot-air temperature (T), air velocity (V) and tempering time (t), their experimental porosity ranges are determined by utilizing single-factor test method. In this paper, we mainly explored the relationships between drying time and the 5 parameters by the quadratic orthogonal rotation combination design [5]. Firstly we introduce the structure of the test-bed, as shown in Fig. 1.

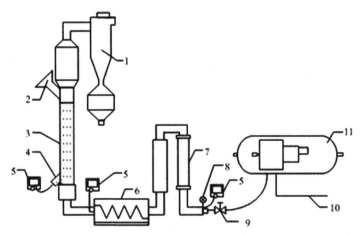

1 cyclone separator, 2 charging port, 3 drying cylinder, 4 discharge outlet
5 digital thermograph, 6 air heater, 7 float flowmeter, 8 pressure gauge, 9 control valve,
10 regulated exhaust pipe, 11 air compressor

Fig. 1. Structure of the drying test bed

In the experimental process, the air compressor is firstly started and the compressed air is stored in the storage tank. Controlled by the valve and air flow meter, the regulated air flow is heated in air heater. And then, the paddy is dried by the hot air flow. When drying the paddy, we continue testing the air flow temperature and paddy moisture content of each paddy layer and stop the drying process once the paddy moisture content in the top of the drying cylinder is below 14 %.

2.3   Determination of Drying Time

When air velocity and temperature are regulated as the predetermined value and keep stable, the paddy can be put into the drying cylinder and timing operation is started. The drying process and timing operation is terminated once the paddy moisture content (w-b) is lower than 14 %. The obtained time period is the total drying time T_t, which equals to the addition of the tempering time and the net paddy drying time T_p.

3   Experiment Scheme

In this test, the experimental parameters are set as hot-air temperature X_1, air velocity X_2, paddy layer thickness X_3, moisture content X_4 and tempering time X_5, their value ranges are $45°C \sim 85°C$, $0.45 \text{ m·s}^{-1} \sim 1.2 \text{ m·s}^{-1}$, $20 \text{ cm} \sim 60 \text{ cm}$, $16 \% \sim 28 \%$, $0 \text{ h} \sim 6 \text{ h}$, respectively. Then we use the quadratic orthogonal rotation combination design to study the relationship between drying time and these 5 parameters, whose coding levels are shown in Table 1.

Table 1. Coding levels of the five parameters

x_j Coding Levels	$X_1(°C)$ hot-air temperature	$X_2(\text{ms}^{-1})$ hot-air velocity	$X_3(\text{cm})$ paddy layer thickness	$X_4(\%)$ moisture content	$X_5(\text{h})$ tempering time
+2	85	1.20	60	28	6
+1	75	1.01	50	25	4.5
0	65	0.83	40	22	3
−1	55	0.64	30	19	1.5
−2	45	0.45	20	16	0

4   Analysis of Experimental Results

4.1   Regression Equation of Paddy Drying Time

We conducted 36 times for the quadratic orthogonal rotation combination design, the obtained results indicate the regression equations between drying time (both T_t and T_p). By using the lack of fit test and the significance test, we exclude the non-significant regression coefficients, as show in the following equations.

$$T_p = 404.9157 - 48.3333x_1 - 61.25x_2 + 112.9167x_3 + 77.0833x_4 - 25x_2x_3$$
$$- 28.0511x_4{}^2 - 26.8011x_5{}^2 \tag{1}$$

$$T_t = 585.1409 - 48.3333x_1 - 61.25x_2 + 112.9167x_3 + 77.0833x_4 + 85.4167x_5$$
$$- 25x_2x_3 - 28.1186x_4{}^2 - 26.8686x_5{}^2 \tag{2}$$

Comparison results of the observed values and the predicted ones show that the regression models have good conformity. In addition, the results of relative error ratios also demonstrate that the regression model of T_p performs better than the one of T_t. In the situation that the average total drying time is 539 min, the average prediction error of the regression model is 30 min, the respect minimum and maximum error are 14 min and 55 min.

Analysis of Parameters' Contribution Rates

Each factor has its own impact degree on the drying time and it can be calculated by the contribution rate method. For net drying time, the respect contribution rates of the five parameters are $\Delta_1 = 0.9550$, $\Delta_2 = 1.34579$, $\Delta_3 = 1.5151$, $\Delta_4 = 1.8821$, $\Delta_5 = 0.1495$. Apparently, we can conclude the impact ranking of the 5 parameters is: moisture content (X_4) > paddy layer thickness (X_3) > tempering time (X_5) > air velocity (X_2) > hot-air temperature (X_1). The result shows that an appropriate tempering process is significant for reducing the energy consumed by the air temperature and velocity, which are usually increased as high as possible in traditional ways. Therefore, adjusting the tempering time is a new way for reducing drying energy consumption and increasing the efficiency of the drying techniques.

For the total drying time, the respect contribution rates of the five parameters are $\Delta_1 = 1.3111$, $\Delta_2 = 1.3470$, $\Delta_3 = 1.5189$, $\Delta_4 = 1.8835$, $\Delta_5 = 2.0296$. Hence the rank of the five parameters is tempering time (X_5) > moisture content (X_4) > paddy layer thickness (X_3) > air velocity (X_2) > hot-air temperature(X_1). This result shows that the tempering time is the primary factor influencing the drying time which means tempering time is effective in reducing the drying energy consumption. However, it doesn't mean that we can prolong the tempering time with impunity, because efficiency of the drying process is also important for the storage and processing of seasonal paddy. Besides, to improve the overall production efficiency, the drying capacity of the drying agent and the amount of the drying paddy are also should be carefully treated, but not only considering the original paddy moisture content.

4.2 Effect Rules of Two Parameters

We choose two of the five parameters and studied their relationships between T_t and T_p, as shown in Figs. 2 to 11. The coding levels of fixed parameters are all set as 0.

Figures 2 and 3 illustrate that hot-air temperature, original paddy moisture content, air velocity and paddy layer thickness have the same effect rules on the total drying time. The drying time increases when monotonously increasing the paddy moisture content or monotonously decreasing the hot-air temperature. Compared with original paddy moisture content, hot-air temperature plays a more important role that improves the efficiency of drying time. When paddy layer is higher than level 0, drying time is

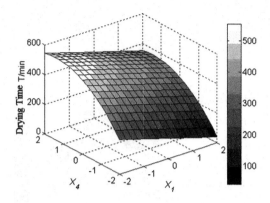

Fig. 2. Effect of hot-air temperature (X_1) and original paddy moisture content (X_4) on drying time

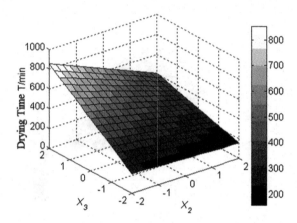

Fig. 3. Effect of air velocity(X_2) and paddy layer thickness (X_3) on drying time

greatly reduced. However, with the same paddy layer thickness, drying time is insensitive to the air velocity. Therefore, compared with air velocity, paddy layer thickness has a greater influence on drying time.

The tempering time with other parameters have different influences on the net drying time and total drying time. As shown in Figs. 4 and 5, with the same hot-air temperature, when the coding level of tempering time is lower than 0, the net drying time decreases with the decreasing of tempering time. In the condition that tempering time is higher than 0, net drying time increases with the decreasing of tempering time. For total drying time, it monotonously decreases with the deceasing of tempering time. Hence we draw the conclusion that an appropriate period of tempering time is good for reducing the net drying time. However, it is worth to point out that, we also should consider the efficiency of the whole process, as the total drying time is limited in practical situations.

As shown in Figs. 6 and 7, the combination of air velocity and tempering time has almost the same effect rules on the total drying and net drying time. Relatively

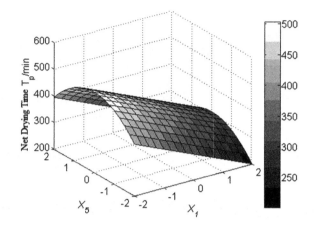

Fig. 4. Effect of hot-air temperature (X_1) and tempering time (X_5) on net drying time

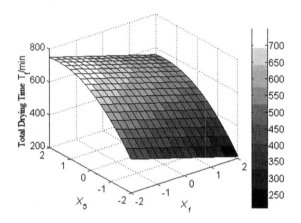

Fig. 5. Effect of hot-air temperature (X_1) and tempering time (X_5) on total drying time

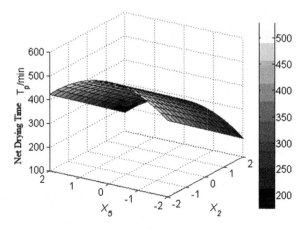

Fig. 6. Effect of air velocity (X_2) and tempering time (X_5) on net drying time

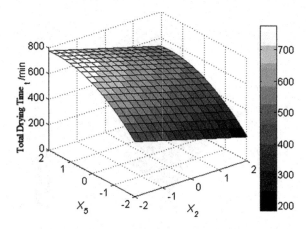

Fig. 7. Effect of air velocity (X_2) and tempering time (X_5) on total drying time

speaking, high air velocity improves the drying ability more effectively because its drying time reduces evidently. For paddy net drying time, with the same other drying parameters, when tempering time is below level 0, the shorter tempering time brings shorter net drying time; When tempering time is higher than level 0, net drying time decreases with the increasing of tempering time. Generally speaking, compared with the air velocity, tempering time has a deeper impact on the total drying time and net drying time.

Figures 8 and 9 show the rule that, with the same parameters, the increasing of paddy layer thickness will greatly increase the total drying time and net drying time. Take level 0 as the critical point, with the same paddy layer thickness, net drying time increases when tempering time slides away from the critical point, but a small variation influences little on the net drying time.

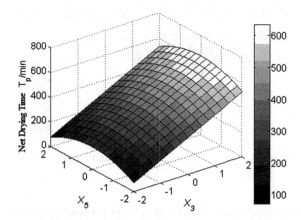

Fig. 8. Effect of paddy layer thickness (X_3) and tempering time (X_5) on net drying time

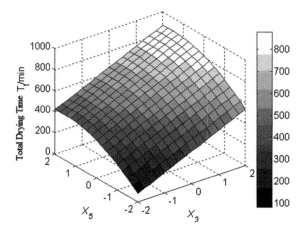

Fig. 9. Effect of paddy layer thickness (X_3) and tempering time (X_4) on total drying time

As shown in Figs. 10 and 11, with the same condition, original paddy moisture content is distinctly quadratic positive related to the net drying time. Higher moisture content means longer net drying time. When tempering time is below level 0, net drying time decreases with the decreasing of tempering time, while net drying time decreases with the increasing of tempering time when tempering time is higher than level 0. For total drying time, the rule is very simple: higher moisture content means longer total drying time.

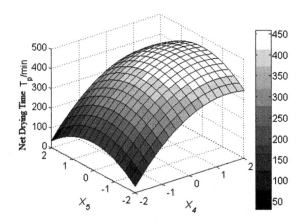

Fig. 10. Effect of original paddy moisture content (X_4) and tempering time (X_5) on net drying time

In above we studied the rules of the relationships between tempering time and drying time (both total and net) in the conditions of two different parameters. We can draw a comprehensive conclusion that, when tempering time is higher than level 0 (3 h), the net

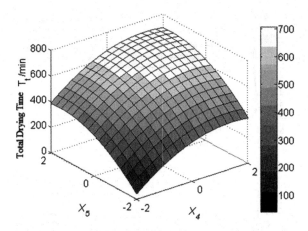

Fig. 11. Effect of original paddy moisture content (X_4) and tempering time (X_5) on total drying time

paddy drying time will be effectively reduced, which means it is beneficial for reducing energy consumption. However, as aforementioned, tempering time will increase the total drying time, which may be not acceptable in some practical applications. It is an advisable way to set up a separate tempering storehouse or provide a transportation technology guaranteeing tempering requirements to improve the paddy drying efficiency.

4.3   Analysis of Drying Characteristics with Different Paddy Layer

With different drying temperatures, the curves of different paddy layer temperatures are different too. We can study the qualitative or quantitative relationships between paddy layer temperature and hot-air temperature or drying time. With different hot-air temperatures, the variation relationships between paddy layer and hot-air temperature are shown in Fig. 12.

As shown in Fig. 12, with the same hot-air temperature, different layers have different changing rules, but the changing rules of the same paddy layer with different hot-air temperature are almost the same with each other. Thus, with given moisture content and drying-air velocity, it can be concluded that it is a two-variable nonlinear function between paddy layer temperature and drying-air temperature.

5   Conclusion and Discussions on Paddy's Influencing Parameters

In previous studies, Zheng et al. [1, 6, 7] took the drying time as the independent variable and studied the relationship between additional crack percentage and drying time. In the proposed work, it is able to determine the drying time with required additional crack percentage with the obtained equations between additional crack percentage and drying parameters. And also, additional crack percentage can be

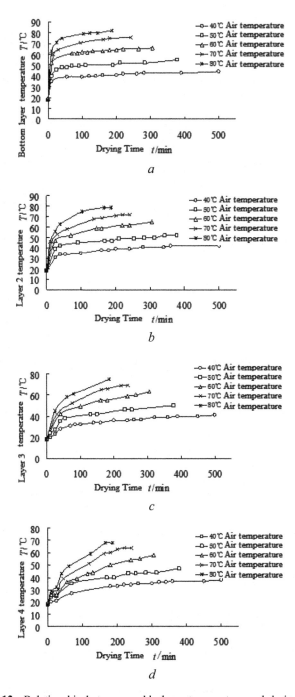

Fig. 12. Relationship between paddy layer temperature and drying time

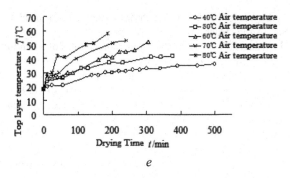

e

Fig. 12. (continued)

predicted with different drying conditions. Chen *et al.* [4] summarized the drying time's changing rules with paddy layer temperature and moisture content. It is also concluded that each paddy layer's temperature and moisture content have their own changing rules with the drying time [4]. In [8], the study shows that, when other drying conditions are the same, the thicker the paddy layer, the longer the drying time; On the other hand, thinner paddy layer and longer stirring interval lead to shorter drying time and higher additional crack percentage. In [10], Cao *et al.* analyzed drying time's changing rules with some indirect parameters, including amount of precipitation, amount of moisture evaporation and drying ability. They concluded that downstream drying technique and low drying temperature are good for paddy with high original moisture content, while upstream drying technique and high drying temperature are good for paddy with low relative low moisture content. Besides, compared with downstream drying and upstream drying technique the down-mix-stream drying technique has advantages of lower additional crack percentage, higher productivity and lower energy consumption [9]. In this paper, we studied drying time's changing rules with different drying conditions and the obtained results have significant importance in practical productions. However, in the situation when time lag is high and relationships between drying time and influencing parameters are nonlinear, there is no consensus to take the drying time as a drying factor or controlling factor. The study in this paper focuses on optimizing the drying process and it lays some essential experimental and theoretical foundations for the research on paddy deep fixed-bed drying's nonlinear dynamic process.

References

1. Zheng, X.Z., An, Y.L., Wang, C.Z.: Experimental study on the drying process based on combined high and low temperature. Chin. Agric. Mechanization **1**, 210–213 (1997)
2. Zhu, W.X., Lian, Z.G., Zhang, Y.X., Cao, C.W.: Study on weighted parameters of characteristic index of drying. Trans. Chin. Soc. Agric. Mach. **31**(1), 72–75 (2000)
3. Zhang, Q.H., Simon, X.: Prediction of performance indices and optimal of rough rice drying using neural networks. Biosyst. Eng. **83**(3), 281–290 (2002)

4. Chen, K.J., Li, J.L., Yang, M.Y., Ji, C.Y.: Drying experiments of paddy in a deep fixed-bed. Trans. Chin. Soc. Agric. Mach. **32**(2), 58–61 (2001)
5. Yi, S.J., li, B.H., Wang, J.W.: Experimental study on the rules of additional crack percentage in paddy drying. J. Agric. Mechanization Research. **1**, 76–77 (1999)
6. Zheng, X.Z., Wang, C.Z.: Study on the relationship of additional crack percentage and drying conditions. Trans. Chin. Soc. Agric. Eng. **15**(2), 194–197 (1999)
7. Zheng, X.Z., Zhou, X.L., Xia, J.Q.: The study on drying condition influencing paddy mill quality. J. Northeast Agric. Univ. **32**(1), 48–52 (2001)
8. Liu, J.W., Xu, R.Q., Bao, Q.B.: Study on natural drying characteristics and quality of paddy. Grain Storage **5**, 37–41 (2001)
9. Wang, G.X.: Study on process of rice concurrent-mixed flow drying. Trans. Chin. Soc. Agric. Eng. **16**(2), 109–202 (2000)
10. Wang, C.W., Huang, X.Y.: Research on grain drying process. J. Beijing Agric. Eng. Univ. **15**(2), 51–57 (1995)
11. Wang, Y.: Contribution of modern grain drying equipment to grain drying. Agric. Machinery Technol. Extension. **2004**(9), 10–11 (2004)
12. Cao, C.W.: Understanding the paddy drying in china and suggestions on development of equipment. Agric. Mach. **10**, 10–12 (2000)
13. Wang, D.Y., Li, C.H.: Experiment study on influence of drying parameters on drying duration of paddy rice in a deep fixed-bed. J. Shenyang Agric. Univ. **39**(2), 213–217 (2008)
14. Wang, D.Y., Li, C.H.: Influence of drying parameters on additional crack percentage of rice in a deep fixed-bed. J. Shenyang Agric. Univ. **36**(4), 482–484 (2005)
15. Zheng, X.Z.: The study on drying condition influcing paddy mill quality. J. Northeast Agric. Univ. **32**(1), 48–52 (2001)
16. Dimitriadis, A.N., Akritidis, C.B.: A model to simulate chopped alfalfa drying in fixed deep-bed. Drying Technol. **17**(6), 1247–1253 (2005)
17. Aregba, A.W., Nadeau, J.P.: Stationary deep-bed drying: a comparative study between a logarithmic model and a non-equilibrium model. TREFLE UMR- CNRS **65**, 114–116 (2005)
18. Chongwen, Cao: Rice drying and development of rice dryers in China. Trans. CSAE **17**(1), 5–9 (2001)
19. Brooker, D.B., Bakker –Arkema, F.W.: Drying and storage of grains and oilseeds (1992)
20. Movagharnejad, K.: Maryam Nikzad modeling of tomato drying using artificial neural network. Comput. Electron. Agric. **59**, 78–85 (2007)
21. Sarker, N.N., Kunze, O.R., Strouboulis, T.: Transinet moisture gradients in rough rice mapped with finite element model and related to fissures after heated air drying. Trans. ASAE **39**(2), 625–631 (1996)
22. Zhang, Q.H., Yang, S.X.: Prediction of performance indices and optimal of rough rice drying using neural networks. Biosyst. Eng. **83**(3), 281–290 (2002)

Feature Extraction and Recognition Based on Machine Vision Application in Lotus Picking Robot

Shuping Tang[1(✉)], Dean Zhao[1,2], Weikuan Jia[1], Yu Chen[1,2],
Wei Ji[1,2], and Chengzhi Ruan[1]

[1] School of Electrical and Information Engineering,
Jiangsu University, Zhenjiang 212013, China
1015412838@qq.com,
jwk_1982@163.com, ruanczhi@163.com
[2] Key Laboratory of Facility Agriculture Measurement
and Control Technology and Equipment of Machinery Industry,
Jiangsu University, Zhenjiang 212013, China
{dazhao, chenyu, jiwei}@ujs.edu.cn,

Abstract. Recently the picking technology of high value crops has become a new research hot spot, and the image segmentation and recognition are still the key link of fruit picking robot. In order to realize the lotus image recognition, this paper proposes a new feature extraction method combined with shape and color, and uses the K-Means clustering algorithm to get lotus recognition model. Before the feature extraction, the existing pulse coupled neural network segmentation algorithm, combined with morphological operation, is used to achieve nice segmentation image, including lotus, lotus flower, lotus leaf and stems. Then in the feature extraction processing, the chromatic aberration method and the moment invariant algorithm are selected to extract the color and shape features of the segmented images, in which principal component analysis algorithm is selected to reduce the dimension of the color and shape features to achieve principal components of lotus, lotus flower, lotus leaf and stems. In the experiment, K-Means clustering algorithm is used to get lotus recognition model and four clustering centers according to above principal components of training samples about lotus, lotus flower, lotus leaf and stems; then the testing experiment is applied to validate the recognition model. Experimental results shows that the correct recognition rate is 90.57 % about 53 testing samples of lotus, and the average recognition time is 0.0473 s, which further indicates that the feature extraction algorithm is applicable to lotus feature extraction, and K-Means algorithm is simple, reliable and feasible, providing a theoretical basis for positioning and picking of lotus harvest robot.

Keywords: Picking robot of lotus · Feature extraction · Invariant moment · Principal component analysis · K-Means clustering

© IFIP International Federation for Information Processing 2016
Published by Springer International Publishing AG 2016. All Rights Reserved
D. Li and Z. Li (Eds.): CCTA 2015, Part I, IFIP AICT 478, pp. 485–501, 2016.
DOI: 10.1007/978-3-319-48357-3_46

1 Introduction

Recently, the picking technology of high value crops has become a new research hot spot, and lotus belongs to high value crops because of its great medical value, high risk of picking and high labor input required [1–3]. Fresh lotus seeds have great medicinal value in calming nerves, having healthy brain, developing intelligence and removing fatigue, but the high risk and low efficiency of manual picking restrict the lotus market development. Therefore the study of lotus picking robot has broad research prospect.

The image segmentation and recognition algorithms are key link of lotus harvesting robot vision system, and the ability to accurately and quickly identify the lotus directly affects the reliability and timeliness of lotus harvesting robot [4–6]. Then many factors increase the difficulty of lotus recognition, for example, the growth of lotus is changeable and mature lotus, lotus leaf, lotus flower and lotus stem often densely distribute in an area of the image, usual recognition algorithm based on RGB components [7, 8] is difficult to distinguish the lotus from the complex background because the colors of lotus leaf and lotus stem are closer than lotus.

However it is obvious about color difference between lotus and lotus flower, and shape difference between lotus and lotus flower and lotus stem, so in this study, the chromatic aberration method and the moment invariant algorithm are selected to extract the color and shape features from the divided images [9–13], in which there are seven moment invariant shape characteristics and one color feature (r-g) extracted from each connected domain in divided image. The moment invariant algorithm is simple and reliable, but has too many extracted characteristic variables, that causes the calculation is too big, and the time of recognition is too long, then the principal component analysis (PCA) algorithm is introduced to reduce the dimensionality of color and shape information in order to accelerate the speed of image recognition [14–17].

K-Means clustering algorithm is simple, reliable and suitable for high dimensional data clustering [18–21]. Due to the higher dimension of sample characteristics and the rapid identification requirement, we choose the most classical K-Means algorithm to cluster the extracted training feature set in the case of ensuring the recognition accuracy and to get four clustering centers of lotus, lotus leaf, lotus flower and lotus stem in order to identify lotus goal.

This paper describes a machine vision recognition system of lotus picking robot, and proposes a new shape and color feature extraction method combined with K-Means clustering algorithm to get lotus recognition model. Before the feature extraction, the existing pulse coupled neural network segmentation algorithm(PCNN), combined with morphological operation [22–25], is used to achieve nice segmentation image, including lotus, lotus flower, lotus leaf and stems. Then the chromatic aberration method and the moment invariant algorithm are selected to extract the color and shape features of the segmented images, in which principal component analysis algorithm is selected to reduce the dimension of the color and shape features to achieve principal components of lotus, lotus flower, lotus leaf and stems.

2   Image Acquisition and Segmentation

2.1   Image Acquisition

Lotus pictures researched in this paper are taken from the lotus pond of Jiangsu university in Jiangsu province. In here, we choose by CCD color light source with 1024×960 pixels resolution to get lotus color images. Then 120 images, used for training and testing experiments, are closely shot with the method of elevation shooting on the side of lotus live action, in which we adopt 15 to 30 degrees elevation through the experience summary, and each picture include lotus and lotus stem with lotus flower or lotus leaf. Moreover all lotus pictures are shot under natural light conditions, and saved by JPG format with the resolution of 320×240 pixels in order to improve the seed of image process. Figure 1 is a lotus real picture.

Fig. 1. Lotus color picture

2.2   Image Segmentation Algorithm Introduction

This paper chooses the existing PCNN segmentation algorithm proposed by Ma Yide professor from Lanzhou University to get the segmentation images. The method uses PCNN segmentation algorithm and morphological image processing algorithm to divide the acquired image [22, 23]. Neuron model of simplified PCNN is shown in Fig. 2.

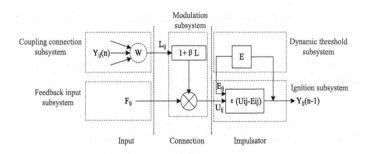

Fig. 2. Neuron model of simplified PCNN

Mathematical model of simplified PCNN neurons describes as follows:

$$F_{ij}(n) = S_{ij} \tag{1}$$

$$L_{ij}(n) = V_L \cdot \sum W_{ijkl} Y_{kl}(n-1) \tag{2}$$

$$U_{ij}(n) = F_{ij}(n)(1 + \beta L_{ij}(n)) \tag{3}$$

$$E_{ij}(n) = e^{-\alpha_E} \cdot E_{ij}(n-1) + V_E \cdot Y_{ij}(n-1) \tag{4}$$

$$Y_{ij}(n) = \begin{cases} 1 & (U_{ij}(n) > E_{ij}(n-1)) \\ 0 & (U_{ij}(n) \le E_{ij}(n-1)) \end{cases} \tag{5}$$

In these equations, F_{ij} is neural input item, S_{ij} is external neural input excitation, which is corresponding to the eigenvalues of pixel (i,j), L_{ij} is connection input item, Y_{kl} is pulse output item, W_{ijkl} is connecting factor of coupling connection field, V_L is the connection power amplification factor of L_{ij}, U_{ij} is internal activity item, β is the connecting factor of internal activity item, E_{ij} is dynamic threshold item, α_E is the iterative damping time constant of E_{ij}, and V_E is amplification factor of dynamic threshold.

In this paper, we set up $V_L \cdot \beta = D$, and $W_{ijkl} = [0.5/D \ 1/D \ 0.5/D, 1/D \ 1 \ 1/D, 0.5/D \ 1/D \ 0.5/D]$. We select Fig. 1 as an example, and after PCNN segmentation and morphological noise reduction, we can get the initial segmentation figure of Fig. 1 as Fig. 3.

Fig. 3. Segmentation Image of Fig. 1 by PCNN

Then we use the corrosion and expansion morphology method to remove the lotus stem connected with lotus and lotus flower in Fig. 3.

3   Image Feature Extraction

Considering the color differences between lotus and lotus flower and the shape differences between lotus, lotus leaf and lotus stem, this paper takes the color and shape feature extraction algorithm to extract the color and shape features of target image.

3.1   Color Feature Extraction

This paper uses color difference method to extract the target color feature based on the difference between lotus and lotus flower.

In the RGB color space, colors can be compared directly, and the color characteristic information can be controlled easily, so we can choose RGB color difference method for feature extraction [26, 27]. The classic color feature extraction method is selected to extract the color features of target image, in which R component and G component are regarded as color feature according the color difference between lotus and lotus flower. We choose $(r - g)$ on behalf of color difference, and (x, y) on behalf of arbitrary pixel values in the image, then the formula of color difference is

$$r - g = \bar{R}(x, y) - \bar{G}(x, y) \tag{6}$$

Where $\bar{R}(x, y)$ is average pixels of red component, $\bar{G}(x, y)$ is average pixels of green component.

3.2   Shape Feature Extraction

Classic Hu invariant moments algorithm combined with PCA algorithm is selected to extract the shape features of segmented image considering about the significant shape differences of lotus, lotus leaf, lotus stem, and changeable growth posture of lotus. Moment invariant algorithm processes the area of the target, and has simple concept, invariant translation, scaling and rotation, and stable and reliable recognition rate, which can avoid the shape deviation caused by changeable growth posture of lotus. PCA algorithm is used to get the invariant feature extraction to obtain comprehensive shape characteristic descriptions of the target image, which avoids the low recognition rate caused by invariant moment data redundancy.

3.2.1   Hu Invariant Moment Feature Extraction

Hu invariant moments is put into use for extracting the shape features of each connected area in target image, the seven invariant moments are as follows:

$$a_1 = \eta_{02} + \eta_{20}$$
$$a_2 = (\eta_{20} - \eta_{02})^2 + 4\eta_{11}^2$$
$$a_3 = (\eta_{30} - 3\eta_{12})^2 + (3\eta_{21} - \eta_{03})^2$$
$$a_4 = (\eta_{30} + \eta_{12})^2 + (\eta_{21} + \eta_{03})^2$$
$$a_5 = (\eta_{30} - 3\eta_{12})(\eta_{30} + \eta_{12})[(\eta_{30} + \eta_{12})^2 - 3(\eta_{21} + \eta_{03})^2] + (3\eta_{21} - \eta_{03}) \qquad (7)$$
$$\quad (\eta_{21} + \eta_{03})[3(\eta_{30} + \eta_{12})^2 - (\eta_{21} + \eta_{03})^2]$$
$$a_6 = (\eta_{20} - \eta_{02})[(\eta_{30} + \eta_{12})^2 - (\eta_{21} + \eta_{03})^2] + 4\eta_{11}(\eta_{30} + \eta_{12})(\eta_{21} + \eta_{03})$$
$$a_7 = (3\eta_{21} - \eta_{03})(\eta_{30} + \eta_{12})[(\eta_{30} + \eta_{12})^2 - 3(\eta_{21} + \eta_{03})^2] + (3\eta_{12} - \eta_{30})$$
$$\quad (\eta_{21} + \eta_{03})[3(\eta_{30} + \eta_{12})^2 - (\eta_{21} + \eta_{03})^2]$$

Where, η_{pq} is normalized central moment, which can ensure the rotation, translation and scaling invariance of Hu invariant moments. We get the normalized center moment through the following steps.

Firstly, we set (x, y) as one pixel point on the connected domain, $f(x, y)$ is the corresponding gray value of (x, y), then the $(p + q)$ moment of the entire connected domain is

$$m_{pq} = \sum_x \sum_y f(x, y) x^p y^q \qquad (8)$$

From the above formula, we can get the image barycentric coordinates is

$$x_0 = m_{10}/m_{00}$$
$$y_0 = m_{01}/m_{00} \qquad (9)$$

Where, m_{00} and m_{01}, m_{10} are zero-order and first-order moments of the target image respectively.

Then, we use barycentric coordinates to get the center distance which is as follows

$$\mu_{pq} = \sum_x \sum_y (x - x_0)^p (y - y_0)^q f(x, y) \qquad (10)$$

From the above formula, we know that central distance has translation invariance.

Then we need normalize the central distance in order to get the normalization central distance which has rotation, translation and scaling invariance. The formula of normalization central distance is as follows:

$$\eta_{pq} = \frac{\mu_{pq}}{\mu_{00}^{1 + (p+q)/2}} (p + q = 2, 3, 4 \cdots) \qquad (11)$$

By the above principle formula, we can get the seven Hu invariant moment shape feature quantities of the lotus, lotus leaf, lotus flower, lotus stem.

3.2.2   PCA Feature Reduction

We can get seven Hu invariant moments which are used to describe the shape feature amounts of the lotus, lotus leaf, lotus flower, lotus stem from the Sect. 3.2.1 feature extraction principle. But the calculated amount of Hu invariant moments is bigger and the dimension of feature amount is too high, PCA algorithm is introduced, to get on the feature extraction, in order to reduce data redundancy and avoid curse of dimensionality. The PCA algorithm gets the linear combination for the obtained seven invariant moments, and extracts the principal components which can able to represent the different shape features of lotus, lotus leaf, lotus flower and lotus stem.

Suppose, the sample size is n, each sample has p invariant moments, so there is a $n \times p$ matrix,

$$
A = \begin{bmatrix}
a_{11} & a_{12} & \cdots & a_{1p} \\
a_{21} & a_{22} & \cdots & a_{2p} \\
\vdots & \vdots & \vdots & \vdots \\
a_{n1} & a_{n2} & \cdots & a_{np}
\end{bmatrix}
\tag{12}
$$

To calculate the correlation coefficient matrix r_{ij},

$$
r_{ij} = \frac{\sum\limits_{k=1}^{n} (a_{ki} - \bar{a}_i)(a_{kj} - \bar{a}_j)}{\sqrt{\sum\limits_{k=1}^{n} (a_{ki} - \bar{a}_i)^2 \sum\limits_{k=1}^{n} (a_{kj} - \bar{a}_j)^2}}
\tag{13}
$$

According to the correlation coefficient matrix R to construct characteristic equation $|\lambda I - R| = 0$, and to calculate the eigenvalue, and sorted them by their value, that is, $\lambda_1 \geq \lambda_2 \geq \cdots \geq \lambda_p$.

By the eigenvalue, to calculate the principal component contribution rate α and the accumulative contribution rate η,

$$
\alpha = \lambda_i \bigg/ \sum_{i=1}^{p} \lambda_i \quad (i = 1, 2, \ldots, p)
\tag{14}
$$

$$
\eta = \sum_{i=1}^{m} \lambda_i \bigg/ \sum_{i=1}^{p} \lambda_i \quad (i = 1, 2, \ldots, p)
\tag{15}
$$

The front m eigenvalues make $\eta \geq 85\%$, to calculate their corresponding eigenvectors, and to calculate the loading l_{ij} of invariant moments $a_1, a_2, \cdots, a_7$ load on the principal components $z_1, z_2, \cdots, z_m$

$$
l_{ij} = p(z_i, a_j) = \sqrt{\lambda_i} \vec{e}_{ij} (i, j = 1, 2, \ldots, p)
\tag{16}
$$

And extract the front m principal components

$$Z = \begin{bmatrix} z_1 \\ z_2 \\ \cdots \\ z_m \end{bmatrix} = \begin{bmatrix} l_{11} & l_{12} & \cdots & l_{17} \\ l_{21} & l_{22} & \cdots & l_{27} \\ \cdots & \cdots & \cdots & \cdots \\ l_{m1} & l_{m2} & \cdots & l_{m7} \end{bmatrix} \begin{bmatrix} a_1 \\ a_2 \\ \cdots \\ a_7 \end{bmatrix} = \begin{cases} l_{11}a_1 + l_{12}a_2 + l_{13}a_3 + \ldots + l_{17}a_7 \\ l_{21}a_1 + l_{22}a_2 + l_{23}a_3 + \ldots + l_{27}a_7 \\ \cdots \\ l_{m1}a_1 + l_{m2}a_2 + l_{m3}a_3 + \ldots + l_{m7}a_7 \end{cases}$$

$$(17)$$

The basic steps of optimized Hu invariant moments algorithm based on PCA are as follow:

Step 1 to extract the $p+q$ moments of each segmented image's connected domain, according to Eq. (8) calculate the barycentric coordinate (x_0, y_0);

Step 2 according to Eq. (9) calculates the center distance, and normalized by Eq. (10);

Step 3 by Eq. (11) to calculate the each eigenvalue of invariant moments corresponding with connected domain, and extract the shape features, then obtained invariant moments' information of all samples;

Step 4 aim at the data sample matrix, to calculate its the correlation coefficient matrix R by Eq. (13);

Step 5 to calculate the eigenvalue of matrix R, and sorted them by their value;

Step 6 according to Eq. (15) determine the front m principal components;

Step 7 by Eq. (16) to calculate the load of principal components, and obtained the principal features Z.

4   K-Means Cluster Analysis

K-means cluster algorithm is one of classic cluster algorithm, it has the advantages of description easy, algorithm simple, operating efficient high, and suitable for large-scale data processing. In this study, using the K-means cluster algorithm can be accurately obtain the lotus clustering center, to achieve the purpose of fast recognition for lotus.

In the clustering process, the selection of parameters is directly influence the reliability and recognition efficiency of the experiment. The number of clusters is K, the K initial clustering centers and the distance parameter d plays a decisive role for K-Means cluster algorithm. The selection of K has a certain human factors, and the selection of K initial clustering centers has a certain randomness, all of these have a directly impact on the result of clustering.

In this study, the city distance is adopted,

$$d = |x - x'| + |y - y'| + |z - z'| + \ldots \qquad (18)$$

where, $(x, y, z, \cdots)$ and $(x', y', z' \cdots)$ respectively represent the cluster center and the coordinate point of sample.

Suppose the number of samples is n, using the color and shape features to cluster by K-means algorithm, the basic steps are as follow.

Step 1 the color and shape features of n samples as inputs;

Step 2 to determine K initial clustering centers $c = \{c_1, c_2, \cdots, c_k\}$ randomly, the samples is divided into K classes;

Step 3 to calculate the city distance between eigenvalue of each sample to the center of clustering, according the principle of the shortest distance, each sample is divided into the sub-classes of corresponding cluster center;

Step 4 to calculate the average value of each sub-class, and regard as the new cluster center;

Step 5 repeat the step 3 to 4, until the cluster center no longer changes, the algorithm stop, and output the cluster graph and cluster center C.

5 Image Recognition

The lotus image recognition algorithm based on machine vision is roughly divided into four steps, that is, image segmentation, feature extraction, recognition model training, and sample recognition. The flow chart of feature extraction is show in Fig. 4, the flow chart of model training is show in Fig. 5, the flow chart of sample recognition is show in Fig. 6.

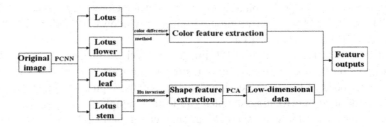

Fig. 4. The flow chart of feature extraction

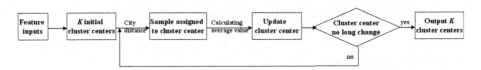

Fig. 5. The flow chart of K-means cluster training

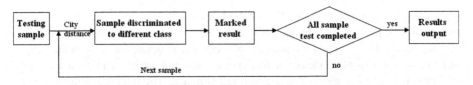

Fig. 6. The flow chart of testing sample recognition

The basic steps of lotus image recognition algorithm are as follow.

Step 1 the captured image is segmented by PCNN algorithm, get the segmented images;

Step 2 to extract the color feature information of segmented images by color difference method;

Step 3 to extract the Hu invariant moments of segmented images, using PCA algorithm to reduce dimension, get the principle component data of sample;

Step 4 the color feature and principle components as the inputs, for the training samples, by K-means clustering algorithm, to determine the four cluster centers, that is, lotus, lotus leaf, lotus flower and lotus stem, and then get the trained recognition model;

Step 5 to calculate the distance between test samples to each cluster center;

Step 6 to judge the distance to lotus cluster center whether is closest, if yes, the sample is the lotus target.

6 Experiments

6.1 Experiment Design

This experiment's operation platform, CPU: Intel(R) Core(TM) 2 Duo E7300 @ 2.66 GHz, RAM: 1.99 GB, operating environment: windows XP and matlab R2012a, graphic card: Intel® G33/G31 ECF.

In the experiment, the captured 120 images, all of these images are captured under natural light. In the image process, the pictures are converted to 320×240 pixels to improve the speed of image processing. After the image segmentation, the segmented images are divided into training samples and testing samples.

This experiment firstly uses combination method of PCNN and morphological processing to segment the captured images, obtained segmented images; respectively uses color difference method to extract the color feature of segmented images, uses Hu invariant moments method based on PCA to extract the shape feature of segmented images; and then uses K-means algorithm to cluster the training samples, obtained K cluster center; finally, classifies the test sample according to the distance to cluster center in order to verify the reliability, recognition precision and operating efficiency of the lotus recognition model.

6.2 Lotus Image Segmentation

All images are segmented by PCNN method described in Sect. 2, and then do some morphological processing, obtained the segmented images. Segmentation images are pictured above Fig. 3. Then we select 3×3 circular structure element to etch Fig. 3 and use 4×4 structural elements to expand corrosion image, and then make '&' operation between expansion image and original image. The final segmentation image as shown in Fig. 7.

Fig. 7. Final segmentation image

From Fig. 7, most lotus stem has been removed, only leaving a small part connected with lotus and lotus flower for the positioning of the fruit and picking, and the outline of segmentation target is clear, the connected area is obvious. In other words, it is easy to extract the target's features. So the result shows that PCNN used to segment lotus image is feasible and reliable.

The captured 120 images, after segmentation, obtained 200 segmented images, among 73 lotus images, 39 lotus flower images, 44 lotus leaf images and 46 lotus stem images. Respectively, selected 20 images from each type segmented images as training sample, the rest of the images as the test sample.

6.3   Lotus Image Feature Extraction

6.3.1   Hu Invariant Moments

For all segmented images, using the invariant moment algorithm to extract the shape feature, and get the invariant moment data of all samples, each sample has seven feature variables $a_1, a_2, \cdots, a_7$. Invariant moments feature of 8 examples are listed in Table 1.

Table 1. Invariant moment data

Sample	a_1	a_2	a_3	a_4	a_5	a_6	a_7
a	1.67E−01	6.15E−04	5.78E−04	5.81E−06	−3.10E−10	−1.30E−07	1.31E−10
b	1.70E−01	5.98E−05	1.12E−03	6.29E−07	−1.62E−12	−5.62E−10	1.66E−11
c	1.03E−01	2.84E−03	1.22E−05	6.61E−07	−1.83E−12	−3.22E−08	−4.06E−13
d	1.99E−01	6.42E−03	1.44E−04	9.72E−06	−3.08E−10	−5.46E−07	−1.93E−10
e	2.33E−01	2.68E−02	1.83E−04	4.98E−05	4.57E−09	6.87E−06	1.31E−09
f	2.15E−01	1.81E−02	2.89E−04	9.39E−05	1.55E−08	1.26E−05	−4.82E−10
g	3.30E+00	1.07E+01	5.46E−01	8.09E−02	−1.82E−03	−1.21E−01	−1.69E−02
h	1.45E+00	2.06E+00	2.87E−02	3.20E−03	−3.03E−05	−4.57E−03	4.36E−06

From Table 1, the results show that the value of invariant moments has a descending trend, except lotus stem, this is caused by the particularity of the shape of lotus stem. The scope of invariant moment of lotus is part overlap with lotus flower and lotus leaf, this will bring some interference for lotus recognition. Thus it can be seen, single from the shape feature to recognize lotus, lotus flower, lotus leaf, is very difficult, so in this study, need to combine with the color feature.

6.3.2   PCA Feature Reduction

Invariant moment information is not directly used to recognize the lotus, lotus flower, lotus leaf and lotus stem. The data information is too redundant, is not convenient to cluster analysis. So in this study, in order to using these invariant moment information, need to dimension reduction. PCA algorithm is introduced, and the invariant moment information is integrated into a few principal components.

We do principle component analysis for 200 segmented sample, and the eigenvalue, contribution rate (CR) and accumulative contribution rate (ACR) are listed in Table 2.

Table 2. Eigenvalue, contribution rates and accumulative contribution rates

Principal component	Eigenvalue	CR, %	ACR, %
z_1	4.99E+00	71.286	71.286
z_2	1.15E+00	16.429	87.714
z_3	8.36E−01	11.943	99.657
z_4	2.06E−02	0.294	99.951
z_5	2.81E−03	0.040	99.992
z_6	5.09E−04	0.007	99.999
z_7	8.09E−05	0.001	100.000

Table 2 shows that, if the amount of information can reach 85 %, the solution will not be influenced, in this study need to extract 3 principal components. That is, the dimension of the shape features is reduced from 7 to 3. These 3 principal components represent the shape feature of lotus, lotus flower, lotus leaf and lotus stem.

6.3.3   Combination of Color and Shape Feature

Combination of one color feature and three shape features, using these four variables to describe the characteristics of the sample. Take Table 1 as an example, the samples corresponding feature are listed in Table 3.

Table 3 shows that, these four variables regard as the inputs of the training and recognition model.

Table 3. Feature variables of example samples

Sample	z_1	z_2	z_3	r−g
a	2.98E−01	−5.36E−02	−6.15E−02	−1.04E+01
b	3.04E−01	−5.41E−02	−6.26E−02	−1.62E+01
c	1.83E−01	−3.43E−02	−3.86E−02	6.70E+01
d	3.53E−01	−6.67E−02	−7.50E−02	2.84E+01
e	4.07E−01	−8.68E−02	−9.42E−02	−5.24E+01
f	3.78E−01	−7.70E−02	−8.48E−02	−4.31E+01
g	2.53E+00	−5.63E+00	−4.92E+00	−1.84E+01
h	1.90E+00	−1.38E+00	−1.22E+00	−1.93E+01

6.4 Training

In the process of lotus recognition, the interference factors mainly include lotus flower, lotus leaf, lotus stem, so in the training, set $K = 4$. The determination of cluster center is randomly, selected the city distance as the criteria of similarity and discriminated. From 200 samples, respectively, choose 20 lotus, 20 lotus flowers, 20 lotus leaves and 20 lotus stems, so total 80 samples as training sample. The 80 training samples can be divided into four classes by K-means cluster algorithm, four cluster centers $\{c_1, c_2, c_3, c_4\}$ are obtained. In order to the clustering figure show in 2 dimension coordinate, for 4 extracted features $(z_1, z_2, z_3, (r - g))$ to do linear combination, get a new feature $(z', (r - g)')$. The transfer formula is

$$\begin{cases} z' = \lg |z_1 . * z_2 . * z_3| \\ (r - g)' = 0.01(r - g) \end{cases} \tag{20}$$

The clustering figure is show as Fig. 6.

Note: Abscissa represents the linear combination of principal components $(\lg |z_1 . * z_2 . * z_3|)$, ordinate represents color feature value after the linear operation $(0.01(r - g))$; peach 'o' area represents lotus, red '*' area represents lotus flower, green '+' area represents lotus leaf, blue '△' area represents lotus stem; '□' represents the cluster center of each cluster area.

The cluster centers' coordinate is

$$C = \begin{cases} c1 & c2 & c3 & c4 \\ -4.1389 & -3.4894 & -3.3214 & 0.0637 \\ -0.1536 & +0.6433 & -0.5244 & -0.1928 \end{cases} \tag{21}$$

Figure 8 shows that, the 80 samples are clustered into four classes, the cluster area is obvious. It is illustrate that K-means cluster algorithm is suitable for distinguish lotus, lotus flower, lotus leaf, lotus stem. Lotus flower area locate at above coordinate axis $(x, 0)$, it well to distinguish the lotus flower and the lotus, lotus leaf, and to avoid the lotus recognition interference caused by lotus flower. Lotus flower area locate at

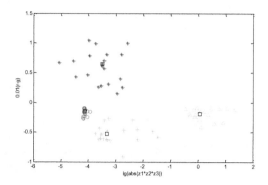

Fig. 8. Clustering chart of K-Means (Color figure online)

right coordinate axis $(-1, y)$, the cluster area is fragmented, it is verify that the difference of lotus stem shape is bigger, the recognition interference is small.

The cluster area of lotus is relatively concentrated, it has obvious boundary with lotus leaf, noninterference with each other. It is further illustrating that K-Mean clustering algorithm for lotus recognition is simple and reliable.

6.5   Testing

For 120 test samples, to calculate the city distance of each sample to each cluster center, according the shortest distance to judge the sample belong to which class. The results of recognition are listed in Table 4.

Table 4. Recognition result of testing images

Indexes	Lotus	Lotus flower	Lotus lesf	Lotus stem
Number of images	53	19	22	26
Correct recognition	48	16	20	26
Misrecognized to lotus	-	3	2	0
Misrecognition rate, %	9.43	0	0	0
Correct recognition rate, %	90.57	84.21	90.91	100
Recognition time, s	0.0473			

Note: Recognition time is the average of total time, include the color feature extraction, shape feature extraction, and test running time of each image.

From Table 4, the results show that, using for testing 53 lotus images, the number of correct recognition image is 48, 2 images are misrecognized to lotus flower, 3 images are misrecognized to lotus leaf, the recognition precision of lotus is 90.57 %.

For 19 lotus flowers, the recognition precision is 84.21 %, 3 images are misrecognized to lotus, because they are un-blooming lotus flower, their shape is very similar with lotus, and color also is very close.

For 22 lotus leaves, the recognition precision is 90.91 %, 2 images are misrecognized to lotus, due to the different shooting angle of the camera, and the growth posture of the lotus is variety.

For 26 lotus stems, the recognition precision is 100 %, because the shape of lotus is difference with other target.

To test 120 images, the average recognition time of each image is 0.0473 s, recognition time include feature extraction and testing time.

7 Conclusion

In order to achieve the automatic harvesting lotus, this paper carries on some related preliminary study of lotus harvesting robot, and a lotus recognition method for vision system of harvesting robot is designed.

Lotus target image recognition based on machine vision, the target image segmentation by PCNN, and obtained the segmented lotus image, lotus flower image, lotus leaf image, and lotus stem image. Due to the lotus has color difference with lotus flower, has the shape difference with lotus leaf and lotus stem, so the feature extraction method of lotus target adopts the combination of color and shape feature. In the color feature extraction, color difference method is used to extract the red and green feature difference, to avoid the interference from lotus flower to the lotus and lotus leaf recognition. In the shape feature extraction, Hu invariant moment algorithm is adopted, combine with PCA algorithm, to reduce the dimension of shape features. The extracted principal components represent the shape feature of lotus, lotus flower, lotus leaf and lotus stem. It reduces the data redundancy, improve the computing speed.

All samples are divided into training sample and testing sample. For training sample, by K-means cluster algorithm to determine four cluster center of lotus, lotus flower, lotus leaf and lotus stem, obtained the lotus recognition model. In the K-means cluster algorithm, city distance is adopted as the criteria of similarity and discriminated. In order to easily observe the cluster figure, make its display in 2d coordinate system, the extracted four features need to be done linear combination. Finally, the city distance of each testing sample to four cluster center has been calculated, according the shortest distance to discriminate sample of belonging.

By the experiment analysis, from image segmentation and feature extraction, to recognition model training and testing sample recognition, this study adopts the method of lotus image processing, it gets the better effect, and the results are practical and reliable. From the lotus correct recognition rate, this study achieves the desired effect. This paper implements the preliminary discussion on the lotus auto-harvesting, next step of study, we will do some improvement on the recognition accuracy and efficiency.

Acknowledgments. This work is supported by the National Natural Science Foundation of China (No. 31571571); Priority Academic Program Development of Jiangsu Higher Education Institutions (PAPD); The Specialized Research Fund for the Doctoral Program of Higher Education of China (No. 20133227110024); Ordinary University Graduate Student Research Innovation Projects of Jiangsu Province (No. KYLX 14_1062).

References

1. Bac, C.W., Henten, E.J., Hemming, J., et al.: Harvesting robots for high-value crops: state-of-the-art review and challenges ahead. J. Field Robot. **31**(6), 888–911 (2014)
2. Li, P., Lee, S.H., Hsu, H.Y.: Review on fruit harvesting method for potential use of automatic fruit harvesting systems. Procedia Eng. **23**(5), 351–366 (2011)
3. Sánchez-González, M.J., Sánchez-Guerrero, M.C., Medrano, E., et al.: Influence of pre-harvest factors on quality of a winter cycle, high commercial value, tomato cultivar. Sci. Hortic. **189**, 104–111 (2015)
4. Xu, L., Zhang, T.: Present situation of fruit and vegetable harvesting robot and its key problems and measures in application. Trans. Chin. Soc. Agric. Eng. **20**(5), 38–42 (2004)
5. Zhao, D.A., Lv, J., Ji, W., et al.: Design and control of an apple harvesting robot. Biosyst. Eng. **110**(2), 112–122 (2011)
6. Xiang, R., Ying, Y.B., Jiang, H.Y.: Development of real-time recognition and localization methods for fruits and vegetables in field. Trans. Chin. Soc. Agric. Mach. **44**(11), 208–223 (2013)
7. Qian, J.P., Yang, X.T., Wu, X.M., Chen, M.X., Wu, B.G.: Mature apple recognition based on hybrid color space in natural scene. Trans. Chin. Soc. Agric. Eng. **28**(17), 137–142 (2012)
8. Shen, B.G., Chen, S.R., Yin, J.J., Mao, H.P.: Image recognition of green weeds in cotton fields based on color feature. Trans. CSAE **25**(6), 163–167 (2009)
9. Dai, J.B.: Algorithms of image feature extraction based on visual information. Jilin University (2013)
10. Chaki, J., Parekh, R., Bhattacharya, S.: Plant leaf recognition using texture and shape features with neural classifiers. Pattern Recogn. Lett. **58**, 61–68 (2015)
11. Banerji, S., Sinha, A., Liu, C.: New image descriptors based on color, texture, shape, and wavelets for object and scene image classification. Neurocomputing **117**(14), 173–185 (2013)
12. Yang, M., Kpalma, K., Ronsin, J.: Shape-based invariant feature extraction for object recognition. In: Kountchev, R., Nakamatsu, K. (eds.) Advances in Reasoning-Based Image Processing Intelligent Systems. Intelligent Systems Reference Library, vol. 29, pp. 255–314. Springer, Heidelberg (2012)
13. Yang, M., Kpalma, K., Ronsin, J.: Shape-based invariant feature extraction for object recognition. Intell. Syst. Ref. Libr. **29**, 255–314 (2012)
14. Diao, Z., Zheng, A., Wu, Y.: Shape feature extraction of wheat leaf disease based on invariant moment theory. In: Li, D., Chen, Y. (eds.) Computer and Computing Technologies in Agriculture V. IFIP Advances in Information and Communication Technology, vol. 369, pp. 168–173. Springer, Heidelberg (2012)
15. Vijayalakshmi, B.: A new shape feature extraction method for leaf image retrieval. In: Mohan, S., Suresh Kumar, S. (eds.) Proceedings of the Fourth International Conference on Signal and Image Processing 2012 (ICSIP 2012). LNCS, vol. 221, pp. 235–245. Springer, Heidelberg (2012)
16. Wang, M., Wang, G.H., Fang, P.Y., Sun, S.J.: Method for vehicle-logo location and recognition based on PCA and invariant moment. Geomat. Inf. Sci. Wuhan Univ. **33**(1), 36–40 (2008)
17. Dray, S., Josse, J.: Principal component analysis with missing values: a comparative survey of methods. Plant Ecol. **216**(5), 657–667 (2014)

18. Chandrasekhar, A.M., Raghuveer, K.: Intrusion detection technique by using k-means, fuzzy neural network and SVM classifiers. In: 2013 International Conference on Computer Communication and Informatics (ICCCI), vol. 33, pp. 1–7. IEEE (2013)

19. George, A.: Efficient high dimension data clustering using constraint-partitioning k-means algorithm. Int. Arab J. Inf. Technol. **10**(5), 467–476 (2013)

20. Li, Z., Hong, T.S., Zeng, X.Y., Zheng, J.B.: Citrus red mite image target identification based on k-means clustering. Trans. Chin. Soc. Agric. Eng. **28**(23), 147–153 (2012)

21. Wang, L.S., Zhang, G.Y., Sha, Y.: Feature extraction and clustering of complex ore image. J. Beijing Inst. Petro-chem. Technol. **18**(4), 36–42 (2010)

22. Wang, Z., Ma, Y., Cheng, F., et al.: Review of pulse-coupled neural networks. Image Vis. Comput. **28**(1), 5–13 (2010)

23. Wang, H.Q., Ji, C.Y., Gu, B.X., An, Q.: In-greenhouse cucumber recognition based on machine vision and least squares support vector machine. Trans. Chin. Soc. Agric. Mach. **43**(3), 163–167 (2012)

24. Wen, C., Yu, H., He, S.: An image segmentation algorithm of corn disease based on the modified bionic pulse coupled neural network. In: 2013 Fourth Global Congress on Intelligent Systems (GCIS), pp. 97–101. IEEE (2013)

25. Sarkar, S., Das, S., Chaudhuri, S.S.: A multilevel color image thresholding scheme based on minimum cross entropy and differential evolution. Pattern Recogn. Lett. **54**(1), 27–35 (2015)

26. Dun, S.K., Wei, H.P., Sun, M.Z.: A new distance color difference formula in RGB color space. Sci. Technol. Eng. **11**(8), 1833–1836 (2011)

27. Zhang, C.L., Zhang, J., Zhang, J.X., Li, W.: Recognition of green apple in similar background. Trans. Chin. Soc. Agric. Mach. **10**, 277–281 (2014)

28. Wang, J.J., Zhao, D.A., Ji, W., Zahng, C.: Apple fruit recognition based on support vector machine using in harvesting robot. Trans. Chin. Soc. Agric. Mach. **40**(1), 148–151 (2009)

29. Jiang, L., Chen, D.F., Pan, C., et al.: Recognition of difformity mature fruits in natural environment. Sci. Technol. Eng. **20**(21), 5135–5138 (2012)

30. Zhang, H.T., Mao, H.P., Qiu, D.J.: Feature extraction for the stored-grain insect detection system based on image recognition technology. Trans. Chin. Soc. Agric. Eng. **25**(2), 126–130 (2009)

31. Deng, J.Z., Li, M., Yuan, Z.B., Jin, J., Huang, H.S.: Feature extraction and classification of Tilletia diseases based on image. Trans. Chin. Soc. Agric. Eng. **28**(3), 172–176 (2012)

32. Xiong, J.T., Zou, X.J., Liu, N., Peng, H.X., Lin, G.C.: Fruit quality detection based on machine vision technology when picking litchi. Trans. Chin. Soc. Agric. Mach. **45**(7), 54–60 (2014)

33. Wang, X.P., He, Z.J.: Accurate recognition of hairs in canned mushroom under different kinds of lighting conditions. Trans. Chin. Soc. Agric. Eng. **30**(4), 264–271 (2014)

Rapeseed (*Brassica napus* L.) Primary Ramification Morphological Structural Model Based on Biomass

Weixin Zhang, Hongxin Cao[(✉)], Wenyu Zhang, Yan Liu, Daokuo Ge, Chunhuan Feng, Weitao Chen, and Chuwei Song

Institute of Agricultural Economic and Information/Engineering Research Center for Digital Agriculture, Jiangsu Academy of Agricultural Sciences, Nanjing 210014, China nkyzwx@126.com, caohongxin@hotmail.com, research@wwery.cn, liuyan0203@aliyun.com, gedaokuo@163.com, 1286234727@qq.com, 1303079141@qq.com, 923903764@qq.com

Abstract. Primary ramification morphogenesis has a significant influence on the yield of rapeseed. In order to quantify the relationship between rapeseed architecture indices and the organ biomass, a rapeseed primary ramification structural model based on biomass were presented. Intended to explain effects of cultivars and environmental conditions on rapeseed PR morphogenesis. The outdoor experiment with cultivars: Ningyou 18 (V1, conventional), Ningyou 16 (V2, conventional) and Ningza 19 (V3, hybrid), and designed treatment of cultivar-fertilizer, cultivar-fertilizer-density, and cultivar tests in 2011–2012 and 2012–2013. The experimental result showing that the leaf blade length of PR, leaf blade width of PR, leaf blade bowstring length of PR, PR length, and PR diameter from 2011 to 2012 were goodness, and their d_a values and *RMSE* values were −1.900 cm, 5.033 cm ($n = 125$); −0.055 cm, 3.233 cm ($n = 117$); 0.274 cm, 2.810 cm ($n = 87$); −0.720 cm, 3.272 cm ($n = 90$); 0.374 cm, 0.778 cm ($n = 514$); 0.137 cm, 1.193 cm ($n = 514$); 0.806 cm, 8.990 cm ($n = 145$); and −0.025 cm, 0.102 cm ($n = 153$), respectively. The correlations between observation and simulation in the morphological indices were significant at $P < 0.001$, but the d_{ap} values were <5 % for the second leaves length and the third leaves length, leaf blade bowstring length, PR length, and PR diameter, which indicated that the model's accuracy was high. The models established in this paper had definite mechanism and interpretation, and the impact factors of N, the ratio of the leaf length to leaf dry weight of primary ramification (*PRRLW*), and the partitioning coefficient of leaf blade dry weight of primary ramification (*PRCPLB*) were presented, enabled to develop a link between the plant biomass and its morphogenesis. Thus, the rapeseed growth model and the rapeseed morphological model can be combined through organ biomass, which set a reference for the establishment of FSPMs of rapeseed.

Keywords: Biomass · FSPMs · Rapeseed · Primary ramification

D. Li and Z. Li (Eds.): CCTA 2015, Part I, IFIP AICT 478, pp. 502–518, 2016.
DOI: 10.1007/978-3-319-48357-3_47

1 Introduction

Rapeseed is a very good oil crop with high economic and nutritional values. In the ten years from 2004 to 2014, the production and consumption of rapeseed have increased significantly, and the growth rates of the plant area and total production of rapeseed were 35 % and 54 %, respectively [1]. Plant morphological structure simulation and visualization is one of the important content of agro-informatics in nowadays, and its latest trend is to establish Functional-structural Plant Models (FSPMs). Some crop morphological structure models had been developed based on GDD [2–7], which described the main stems, branches, leaves, leaf sheaths, and internodes of the morphological model of the crops, and realized the quantitative simulation of the crop morphogenesis process. With the maturity of the research conditions, crop morphological structure and physiological ecological process will become the new focus. In order to analyze the relationships between the morphological parameters of the organs of crops and biomass, some related studies had been reported [8–11]. Rapeseed morphology directly influences its biomass production. In recent years, studies of the rapeseed morphological structure model had also been proposed [12–14]. A rapeseed leaf geometric parameters model based on biomass presented quantified the relationship between biomass and rapeseed leaf geometric parameters, a leaf curve model based on biomass for rapeseed was established, which described the relationships between the leaf curve and the corresponding leaf biomass for rapeseed on main stem, and a morphological structure model of leaf blade space based on biomass at pre-overwintering stage in rapeseed plant presented revealed the relationships between biomass and rapeseed architecture indices. These studies laid a good foundation for FSPMs of rapeseed. By combining LEAFC3-N with the FSPMs, rapeseed functional structural plant model was established, which could respond to the environmental conditions [15, 16].

However, the rapeseed primary ramification morphological structural model based on biomass has not been reported. Vegetative organs of rapeseed including leaf, stem, ramification, and root [17]. Leaf is the crucial organs of photosynthesis, ramification leaves play gradually an important role in the rapeseed mid and late growth stage, and the ramification becomes a vital source. Therefore, how to accurate quantitative description of the rapeseed morphological variation is vitally important and difficult.

The objectives of this paper were to link primary ramification architectural parameters of rapeseed plant with biomass, by analyzing field experimental data from 2011–2012, and 2012–2013, to develop finally the rapeseed primary ramification morphological structural model, and to lay a foundation for rapeseed morphological structure model and visualization.

2 Experiments and Methods

2.1 Experimental Samples

Three varieties Ningyou 18 (V1) (conventional), Ningyou 16 (V2) (conventional), and Ningyou 19 (V3) (hybrid) were used in experiments, and they all belong to *brassica napus*. Canopy morphology structure of the three cultivars had following traits: V1 with the overwintering half-vertical cultivars, and medium height, had higher rankof branch, compacter in plant type; V2 with the overwintering half-vertical cultivars, medium height, andcompact plant type; V3 with the overwintering half-vertical cultivars, had broader and thicker leaves, the leaves light green, and edge of leaves with saw teeth.

2.2   Experimental Methods

Experimental Conditions: Field experiment was conducted at the experimental site (32.03° N, 118.87° E) of JAAS, China during the 2011 to 2013 rapeseed growing period. The soil type is the yellow-brown soil, and its basic nutrient status: organic carbon, 13.8 gkg^{-1}; available Nitrogen, 58.95 $mgkg^{-1}$; available Phosphorus, 29.25 $mgkg^{-1}$; available Potassium, 109.05 $mgkg^{-1}$; and pH 7.84. Three experiments were designed to implement.

Experiment on Cultivar and Fertilizer: By split-plot design, fertilization levels (N 180 $kghm^{-2}$, P_2O_5 120 $kghm^{-2}$, K_2O 180 $kghm^{-2}$ borax 15 $kghm^{-2}$ and CK) was assigned to the whole-plot, and cultivars (V1, V2, V3) to the sub-plot. Six treatments were repeated three times, and the 18 subplots were arranged randomly. The area of each plot was 7.0 m × 5.7 m = 39.9 m^2, the density design is 30 cm of row spacing and 17–20 cm of distance between plants, respectively. The 13 rows were planted in each subplot, and a blank line were stayed in the inter-plot. The sowing date and the transplanting date was October 15 and November 4 respectively, for both 2011 and 2012. The total amount of N-fertilizer application was 3.26 kg in the fertilization area, and the basal: seedling: winter ratio was 5:3:2 respectively. Basalrate of N + P_2O_5 + K_2O-compound fertilizer was at 16.3 kg (mass fraction ≥ 25 %), rate of calcium superphosphate was at 7.3 kg (mass fraction 12 %), and rate of agricultural potassium sulphate was at 6.4 kg (mass fraction 33 %). the manure for seedling and winter dressing were all urea and its application amount was 1.41 kg, 2.12 kg, respectively (mass fraction of total N ≥ 46.2 %). Other management activities followed local production practice.

Experiment on Cultivar, Fertilizer, and Density: By split-plot design, fertilization levels (N, P_2O_5, K_2O each of 180 $kghm^{-2}$, 90$kghm^{-2}$ $kghm^{-2}$ and CK) was assigned to the whole-plot and cultivars (V1, V3) and density levels (D_1 (6 × 10^4 planthm^{-2}), D_2 (1.2 × 10^5 planthm^{-2}) and D_3 (1.8 × 10^5 planthm^{-2})) to the sub-plot. Eighteen treatments were repeated three times, and the 54 plots were arranged randomly. The area of each plot was 3.99 m × 3.5 m = 13.97 m^2, the density design is 42 cm of row spacing and the distance between plants was calculated by density. The 9 rows were planted in each subplot, and a blank line was stayed in the inter-plot. The sowing date and the transplanting date was October 8 and November 9 respectively, for both 2011 and 2012.

The basal: seedling: winter ratio of fertilizer application was 5:3:2 respectively. Folia application of borax was at a rate of 15 $kghm^{-2}$ after bolting of rapeseed. Other management activities followed local production practice.

Experiment on Cultivar Experiment: The experiments were a randomized complete block design, with three cultivars (V1, V2, V3), under the same fertilization level, N90 (N, P_2O_5, K_2O each of 90 kghm^{-2}) and density level, D_2 (1.2 × 105 planthm^{-2}), 3 replications, and the 9 subplots. The area of each plot was 7.98 m × 3.5 m = 27.93 m^2, the density design is 42 cm of row spacing. Other treatments with the Cultivar, fertilizer and density experiment.

Measurements: We selected 3 plants with similar growth status in each treatment, and determined the blade length, blade width, and blade bowstring length at various leaf ranks on primary ramification more than 2.5 cm, and the length and the diameter of the primary ramification. Then, leaves were separated from the ramification, and into the paper bags, then put in oven, the temperature of green removing in 105 °C for 30 min, in 80 °C until reaching a stable weight.

Blade: Leaf blade length: measuring the length of the blade straight state from the leaf tip to the leaf base; Leaf blade width: measuring the maximum length of the leaf width value (in the middle of the blade), average value was gained by multiple measurements. Leaf blade bowstring length: as the elongation of the blade, due to gravity and other effects, the leaf is deformed, and bends into an arc downwardly. So the leaf blade bowstring length is from the leaf base to the leaf tip of the linear distance of space, average value was gained by multiple measurements.

Primary Ramification (PR): Length of PR: the length of the straight state which is the distance from the basal of the PR to the top of the PR; Diameter of PR: measuring the base, middle and top of PR several times using the vernier caliper and average value was gained by multiple measurements.

Data Analysis: We used the MS EXCEL 2007 and SPSS version 19 to analyzed the experiment data. Part of the data of different cultivars and fertilizer levels were for the modeling and parameter determination in 2011–2012. The remaining independent data were for model testing and inspection.

Model Validation: We used the root mean squared errors (*RMSE*), the correlation (*r*), the average absolute difference (d_a), the ratio of d_a to the average observation (d_{ap}), and 1:1 chart of measured values and simulated values properties to validate the models developed in this paper (Cao et al. 2012). The smaller *RMSE*, d_a, and d_{ap} values, the better consistency of simulated and observed values, and the deviation will be small, the simulation results of the model proved to be accurate and reliable. So the d_a, and d_{ap} are defined as:

$$d_i = X_{Oi} - X_{Si}$$

$$d_a = \frac{1}{n}\sum_{i=1}^{n} d_i$$

$$\bar{X}_o = \frac{1}{n}\sum\nolimits_{i=1}^{n} X_{oi}$$

$$d_{ap} = d_a/\bar{X}_o \times 100\%$$

With i = sample number, X_{oi} = observed values, X_{si} = simulated values, n = total number of measurements.

3   Results and Discussion

3.1   Model Description

Leaf is an important organ of photosynthesis, and the blades of PR are the main organ of photosynthesis in the mid and late period of rapeseed growth stages, which directly determines the photosynthetic capacity and the final yield. The morphology of the rapeseed leaves have characteristics with complexity, variability, and difficult to obtain, so it is an important part of the rapeseed plant model. According to the status of the petiole, the leaf for *brassica napus* can be divided into long-petiole, short-petiole, and sessile leaf on the main stem, and the leaves of PR are similar to the sessile leaf [18].

3.2   Leaf Blade Length Model

The production of the effective PR usually occurs in the axillary buds above the tenth leaf in the upper on the main stem. According to observation in 2011 to 2012 experiment, the leaf blade length with the leaf dry weight was close to proportional increasing trend (Fig. 1). Hereby, the models can be expressed as follows:

$$PRLL_{jk}(i) = DWPRLB_{jk}(i) \times PRRLW_{jk}(i) \tag{1}$$

$$DWPRLB_{jk}(i) = PRCPLB_{jk}(i) \times DW_{SP}(i) \tag{2}$$

$$DW_{SP}(i) = MDW_{SP}(i) \pm SDW_{SP}(i) \tag{3}$$

$$MDW_{SP}(i) = DW_{CP}(i)/DES \tag{4}$$

where, $PRLL_{jk}(i)$ is the kth leaf blade length of the jth primary ramification (cm), $DWPRLB_{jk}(i)$ is the kth leaf dry weight of the jth primary ramification (cm), $PRRLW_{jk}(i)$ is the ratio of the kth leaf length of the primary ramification to leaf biomass (cm g^{-1}), $PRCPLB_{jk}(i)$ is the ratio of the kth leaf biomass of the primary ramification to the biomass of upper plant part (g g^{-1}), $MDW_{SP}(i)$ is the mean dry weight of per plant (g plant^{-1}), $DW_{SP}(i)$ is the dry weight of per plant (gplant^{-1}), $DW_{CP}(i)$ is the biomass in canopy per unit area (g m^{-2}), $SDW_{SP}(i)$ is the standard error of dry weight of per plant

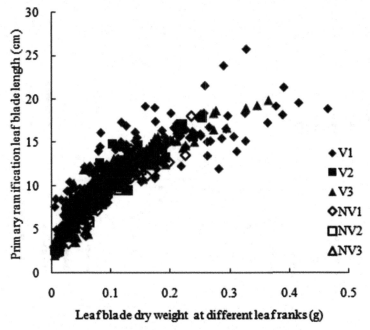

V1-Ningyou18 with fertilizer; V2-Ningyou16 with fertilizer;
V3-Ningza 19 with fertilizer;NV1- Ningyou18 with no fertilizer;NV2-
Ningyou16 with no fertilizer;NV3- Ningza19 with no fertilizer (the same as follows)

Fig. 1. Variation of the primary ramification leaf length by the leaf dry weight for different treatments in 2011–2012

(g plant^{-1}) (determined by experiment), and *DES* is the plant number unit area (plant m^{-2}).

The data in the 2011 to 2012 and 2012 to 2013 experiment showed that the number of the ramification leaf has a definite relationship with ramification rank. Generally effective ramification was on the middle and upper part of the main stem, all can reach four leaves, and some ramifications was up to six or more but relatively less. Therefore, the first four leaves were studied in this study, and the other blades were not measured as it is difficult to determine the relationship with biomass.

The data in the 2011 to 2012 experiment showed that the values of $PRRLW_{jk}(i)$ with the leaf rank on primary ramification were close to quadratic function. The significant $R = 0.768$ ($n = 74$, $R_{(72,0.001)} = 0.375$, $P < 0.001$) and $R^2 = 0.589$ for the first leaf; $R = 0.584$ ($n = 77$, $R_{(75,0.001)} = 0.367$, $P < 0.001$) and $R^2 = 0.341$ for the second leaf; $R = 0.489$ ($n = 32$, $R_{(30,0.01)} = 0.452$, $P < 0.01$) and $R^2 = 0.240$ for the third leaf; $R = 0.557$($n = 36$, $R_{(34,0.001)} = 0.525$, $P < 0.001$) and $R^2 = 0.310$ (Fig. 2) for the fourth leaf (Eq. (5), Table 1).

The F-values, t-values, c_1, d_1, e_1, c_2, d_2, e_2, c_4, and e_4 all were significant at $P<0.001$, apart from c_3, d_3, d_4, and e_3 (Table 1).

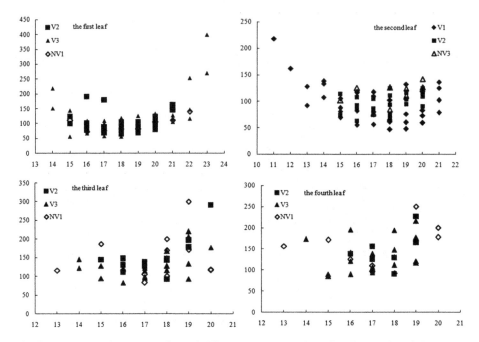

Fig. 2. Variation of *PRRLW* values of different treatments of the first four leaf blade length by primary ramification rank in 2011–2012.

$$PRRLW_{jk}(i) = \begin{cases} c_1 PRR_{j1}(i)^2 + d_1 PRR_{j1}(i) + e_1, & \text{the first leaf, } 14 \le j \le 23 \\ c_2 PRR_{j2}(i)^2 + d_2 PRR_{j2}(i) + e_2, & \text{the second leaf, } 11 \le j \le 21 \\ c_3 PRR_{j3}(i)^2 + d_3 PRR_{j3}(i) + e_3, & \text{the third leaf, } 13 \le j \le 20 \\ c_4 PRR_{j4}(i)^2 + d_4 PRR_{j4}(i) + e_4, & \text{the fourth leaf, } 13 \le j \le 20 \end{cases} \quad 1 \le k \le 4$$

(5)

The data in the 2011 to 2012 experiment showed that $PRCPLB_{jk}(i)$ and $PRRLB_{jk}(i)$ with primary ramification rank were linear function model and exponential function model, respectively. R19 was the inflexion under different treatment levels, and the fitting precision for $PRCPLB_{jk}(i)$ in various treatments were: fertilizer with cultivars significant $R = 0.603$ ($n = 56$, $P < 0.001$, $R_{(54,0.001)} = 0.428$), $R^2 = 0.363$ and $R = 0.568$ ($n = 34$, $P < 0.001$, $R_{(32,0.001)} = 0.539$), $R^2 = 0.323$; no fertilizer with cultivars significant $R = 0.567$ ($n = 39$, $P < 0.001$, $R_{(37,0.001)} = 0.507$), $R^2 = 0.321$ and $R = 0.599$($n = 32$, $P < 0.001$, $R_{(30,0.001)} = 0.554$), $R^2 = 0.359$ (Fig. 3). The fitting precision for $PRRLB_{jk}(i)$ in various treatments were: fertilizer with cultivars significant $R = 0.592$ ($P < 0.001$, $n = 159$, $R_{(157,0.001)} = 0.264$), $R^2 = 0.351$; no fertilizer with cultivars significant $R = 0.466$ ($n = 102$, $P < 0.001$, $R_{(100,0.001)} = 0.321$), $R^2 = 0.217$ (Fig. 4, Eqs. (6)– (8), Table 2).

The F-values, t-values, all model parameters apart from B_5 were at P < 0.001 (Table 2).

Table 1. The determination of model parameters and significance test

Eq.	n	Parameter symbolic	Unstandardized coefficients	t
$PRRLW_{j1}(i) = c_1PRR_{j2}(i)^2 +$ $d_1PRR_{j2}(i) + e_1$	74	c_1	6.364	8.554[***]
		d_1	−97.389	−7.69[***]
		e_1	451.118	8.752[***]
$PRRLW_{j2}(i) = c_2PRR_{j2}(i)^2 +$ $d_2PRR_{j2}(i) + e_1$	77	c_2	2.446	5.563[***]
		d_2	−36.179	−5.913[***]
		e_2	221.418	10. 806[***]
$PRRLW_{j3}(i) = c_3PRR_{j3}(i)^2 +$ $d_3PRR_{j3}(i) + e_3$	32	c_3	3.732	2.074[*]
		d_3	−40.823	−1.616
		e_3	230.506	2.684[*]
$PRRLW_{j4}(i) = c_4PRR_{j4}(i)^2 +$ $d_4PRR_{j4}(i) + e_4$	36	c_4	5.608	3.068[***]
		d_4	−46.053	−2.528[*]
		e_4	214.337	4.933[***]
$PRLW_{jk}(i) = c_5PRL_{jk}(i) + d_5$	262	c_5	0.528	36.285[***]
		d_5	−0.941	−5.356[***]
$PRBL_{jk}(i) = c_6PRL_{jk}(i) + d_6$	262	c_6	0.703	25.536[***]
		d_6	1.120	3.369[***]
$PRSL_{ji} = c_7PRSDW_{ji}^{d7}$	106	c_7	37.098	20.956[***]
		d_7	0.571	12.563[***]
$PRSD_{ji} = c_8PRSDW_{ji} + d_8$	96	c_8	0.082	5.619[***]
		d_8	0.508	30.897[***]

[***], [**], and [*] denotes $P < 0.001$, $P < 0.01$, and $P < 0.05$, respectively.

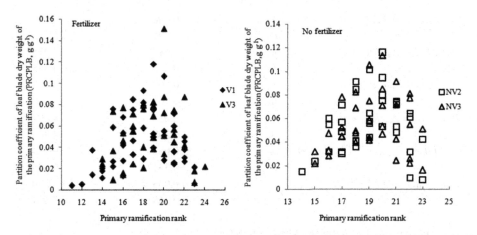

Fig. 3. Variation of the $PRCPLB_{ji}$ values by primary ramification rank in 2011–2012

$$PRCPLB_{ji} = \begin{cases} A_3 PRR_{ji} + B_3 & 11 \leq j \leq 19 \\ A_4 PRR_{ji} + B_4 & 19 \leq j \leq 24 \end{cases} \quad \text{Fertilizer} \tag{6}$$

$$NPRCPLB_{ji} = \begin{cases} A_5 PRR_{ji} + B_5 & 11 \leq j \leq 19 \\ A_6 PRR_{ji} + B_6 & 19 \leq j \leq 23 \end{cases} \quad \text{No fertilizer} \tag{7}$$

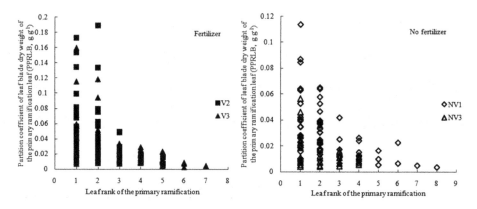

Fig. 4. Variation of the $PPRLBjk(i)$ values by leaf rank on primary ramification in 2011–2012

$$PPRLB_{jk}(i) = \begin{cases} A_7 \times e^{B_7 PRLR_{jk}(i)}, & 11 \leq j \leq 13 \quad \text{Fertilizer} \\ A_8 \times e^{B_8 PRLR_{jk}(i)}, & 11 \leq j \leq 23 \quad \text{No fertilizer} \end{cases} \tag{8}$$

Table 2. The determination of model parameters and significance test

Eq.	n	Parameter symbolic	Unstandardized coefficients	t
$PRCPLB_{ji} = A_3 PRR_{ji} + B_3$	56	A_3	0.008	5.549***
		B_3	−0.085	−3.513***
$PRCPLB_{ji} = A_4 PRR_{ji} + B_4$	34	A_4	−0.015	−3.915***
		B_4	0.368	4.515***
$PRCPLB_{ji} = A_5 PRR_{ji} + B_5$	39	A_5	0.009	4.187***
		B_5	−0.103	−2.778*
$PRCPLB_{ji} = A_6 PRR_{ji} + B_6$	32	A_6	−0.016	−4.097***
		B_6	0.398	4.803***
$PRCLB_{jk}(i) = A_7 e^{B7PRLPjk}$ (i)	159	A_7	0.053	−9.214***
		B_7	−0.341	9.645***
$PRCLB_{jk}(i) = A_8 e^{B8NPRLPjk}$ (i)	102	A_8	0.033	−5.270***
		B_8	−0.258	7.554***

3.3 Maximum Leaf Blade Width Model

The experiment on cultivar and fertilizer in the 2011 to 2012 showed that the $PRLW_{jk(i)}$ with the leaf length were described by a growth function (Eq. (9)). $R = 0.914$, $n = 262$, $P < 0.001$, $R_{(260,\ 0.001)} = 0.206$, $R^2 = 0.835$ (Fig. 5).

The *F*-values, *t*-values, c_5, d_5 all at $P < 0.001$ (Table 1).

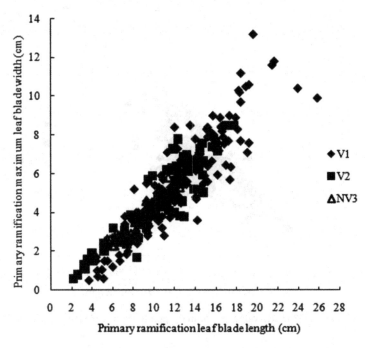

Fig. 5. Variation of $PRLW_{jk}(i)$ values by leaf length in 2011–2012

$$PRLW_{jk}(i) = c_5 PRL_{jk}(i) + d_5, \quad 11 \leq j \leq 23 \tag{9}$$

3.4 Leaf Blade Bowstring Length Model

Apparently, the maximum $PRBL_{jk}(i) = \text{PRLL}_{jk}(i)$. The experiment on cultivar and fertilizer in the 2011 to 2012 showed that variation of leaf bowstring length with the leaf length could be represented by the linear function. $R = 0.846, P < 0.001, \text{n} = 262,$ $R_{(260, 0.001)} = 0.206, R^2 = 0.715$ (Fig. 6).

The *F*-value, *t*-values, c_6, d_6 all were t at $P < 0.001$ (Table 1).

$$PRBL_{jk}(i) = c_6 PRL_{jk}(i) + d_6, \quad 11 \leq j \leq 23 \tag{10}$$

3.5 Stem Length Model of Primary Ramification

From the experimental data in the 2011 to 2012 we can see that the *j*th stem length of primary ramification (cm), $PRSL_{ji}$, changes with the dry weight could be described by a power function. $R = 0.776, \quad n = 106, \quad P < 0.001, \quad R_{(104, 0.001)} = 0.314, \quad R^2 = 0.603$ (Fig. 7).

The *F*-value, *t*-values, c_7, d_7 all were at $P < 0.001$ (Table 1).

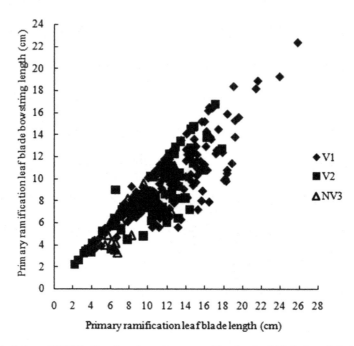

Fig. 6. Variation of $PRBL_{jk}(i)$ values by primary ramification leaf blade length in 2011–2012

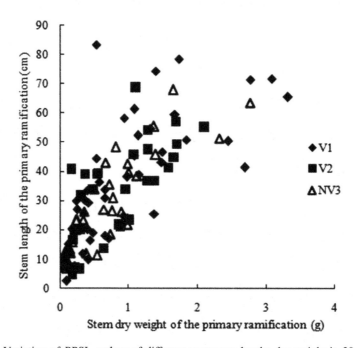

Fig. 7. Variation of $PRSL_{ji}$ values of different treatments by the dry weight in 2011–2012

$$PRSL_{ji} = c_7 PRSDW_{ji}^{d_7}, \quad 11 \leq j \leq 23 \tag{11}$$

3.6 Stem Diameter Model of Primary Ramification

The experimental data in the 2011 to 2012 showed that the *j*th stem diameter of primary ramification, $PRSD_{ji}$, changed with the dry weight could be described by a linear function with significant $R = 0.501$, $n = 96$, $P < 0.001$, $R_{(94,\ 0.001)} = 0.331$, $R^2 = 0.251$ (Fig. 8).

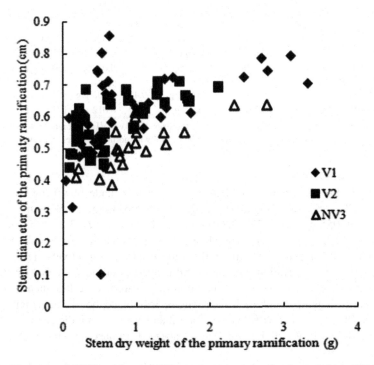

Fig. 8. Variation of $PRSD_{ji}$ values of different treatments by the dry weight in 2011–2012

The *F*-value, *t*-values, c_8, d_8 all were at $P < 0.001$ (Table 1).

$$PRSD_{ji} = c_8 PRSDW_{ji} + d_8, \quad 11 \leq j \leq 23 \tag{12}$$

3.7 Validation

We used the independent experimental data to validate the biomass-based rapeseed plant primary ramification morphological structure model proposed in this study. The *RMSE*, and the d_a in rapeseed primary ramification morphological parameters, leaf

Table 3. Model validation results in 2012–2013

Model parameters		Statistic parameters of measured values and simulated values					
		r	d_a	d_{ap} (%)	RMSE	n	Sig.
LL_1(cm)		0.6520***	−1.900 cm	21.11	5.033 cm	125	$r_{(123,0.001)} = 0.291$
LL_2 (cm)		0.6846***	−0.055 cm	0.70	3.233 cm	117	$r_{(115,0.001)} = 0.300$
LL_3 (cm)		0.6169***	0.274 cm	4.54	2.810 cm	87	$r_{(85,0.001)} = 0.347$
LL_4 (cm)		0.5275***	−0.720 cm	11.56	3.272 cm	90	$r_{(88,0.001)} = 0.341$
LW of PR (cm)		0.9071***	0.374 cm	11.90	0.778 cm	514	$r_{(512,0.001)} = 0.146$
LBL of PR (cm)		0.8555***	0.137 cm	2.21	1.193 cm	514	$r_{(512,0.001)} = 0.146$
PRL (cm)		0.8325***	0.806 cm	2.73	8.990 cm	145	$r_{(143,0.001)} = 0.273$
PRD (cm)		0.4476***	−0.025 cm	4.64	0.102 cm	153	$r_{(151,0.001)} = 0.264$
V1–V3$PRCPLB_{ji}$ (gg^{-1})	$11 \le j \le 19$	0.9091***	−0.004 gg^{-1}	10.81	0.021 g/g	50	$r_{(48,0.001)} = 0.451$
	$j > 19$	0.5079**	−0.003gg^{-1}	5.41	0.021 g/g	26	$r_{(24,0.01)} = 0.496$
NV2– NV3$PRCPLB_{ji}$ (gg^{-1})	$11 \le j \le 1$	0.6364***	0.003gg^{-1}	5.85	0.030 g/g	52	$r_{(50,0.001)} = 0.443$
	$j > 19$	0.6500***	−0.013 gg^{-1}	29.55	0.028 g/g	30	$r_{(28,0.001)} = 0.570$
V2–V3 $PRRLB_{jk}(i)$ (gg^{-1})		0.7440***	−0.0001 gg^{-1}	0.458	0.008 g/g	352	$r_{(350,0.001)} = 0.175$
NV1–NV3 $PRRLB_{jk}(i)$ (gg^{-1})		0.2385**	0.004 gg^{-1}	17.39	0.012 g/g	185	$r_{(183,0.001)} = 0.190$

N0V1–V1 with no fertilizer, N90V1–V1 with normal fertilizer, N180V1–V1 with high fertilizer; N90V2–V2 with normal fertilizer; N0V3–V3 with nofertilizer, N90V3–V3 with normal fertilizer, N180V3–V3 with high fertilizer

blade length, the maximum leaf blade width, the leaf blade bowstring length, stem length of PR, and stem diameter of PR were 5.033 cm, −1.900 cm (n = 125); 3.233 cm, −0.055 cm (n = 117); 2.810 cm, 0.274 cm (n = 87); 3.272 cm, −0.720 cm (n = 90); 0.778 cm, 0.374 cm (n = 514); 1.193 cm, 0.137 cm (n = 514); 8.990 cm, 0.806 cm (n = 145); and 0.102 cm, −0.025 cm (n = 153), respectively. The r values in rapeseed primary ramification morphological properties all at $P < 0.001$ or $P < 0.01$, but the ratio of d_a to the average observation (d_{ap}) values were less than 5 % for the second leaves length, the third leaves length, leaf blade bowstring length, the PR length, the PR diameter, $PRRLB$ values for V2 and V3, which indicated these model's accuracy is high (Table 3). The 1:1 line in rapeseed primary ramification were represented in Fig. 9.

The r values in rapeseed $PRCPLB$ all at $P < 0.001$ or $P < 0.01$, but the ratio of d_a to the average observation (d_{ap}) values were between 5 %−10 % for $PRCPLB$ with fertilizer ($j > 19$) and $PRCPLB$ no fertilizer ($11 \le j \le 19$), which indicated that these model's accuracy is good. The dap values of more than 10 % which indicated that the model had a lower accuracy, but the d_a values and the $RMSE$ of $PRCPLB$ and $PRRLB$ were small (Table 3). The 1:1 chart of measured values and simulated values in $PRCPLB$ and $PRRLB$ are represented in Fig. 9.

Notably, we had seen that the first leaf blade length and $PRRLB$ value(no fertilizer) models had obvious errors from Table 3, and Fig. 9, which showed that the two models still needed to be improved and perfected in the further.

3.8   Discussion

The study on FSPMs of rapeseed has important theoretical and application value for selection ideal plant type and regulation of plant type. One of the most important methods to establish the Function-Structural Model of rapeseed is to combine the rapeseed growth model with the morphological structure model by biomass. This paper established the relationships between morphological parameters of rapeseed and organ biomass, realized the organic combination of rapeseed growth model and rapeseed morphological model. It lays the basis for the establishment of FSPMs of rapeseed.

Studies on crop morphological structural models, such as rice, wheat, cotton, corn have been many reported. Chang *et al.* [19] constructed the simulation model of leaf elongation process in rice, analyzed the variation of leaf blade geometric morphology indices of rice with the growth process and environment conditions, and provided facilitate to digital and visualization of rice. Zhang *et al.* [20] established a process-based model with the methods of system analysis and dynamic modeling

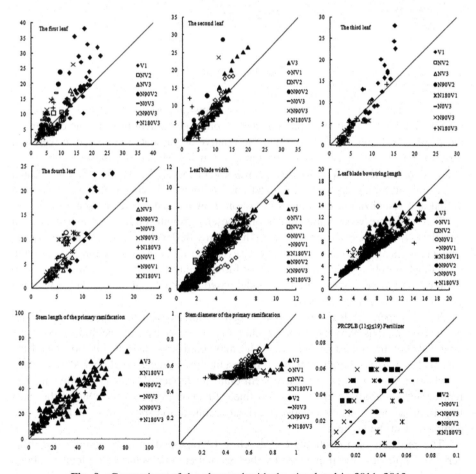

Fig. 9. Comparison of the observed with the simulated in 2011–2013

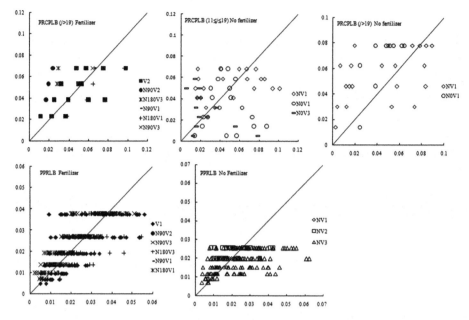

Fig. 9. (continued)

techniques. Fournier *et al.* [21] by using L-systems proposed a detailed description of the relationship between leaf functions and chlorophyll content of leaf. The concept of the relative leaf area index (LAI) and the relative accumulated temperature were put forward, and fitted parameters of the model by using MATLAB, the dynamic simulation model of leaf area index of corn was established [22]. By the computer model based on combining crop physiological and ecological process with visualization, a digital and visualization techniques of cotton growing system was presented [23].

Through there were some studies on rapeseed growth models and morphological models [24, 25], there are many research on growth-development law and structural characteristics of rapeseed [15, 26]. However, the combination of rapeseed growth model and morphological model is lack of further study. Yue [6] conducted a rapeseed ramification morphological structure model based on GDD, but the combined with the growth model was not mentioned.

In this paper, the leaf length models can be linked with the rapeseed dry matter production models through biomass, and the models also can be linked with the rapeseed growth models by biomass. The maximum leaf blade width model and the leaf blade bowstring length model are represented by a linear function with leaf blade length well. Stem length model and stem diameter model of primary ramification are represented by a power function and a linear function, respectively. One of the most important factors is the angle between primary ramification and main stem, and we will complete it in the coming.

The research on morphological structure of rapeseed primary ramification hasa certain complexity, and influenced by the external environment and difficult to obtain morphological indices accurately, tends to appear some experimental error. We will improve measurement methods, such as digital camera, and 3D laser scanner etc., in order to get the whole morphology of plant leaves and ramifications, exploration of the way to enhance the model accuracy.

4   Conclusions

In this paper, we developed a rapeseed PR structural model, which indented to explain the response mechanism of the PR morphogenesis to environmental conditions and varieties. Validation of the model with the independent experiment data indicated a good fitness between the simulated and observed in rapeseed.

The *PRRLW* was first put forward, and the relationships between rapeseed plant primary ramification morphological structure and the organ biomass to be established by *PRRLW*. It is a morphological structural parameter of rapeseed with a biological significance, and enhanced mechanism of this study.

Thus, the rapeseed plant primary ramification morphological structure model in this paper is feasible. We expect the proposed to be useful for other morphological indices of rapeseed in the further.

Acknowledgment. This work was supported by NFSC (31171455, 31201127, and 31471415), Jiangsu Province Agricultural Scientific Technology Innovation Fund [CX(12)5060], National High-Tech Research and Development Program of China (2013AA102305-1), and JAAS Basic Scientific Research Work Special Fund [ZX(15)4001, ZX(15)2008].

References

1. Foreign Agricultural Service. Oilseeds: World Markets and Trade (2014)
2. Xinyu, G., Xuyang, D., Wengang, Z., et al.: Study on the 3-D visualization of leaf morphological formation in corn. J. Maize Sci. **12**(special column), 27–30 (2004). (in Chinese with English abstract)
3. Zihui, T.: Study on simulation model of morphological development in wheat plant. Master degree dissertation of Nanjing Agricultural University (2006) (in Chinese)
4. Chunlin, S.: Study on morphological model and visual growth in rice. Ph.D. dissertation of Nanjing Agricultural University (2006) (in Chinese)
5. Juan, Z., Shuang, J., Binglin, C., et al.: Study on morphological model of stem, branch and leaf in cotton. Cotton Sci. **21**(3), 206–211 (2009). (in Chinese with English abstract)
6. Yanbin, Y.: The morphological structural model and visualization of rapeseed (*Brassica napus* L.) plant. Master degree dissertation of Nanjing Agricultural University (2010) (in Chinese)
7. Aizhen, S., Huojiao, H., Hongyun, Y., et al.: A visualization simulation of leaf shape elongation process in rice based on accumulated temperature change. Acta Agricuturae Universitatis Jiangxiensis **34**(5), 1058–1063 (2012). (in Chinese with English abstract)
8. Youhong, S., Yan, G., Baoguo, L., et al.: Virtual maize modelII. Plant morphological constructing based on organ biomass accumulation. Acta Ecolegica Sinica **23**(12), 2579–2586 (2003). (in Chinese with English abstract)

9. Mengzhen, K., Evers, J.B., Vos, J., et al.: The derivation of sink functions of wheat organs using the Greenlab model. Ann. Bot. **101**(8), 1099–1108 (2008)

10. Liu Yan, L., Jianfei, C.H., et al.: Main geometrical parameter models of rice blade based on biomass. Scientia Agricultural Sinica **42**(11), 4093–4099 (2009). (in Chinese with English abstract)

11. Hongxin, C., Yan, L., Yongxia, L., et al.: Biomass-based rice (*Oryza sativa* L.) above ground architectural parameter models. J. Integr. Agric. **11**(10), 101–108 (2012)

12. Hongxin, C., Wenyu, Z., Weixin, Z. et al.: Biomass-based rapeseed (*Brassica napus* L.) leaf geometric parameter model. In: Proceedings of the 7th International Conference on Functional-Structural Plant Models, Saariselkä, Finland, 9–14 June 2013 (2013) http://www.metal.fi/fspm2013/proceedings

13. Wenyu, Z., Weixin, Z., Daokuo, G., et al.: A biomass-based leaf curve model for rapeseed (Brassica napus L.). Jiangsu Agric. Sci. **30**(6), 1259–1266 (2014). (in Chinese with English abstract)

14. Weixin, Z., Hongxin, C., Yan, Z., et al.: Morphological structure model of leaf space based on biomass at pre-overwintering stage in rapeseed (*Brassica napus* L.) plant. Acta Agronomica Sinica **41**(2), 318–328 (2015). (in Chinese with English abstract)

15. Groer, C., Kniemeyer, O., Hemmerling, R. et al.: A dynamic 3D model of rape (*Brassica napus* L.) computing yield components under variable nitrogen fertilization regimes (2007). http://algorithmicbotany.org/FSPM07/proceedings.html

16. Müller, J., Braune, H., Wernecke, P. et al.: Towards universality and modularity: a generic photosynthesis and transpiration module for functional structural plant models (2007). http://algorithmicbotany.org/FSPM07/proceedings.html

17. Oil crops research institute, Chinese academy of agricultural sciences. Chinese rape cultivation. Agricultural Press, Beijing (1987) (in Chinese)

18. Houli, L.: Practical rapeseed cultivation. Shanghai Science and Technology Press, Shanghai (1987). (in Chinese)

19. Chang Liying, G., Dongxiang, Z.W., et al.: A simulation model of leaf elongation process in rice. Acta Agronomica Sinica **34**(2), 311–317 (2008). (in Chinese with English abstract)

20. Wenyu, Z., Liang, T., Xiangcheng, Z., et al.: Dynamic simulation of wheat stem-sheath angle based on process. Chin. J. Appl. Ecol. **22**(7), 1765–1770 (2011). (in Chinese with English abstract)

21. Fournier, C., Andrieu, B.: A 3D architectural and process-based model of maize development. Ann. Bot. **81**, 233–250 (1998)

22. Xinlan, L., Xianglan, C., Yunsheng, Y., et al.: Study on dynamic simulation model of leaf area index of corn in Northeast. Jiangsu Agric. Sci. **40**(1), 91–94 (2012). (in Chinese)

23. Xuebiao, P., Lizhen, Z., Chao, C., et al.: Digital and visualization techniques of cotton growing system. China Sci. Technol. Achievements **10**, 24 (2011). (in Chinese)

24. Liangzhi, G.: The basis of agricultural model. Tianma book co., LTD, Hong Kong (2004). (in Chinese)

25. Weixing, C., Weihong, L.: Crop System Simulation and Intelligent Management, pp. 149–166. Higher Education Press, Beijing (2003). (in Chinese)

26. Hongxin, C., Chunlei, Z., Guangming, L., et al.: Researches of simulation models of rape (*Brassica napus* L.) growth and development. Acta Agronomica Sinica **32**(10), 1530–1536 (2006). (in Chinese with English Abstract)

Effective Wavelengths Selection of Hyperspectral Images of Plastic Films in Cotton

Hang Zhang[1,2,3], Xi Qiao[1,3], Zhenbo Li[1,3], and Daoliang Li[1,3(✉)]

[1] College of Information and Electrical Engineering,
China Agricultural University, Beijing 100083, China
dliangl@cau.edu.cn
[2] College of Computer and Information Engineering,
Tianjin Agricultural University, Tianjin 300384, China
[3] Key Laboratory of Agricultural Information Acquisition Technology,
Ministry of Agriculture, Beijing 100083, China

Abstract. This research was conducted to investigate the application of detecting plastic films in cotton using visible and near-infrared hyperspectral imaging. A line-scan hyperspectral imaging system (326–1100 nm) was used to detect plastic films mixed with cotton which was an important quality issue. Hyperspectral reflectance images were acquired and difference spectra of cotton and plastic films were extracted and analyzed to determine the dominant wavelengths. Also, as one of the most commonly used methods for dimensionality reduction, principal component analysis (PCA) was chosen to process the hyperspectral images. Afterwards, effective wavelengths were selected by analyzing the first three principal components (PCs) and six single-band images at 473.24 nm, 497.29 nm, 530.6 nm, 670.81 nm, 674.71 nm, and 955.68 nm were extracted respectively. Finally, the selected wavelengths were validated to prove the effectiveness. The results indicated that the selected wavelengths could be able to detect plastic films in cotton instead of the whole wavelengths.

Keywords: Hyperspectral imaging · Difference spectra · Principal component analysis · Wavelengths selection

1 Introduction

The quality of cotton products is easily and seriously affected by foreign matter mixed into cotton during spinning, weaving, and dyeing. As one kind of foreign matter, plastic films are widely used to preserve soil temperature and moisture in China when growing cotton and make it a problem that a great many plastic films might be mixed with cotton when harvesting. They are easily to be broken into lots of very small pieces during spinning [1], which make removal of them difficult and make breakability of cotton yarn increase. Therefore, it is crucial for us to detect and remove plastic films in cotton rapidly and accurately.

© IFIP International Federation for Information Processing 2016
Published by Springer International Publishing AG 2016. All Rights Reserved
D. Li and Z. Li (Eds.): CCTA 2015, Part I, IFIP AICT 478, pp. 519–527, 2016.
DOI: 10.1007/978-3-319-48357-3_48

For the past two decades, computer vision technique has been applied to identify bark in cotton and determine the gravimetric bark content in cotton [2], conduct automated visual inspection of cotton quality [3], measure interlace of intermingled and false-twist textured yarns [4], detect structural defects in textiles [5], inspect and classify different types of foreign fibers [6, 7], etc. Furthermore, research has also been conducted on plastic films. Fang et al. [1] used an online visual detection machine to acquire images of plastic films and proposed a new method to identify plastic films. The result showed that the prosed method improved the detection rate of plastic films and reduced the negative identification accuracy. Nevertheless, the average identification accuracy was still not very high (43.33 %). In order to detect white foreign fibers (such as white plastic bags, white cotton cloth, transparent plastic films, etc.), Hua et al. [8] constructed a linear laser imaging system and carried out a series of experiments. The results indicated that it's still difficult to detect transparent plastic films although the contrast between object and background was increased effectively. Liu et al. [9] used linear laser cross-sectional imaging to distinguish typical white contaminants from cotton. 12 types of white contaminants including semitransparent plastic mulch and white plastic film were used in their experiment. The result indicated the detection rate was 97 % for white plastic films and 95 % for semitransparent plastic mulch.

As a promising technique, hyperspectral imaging has been used for detection of bruises on apples [10, 11], canker lesions on citruses [12], sprout damage and Fusarium head blight in wheat [13, 14], common defects on oranges [15], etc. However, very few researches have been reported for the application of this technique to evaluate cotton quality. To our knowledge, Guo et al. [16] firstly developed a hyperspectral imaging system in reflectance mode (422 nm–982 nm) to detect different trash on the surface of cotton including fine, colorless, light color, and white foreign matter. Nevertheless, the transparent polyethylene mulching films' recognition rates of training sets and test sets were only 41.7 % and 53.8 % respectively.

Furthermore, there are still some disadvantages for hyperspectral imaging technique to consider, such as higher cost and lower speed compared to conventional machine vision systems. Therefore, it is still important to investigate hyperspectral imaging for detecting plastic films in cotton and determine the effective wavelengths that can best characterize the features of plastic films.

2 Materials and Methods

2.1 Samples

Cotton samples and plastic films with different sizes and shapes were collected from a local cotton processing plant. The cotton samples were made to flat layers and the length and width of the cotton layer were 255 mm and 175 mm, respectively. Some plastic films were put on the surface of the cotton layer randomly and the others were put into the cotton layer at the depth of 1–6 mm. The image of the plastic films was shown in Fig. 1.

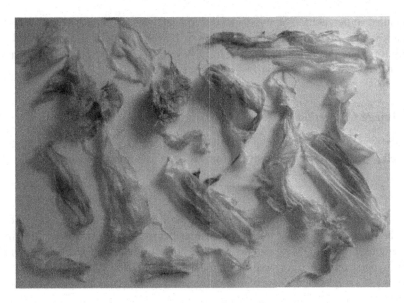

Fig. 1. Image of the plastic films.

2.2 System Setup

The assembled hyperspectral imaging system (Fig. 2) consisted of a prism-grating-prism imaging spectrograph (ImSpector V10E), a standard 23 mm C-mount zoom lens, an EMCCD camera, two 150-watt halogen lamps, an electric displacement platform, an image capture software, and a computer with a frame grabber. The scanning speed was adjusted at 0.8 mm/s. The object distance was set to 45 cm throughout the test. The exposure time was 20 ms with a frame rate of 16.47 fps to build hyperspectral images

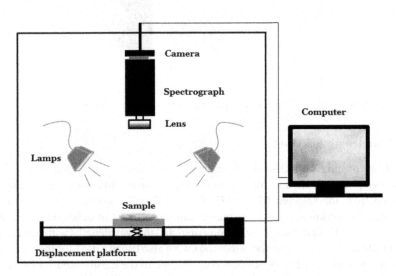

Fig. 2. Schematic of the hyperspectral imaging system.

(also called hypercube). The hyperspectral images were acquired at 1000 wavelength channels in the wavelengths from 326.7 nm to 1098 nm with an increment of 0.78 nm line by line. The image acquisition parameters were set by the Spectral Image and raw hyperspectral images with 1004 × 1420 pixles were obtained.

2.3   Image Correction

In order to correct the raw images, a dark image and a white reference image were acquired seperately. The equation below was used to correct the raw image:

$$I_C = I_R - I_D/I_W - I_D \tag{1}$$

In the equation, I_R is the raw image, I_W is the white reference image, I_D is the dark image and I_C is the corrected image. After image correction, the images at the first 170 wavelengths and last 80 wavelengths were cropped for their high level of noise and the size of hyperspectral images was then reduced to 600 × 400 pixels at 750 wavelength channels. A sample data cube of the corrected hyperspectral images was shown in Fig. 3.

Fig. 3. Data cube of a hyperspectral image of plastic films in cotton.

2.4   Wavelength Selection

The regions of interest (ROIs) of hyperspectral images were used to extract spectra of cotton and plastic films and mean spectra that used for following processing were computed from the selected ROIs [17]. Then, difference spectra were calculated by subtracting the cotton spectra from plastic films spectra. The difference between the spectra of cotton and plastic films was compared and the wavelengths with local maxima were selected. Multivariate analysis methods were also usually used for the reduction of redundant information of hyperspectral data. As one of the mostly used statistical tools, principal component analysis (PCA) was adopted for dimensionality reduction by transforming original data (usually high correlated) into a new set of uncorrelated images that were called principal components (PCs) images. PCA loadings or eigenvectors were different among PCs images and the local maxima and/or local minimum corresponding wavelengths could be chosen as the determinative wavelengths. Most of the information could be carried by the images with selected effective wavelengths and might be more efficient for the following data processing [18].

3 Results and Discussion

3.1 Difference Spectra Between Cotton and Plastic Films

8 ROIs chosen from different locations on cotton and plastic films of every hyperspectral images in the 450–1035 nm region were selected to extract the spectral characteristics. The elliptic ROI areas with different sizes (about 80 to 200 pixels) were selected as representative areas of spectral reflectance. The extracted spectra of cotton and plastic films were averaged respectively and the average reflectance intensities from images in cotton were higher than those in plastic films. Then, the mean spectra of cotton were subtracted from the mean spectra of plastic films to obtain the difference spectra. As shown in Fig. 4, there were two local maxima at 497.29 nm and 670.81 nm indicating that the reflectance difference values of cotton and plastic films in these two wavelengths had large intensities and the two wavelengths could be selected as the dominant wavelengths. Thus, the images obtained from the two wavelengths could be used for further analysis.

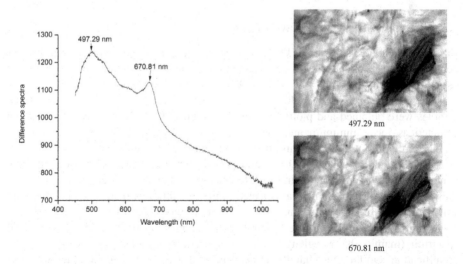

Fig. 4. Difference spectra of cotton and plastic films and hyperspectral reflectance images at 497.29 nm and 670.81 nm.

3.2   PCA on the Whole Wavelengths

The hyperspectral images after correction were subjected to PCA using the wavelengths in the spectral ranges (450–1035 nm). The representative PC1 to PC3 obtained from PCA accounted for over 96 % of the total variance of all bands for the images of plastic films, thus the first three PC images could be an alternative to substitute the raw images for image processing and data analysis. Also, PCA could be used to conduct dimensionality reduction of hyperspectral images.

As shown in Fig. 5, the resultant PC images indicated that the major features of plastic films became more evident, especially PC1 images. PC1 images represented the gray value information of the sample and provided more features of plastic films to make them be clearly identified. The plastic films in the PC2 images were brighter than those in the PC1 images. PC3 images were much noisy than the PC1 and PC2 images. PC2 images and PC3 images gave some other useful features for detection as well. The rest PCs depicted other features. However, starting from PC4 image, the transformed images did not give more meaningful information to detect plastic films.

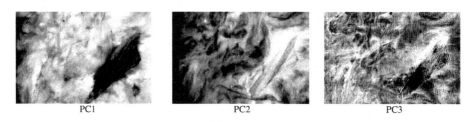

Fig. 5. The first three PC images obtained using the whole wavelengths

3.3   Selection of Effective Wavelengths

Based on the analysis results of PCA, several wavelengths of the hyperspectral images could be selected as the effective wavelength and the corresponding images could be used to represent the whole hyperspectral images.

The average weighing coefficients (eigenvectors) for the first three PCs of the samples were obtained and plotted in Fig. 6. A total of five effective wavelengths that depicted as local minimum and maximum were selected, which were centered at 473.24 nm, 530.6 nm, 670.81 nm, 674.71 nm, and 955.68 nm, respectively.

As aforementioned, 497.29 nm and 670.81 nm were the other two effective wavelengths that were selected in the difference spectra analysis. It could be found that the wavelength of 670.81 nm was selected in the two different methods. Therefore, six wavelengths were chosen for analysis. It's quite obvious that five of the six selected wavelengths were in the visible spectral region suggesting that the visible region of spectrum (mainly the wavelengths in red and green regions) was critical for the identification. On the other hand, one near-infrared wavelength (955.68 nm) was also selected as the effective wavelength indicating that the near-infrared spectral region might have the possibility to detect plastic films in cotton. Further research could be conducted in the whole near-infrared spectral region to verify the feasibility.

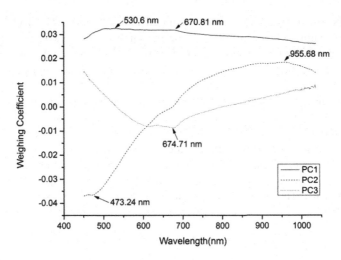

Fig. 6. Weighing coefficients (eigenvectors) of first three PCs obtained from principal component analysis.

3.4 Validation of Selected Wavelengths

PCA was then conducted on the six effective wavelengths (473.24 nm, 497.29 nm, 530.6 nm, 670.81 nm, 674.71 nm, and 955.68 nm). Figure 7 showed the first three PC images. As illustrated in Fig. 7, PC1 and PC2 images were similar to those PC images previously obtained based on analysis of the full wavelength range although PC3 image was not quite the same to the corresponding one shown in Fig. 5. Also, the first two PCs explained almost 98 % (PC1: 92.09 % and PC2: 5.92 %) of the variance indicating that the selected wavelengths could be efficiently used for plastic films detection in cotton. And the selected wavelengths could also be used to construct a multispectral imaging system to detect plastic films in cotton in real time.

PC1   PC2   PC3

Fig. 7. The first three PC images obtained using the six effective wavelengths

4 Conclusion

Hyperspectral reflectance images was acquired for the detection of plastic films in cotton and a new wavelength selection method was proposed by combining difference spectra analysis with principal component analysis.

Hyperspectral images were firstly acquired and corrected. The ROIs were then selected and subjected to difference spectra analysis and two local maxima at 497.29 nm and 670.81 nm were chosen because the reflectance difference values of cotton and plastic films in these two wavelengths had large intensities. Then, the hyperspectral images were subjected to PCA and the effective wavelengths were obtained by analyzing the first three PCs. Afterwards, six single-band images at 473.24 nm, 497.29 nm, 530.6 nm, 670.81 nm, 674.71 nm, and 955.68 nm were extracted respectively. Finally, the selected wavelengths were validated by using PCA and the effectiveness was proved. Furthermore, multispectral imaging systems could be developed using the selected wavelengths to lower the cost and enhance the speed.

Acknowledgments. This work was supported by the National Natural Science Foundation of China (31228016).

References

1. Fang, J., Jiang, Y., Yue, J., et al.: A hybrid approach for efficient detection of plastic mulching films in cotton. Math. Comput. Model. **58**(3–4), 834–841 (2013)
2. Lieberman, M., Bragg, C.K., Brennan, S.N.: Determining gravimetric bark content in cotton with machine vision. Text. Res. J. **68**(68), 94–104 (1998)
3. Tantaswadi, P., Vilainatre, J., Tamaree, N., et al.: Machine vision for automated visual inspection of cotton quality in textile industries using color isodiscrimination contour. Comput. Ind. Eng. **37**(1–2), 347–350 (1999)
4. Millman, M.P., Acar, M., Jackson, M.R.: Computer vision for textured yarn interlace (nip) measurements at high speeds. Mechatronics **11**(8), 1025–1038 (2001)
5. Abouelela, A., Abbas, H.M., Eldeeb, H., et al.: Automated vision system for localizing structural defects in textile fabrics. Pattern Recogn. Lett. **26**(10), 1435–1443 (2005)
6. Yang, W., Li, D., Zhu, L., et al.: A new approach for image processing in foreign fiber detection. Comput. Electron. Agric. **68**(1), 68–77 (2009)
7. Li, D., Yang, W., Wang, S.: Classification of foreign fibers in cotton lint using machine vision and multi-class support vector machine. Comput. Electron. Agric. **74**(2), 274–279 (2010)
8. Su, H.C.: White Foreign Fibers Detection in Cotton Using Line Laser. Trans. Chin. Soc. Agric. Mach. **43**(2), 181–185 (2012). Nongye Jixie Xuebao
9. Liu, F., Su, Z., He, X., et al.: A laser imaging method for machine vision detection of white contaminants in cotton. Text. Res. J. **84**(18), 1987–1994 (2014)
10. Lu, R.: Detection of bruises on apples using near–infrared hyperspectral imaging. Trans. Asae **46**(2), 523–530 (2003)
11. Xing, J., Bravo, C., Jancsók, P.T., et al.: Detecting bruises on 'golden delicious' apples using hyperspectral imaging with multiple wavebands. Biosyst. Eng. **90**(1), 27–36 (2005)
12. Qin, J., Burks, T.F., Kim, M.S., et al.: Citrus canker detection using hyperspectral reflectance imaging and PCA-based image classification method. Sens. Instrum. Food Qual. Saf. **2**(3), 168–177 (2008)
13. Xing, J., Symons, S., Shahin, M., et al.: Detection of sprout damage in Canada Western Red Spring wheat with multiple wavebands using visible/near-infrared hyperspectral imaging. Biosyst. Eng. **106**(2), 188–194 (2010)

14. Barbedo, J.G.A., Tibola, C.S., Fernandes, J.M.C.: Detecting Fusarium head blight in wheat kernels using hyperspectral imaging. Biosyst. Eng. **131**, 65–76 (2015)
15. Li, J., Rao, X., Ying, Y.: Detection of common defects on oranges using hyperspectral reflectance imaging. Comput. Electron. Agric. **78**(1), 38–48 (2011)
16. Guo, J., Ying, Y., Cheng, F., et al.: Detection of foreign materials on the surface of ginned cotton by hyper-spectral imaging. Trans. Chin. Soc. Agric. Eng. **28**(21), 126–134 (2012). Nongye Gongcheng Xuebao
17. Qin, J., Burks, T.F., Zhao, X., et al.: Multispectral detection of citrus canker using hyperspectral band selection. Trans. Asabe **54**(6), 2331–2341 (2011)
18. Wold, J.P., Jakobsen, T., Krane, L.: Atlantic salmon average fat content estimated by near-infrared transmittance spectroscopy. J. Food Sci. **61**(61), 74–77 (1996)

Research and Experiment on Precision Seeding Control System of Maize Planter

Nana Gao, Weiqiang Fu, Zhijun Meng$^{(\boxtimes)}$, Xueli Wei, You Li, and Yue Cong

Beijing Research Center of Intelligent Equipment for Agriculture, Shuguang Park No. 11, Banjing, Haidian District, Beijing, China
{gaonn, fuwq, mengzj, weixl, liy, congy}@nercita.org.cn

Abstract. A precision seeding control system was developed based on GPS, which could achieve real-time adjustment of seeding speed to ensure seed distribution uniformity. Parameters including seeding rate, row spacing and hole number of seed metering device were entered by users which were transferred by CAN bus to the controller. Its electronic speed control system could work out rotational speed of sowing axis matching with operating speed of tractor automatically. Seed metering device was driven by the hydraulic motor. The field experiment results demonstrated that rate of spacing of normally sown seeds and variable coefficient of this system comparing to traditional land wheel driven planter was relatively 94.1 %, 24.6 % and 89.1 %, 33.1 % while operating speed was 5.9 km/h. If the variation range of seeding spacing shrinks 40 % according the standard request, rate of spacing of normally sown seeds of precision seeding control system was 89.7 % exceeded the traditional land wheel 23 %.

Keywords: Precision seeding · Control system · GPS speed measurement · Hydraulic motor · Maize planter · Distribution uniformity of seeds

1 Introduction

Precision seeding is sowing a certain number of seeds into predetermined location of soil that is to say, the 3-dimensional spatial coordinate position of seeds is constituted by row space, seed spacing and seed depth. At present, land wheel was commonly used to drive seed metering device of maize planer and groups of sprockets were applied to change the transmission ratio and adjust seed spacing on maize planters. High resistance and skid of land wheel during seeding operation lead to lower rate of spacing, especially when planters were working in high-speed, heavy straw coverage and none-plowing soil. Adjusting sprockets manually can only change seed space step by step within a certain range, it cannot meet the requirement of precision seeding [1–3]. Practice has proved that there was great different between calibration seeding rate and actual seeding rate. The difference resulted seeding rate decreasing, number of seedlings reducing and lower production which was caused by soil texture, soil moisture, tillage quantity and skid of land wheel, and so on.

© IFIP International Federation for Information Processing 2016
Published by Springer International Publishing AG 2016. All Rights Reserved
D. Li and Z. Li (Eds.): CCTA 2015, Part I, IFIP AICT 478, pp. 528–535, 2016.
DOI: 10.1007/978-3-319-48357-3_49

Precision seeding control systems have been developed for a long time to save seeds, improve operating speed and make sure seeding uniformity. For example, Germany Amazone Company produced precise variable seeder which could adjust speed infinitely according to instructions from airborne computer with DGPS [4, 5]. A poleless speed controller was designed by Shandong Agricultural University with their seed metering which can adjust rotation speed of sowing axis based on forward speed in PID control mode [6–8]. The electro-hydraulic propritional control hill distance system developed by HAU used PLC to control proportional flow valve and adjusted motor speed in real time based on land wheel speed [9]. However, groups of sprockets and geared head are widely used to change the transmission ratio and adjust seed spacing on maize planters. There is no mature seeding control system to apply in production [10].

In conclusion, a precision seeding control system of maize planter based on GPS was developed by Beijing Research Center of Intelligent Equipment for Agriculture of which hydraulic motor replaced land wheel to drive seed metering. The system realized real-time adjustment of seeding speed to ensure the distribution uniformity of seeds. Performance experiments of the system were conducted in the field.

2 Materials and Methods

2.1 System Organization and Working Principle

Precision seeding control system of maize planter includes a mechanical seeding system, speed measurement device, hydraulic system, and electronic speed control system, shown in Fig. 1. The mechanical seeding system is consisted of seed metering device and its auxiliary mechanism. Speed measurement device based on high precision GPS and its rate accuracy is 0.1 m/s, the time interval of measurement 1 s. Hydraulic system includes hydraulic motor and proportional valve the function of which is driving seeding axis to match working speed of planter. Electronic speed control system is the most important controller of the system which received working

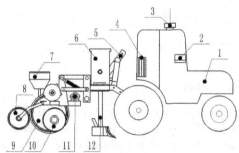

1.tractor 2.touch screen 3. speed measuring device 4.controller 5.trifilar suspension 6.fertilizer box 7.seed box 8.press wheel 9.land wheel 10.seed opener 11.hydrualic motor 12.fertilizer opener

Fig. 1. Structure diagram of precision seeding control system of maize planter

speed by CAN bus from GPS, computed rotational speed of hydraulic motor based on parameters from user and GPS, and controlled hydraulic motor rotation. There is a close-cycle control system with hydraulic motor speed feedback measured by a photoelectric encoder. Speed of seed metering axis will adjust in real time while working speed changing which make sure seed spacing adjusting accuracy and its adjustment range expanded.

2.2   System Seeding Space Control

The system requires users to input parameters of planter and objective seeding rate in human-computer interaction interface. Figure 2 is the seeding setting module of system. High precision GPS measures working speed in real time while operation starting, and the speed information and other parameters transfer to the controller for calculating rotational speed of motor. The relation between motor speed and other parameters is as follows:

$$n_m = 60v_m/ \left(S \cdot Z \cdot i \right) = 6 \times 10^{-3} C \cdot N \cdot v_m/ \left(Z \cdot i \right)$$

Where n_m is output rotational speed of motor/r·min^{-1}, v_m is working speed measured by GPS/m·s^{-1}, C is row space/m, S is seed spacing/m, Z is hole number of seed metering device, i is transmission ratio of motor axis and sowing axis, N is the number of seeds per ha.

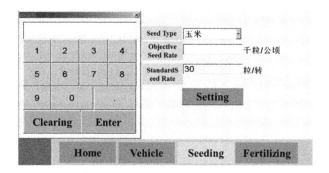

Fig. 2. Seeding setting module

2.3   Experiment Methods

Experiments were conducted in spring 2015 in Xiaotangshan National Precision Agriculture Research Demonstration Base. Sowing uniformity of the system on 2BQX-6 maize planter was measured in different working speed and different seeding rate. Two kinds of treatment were arranged: sowing on surface and normal sowing. Sowing on surface can make sure seeds maximally exposed on soil surface, which should decrease the action of depth roller and press wheel to minimize. Normal sowing is seeding to suitable depth according to maize demand. Experiments mainly surveyed seeding uniformity index included rate of spacing of normally sown seeds, rate of miss (multiple), and variable coefficient based on JB/T 10293-2013 Specifications for single seed drills (precision drills).

3 Results and Discussion

3.1 Seeding Uniformity in Different Working Speed

In this experiment, setting the same row space, hole number of seed metering device, and target seeding rate, sowing on surface, test results for low speed and high speed respectively repeated three times were presented in Table 1.

Table 1. Test results of seeding uniformity on different speed

No.	1	2	3	4	5	6
Working speed/km/h	5.7			3.2		
Rate of spacing/%	83.3	95.8	82.6	91.7	87.0	91.3
Rate of multiple/%	11.1	0.0	9.7	8.3	8.7	1.0
Rate of miss/%	5.6	4.2	7.7	0.0	4.3	7.7
Standard deviation	9.2	5.4	6.8	5.2	8.5	7.2
CV/%	38.5	25.0	31.9	24.8	38.3	34.3

3.2 Seeding Uniformity in Different Seeding Rate

In this experiment, setting the same row space, hole number of seed metering device, and keeping working speed at 5.6 km/h, sowing on surface, test results for target seeding rate 75000 per ha and 67500 per ha respectively repeated three times were presented in Table 2.

Table 2. Test results of seeding uniformity on different seeding rate

No.	1	2	3	4	5	6
Sowing rate/ha.	75000			67500		
Rate of spacing/%	83.3	86.4	91.1	95.0	85.0	81.0
Rate of multiple/%	12.5	9.1	4.3	0	10.0	11.5
Rate of miss/%	4.2	4.5	4.6	5.0	5.0	7.5
Standard deviation	7.7	8.1	5.6	4.8	5.9	9.7
CV/%	35.4	35.8	26.8	19.7	23.6	39.6

The results of different speed and different seeding rate showed that it was up to standard request that rate of spacing of normally sown seeds of the precision seeding control system. Meanwhile, there was no significant difference between the different levels of working speed and seeding rate. Sowing on surface always led to bouncing while seeds touching soil and influencing seed space greatly. In view of the disadvantage of sowing on surface, sowing on surface and normal sowing comparison experiments were conducted.

3.3   Different Sowing Method Comparison Experiments

In this experiment, setting the same seeding rate and keeping the same working speed, test results of distribution uniformity of seed on different sowing method, sowing on surface and normal sowing respectively repeated three times were presented in Table 3. The results showed that sowing on surface leads to lower uniformity index because of seed bouncing. However, rate of spacing, rate of miss and rate of multiple were qualified when seed was sown into soil layer as standard requested. To sum up, the precision seeding control system met specifications for precision seed drills.

Table 3. Test results of seeding uniformity on different seeding method

No.	1	2	3	4	5	6
Sowing method	Sowing on surface			Normal sowing		
Rate of spacing/%	90.0	82.4	84.2	95.0	94.7	98.0
Rate of multiple/%	5.0	9.8	10.0	0.0	0.0	2.0
Rate of miss/%	5.0	7.8	5.8	5.0	5.3	0.0
Standard deviation	5.2	8.9	9.4	5.8	6.4	3.5
CV/%	18.3	32.6	37.2	22.4	26.8	14.3

3.4   Different Seeding Driven Comparison Experiments

Performance testing of precision seeding control system indicated that the major index such as rate of spacing, CV, were above standard request. Moreover, the performance of system was close to bench test [11–14]. To check out the system advantage, comparison experiments on seeding driven by traditional land wheel and system were conducted. In this experiment, using 2BQX-6 maize planter and keeping parameters consistent except seeding driven device, test results for seeding driven by land wheel and control system respectively repeated three times were presented in Table 4.

Table 4. Test results of seeding driven by land wheel and control system

No.	1	2	3	4	5	6
Seeding driving device	Driven by land wheel			Driven by precision seeding control system		
Rate of spacing/%	81.2	85.7	91.0	100.0	94.7	95.0
Rate of multiple/%	14.3	6.8	4.5	0.0	0.0	5.0
Rate of miss/%	4.5	7.5	4.5	0.0	5.3	0.0
Standard deviation	8.3	7.4	7.9	3.3	5.3	5.0
CV/%	37.9	29.5	34.1	13.4	20.2	19.4

The results showed that seeding uniformity of system was apparently higher than land wheel, and CV was obvious lower than it. Therefore, the system has significant function on enhancing seeding precision.

3.5   Analysis on Rate of Spacing

Normal seed spacing is distance of two adjacent seeds in the soil between half theory seed spacing and one-and-a-half seed spacing in JB/T 10293-2013 Specifications for single seed drills (precision drills). If rate of seed spacing is equal or greater than 80 %, and rate of multiple, rate of miss is respectively equal or less than 15 % and 8 % while theory seed spacing between 20 and 30 cm, CV equal or less than 30 % is qualified operation [15].

The system is based on high precision GPS measuring speed and hydraulic motor driving seed metering device which has high accuracy speed measurement and high stability rotational speed control. It can improve uniformity and distribution stability of seed spacing. If the range of normal seed spacing shrinks 40 %, that is new normal seed spacing between 0.7 times and 1.3 times theory seed spacing as qualified. Figures 3 and 4 demonstrated the distribution of seed spacing driven by land wheel and the precision seeding control system respectively. Rate of spacing of normally sown seeds measured according to standards were 89.1 % and 94.1 %, driven by land wheel and control system respectively. Meanwhile, based on the new normal seed spacing standard, the corresponding rate of spacing were 66.7 % and 89.7 % respectively. It can be seen that precision seeding control system can greatly keep seed spacing stability and increase seed distribution uniformity. It is helpful for maize to enhance ventilation and light transmission, be beneficial to root growth, and improve lodging-resistance capability [16–18].

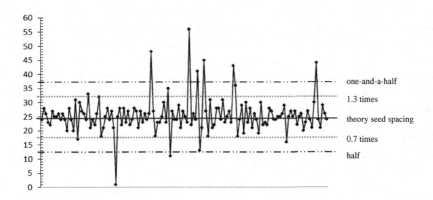

Fig. 3. Distribution of seed spacing driven by traditional land wheel

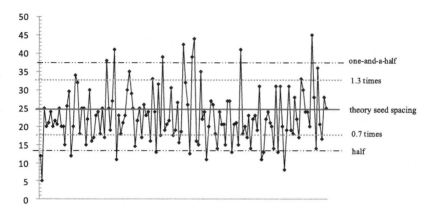

Fig. 4. Distribution of seed spacing driven by precision seeding control system

4 Conclusions

Precision seeding control system consists of a mechanical seeding system, speed measurement device, hydraulic system, and electronic speed control system. The system was adopted to solve land wheel skid and seed spacing adjusting problem during seeding operation. The system experiments indicated that different levels of working speed and seeding rate showed no significant difference on rate of spacing, rate of miss (multiple), et al. Rate of spacing of normally sown seeds measured according to standards were 89.1 % and 94.1 %, driven by land wheel and control system respectively. Meanwhile, after shrinking the range of normal seed spacing 40 %, the corresponding rate of spacing was 66.7 % and 89.7 % respectively. The system provides technology and experiment reference for precision seeding system, whereas there are some improvements in starting and stopping stage of seeding operation. Seeding information offered by users transferred from human-computer interaction interface to seed metering axis goes through GPS device, seeding control terminal, rotational speed controller and hydraulic motor. There is delayed starting and stopping of seed metering about 5 m respectively during an operation speed between 5 km/h and 6 km/h. More research on method and precision of switching between land wheel and control system for seeding driven is likely to solve the problem of delayed, and this will provide assistance to the precision seeding control system enlarging application range in kinds of planters.

Acknowledgment. Fund for this research was provided by the National 863 Program (2012AA101901), precise operation technology and equipment for large-scale production of food crops.

References

1. Mei, T., Li, Z., Wang, X., et al.: Study on pneumatic precision seeder in China. Agric. Equip. Veh. Eng. **51**(4), 17–21 (2013)

2. Sun, Q., Wang, Y., Wang, D., et al.: Experimental study on pneumatic precision seed metering mechanism of field breeding planter. Trans. Chin. Soc. Agric. Eng. (Trans. CSAE) **28**(Supp. 2), 59–64 (2012)
3. Gong, L., Yuan, Y., Shang, S., et al.: Design and experiment on electronic control system for plot seeder. Trans. CSAE **27**(5), 122–126 (2011)
4. Zhang, M., Wu, C.: The survey and the prospect in china of the pneumatic drills. Chin. Agric. Mechanization **2**, 70–72 (2008)
5. He, Y.: Precision Agriculture, pp. 178–192. Zhejiang University Press, Hangzhou (2010)
6. Zhao, L., Zhang, Y., Song, W., et al.: Improved application of stepless speed control technology in seeding machine. Comput. Syst. Appl. **23**(6), 250–254 (2014)
7. Wang, C., He, R.: Performance detection of precision seed-metering device based on single chip microprocessor. Sci. Technol. Eng. **33**(12), 8302–8329 (2011)
8. Li, J., Zhao, L., Bi, J., et al.: Design of intelligent control system for two-row precise seeding of wheat. Trans. CSAE **5**(28), 134–139 (2012)
9. Fu, W., Luo, X., Zeng, S., et al.: Design and experiment of electro-hydraulic proportional control hill distance system of precision rice hill-drop drilling machine for dry land. Trans. CSAE **31**(9), 25–31 (2015)
10. Tang, X.: Research on the Planting Drive System Based on a Tractor Front-Wheel Speed. Hebei Agricultural University, Baoding (2009)
11. Wang, J., Tang, H., Zhou, W., et al.: Improved design and experiment on pickup finger precision seed metering device. Trans. CSAM **9** (2015)
12. Liu, J., Cui, T., Zhang, D., et al.: Experimental study on pressure of air-blowing precision seed-metering device. Trans. CSAE **27**(12), 18–22 (2011)
13. Yu, J., Wang, G., Xin, N., et al.: Simulation analysis of working process and performance of cell wheel metering device. Trans. CSAM **142**(12), 83–87 + 101 (2011)
14. Shi, S., Zhang, D., Yang, L., et al.: Design and experiment of pneumatic maize precision seed-metering device with combined holes. Trans. CSAE **30**(5), 10–18 (2014)
15. People's Republic of China Ministry of Industry and Information Technology. JB/T10293-2013 Specifications for single seed drills (precision drills). Machinery Industry Press, Beijing (2014)
16. Cui, T., Zhang, D., Yang, L., et al.: Design and experiment of collocated-copying and semi-low-height planting-unit for corn precision seeder. Trans. CSAE **28**(S2), 18–23 (2012)
17. Li, D., Jiang, X., Tong, L.: Effect of planting density on root-shoot growth and water utilization efficiency of seed corn. J. Drainage Irrig. Mach. Eng. **32**(12), 1091–1097 (2014)
18. Liu, Z., Xiao, J., Yu, J., et al.: Effects of varieties and planting density on plant traits and water consumption characteristics of spring maize. Trans. CSAE **28**(11), 125–131 (2012)

Study on Time and Space Prediction Model About Rice Yield in Hei Longjiang Province

Guowei Wang[✉], Hongyan Hu, Hao Zhang, and Yu Chen

College of Information Technology, Jilin Agricultural University,
Changchun 130118, China
{41422306,1033823087,496837382,123929697}@qq.com

Abstract. Predicting rice yield plays a significant role in preparing production plan and relevant decision-making for following year. The paper adopts ARIMA time series algorithm taking rice yield per hectare collected between 1991 and 2010 from different prefecture-level cities of Heilongjiang Province as the object of time series processing, to analyze the rice yield recorded between 1991 and 2009 of different prefecture-level cities and to build prediction model of rice yield. By using the model, the rice yield of different prefecture-level cities in 2010 is predicted. The predicted and actual space distribution of rice yield is obtained using ArcGIS software. And by conducting spatial data analysis, the space distribution diagram is mapped out. It is analyzed that the accuracy of predicted model on average attains over 95 %, featuring good prediction effect.

Keywords: ARIMA model · Rice · GIS · Yield prediction

1 Introduction

Heilongjiang is the national important commodity grain base; among which rice is one of the three major crops. Predicting rice production in Heilongjiang province exerts significant impact on the national macro policy. Foreign production prediction method focuses on statistical dynamics growth simulation model, meteorological yield prediction model and remote sensing technology prediction model; while domestic production prediction is based on the mathematical model; Liu Qianpu uses the space-time regression prediction model to predict the output of grain of Henan province and municipality, and Li Bingjn predicts short-term grain output using grey linear regression model; Chen Xiangfang proposes a kind of regression tree based on multivariate time series prediction model, to predict the yield of cucumber; Xu Xingmei proposes the model based on clustering analysis and scheduling algorithm to predict corn production. By removing the noise data and reducing the data dimension [7], the paper builds space-time prediction model of rice yield and the rice yield collected in 2010 of Heilongjiang province was predicted by using this model. Such aspects as yield data

D. Li and Z. Li (Eds.): CCTA 2015, Part I, IFIP AICT 478, pp. 536–545, 2016.
DOI: 10.1007/978-3-319-48357-3_50

processing, selection of predictors, establishment of model and stability of model is noticed during the process of prediction, which has improved the accuracy of the prediction.

To sum up, it has yet reported that the comprehensive utilization of gis and time series analysis are used for space-time prediction of rice production in Heilongjiang province.

2   ARIMA Model

The ARIMA model is difference autoregressive moving average model in its full name; ARIMA (p, d, q) is called difference autoregressive moving average model, with AR being autoregression; p autoregression item, MA moving average number, d the difference time conducted when time series become reliable. The so-called ARIMA model refers to the model established by converting non-stationary time series into the stationary time series and then performing regression of the dependent variable lag value and the present value of the random error and lag values, so as to convert the non-stationary series into stationary series.

The model is generally referred to as the ARIMA (p, d, q); the model parameter–p, d, q are nonnegative integers, meaning autoregression, the order of integration, and all parts of the moving average model. Baucus ARIMA model serves as an important part of a Jenkins method of time series model.

2.1   Stationarity Test

Usually it is based on the following functions to determine sequences' stationary.
Mean function:

$$\mu_n = EX_n = \int_{-\infty}^{\infty} x dF_n(x) \tag{1}$$

Variance function:

$$DX_n = E(X_n - \mu_n)^2 = \int_{-\infty}^{\infty} (x - \mu_n)^2 dF_n(x) \tag{2}$$

Auto-covariance function:

$$\gamma(n, n+k) = E(X_n - \mu_n)(X_{n+k} - \mu_{n+k}) \tag{3}$$

Autocorrelation function:

$$\rho(n, n+k) = \frac{\gamma(n, n+k)}{\sqrt{DX_n \cdot DX_{n+k}}} \tag{4}$$

2.2 Pure Randomness Test

Null hypothesis: delay between periods no more than m phase sequence values are independent of each other.

$$H_0 : \rho_1 = \rho_2 = \rho_3 = \cdots = \rho_m = 0, \forall m \geq 1 \tag{5}$$

The test statistic:

$$Q_{LB} = n(n+2) \sum_{k-1}^{m} \left(\frac{\hat{\rho}_k^2}{n-k}\right) \sim \chi^2(m) \tag{6}$$

$Q_{LB} = n(n+2) \sum_{k-1}^{m} \left(\frac{\hat{\rho}_k^2}{n-k}\right) \sim \chi^2(m)$, reject the null hypothesis, and consider the sequence as a purely random sequence, can be modeled

$Q_{LB} = n(n+2) \sum_{k-1}^{m} \left(\frac{\hat{\rho}_k^2}{n-k}\right) \sim \chi^2(m)$, accept the null hypothesis, consider the sequence as pure random sequence and model terminal.

In general, take m = 6, 12, 18.

2.3 Processing of Outliers

If X_{t+1} is an outlier, we can use $\hat{X}_t$ to correction X_{t+1}, $\hat{X}_t = 2X_t - X_{t-1}$ [8].

3 Rice Yield Prediction Model

3.1 Data Collection

This paper collected the rice yield data in municipalities of Heilongjiang Province from 1991 to 2010, taking rice yield from 1991 to 2009 as training set, taking rice yield in 2010 as test set.

3.2   Data Processing

In the case of Rice yield per unit area of Mudanjiang city, after the treatment of abnormal, getting a visual distribution map of the rice yield, as shown in Fig. 1:

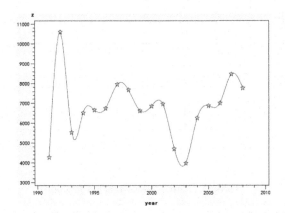

Fig. 1. Rice yield distribution maps in Mudanjiang city from 1990 to 2010

Analysis of pictorial diagram, the rice yield doesn't move smoothly enough. After a white noise inspection, we find that the sequence exists in white noise. Quadratic differential on the sequence as shown in Fig. 2:

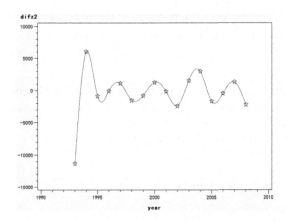

Fig. 2. Rice yield distribution after the second differential pictorial diagram

To make the group number of autocorrelation and partial autocorrelation, Table 1 for autocorrelation function, Table 2 for the partial autocorrelation function:

Table 1. Autocorrelation function

				Autocorrelations		
L **ag**	*Covaria* *nce*	*Correlat* *ion*		*-1 9 8 7 6 5 4 3 2 1 0 1 2 3 4 5 6 7 8 9 1*		*Std* *Error*
0	12205647	1.00000	I	|**************|		0
1	-4949965	-0.40555	I	. ******| .	I	0.250000
2	-644815	-0.5283	I	. *| .	I	0.288199
3	1496268	0.12259	I	. |** .	I	0.288803
4	-1120015	-0.09176	I	. **| .	I	0.292038

Table 2. Partial autocorrelation function

		Partial Autocorrelations
L **ag**	*Correlation*	*-1 9 8 7 6 5 4 3 2 1 0 1 2 3 4 5 6 7 8 9 1*
1	-0.40555	I . ******| . I
2	-0.26007	I . *****| . I
3	-0.01268	I . | . I
4	-0.06155	I . *| . I

Pictorial diagram can be seen that the production distribution has no obvious cyclical rice production, and the quadratic differential autocorrelation function and partial autocorrelation function can be seen that rice yields in Mudanjiang city Heilongjiang province in 1991–2008 were stationary series.

3.3   Determine the Order Number

Based on the BIC criterion

$$BIC(n) = \ln \hat{\sigma}_\varepsilon^2(n) + \frac{n}{N} \ln NZ$$

Inside, n is the number of parameters. If an order number n_0' meet

$$BIC(n_0') = \min_{1 \le n \le M(N)} BIC(n)Z,\ M(N) \text{ is equal to } [\sqrt{N}] \text{ or } [\tfrac{N}{10}],\ n_0' \text{ is the best order.}$$

After the calculation, p = 3, q = 1. Therefore, rice yield prediction model was ARIMA (3, 2, 1), To examine the Bic sequence analysis by SAS software, as shown in Table 3.

Table 3. BIC order determination results map

Minimum information criterion						
Lags *MA0*	*MA1*	*MA2*	*MA3*	*MA4*	*MA5*
AR0	13.1106	13.03749	10.64886	−22.9253		
AR1	13.13289	13.08792				
AR2	9.877755	−22.5844	−19.9156	−24.3156		
AR3	−22.8096	−26.8701	−25.0403			
AR4	−25.6454	AR4				
AR5						

Minimum Table Value: BIC(3,1) = −26.8701

4 Yield Prediction

According to the rice yield prediction model, To predict rice yield in cities of Heilongjiang province in 2010. The forecast output and the actual output are shown in Table 4. The distribution maps of Prediction and actual yield in Mudanjiang city in 2010 is shown in Fig. 3.

Table 4. The yield of rice in Heilongjiang province in 2010 actual yield is compared with the predicted values.

The prefecture level city	Prediction of yield (kg/HA)	The actual yield (kg/HA)
Daqing	9326.15	9094.245
Harbin	8927.8238	8842.127
Hegang	6226.5706	6174.625
Heihe	6672.7103	5387.501
Jixi	7851.6911	7774.634
Jia Musi	8225.7806	7517.753
Mudanjiang	7721.8741	7839.623
Qigihar	7154	6725.376
Shuangyashan	7396.7602	7567.996
Suihua	8857	8959.85
Yichun	8068	8046.453

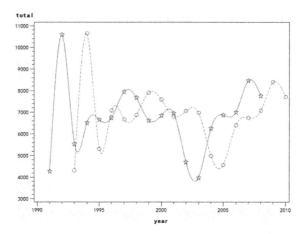

Fig. 3. The distribution maps of prediction and actual yield n Mudanjiang city in 2010

5   Spatial Analysis of Rice Yield

By using ArcGIS software, To establish the spatial distribution map of The yield of rice in Heilongjiang Province in 2010 cities that the actual and the predicted. As shown in Figs. 4, 5.

Fig. 4. The actual yield distribution in space

Fig. 5. Prediction of yield distribution in space

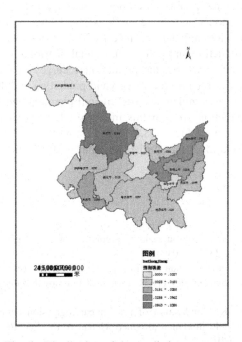

Fig. 6. Distribution of the prediction error space

Using the functions of GIS spatial analysis, Error analysis for the actual yield of rice and predicted values in 2010 Heilongjiang province. The spatial distribution map of error is shown in Fig. 6.

According to the above analysis, Getting the prediction error of Rice yield of municipalities in Heilongjiang province. As shown in Table 5.

Table 5. Prediction error of rice yield of Heilongjiang province in 2010

Cities	Prediction error	Cities	Prediction error
Daqing	0.0255	Mudanjiang	0.01502
Harbin	0.009692	Qigihar	0.063732
Hegang	0.008413	Shuangyashan	0.022626
Heihe	0.238554	Suihua	0.011479
Jixi	0.009911	Yichun	0.002678
Jia Musi	0.094181		

It can be seen from Table 5 that Yichun has the least prediction error of 0.26 %, with Heihe having the largest prediction error of 23.86 %. By carrying out calculation, the prediction error of different cities on average is 4.5 %.

6   Conclusions

The paper establishes time prediction model of rice production in Heilongjiang province using ARIMA model and gis technology; By using the model, the prediction value of rice yield in Heilongjiang province in 2010 is obtained; and the actual production, predicted production and predict the spatial distribution of prediction error is acquired. By conducting analysis, the prediction accuracy of the model on average reaches over 95 % that can be used for rice yield prediction and that provides relevant scientific reference for government sectors' overall planning and decision-making.

However, the prediction error of this model used for prediction of rice yield of Heihe reaches a maximum value of 23.8 %, indicating the model is not well designed taking all factors into account that further research is still required.

References

1. Wang, H.: Statistics and decision making, prediction of grain yield of mixture time series model based on (12), 23–25 (2013)
2. Liu, Q.: The grain yield of Henan province prediction. Spatio-Temporal Regression Model Based Anhui Agric. Sci. **39**(28), 17631–17633 (2011). doi:10.3969/j.issn.0517-6611.2011.28.189
3. Li, B., Li, Q., Lu, X.: Application of gray linear regression combination model in the prediction of grain yield in Henan province. Henan Agric. Sci. **10**, 44–47 (2009). doi:10.1016/j.jhin.2010.02.010

4. Li, J.: Prediction of grain production in Henan province, the grey system model. J. Henan Univ. Technol. (Soc. Sci. Publishing) **5**(4), 1–3, 7 (2009). doi:10.3969/j.issn.1673-1751. 2009.04.001
5. Chen, X., Chen, M., Feng, G., et al.: Computer engineering and design model for the prediction of multi variable time series regression tree yield of cucumber **33**(1), 407–411 (2012). doi:10.3969/j.issn.1000-7024.2012.01.080
6. Xu, X., Zhou, C., Chen, G., et al.: Clustering analysis and sequential algorithm of corn production model test. J. Jilin Agric. Univ. **34**(6), 688–691, 704 (2012)
7. Su, H., Zhu, C., Wen, C.: Model in corn fertilization forecasting. J. Jilin Agric. Univ. **32**(3), 312–315 (2010)
8. Xiao, Z., Guo, M.: Time Series Analysis and Application of SAS. Wuhan University Press, Wuhan (2009)
9. Ana, S., Leonid, S., Aleksandar, D., Slovodanka, D.-K.: GinisWeb-the tool for GIS applications development based on Web. In: 10th International Electrotechnical Conference MEleCon 2000, vol. 1, pp. 331–333
10. ESRI Shapefile Technical Description, GIS by ESRI (1998)

Research on the Digital Machine for Killing the Larva of Longicorn Beetle with Microwave Based on the Arduino

MingXi Shao[2], XiuMei Zhang[1], BingGuo Liu[1],
ChangYong Shao[2,3(✉)], AiSheng Ma[1], ShouSheng Zhang[1],
Sheng Liu[1], YuJie Liu[4], LiJing Zhao[3], and Lin Dong[1]

[1] Weifang University of Science and Technology, Shouguang 262700, China
{2676377107,837647697,582003674,847281661}@qq.com,
lbgwhm@icloud.com, sg_xiaoma@163.com
[2] College of Engineering, China Agricultural University, Beijing 10083, China
390606109@qq.com, shaochangyong@cau.edu.cn
[3] Shandong Province Seeds Group Co., Ltd., Jinan 250100, China
zhaolijing008@163.com
[4] Shandong Vocational College of Science & Technology, Weifang 261053,
China
lyj_1215@163.com

Abstract. Longicorn beetle is an important wood-boring insect. In order to probe into the effective eco-friendly method controlling longicorn beetle, this document recommends a new dual optical digital machine for killing the larva of longicorn beetle with microwave based on the arduino. This machine uses the arduino with temperature controller as the core, touch switch, and a temperature sensor as important parts. By using arduino and sensor measurement technology, a kind of economical close-loop grinding control system is formed to control the temperature with the control of duty cycle and frequency of pulses. The results of the operation showed that the temperature controlling, microwave power is adjustable, man - machine dialogue is woven and seized. It is reliable to be easy to operate and high in automation.

Keywords: Pulse microwave · The microwave oven · The arduino · DS18B20 temperature sensor · Silicon controlled rectifier

1 Introduction

Longicorn beetle is a kind of widespread occurrence of pests worldwide, is the floorboard of the superfamily coleoptera leaf beetle cerambycidae insects and their host variety, damage is extensive, such as forestry, garden trees and fruit trees were infect it, next, because it has a way of life of concealed, natural enemy species less, low interference by natural factors, etc., so control is very difficult, often cause harm in a cabin myself, even the deforestation. Longicorn larvae of vermicular, can crawl into life within two years, at the beginning of hatched larvae in the bark of feeding, with the increase of age, namely greater in phloem and xylem in feeding, trees conducting tissue

D. Li and Z. Li (Eds.): CCTA 2015, Part I, IFIP AICT 478, pp. 546–555, 2016.
DOI: 10.1007/978-3-319-48357-3_51

damage, leading to death. Sawyer neutralising the commonly used method at present is mainly chemical method, namely: by dimethoate and dichlorvos highly toxic pesticides into wormholes, purpose of pesticides. The shortcomings of this approach is that as the longicorn larvae to become resistant to treatment, result in applying pesticide concentration increasing, the environmental pollution is more and more serious, at the same time, also cause serious damage to human and animal health. Therefore, attempts to make the field technicians physical methods neutralising the longicorn, but so far, there is still no good insecticidal equipment or physical insecticidal method can achieve the effect of neutralising the sawyer, high efficiency, high killing rate.

Microwave is a kind of can make the material in the inland waters molecules to vibrate and generate heat, therefore, in our already use microwave oven to heat food in the kitchen, and hotel, the hotel has to be used for drying, sterilization equipment, the effect is obvious.

Currently used in the microwave oven is infrared temperature detection device, the thermal stability of the sensor of the device, though, but as a result of microwave heating from the inside out, at this time of the infrared temperature sensor, temperature is only the surface temperature were collected and the internal temperature of the heated material difference is very big. At the same time, because now the microwave equipment for microwave magnetron is realized by using a timer to control, but this method reduces the life of a microwave magnetron, also have impact on control system.

Based on the above problems, in this paper, after studying, put forward a new kind of kill longicorn larvae of automatic control system. This design USES the arduino as the control core, DS18B20 temperature acquisition module for the temperature control unit for soil temperature and the temperature detection, closed-loop control to realize interpolation, interpolation signal transmission to the SCR to magnetron microwave output control.

2   Microwave Kill the Larva of Longicorn Beetle Composition of the System

Microwave kill the larva of longicorn beetle system main structures is shown in Fig. 1:

(1) Magnetron and cavity, adopts the structure combining a snap button.
(2) Install the ring frame 12 groups of magnetron pulse current of capacitor and transformer and support. Type a panasonic m210-2 M1, maximum output power 900 w. Capacitor using the treasure chang 2100 v 1 u f microwave special capacitance, 12 groups of transformer the GAL - 700 - e - 1 s. 36 sets of SCR control on rectangular frame is divided into three groups of 12 groups of magnetron.
(3) This system USES through the 12 v supply to the output transformer for power supply way of wind cold short of 12 groups of fans.
(4) The control of this system adopts the arduino and touch key as the core controller to realize automatic control.

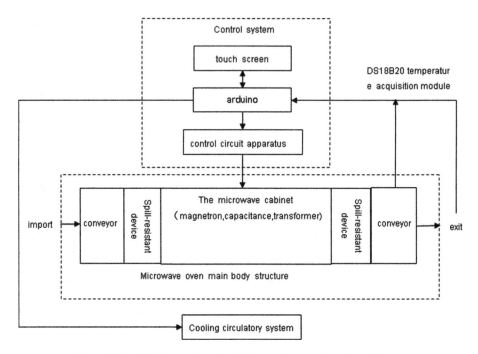

Fig. 1. The digital machine for killing the larva of longicorn beetle

(5) Temperature detection and feedback system USES DS18B20 temperature acquisition module, the detection of light soil and the internal temperature of the trunk, closed-loop control to realize interpolation.

(6) Because the microwave will cause certain damage to human body, the closed loop type is adopted stainless steel sheet to prevent overflow, microwave scattering and in use process, made up of closed loop cavity surrounded the trunk or the roots of the soil.

3   Microwave Kill the Larva of Longicorn Beetle of Electrical Control System

Microwave kill the larva of longicorn beetle control system is shown in Fig. 2, through the picture you can see, the control unit is mainly the arduino and relevant input and output unit, temperature display, SCR, temperature sensor and controlled object.

3.1   Microwave Kill Longicorn Larvae System Hardware Design

3.1.1   Arduino Port Circuit Design

According to the control requirements of the system, need eight digital output port and four analog input ports and three analog output port. In order to meet this requirement,

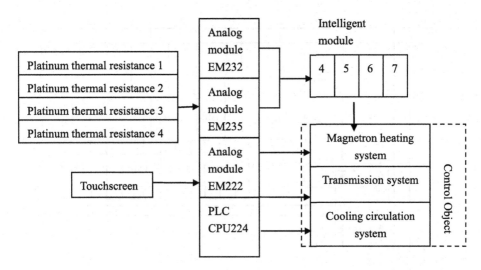

Fig. 2. Control system

choose the Arduino UNO platform, this platform is the core of ATmega328 processing chip, a total of 14 digital input/output: 6 road PWM output port, no. 6 analog input port, all the way ICSP header, a 16 MHz crystal oscillator, a power switch.

A USB interface and a reset button. In the AREF new added two pin: SCL and SDA, support the I2C interface; Increase IOREF and reserve a pin, control panel can be compatible with 5 V and 3.3 V supply voltage.

Device name	The number	Rated power (KW)
Magnetron	90	0.9
Intelligent thyristor	48	0.05
Cooling tower fan	1	1.1
Water pump	2	2.2
Elevator	1	0.75
Oil pump	1	0.37
Conveyor	1	2.2

3.1.2 Microwave Kill Longicorn Larvae Circuit Design of the System

Choose intelligent thyristor control insecticidal machine heating unit, this system USES MJYS - JL - 450 type intelligent thyristor, thyristor module of the input signal of current 4 ma–20 ma, or 0 to 5 v voltage, control principle are shown in Fig. 3 below.

The system control method is as follows:

(1) use the arduino form PWM pulse signal timing control function, low level is the input signal is 4 ma thyristor, magnetron no microwave power is zero at this time; High electricity at ordinary times, silicon controlled rectifier input signal to 20 ma, microwave magnetron, at this time a single power 900 w.

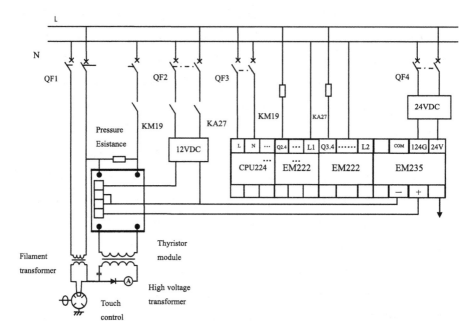

Fig. 3. Microwave kill longicorn larvae circuit

(2) power regulation. Through the change of the PWM duty cycle of the signal, when the duty ratio is less than 1, can activate the microwave magnetron output. So you can through the PWM pulse width modulation pulse rate and silicon controlled rectifier input signal to change the output power of microwave insecticidal machine.

3.1.3 Microwave Kill Longicorn Larvae Temperature Control Circuit Design of the System

In the system of the temperature feedback collection unit adopts the digital temperature sensor DS18B20, the module will be collected the temperature of the analog signal into digital signal, the sensor includes a NTC thermometer element and a resistive touch wet element. DS18B20 temperature detection range in $-55 \sim +125$ °C, in $-10 \sim +85$ °C accuracy of +/−5 °C. In each bucket cover the bottom with a DS18B20, soil temperature signal collected and processed output PWM pulse signal, control thyristor module, change the power magnetron, achieve control of the temperature.

3.2   Microwave Kill Root Nematode System Software Design

This system adopts the aiduino as processing core, the main program includes the main program, subprogram, temperature acquisition and practical program, microwave control instruction, fault alarm procedures, etc.

3.2.1   Microwave Kill Root Nematode System Temperature and Pulse Microwave Control Program

The core of this system is for the control of temperature and microwave power, the soil temperature DS18B20 temperature sensor acquisition and processing, the input signal to the arduino Atmel AVR ATmega328 processing, after processing by its send a pulse signal to the thyristor module change magnetron power to implement automatic regulation for temperature. This system has two kinds of manual and automatic operation mode. Automatic mode, the use of pid algorithm, and calculate the duty ratio, through

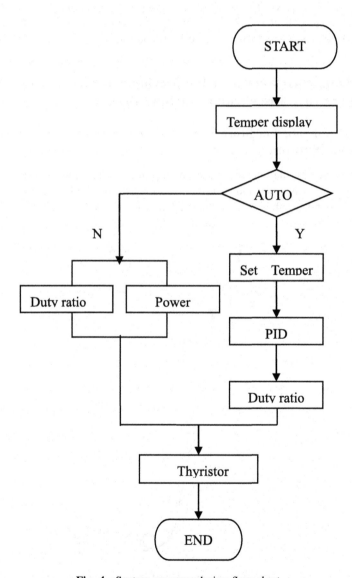

Fig. 4. System program design flow chart

the arduino central processing unit and output control signal, through intelligent thyristor adjust the duty ratio of pulse signal, change the power of microwave magnetron to for temperature control. When malfunction or insecticidal effect is not ideal the need for manual operation, the operator can be manually opened according to the actual temperature heating, namely in setting an interrupt program, through the manual button at the bottom of the screen to switch to the program change PMW duty ratio to change the role of the magnetron power. This system program design flow chart as shown in Fig. 4:

This system adopted by the arduino as the core control unit, by controlling the output power of microwave pulse duty ratio control, and adopted the DS18B20 temperature sensor to collect signal and output control signal so that a complete output, execution, feedback, to perform a closed-loop control.

4 Kill Longicorn Larvae of the Mechanical Mechanism of Microwave Insecticidal Machine Design

4.1 Neutralising the Longicorn Larvae of Microwave Insecticidal Machine Structure

The device including the telescopic rod, the top of the telescopic rod connected to the hull, microwave transmitter and controller installed in the shell, shell on the inside of the board to open hole, the inside of the shell plate respectively hinged on both ends the

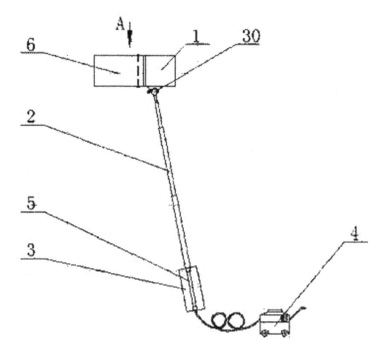

Fig. 5. Mechanical structure

first arc baffle plate and a second reflex arc plate, the first arc baffle in the hinge parts setting first contact pressure, the second arc baffle in the hinge position setting the second contact pressure, the first arc the outer end of the baffle plate and a second reflex arc plate edge set opening snap button, the first baffle plate and a second reflex arc arc plate forming a circular cavity microwave irradiation after closure, hole and cavity microwave irradiation are interlinked, first arc baffle wall by first fittings installed first microwave leak proof brush, the second arc baffle wall brush through the second fittings installed the second microwave leak proof. The invention by microwave irradiation sawyer will kill, makes the killing rate of 100 %, and characterized by the good insecticidal effect. The mechanical structure as follows in Fig. 5:

1. The cavity 2. Connecting rod 3. The control panel 4. Variable pressure control panel 5. Wire 6. Microwave reflector 30. Universal joint.

4.2 Structure of Microwave Irradiation

Mentioned in the present invention arc baffle the end of the opening and closing the structure of the snap button as shown in Fig. 6, the structure for optimizing structure, opening and closing of the snap button for concrete structure: the first reflex arc plate convex block set the end of July 23, 23 on both sides of the convex block with 29 arc grooves, the second arc baffle grooves set the end of June 22, respectively on the 22 of the two grooves inside open half spherical tank 24, 24, respectively, on both sides of the opening half spherical tank is set to half spherical tank 24 raised near the center of the block is 26, each half spherical tank are installed within 24 28, spring 28 end connected to the spherical bead 25, 25 set on spherical beads baffle ring 27, baffle ring located in the half spherical tank within 24, 27 and 26 limit stop. When the contact pressure affected by external force, fit two curved reflector, two spherical bead 25 at arc grooves in the 29, implement two arc fit reflex plate.

Microwave insecticidal machine when using this design, the two pressure contact and collision tree trunk to make two reflex arc plate automatically closed, there will be a wormhole trunk as well in the closed two curved reflector formed the circular microwave irradiation cavity, at this point, the four microwave leak proof brush also closed with arc baffle ring shape formation, the two curved reflector closed after the formation of the circular cavity microwave irradiation to seal. Open microwave transmitter irradiation longicorn larvae, under the action of microwave irradiation, generate a lot of heat and moisture in the trunk at the same time, the longicorn larvae molecules under the action of microwave electromagnetic field produced deformation and vibration, make the function of the cell membrane produces change, until the cell membrane rupture, produce non-thermal effect, superposition of non-thermal effect and the heat generated by the trunk longicorn larvae were culling.

A

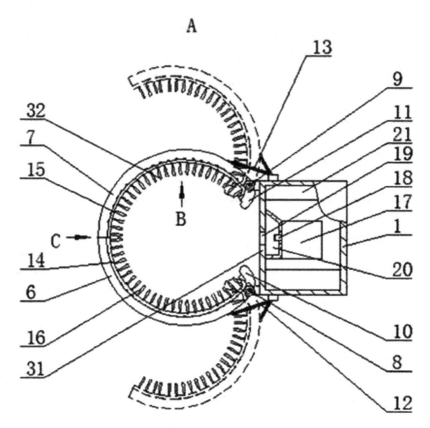

Fig. 6. The structure of the snap button

5   The Insecticidal Effect of Microwave Neutralising the Longicorn Larvae System

This design of neutralising the longicorn larvae microwave insecticidal machine, when use pull telescopic rod according to the height of trees to the length of the need, the microwave irradiation cavity into the trunk of a stem borer pest lateral, microwave irradiation cavity around the trunk and the cavity is closed, intermittent open microwave transmitter, the microwave irradiation cumulative time reaching 3 to 4 min, neutralising the longicorn larvae reached 100 %, and trees 45 to 60 days, the continuous observation trees to normal growth, and before using microwave extermination are exactly the same, without any exception. The effect is according to the statistical data from experiment base. The applicant in the five years of poplar 1000 f to do the test, the poplar sawyer about 70 % less trees are infection, the tree appear a large number of wormholes, peeling bark, longhorn beetle larvae into the depths of the trunk for 15 to 20 cm, the wormhole distribution within the limits of the upper 20 cm, was disseminated trees can be divided into two groups. Using the present invention insecticidal machine will fit with the trunk of the wormhole surrounded in microwave irradiation cavity, the first set of open microwave transmitter 60 s to stop and check the longicorn larvae death in about 31 % and 60 s to stop and open the microwave transmitter, check the longicorn larvae death in about 70 % and the third open microwave transmitter illuminate is 60 s, check the longicorn larvae insect body burst, all died. Continuous observation was light trees 45 to 60 days, tree growth all normal. By microwave irradiation at the same time, the second group of 80 s to stop and check the longicorn larvae death cases, 3 times, total radiation light 80 s after the stop check every time, when the third time after irradiation, check all longicorn larvae were killed, 45 to 60 days continuous observation by microwave irradiation trees, tree growth situation is the same as the first group, without any exception. The irradiation method, in the trunk of the wormhole within the scope of the upper 20 cm mobile microwave transmitter, repeat the above test method, inspection longhorn beetle larva mortality rates were 100 %, and no impact for the growth of trees.

References

1. Sun, F., Xing, Z.B.: Microwave technology principle and its development and application. Sci. Technol. Innov. Appl. **6**, 3–4 (2014)
2. Xue, D., Cheung, Jiang, H., et al.: Food non-thermal effect of microwave processing research. J. Chin. Food **2013**(13), 143–147 (2013)
3. Li, J.: Athermal biological effect of microwave on food microorganism and microwave sterilization technology. J. Southwest Univ. Natl. (Nat. Sci. Edn.) **6**(32), 1219–1222 (2006)
4. Cao, C.: Status of microwave vacuum drying technology. J. Dry. Technol. Equip. **3**(2), 5–9 (2004)
5. Zhou, W.L.: Electromagnetic pulse sterilization study. J. Microw. **3**(16), 318–321 (2007)
6. Nie, S.W., Hu, Y.-L., Huang, H.Y., et al.: Grain insect-resistant mouldproof dedicated microwave oven design. J. Food Mach. **1**, 114–117 (2014)
7. Wang, D.: The power control circuit design of microwave heating system. J. Electron. **11**, 57–58 (2012)
8. Xi, G.Q., Song, J., Yin, Q., Li, K., Xu, J.: Tobacco Science & Technology. Zhengzhou Tobacco Research Institute of CNTC, Zhengzhou 450001, China 2. Xuzhou Zhongkai Electromechanical Equipments Co., Ltd., Xuzhou 221005, Jiangsu, China
9. Hong, X.-S.C., Wei, C.: Insects juvenile hormone analogue "double oxygen wei" on the biological effect of tobacco. J. Tob. Sci. Technol. (12) (2006)
10. Jing, W.D., Ke, L.: Archives Science Study, Microwave insecticidal sterilization performance comparison study effect and side effects
11. Lin, T.: Insects the heating plate temperature measurement and control system design based on MCU. Northwest Agriculture and Forestry University of Science and Technology (2009)

Risk Assessment of Water Resources Shortage in Sanjiang Plain

Qiuxiang Jiang[1,2(✉)], Yongqi Cao[3], Ke Zhao[1], and Zhimei Zhou[1]

[1] College of Water Conservancy and Architecture,
Northeast Agricultural University, Harbin 150030, China
jiangqiuxiang914@163.com
[2] Postdoctoral Mobile Research Station of Agricultural and Forestry Economy
Management, Northeast Agricultural University, Harbin 150030, China
[3] College of Mechanics and Architecture,
Northwestern Polytechnical University, Xi'an 710129, China

Abstract. In view of the problems existing in the development and utilization of water resources in Sanjiang Plain, this paper made a study on the risk assessment of water resources shortage in Sanjiang Plain. From the perspective of water resources shortage risk, analytic hierarchy process was applied to establish assessment index system and evaluation criteria, and then make a comprehensive assessment and regional difference analysis of water resources shortage risk in Sanjiang Plain. Research results showed that Shuangyashan City belonged to high degree of water resources shortage risk, while Jixi, Hegang, Jiamusi, Qitaihe, Muling and Yilan belonged to extremely high water resources shortage risk, which provided a theoretical principle for sustainable and efficient utilization of regional water resources.

Keywords: Water resources · Shortage risk · Analytic hierarchy process · Sanjiang plain

1 Introduction

Water resources are the natural resources for human survival, but also an important resource for the economic development of a country or a region. However, China is a drought and water shortage nation in the world. Fewer amounts, uneven space distribution and serious pollution of water resources make water resources shortage in China more serious. The shortage of water resources in a certain extent has seriously restricted the sustainable development of social economy in China [1]. Therefore, the identification of water resources shortage risk factors, evaluation of its degree to reduce economic losses caused by the shortage, and the optimal allocation of water resources have important practical significance for achieving water resources sustainable utilization.

Water shortage risk assessment is a comprehensive evaluation of multiple factors. Currently, the evaluation methods include analytic hierarchy process (AHP), fuzzy mathematics method, projection pursuit clustering, etc. The calculation process and the advantages and disadvantages of various methods were studied and compared by many former scholars [2, 3]. Due to the impact of many uncertainties, any single method is

© IFIP International Federation for Information Processing 2016
Published by Springer International Publishing AG 2016. All Rights Reserved
D. Li and Z. Li (Eds.): CCTA 2015, Part I, IFIP AICT 478, pp. 556–563, 2016.
DOI: 10.1007/978-3-319-48357-3_52

difficult to evaluate comprehensively. If AHP was combined with fuzzy mathematical method, it could make the abstract evaluation process and algorithm specific, and the evaluation results would be distinct. Therefore, based on the use of AHP to identify the source of water resources shortage risk, membership function was used to evaluate the risk degree of each index, and realize the risk assessment of regional water resources shortage.

2 Research Area

Sanjiang Plain lies in the northeast region of Heilongjiang Province, China, including the triangle confluence region of the Heilongjiang, Songhua and Wusuli Rivers, and Woken River and Muling River Basin and Xingkai Lake Plain. Geographical position of Sanjiang Plain is $43°49'55'' \sim 48°27'20''$ north latitude, $129°11'20'' \sim 135°5'10''$ east longitude, north to the Heilongjiang River, south to the Xingkai Lake, east to the Wusuli River, west to the southeast part of Xiaoxinganling mountains. Its east-west width is 430 km, and north-south long is 520 km, with a total area of 1.09×10^5 km^2 and 24 % of Heilongjiang Province total area. The administrative area of Sanjiang Plain includes Hegang City, Jiamusi City, Shuangyashan City, Qitaihe City, Jixi City and Muling City and Yilan County. The Sanjiang Plain region is essential for Heilongjiang Province to ensure grain produce capacity and food security and is also a significant marketable grain base China. Low water resources pollution emissions compliance rate, unreasonable development and utilization, low industrial water reuse rate and low water resources use rate [4], coupled with increasingly severe extreme weather and water waste caused the serious shortage of water resources. Therefore, it is of great significance to construct the risk assessment model of water resources shortage, and to analyze and evaluate the situation of water shortage in Sanjiang Plain.

3 Water Shortage Risk Assessment

3.1 Evaluation Index System Construction

In order to comprehensively and systematically analyze water resources shortage, risk index and main risk sources of each city in Sanjiang Plain area, and contrast risk differences in different areas, a series of evaluation index were needed to construct a unified and complete evaluation index system. Each evaluation index should reflect the water resources shortage and shortage risk from different angles. It should be able to connect with each other, complement each other, and will be independent and representative. Combined with the characteristics of resource utilization and socioeconomic development in the study area, the risk assessment index system of water resources shortage in Sanjiang Plain was constructed. The index system is made up of target layer, state layer and index layer, in which the target layer includes one target factor (I), the state layer includes five state factors (A to E), and the index layer includes 18 indicators (A1 to E3) shown in Table 1. In the risk assessment index system of water resources shortage, the state layer indicates the risk of regional water resources shortage from different angles: water resources quantity reflects the quantity and background of

Table 1. Water resources shortage risk evaluation index

Target layer	State layer	Index layer	Index measurement method
I Water Shortage risk index	A Water resources	A1 Per capita water resources	Total water resources/total population
		A2 Per square water resources	Total water resources/land area
		A3 Surface water coefficient	Surface water resources/precipitation
	B Social needs	B1 Population density	Total population/land area
		B2 GDP per water consumptions	Gross Regional Product/total water demand
		B3 Per square water demand modulus	Water demand/land area
		B4 Per capita water demand	Total water demand/total population
	C Water source reserve	C1 Water retention coefficient	Total water resources/precipitation
		C2 Reservoir storage ratio	Reservoir water storage/total water resources
		C3 Groundwater resource coefficient	Total groundwater/total water resources
	D Water supply	D1 Per capita water supply	Total water supply/total population
		D2 Water supply rate	Water supply/total water resources quantity
		D3 Groundwater supply ratio	Total groundwater supply/total water supply
		D4 Surface water supply ratio	Total surface water supply/total water supply
	E Ecological environment	E1 Forest coverage	Forest area/total land area
		E2 Ecological water consumption rate	Ecological water consumption/annual average water resources

water resources in evaluated area; social demand mainly reflects the intensity of water demand and social economic scale born by regional water resources; water resource reserve status mainly reflects the ability of relieving regional water shortage; water supply indicators mainly reflect the regional water supply capacity; ecological environment indicators mainly reflect the status of regional ecological balance [5].

3.2   Calculation of Index Weight Based on AHP

There are five steps in the weight calculation of analytic hierarchy process: systemizing problem, valuing index, constructing judgment matrix, determining index weight, and testing judgment matrix consistency [6]. The judgment matrix is an n × n comparison matrix composed by quantification of the comparison of n sub index. In this paper, the relative importance of quantitative value is obtained by using 1–5 scaling method [7], when strong factors were compared with weak factors, the numerical 1–5 were used to characterize the degree of importance, respectively, indicating equal, slightly important, obviously important, strongly important, extremely important. When weak factors were compared with strong factors, the reciprocal of the numerical 1–5 were used to represent. The judgment matrix of this study includes the target -state level (A-E), the state-index layer (A1–A3、B1–B4、C1–C3、D1–D4、E1–E2), which has 6 judgment matrixes.

The index weight was calculated by using the characteristic root of the judgment matrix, and the maximum eigenvalues and eigenvectors of the matrix were calculated by using the MATLAB software. After single ordering the 6 judgment matrix, normalized weight value was obtained by the multiply of each index weight with the corresponding state level factor. The CR value in consistency test of judgment matrix is calculated by the following formula [8]:

$$\begin{cases} CR = CI/RI \\ CI = (\lambda_{max} - n)/(n - 1) \end{cases} \tag{1}$$

Where CI is the consistency index of judgment matrix; λ_{max} is the maximum eigenvalue of judgment matrix; n is the dimensionality of judgment matrix; RI is the random consistency index related to the dimensionality of judgment matrix, when n was 3, 4, 5, 6, 7, 8, 9, 10, RI = 0.52, 0.90, 1.12, 1.26, 1.36, 1.41, 1.46, 1.49. When CR is less than 0.1, it is considered that the judgment matrix is in good agreement, and the weight is rational; elsewise it will need to adjust the judgment matrix until it is satisfied.

In this paper, the CR values from the consistency test of each index layer were all less than 0.1, which indicated that the judgment matrix satisfied the consistency, and the weight distribution was reasonable. The calculation results were shown in Table 2.

Table 2. Calculation results and consistency test of the weight of the index system

State factor weight	Conformance test(CR)	Index factor weight				Conformance test(CR)
		w1	w2	w3	w4	
A 0.4082	0.0498	0.1750	0.1750	0.0582	–	0
B 0.2562		0.0202	0.0349	0.0639	0.1372	0.0258
C 0.1048		0.0419	0.0419	0.210	–	0
D 0.1860		0.0700	0.0700	0.0230	0.0230	0
E 0.0448		0.0075	0.0373	–	–	0

3.3   Membership Evaluation Construction

After the establishment of the risk evaluation index system of water resources shortage, we need to connect each index value with water resource shortage risk index, and the common method solving this problem is fuzzy mathematics evaluation method. By dividing the numerical interval, the index value was converted into the risk degree, and the relationship between the index value and risk index was established. By

Table 3. Grade standard of risk membership degree of water resources shortage index system

Status factor	Index factor	Unit	Risk membership				
			$V_1(1)$	$V_2(2\sim3)$	$V_3(4\sim5)$	$V_4(6\sim7)$	$V_5(8)$
A Water resources	A1 Per capita water resources	m^3	>2000	$2000\sim1600$	$1600\sim1200$	$1200\sim800$	<800
	A2 water resources per square	m^3	>4.5	$4.5\sim3.5$	$3.5\sim2.5$	$2.5\sim1.5$	<1.5
	A3 Surface water coefficient		>0.4	$0.4\sim0.3$	$0.3\sim0.2$	$0.2\sim0.1$	<0.1
B Social needs	B1 Population density	capita/km^2	<25	$25\sim50$	$50\sim100$	$100\sim300$	>300
	B2 GDP per water consumptions	yuan/m^3	>120	$120\sim100$	$100\sim60$	6	<20
	B3 Per square water modulus	m^3	<2	$2\sim5$	$5\sim10$	$10\sim20$	>20
	B4 Per capita water demand	m^3	<300	$300\sim400$	$400\sim500$	$500\sim600$	>600
C Water source reserve	C1 Water retention coefficient	–	>0.6	$0.6\sim0.45$	$0.45\sim0.3$	$0.3\sim0.1$	<0.1
	C2 Reservoir storage ratio	–	>0.8	$0.8\sim0.5$	$0.5\sim0.3$	$0.3\sim0.1$	<0.1
	C3 Groundwater resource coefficient	–	>0.4	$0.4\sim0.35$	$0.35\sim0.3$	$0.3\sim0.25$	<0.25
D Water supply	D1 Per capita water supply	m^3	>800	$800\sim600$	$600\sim400$	$400\sim200$	<200
	D2 Water supply rate	%	>0.6	$0.6\sim0.4$	$0.4\sim0.2$	$0.2\sim0.1$	<0.1
	D3 Groundwater supply ratio	%	>0.05	$0.05\sim0.01$	<0.01	–	–
	D4 Surface water supply ratio	%	>0.95	$0.95\sim0.9$	–	–	–
E Ecological environment	E1 Forest coverage	%	>50	$50\sim40$	$40\sim20$	$20\sim10$	<10
	E2 Ecological water consumption rate	%	<1	$1\sim2$	$2\sim3$	$3\sim5$	>5

transforming the measured values of each and every index in evaluation area into the risk degree, risk index was obtained by weighted sum of the risk of all the indexes in the evaluation area [9]. In this paper, the membership of risk is $1\sim8$, the higher the number, the higher the membership of risk. Based on water resources shortage risk level, the membership degree is divided into 5 levels ($V_1 \sim V_5$) as shown in Table 3.

3.4   Evaluation Results and Analysis

In accordance with the above evaluation process, the water resources shortage risk assessment of Jixi City, Hegang City, Shuangyashan City, Jiamusi City, Qitaihe City, Muling City, Yilan County and the whole Sanjiang Plain was conducted. The risk membership of index in each city was shown in Table 4, and the comprehensive risk index and risk level were shown in Table 5.

Table 4. Index risk membership of evaluation area

City	A1	A2	A3	B1	B2	B3	B4	C1	C2	C3	D1	D2	D3	D4	E1	E2
Jixi	1	8	1	5	7	1	8	2	8	3	1	1	1	1	4	1
Hegang	1	8	1	4	8	1	8	1	8	8	1	1	1	1	2	1
Shuangyashan	1	8	1	4	7	1	8	1	8	1	1	1	1	1	4	1
Jiamusi	1	8	3	4	8	1	8	3	8	3	1	1	1	1	6	1
Qitaihe	6	8	4	6	7	1	8	5	8	3	6	1	1	1	3	1
Muling	1	8	4	3	8	1	8	4	8	3	7	1	1	1	1	1
Yilan	1	8	3	5	8	1	8	5	8	6	1	1	1	1	4	1
Sanjiang Plain	1	8	8	5	7	1	8	8	8	4	1	1	1	1	4	1

Table 5. Risk index and risk rank of evaluation area

City	A	B	C	D	E	Total risk index	Risk level
Jixi	1.6332	1.5068	1.0490	0.1860	0.0673	4.4423	Extremely high risk
Hegang	1.6332	1.5215	2.0571	0.1860	0.0523	5.4501	Extremely high risk
Shuangyashan	1.6332	1.4866	0.5871	0.1860	0.0673	3.9602	High risk
Jiamusi	1.7496	1.5215	1.0909	0.1860	0.0823	4.6303	Extremely high risk
Qitaihe	2.6828	1.5270	1.1747	0.5360	0.0598	5.9803	Extremely high risk
Muling	1.8078	1.5013	1.1328	0.6060	0.0448	5.0927	Extremely high risk
Yilan	1.7496	1.5417	1.8047	0.1860	0.0673	5.3493	Extremely high risk
Sanjiang plain	2.0406	1.5068	1.5104	0.186	0.0673	5.3111	Extremely high risk

The evaluation index in target layer of water resources shortage risk are between $1 \sim 8$. According to the cluster analysis of risk value, the water resources shortage risk was divided into 4 levels: extremely high risk (>4.0), high risk ($4.0 \sim 3.75$), medium risk ($3.75 \sim 3.5$), low risk (<3.5).

The results showed that the whole area of Sanjiang Plain had extremely high risk. Water resources shortage risk mainly came from three aspects: small amount of water resources, high social demand and shortage of water reserves. Low risk of water supply and water environment eased the water shortage risk level to a certain extent. The water shortage risk rank of seven administrative regions in Sanjiang Plain from high to low was as follows: Qitaihe City, Hegang City, Yilan County, Muling City, Jiamusi City, Jixi City, Shuangyashan City.

Jixi City, Hegang City, Qitaihe City, Jiamusi City, Muling City and Yilan County belonged to extremely high water resources shortage risk, and the main reasons were small amount of water resources, high social demand and shortage of water reserves. Shuangyashan City belonged to high water shortage risk and the risk index was close to the extremely high risk, and the main reason was that the region had less rainfall, less reservoir storage and high industrial water consumption. Therefore, the shortage of water resources in Sanjiang Plain is very serious.

4 Conclusions

In order to make risk assessment of water resources shortage in Sanjiang Plain, the study made full use of the uncertainty information contained in the object of study, used numbers to process and express, and quantified decision makers' subjective identification and judgment to simplify complex problem and to analyze the details by hierarchy method. Combining quantitative and qualitative analysis with quantitative data, simply calculating, judging and comparing various factors in each level were carried out, and the weight value of each factor was obtained. By calculating the ordering weights of individual factors relative to the overall goal of the whole system, the overall risk condition of these factors was analyzed, namely, regional water resources shortage risk level. The model constructed in this paper based on AHP, with the advantages of simple structure, simple calculation, clear concept, easy operation, simple and effective modeling method could be applied to regional natural resources shortage risk grade evaluation.

After the evaluation of the analytic hierarchy process, the results showed that Jixi City, Hegang City, Qitaihe City, Jiamusi City, Muling City, Yilan County belonged to the extremely highly water shortage risk, the main reasons were small amount of water resources, high social demand and shortage of water reserves. Shuangyashan City belonged to high water shortage risk and the risk index was close to the extremely high risk. According to the evaluation results, for the region with both extremely high and high shortage risk, the demand of water resources will increase along with population growth and economy development. Therefore it is necessary for us to integrate regional characteristic, make full use of the rich water resources of Heilongjiang River and Xingkai Lake, strengthen water resources management, improve water use efficiency, and then as soon as possible to realize the transformation of economic structure from water-intensive economic structure to water-saving economic structure. At the same time, the government needs to increase investment in water conservancy projects, pay attention to the deep development of water resources, and then guarantee the sustainable water supply ability to socioeconomic development in Sanjiang Plain.

Acknowledgment. The study was supported by National Natural Science Foundation of China (51209038), Public Science and Technology Research Funds Projects of the Ministry of Water Resources (201301096), Foundation of Heilongjiang Educational Committee (12531009), Heilongjiang Postdoctoral Fund (LBH-Z13049), Doctoral Foundation of Northeast Agricultural University (2012RCB58) and "Young Talents" Project of Northeast Agricultural University (14QC47).

References

1. Cui, X., Di, W., Zu, P., et al.: AHP assessment model application in water shortage. J. Math. Pract. Theory **44**(6), 270–273 (2014). (in Chinese)
2. Han, Y.: The theory and practice of water resource shortage risk management. The Yellow River Water Conservancy Press, Zhengzhou (2008). (in Chinese)
3. Li, J., Li, L., Liu, Y., et al.: Framework for water scarcity assessment and solution at regional scales: a case study in beijing-tianjin-tangshan region. Prog. Geogr. **29**(9), 1041–1048 (2010). (in Chinese)
4. Jiang, Q.: Study on Carrying Capacity Evaluation and Dynamic Simulation of Sustainable Utilization of Water and Land Resources in Sanjiang Plain. Northeast Agricultural University, Harbin (2011). (in Chinese)
5. Zhao, J., Zhang, Z., Xu, Y.: Risk assessment of water shortage of han river drainage in hubei province. J. Hubei Univ. Arts Sci. **34**(11), 43–47 (2013). (in Chinese)
6. Huang, Y., Chen, Z., Zhang, L.: Application of AHP and fuzzy evaluation to water-saving potential in irrigation district. Guangdong Water Resour. Hydropwer **5**, 52–55 (2014). (in Chinese)
7. Chen, S.: Engineering Fuzzy Set Theory and its Application. National Defence Industry Press, Beijing (1998). (in Chinese)
8. Wang, J., Wang, X., Lu, X.: Evaluation of the influence on the economy of water source areas in Shanxi province of middle route of south-to-north water diversion project based on the analytic hierarchy process. J. Qufu Normal Univ. **36**(1), 104–108 (2010). (in Chinese)
9. Ruan, B., Han, Y., Hao, W., et al.: Fuzzy comprehensive assessment of water shortage risk. J. Water Conservancy **36**(8), 906–912 (2005). (in Chinese)

Analysis of Soil Fertility Based on FUMF Algorithm

Hang Chen[1,2], Guifen Chen[2(✉)], Yating Hu[2], Liying Cao[2],
Lixia Cai[2], and Sisi Yang[1]

[1] Institute of Scientific and Technical Information of Jilin,
Changchun 130033, China
chenhang0811@163.com, 1046235989@qq.com
[2] College of Information Technology,
Jilin Agricultural University,
Changchun 130118, China
chenhang0811@163.com, guifchen@163.com,
{65447539,45515189,419513823}@qq.com

Abstract. The soil nutrition is an important indicator of soil fertility. The method K-means and FCM are always used to evaluating the soil fertility, but the cluster number need to be set, and the outlier couldn't be eliminated accurately, and there is the deviation between the real result and the soil fertility. So the paper applied the FUMF to analysis the soil nutrient data of Nong An county for eight years, 2005–2012. The result show that the low fertility soils gradually decreased from 2005 to 2012 by precision fertilization, and the moderate and high fertility soil was rising, the overall soil fertility of Nong An had improved significantly. The analysis result was consistent with the actual situation, The FUMF algorithm is proved that was an effective evaluate method of the soil fertility evaluation. It has the practical significance to analyze the large number of soil fertility of high complexity and interactive, it also provided the technical support for precision fertilization decision-making.

Keywords: Fuzzy clustering · Soil fertility · Fertility analysis · Precision fertilization · FUMF

1 Introduction

Soil nutrient content is an important symbol of fertility and productivity of arable land, also it is an important indicator of soil fertility evaluation. With the arrival of precision agriculture era, spatial variability and correlations of wide variety agricultural data which have complex links relationship are more significantly. The attendant massive, diverse and dynamic changes, incomplete, uncertain and a series of characteristics.

Since the 1990s, Data Mining and geographic information systems technology in the agricultural sector has been increasingly widely used. DM and GIS technology can effectively statistics and analysis of massive, complex data. DM Clustering algorithms can dig out the knowledge of soil fertility evaluation from soil nutrient data analysis.

© IFIP International Federation for Information Processing 2016
Published by Springer International Publishing AG 2016. All Rights Reserved
D. Li and Z. Li (Eds.): CCTA 2015, Part I, IFIP AICT 478, pp. 564–573, 2016.
DOI: 10.1007/978-3-319-48357-3_53

Li et al. put forward the application of clustering analysis which is in site classification and soil fertility evaluation [3]. Zheng et al. improved rough K- means algorithm, and put forward the rough K- means clustering algorithm based on density weighted [5]. Chen et al. put forward a weighted spaces fuzzy dynamic clustering algorithm, and proved the validity of method in evaluation of soil fertility [6]. But conventional K-means, FCM and other clustering algorithms have some limitations on soil fertility evaluation. Such as K- means is hard clustering algorithm that can only get a hard divide. Although FCM can get fuzzy clustering divide, both algorithms require artificially set the number of clusters. So it can not eliminate outlier accurately or solve the problem of soil fertility data including complex, dynamic, and interactive fuzzy. Whatever, the clustering results presence of a certain error with the real fertility. For this reason the paper use FUMF algorithm to analyze and evaluate soil fertility.

National measuring territories precise fertilizer projects in Jilin Province for over 10 years. During this period a large number of soil samples were collected and sample of soil nutrients were determined and analyzed. All of this could lay the foundation for soil fertility status by using DM and GIS technology. Thus, this paper use large amounts of data by successive years of soil testing precise fertilizer projects that from Nong An county in Jilin Province. Then, we use GIS and Matlab technical conducted a rapid unsupervised multiscale fuzzy clustering for soil nutrient data from 2005 to 2012. The results show that FUMF algorithm is an effective method for soil fertility evaluation and has practical significance when analyze large amounts of high complexity, strong interaction soil fertility factors. So, it is can provide a technical support for the precise fertilization decision.

2 Experiments and Methods

2.1 The Situation of Research Area

Nong'an is located in Songliao Plain, Changchun, Jilin. specific in northwest of Changchun city away from 60 km, north latitude 43° 54'–44° 56', longitude 124° 32'– 125° 45'; The zone is in the temperate semi-humid continental monsoon climate. So, monsoon features obviously, four distinct seasons, abundant sunshine, less rainfall and the annual average temperature of 4.6 °C, annual average sunshine hours 2590 h, the average annual rainfall 507 mm; On the one hand, there are diverse landforms such as high mesa, mesa, two terraces, a terrace, floodplain, sand dunes, depressions, gullies and so on. Thus, most soil is chernozem, meadow soil and black soil; On the other hand, they grow corn sorghum, wheat, millet and soybean and other crops production as the mainstay. It is arguably one of the country's important commodity grain production bases and its total grain production ranked first in the major grain-producing counties.

2.2   Collection and Analysis of Sample Data

On the basis of field research, we are cooperation with cropland capacity survey quality evaluation office and considering soil types, land use, topography, cropping patterns, management measures and production level and other factors according second national soil survey. After that, we determine the sampling point through DGPS and RS systems. Then we can comprehensive analyze the survey plots of soil testing precision fertilizing work from 2005 to 2012. It collected 23,976 samples, sampling map of soil nutrients in Figs. 1 and 2.

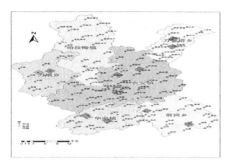
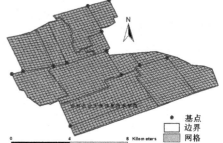

Fig. 1. Nong an soil fertility sampling map   **Fig. 2.** Part of the grid sampling map from Nong an

Samples were collected depth from 0 to 20 cm. Random multi-point sampling within the same plots. Whatever, after mixing the soil by quartering, we take 1.5 or 2 kg bagging spare. Then, take it back to the laboratory for spare through dry naturally pulverized and sieving. Ultimately, index measuring soil nitrogen, phosphorus and potassium and other nutrients in which a total of 26 kinds of soil types. So, in this paper we analyze the collected 23976 data and calculate maximum, minimum, average value of nitrogen (N), phosphorus (P), potassium (K) according to the different soil types. The data of 2010 is shown in Table 1:

Table 1. Soil fertility data

Town name	Country name	Plot name	Alkalystic N (mg/kg^{-1})	Available P (mg/kg^{-1})	Available K (mg/kg^{-1})	Latitude	Longitude
Nong An	Xi Haolai	Jia Bei	154	28	208	44.58417	125.28982
Nong An	Xi Haolai	Jia Bei	136	31.3	217	44.49895	125.25127
Nong An	Xi Haolai	Jia Bei	132	12.8	217	44.4988	125.2509
Nong An	Xi Haolai	Er Jiedi	132	16.3	198	44.49926	125.25072
Nong An	Xi Haolai	Er Jiedi	121	21.4	208	44.5004	125.24821
Nong An	Xi Haolai	Gong Lubei	118	31.8	227	44.50205	125.24968
Nong An	Xi Haolai	Lu Nan	114	37.9	227	44.50206	125.24738
Nong An	Xi Haolai	Er Jiedi	114	28	237	44.5024	125.24905

(continued)

Table 1. (*continued*)

Town name	Country name	Plot name	Alkalystic N (mg/kg^{-1})	Available P (mg/kg^{-1})	Available K (mg/kg^{-1})	Latitude	Longitude
Nong An	Xi Haolai	Yi Jiedi	114	23.3	217	44.5054	125.24378
Nong An	Xi Haolai	Er Jiedi	114	21.3	208	44.50712	125.24472
Nong An	Xi Haolai	San Jiedi	114	19.1	169	44.50855	125.24402
Nong An	Xi Haolai	San Jiedi	121	24.2	160	44.50305	125.23638
Nong An	Xi Haolai	Er Jiedi	118	24.8	140	44.50648	125.24862
Nong An	Xi Haolai	Er Jiedi	118	31.3	179	44.50616	125.24878
Nong An	Xi Haolai	San Jiedi	118	43.6	140	44.50523	125.24845
Nong An	Xi Haolai	San Jiedi	118	32.8	160	44.50502	125.24878
Nong An	Xi Haolai	San Jiedi	110	39.3	169	44.51012	125.2503
Nong An	Xi Haolai	San Jiedi	132	32.8	188	44.50512	125.25013
Nong An	Xi Haolai	San Jiedi	129	37.8	179	44.50522	125.25045
Nong An	Xi Haolai	Er Jiedi	121	23.3	179	44.50755	125.25172
Nong An	Xi Haolai	San Jiedi	114	34.1	227	44.51202	125.2518
Nong An	Xi Haolai	San Jiedi	114	36.8	237	44.5058	125.24893
Nong An	Xi Haolai	Er Jiedi	114	22.7	246	44.50618	125.24475
Nong An	Xi Haolai	Gong Lubei	114	30	217	44.50412	125.23805
Nong An	Xi Haolai	Er Jiedi	118	35.8	208	44.50195	125.2379
Nong An	Xi Haolai	Gong Lubei	129	21.4	237	44.50885	125.24368
Nong An	Xi Haolai	Gong Lubei	140	30.8	227	44.51057	125.24563
Nong An	Xi Haolai	Lu Bei	114	36.3	217	44.51185	125.24018
Nong An	Xi Haolai	Er Jiedi	118	30.8	208	44.51535	125.23695
Nong An	Xi Haolai	San Jiedi	118	36.3	198	44.50502	125.2354
Nong An	Xi Haolai	San Jiedi	118	42.8	227	44.50268	125.23857
Nong An	Xi Haolai	Er Jiedi	103	32.1	198	44.51077	125.25425
Nong An	Xi Haolai	San Jiedi	132	15.8	227	44.50702	125.2542
Nong An	Xi Haolai	Er Jiedi	165	10.8	237	44.51245	125.25438
Nong An	Xi Haolai	Gong Lubei	147	30.8	217	44.51385	125.2544
Nong An	Xi Haolai	Bei Dapian	143	21.3	198	44.5057	125.25578

According to soil grading standards of second soil survey, soil nutrients are divided into six levels, such as shown in Table 2.

Table 2. Soil nutrient grading standards

Project	Level 1	Level 2	Level 3	Level 4	Level 5	Level 6
Alkalystic N(mg/kg)	>150	120–150	90–120	60–90	30–60	≤30
Available P(mg/kg)	>40	20–40	10–20	5–10	3–5	≤3
Available K(mg/kg)	>200	150–200	100–150	50–100	30–50	≤30

According to preliminary results of the analysis, we begin to accurate classification of soil fertility through data mining.

3   Results and Discussion

3.1   Fast Unsupervised Multiscale Fuzzy Clustering (FUMF) Algorithm

First of all, we clustering the N, P, K three indicators of 23976 data by FUMF algorithm, the purpose is to eliminate the isolated samples point of each index. Then these three indicators data were normalized. Finally, set the parameters of weighted dimensional data for clustering analysis by FUMF.

We can accelerate UMF algorithm through nearest neighbor criterion and get FUMF. Well, FUMF method is divided into two stages:

The first stage: re-expression data by using the nearest neighbor criterion, the data is divided into $\bar{n}$ disjoint subsets S_j, Each subset's data represented by its representative point C_j which is as a whole.

The second stage, implementation of weighted UMF algorithm.

FUMF algorithm is as follows:

Step 1. The re-expression data, initialize the $m - 1$, $c_m = \{x_1\}$, $i = 2$ to $N : d(x_i, c_k) = \min_{1 \leq j \leq m} d(x_i, c_j)$, If $d(x_i, c_k) > \Theta$ and $m < q$ then $m = m + 1$ $c_m = \{x_i\}$ Else $c_k = c_k \cup \{x_i\}$.

Step 2. Clustering UMF, set $j = 1$, set a threshold $\varepsilon > 0$ and $v^{(0)} = c_j$, then using the updated formula 1:

$$v^{(l+1)} = \frac{\sum_{k=1}^{\bar{n}} n_k \cdot c_k \cdot \tilde{d}(v^{(l)}, c_k)}{\sum_{k=1}^{\bar{n}} n_k \cdot \tilde{d}(v^{(l)}, c_k)} \tag{1}$$

Calculate convergence point of c_j, denote as p_j. If $j < \bar{n}$, then $j = 1 + j$, repeat step 2.

Step 3. If $\|p_a - p_b\| \leq \varepsilon$, The S_a and S_b of the data points into a class; otherwise, divided into different classes.

3.2   Soil Nutrient Content Analysis

Through statistical analyze 23,976 samples of soil nutrient content, we summarizes the changes of soil nutrients from early, metaphase and anaphase data. As shown in Table 3:

Table 3. Nong an Part of the township of soil nutrient content in different years descriptive statistics

Year	Index	Alkalystic N (mg kg^{-1})	Available P (mg kg^{-1})	Available K (mg kg^{-1})
2005	Range	9.0–193.0	2.0–145.0	8.0–960.0
	Mean	105.79	18.7	138.1
	CV (%)	21.83	67.15	51.15
	Number of samples	2297	2297	2297

(*continued*)

Table 3. (*continued*)

Year	Index	Alkalystic N (mg kg^{-1})	Available P (mg kg^{-1})	Available K (mg kg^{-1})
2009	Range	19.0–211.0	2.1–96.9	40.0–382.0
	Mean	131.84	22.25	184.22
	CV (%)	19.35	54.74	25.87
	Number of samples	5115	5115	5115
2012	Range	88.0–213.0	5.5–82.6	15.0–413.0
	Mean	135.98	24.68	168.47
	CV (%)	13.77	41.97	28.19
	Number of samples	6329	6329	6329

3.3 FUMF Analysis

Taking into account the soil sampling N, P, K three indicators' observed values are different. Data will inevitably be contaminated during sampling that resulting in some isolated points Therefore before cluster analysis of soil nutrient, we need to pre-processing the data set. Pretreatment divided into the following steps:

(1) Executing clustering algorithm for N, P, K three indicators respectively. If it contains a small number of data points when clustering, indicating this category may be constituted by isolated point. In the experiment, we analyze categories which data points lower than 20 and delete those isolated points which beyond the normal range of values.

(2) Because of three indicators of N, P, K have differences in dimension as raw data. Therefore, each of these three indicators were normalized so that the mean of each index is 0 and variance is 1.

After process the raw data, each sample as a data point for clustering. Due to Evaluation of soil fertility mainly depends on the content of P indexes, and P indexes are generally lower than the value of N, K. Thus, we should weighted N, P, K as 1:10:1 before performing clustering algorithm.

(3) Parameter settings: the convergence of the scale parameter is 0.15; convergence precision is 10^{-5}; maximum number of iterations is 100; fuzzy factor is m = 2; data reduction parameters is 0.8; convergence scale parameter 0.14 multiplied mean value; After performing clustering algorithm to pretreatment and weighted data, using inverse transform to get clustering results.

3.3.1 The Initial Precision Fertilization Clustering Results

In this paper, we collected 2297 samples from 27 towns in 2005 to establish the experimental data set(remove isolated points of 38 when prepossessing), all of the data come from Bajilei, Bangchai, Binghe, Fuquanlong, Gaojiadian, Halahai and so on. Then we clustering by FUMF. The clustering results shown in Table 4 and Fig. 3(a 2005).

Table 4. Clustering results in 2005

Category	Available N (mg kg^{-1})	Available P (mg kg^{-1})	Available K (mg kg^{-1})	The amount of data	Fertility
1	105.30	11.30	119.05	824	Low
2	102.83	13.87	122.90	623	Medium
3	105.93	19.50	134.85	427	Medium
4	112.61	64.58	140.52	21	High
5	105.14	54.18	448.18	2	High
6	102.83	32.72	135.60	95	High
7	126.20	47.54	399.26	6	High
8	108.14	38.36	158.07	26	High
9	111.16	40.29	144.33	38	High
10	107.81	44.89	138.15	51	High
11	132.25	60.23	125.78	4	High
12	101.39	29.55	131.55	132	Medium
13	103.13	27.59	129.29	2	Medium
14	107.73	54.61	107.13	5	High
15	132.78	60.18	264.45	3	High

3.3.2   The Middle Precision Fertilization Clustering Results

Experimental data sets with 5115 samples from 23 towns in 2009 (remove isolated points of 24 when prepossessing). The data come from Bajielei, Dehui, Gaojiadian, Halahai, Helong and so on. And clustering results Table 5 and Fig. 3(b 2009).

Table 5. Clustering results in 2009

Category	Available N (mg kg^{-1})	Available P (mg kg^{-1})	Available K (mg kg^{-1})	The amount of data	Fertility
1	123.68	43.40	179.97	102	High
2	131.79	13.81	187.19	1499	Low
3	133.58	22.55	185.53	1156	Medium
4	135.92	17.73	189.62	957	Medium
5	132.67	15.66	187.31	147	Medium
6	134.86	19.69	186.40	441	Medium
7	133.14	36.41	186.40	272	High
8	134.98	30.84	185.22	398	High
9	132.40	62.92	178.31	6	High
10	138.88	76.07	152.41	6	High
11	148.39	67.57	224.91	10	High
12	111.89	81.76	163.05	16	High
13	109.95	72.23	136.23	3	High
14	116.30	53.27	144.66	50	High
15	112.51	57.82	140.77	14	High
16	123.68	43.40	179.97	11	High
17	131.79	13.81	187.19	3	High

3.3.3 The Late Precise Fertilization Clustering Results

Experimental data sets with 6329 samples from 17 towns in 2012 (remove isolated points of 17 when prepossessing). All of the data come from Helong, Qiangang, Bajilei, Fuquanlong, Gaojiadian, Halahai, Huajia and so on. Then we clustering by FUMF. The clustering results shown in Table 6 and Fig. 3(c 2012).

Table 6. Clustering results in 2012

Category	Available N (mg kg^{-1})	Available P (mg kg^{-1})	Available K (mg kg^{-1})	The amount of data	Fertility
1	131.79	20.12	170.56	742	Medium
2	134.85	37.22	160.72	561	High
3	133.05	21.75	166.18	399	Medium
4	137.10	31.71	162.83	666	High
5	133.89	25.76	166.14	699	Medium
6	133.15	28.58	166.18	593	Medium
7	134.18	14.20	162.77	1312	Low
8	133.13	23.58	164.96	260	Medium
9	133.80	18.54	169.35	460	Medium
10	135.91	15.77	165.41	316	Medium
11	135.36	30.17	163.07	58	Medium
12	134.90	47.52	161.72	204	High
13	140.12	63.81	133.31	25	High
14	134.71	45.89	170.48	4	High
15	136.68	35.46	160.15	4	High
16	131.79	20.12	170.56	9	High

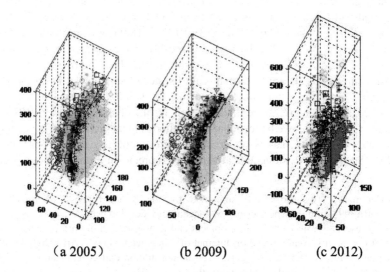

(a 2005)   (b 2009)   (c 2012)

Fig. 3. Soil nutrient spatial clustering map

3.4   Clustering Analysis

When compared the clustering results from 2005, 2009 to 2012, we can derive trend of soil fertility that soil fertility tend towards equilibrium and rise after precision fertilization.

Table 7. Clustering analysis table and soil fertility

Year	The number of samples	Classification number			The percentage of each category in the total sample		
		High	Medium	Low	High	Medium	Low
2005	2259	251	1184	824	11.11 %	52.42 %	36.48 %
2009	5091	891	2701	1499	17.50 %	53.05 %	29.44 %
2012	6312	1473	3527	1312	23.34 %	55.88 %	20.79 %

4   Conclusions

Through the clustering results we analyzed, soil fertility tend towards equilibrium and rise after precision fertilization, so it can reflect the trend of soil fertility better. The results can be seen from Table 7:

(1) We clustering according to the parameters which is set by clustering algorithm. The data is from 2005, 2009 and 2012 and the number of samples are 2259, 5091 and 6312. Then we can derived its cluster classification results are consistent with the actual high fertility, the fertility and low fertility referring soil grading standards. So, it could prove that FUMF algorithm is an effective method to soil fertility evaluation.

(2) The data were compared from 2005, 2009 and 2012, the clustering results show that high fertility soils were increased from 8.16 % to 13.99 % and 15.30 %; The second soil fertility were increased from 53.64 % to 62.80 % and 65.80 %; low soil fertility dropped 22.60 % from 38.10 % to 18.93 % respectively. It is shown that from 2005 to 2012, the low soil fertility decreases and other soil fertility increase after precise fertilization. So, soil fertility has improved significantly.

(3) The analysis results are consistent with the actual situation, it is not only shows FUMF algorithm is an effective method for soil fertility evaluation, but also proved that after precise fertilization soil fertility has improved significantly in general. Therefore, we believe that the method is meaningful by using data mining to analyze fertility factors of large the high complexity and, strong interaction data. So it can provide technical support for precision fertilization decisions.

Acknowledgment. This work was supported by the National "863" High-tech Project (2006AA10A309), Jilin province science and technology development projects(key science and technology research project): "The development and demonstration of corn production's monitoring and traceability system based on Internet of things technology" (20140204045NY), Jilin province agricultural committee projects: "The demonstration and generalization of Corn precision operation system based on Internet of things".

References

1. Zeitouni, K.: A survey of spatial data mining methods databases and statistics point of views. In: Becker, S. (ed.) Data Warehousing and Web Engineering, pp. 229–242. IRM Press, London (2002)
2. Sharma, L.K., Vyas, O.P., Tiwary, U.S., Vyas, R.: A novel approach of multilevel positive and negative association rule mining for spatial databases. In: Perner, P., Imiya, A. (eds.) MLDM 2005. LNCS (LNAI), vol. 3587, pp. 620–629. Springer, Heidelberg (2005). doi:10. 1007/11510888_61
3. Li, L., Li, L.: Application of clustering analysis in site classification and soil fertility evaluation. In: Proceedings of 2010 Third International Conference on Education Technology and Training (2010)
4. Guo, X., Liu, X., Li, X.: Improvement and analysis of hierarchical clustering algorithm. Comput. Appl. Softw. **25**(6), 243–244 (2005)
5. Zheng, C., Miao, D., Wang, R.: Improved K-means clustering algorithm based on rough density weighted. Comput. Sci. (2009)
6. Cheng, G., Cao, L., Wang, G.: Application of weighted spatial fuzzy dynamic clustering algorithm in soil fertility evaluation of the black soil zone. Chin. Agric. Sci. **42**(10), 3559–3563 (2009)
7. Zhang, J., Liu, X.: Research and application of K-means algorithm based on clustering Analysis. Appl. Comput. **24**(5), 166–168 (2007)
8. Li, Y., Yin, P.: Speculative multithreading partitioning algorithm based on fuzzy clustering. J. Comput. **37**(3), 580–592 (2014)
9. Yang, Y., Guo, S.: FCM image segmentation algorithm based on kernel function and spatial information. Jilin Univ. (Eng. Sci.) **41**(2), 283–287 (2011)
10. Liu, F., Shi, X., Yu, D.: Research of mapping of soil properties in Taihu based on geostatistics and GIS–In total soil nitrogen in cartography example. J. Soil **41**(1), 20–27 (2004)
11. Zhao, Y., Shi, X., Yu, D.: Discussion of mall-scale spatial variability of soil nutrients and its influencing factors– In urban areas in Wuxi city, Jiangsu province **37**(2), 214–219 (2006)
12. Huang, S., Jin, J., Yang, L.: Research Grain Crop spatial variability of soil nutrients and the partition management technology in Region of County. J. Soil **40**(1), 79–88 (2003)
13. Yang, Y., Shi, X., Yu, D.: Research of region scale soil nutrients spatial variability and influencing factors. Geogr. Sci. **28**(6), 788–792 (2008)
14. Guo, X., Fu, B., Ma, K.: Research on spatial variability of soil fertility based on GIS and geostatistics—case by Hebei Zunhua. J. Appl. Ecol. **11**(4), 557–563 (2000)
15. Sun, Y., Lu, Y.: High-dimensional data flow subspace clustering algorithm based on grid. Comput. Sci., 199–203 (2007)
16. Shan, S., Yan, Y., Zhang, X.: Clustering algorithm based on K most similar clustering subspace. Comput. Eng., 4–6 (2009)
17. Qiu, B., Zheng, Z.: Clustering algorithm based on the local density and dynamically generated mesh. Comput. Eng. **31**(2), 385–387 (2010)

Modeling and Optimization of Agronomic Factors Influencing Yield and Profit of a Single-Cropping Rice Cultivar

Weiming Liu[1(✉)] and Zuda Bao[2]

[1] Taizhou Vocational College of Science & Technology,
Taizhou 318020, Zhejiang, China
15157622288@139.com
[2] Seed Management Station, JiaoJiang, TaiZhou,
Taizhou 318000, Zhejiang, China

Abstract. The effect of agronomic factors including seedling age, transplanting density and net nitrogen ratio on the yield and profit of a single-cropping hybrid late rice (Oryza sativa) cultivar 'Yongyou 12' was investigated. The orthogonally rotational combination experimental design was used in setting field parameter test. Regression models assuming yield or profit as the objective function were constructed, and the impact from each individual or combinations of the three agronomic factors on the yield and profit of the crop was estimated. A computer simulation was performed to select the optimized agronomic scheme of growing rice 'Yongyou 12' under the field conditions used in this study.

Keywords: Japonica/indica hybrid rice · Rice 'Yongyou 12' · Agronomic factors · Yield · Profit · Mathematical models · Optimization

Since the introduction of the first regression experimental design in the late-1950s, several complex versions have been developed, including the regression for rotary combinational design and regression for d-optimal designs. In agriculture, regression designs are used to develop the optimal agronomic factor schemes for improving agronomic traits. The simplex-lattice and rotation combination designs are used to optimize planting schemes for sweet potato [1–3], and the secondary saturation and D-optimal design for soybean, cassava and Dioscorea zingiberensis [4–6], and the general rotation combination design for the intercropping system of spring maize and sweet potato [7].

Rice (Oryza sativa) cultivar 'Yongyou 12' is a three-line japonica-type hybrid. In the past few years, the cultivar has been recommended in Tai Zhou, and is very promising for future plantings in this area. According on the Zhejiang Rice Variety Certificate 2010015, rice 'Yongyou 12' is a late three-line indica-japonica hybrid cultivar. The cultivar has the following traits: thick and strong stem, high lodging resistance, highly photoperiodic sensitive, long growth season, medium tillering capacity, large panicles with many grains per panicle, high yielding ability, medium-grain-quality, moderate resistance to rice blast disease (Magnaporthe grisea) and rice stripe virus disease, moderate susceptibility to rice bacterial leaf blight

© IFIP International Federation for Information Processing 2016
Published by Springer International Publishing AG 2016. All Rights Reserved
D. Li and Z. Li (Eds.): CCTA 2015, Part I, IFIP AICT 478, pp. 574–579, 2016.
DOI: 10.1007/978-3-319-48357-3_54

(Xanthomonas oryzae), and susceptible to brown planthopper (Nilaparvata lugens). The cultivar is recommended in the area South of Qiantang River in Zhejiang Province. In this study, the rotating regression design was used to set up a field experiment to evaluate effects of three agronomic factors on hybrid rice 'Youngyou 12' which was planted as a single-season late rice cultivar. Mathematical models were constructed and optimal agronomic schemes were identified.

1 Materials and Methods

Rice 'Yongyou 12' was the cultivar, and seedling age (x_1, days), transplanting density (x_2, clusters/hm2) and net nitrogen ratio (x_3, kg/hm2) were the decision variables. The experimental design for field condition parameter test was a quadratic regression orthogonal rotational combing design with three factors [8–10]. Treatment levels and experimental layout for the three agronomic factors (i.e., the three decision variables) and the linear coding were contained in Table 1. Yield (y_1, kg/hm^2) and net profit (y_2, yuan/hm^2) were assumed as objective function in regression modeling.

Table 1. Decision variables and the treatment levels

Linear coding	Seedling age (x_1, days)	Transplanting density (x_2, cluster/hm^2)	Net N ratio (x_3, kg/hm^2)
+1	30	202500	201.15
−1	18	127500	51.15
0	24	165000	126.15
+1.682	34	228075	252.3
−1.682	14	101925	0

In Table 1, seeds were sown on May 26th and 30th, June 4th, 10th, and 14th for seedling age groups of 34, 30, 24, 18 and 14 days, respectively. All seedlings were transplanted on June 28th. Each plot had an area of 16 m^2 (3.2 m × 5 m); a layer of plastic mulch was installed in between plots to prevent unintended fertilizer contamination from adjacent plots. A base fertilizer was applied at a rate of 750 kg/hm^2 of calcium superphosphate and 60 kg/hm^2 of potassium chloride in each plot. The field experiment was conducted in 2011, in Shanjia, Shuzhou county of Taizhou. The whole experimental field with different plots was managed consistently except for the treatment factors.

2 Results and Analysis

2.1 Effect of the Agronomic Factors on Yield

Analysis of yields from different treatments of the three agronomic factors generated the following quadratic regression model:

$\hat{y}_1 = 11844.75 - 25.50x_1 + 158.40x_2 + 835.05x_3 + 54.75x_1^2 - 83.4x_2^2 - 476.25x_3^2 + 145.35x_1x_2 + 66.6x_1x_3 - 42.9x_2x_3$, in which the interval constrain is in the range of $-1.682 \le x_i \le 1.682$, $i = 1, 2, 3$.

In this model, x_3 and x_3^2 both were at the extremely significant level, x_2, x_1x_2 were at a significant level ($\alpha = 0.25$), and the rest of the variables were not significant at the $\alpha = 0.25$ level. To improve the accuracy of model, variables that were not significant ($\alpha = 0.25$) were removed, and a simplified yield regression model was constructed as: $\hat{y}_1 = 11844.75 + 158.40x_2 + 835.05x_3 - 476.25x_3^2 + 145.35x_1x_2$, where the interval constrain is in the range of $-1.682 \le x_i \le 1.682$, $i = 1, 2, 3$. There was an extremely significance level for the simplified model as a whole.

Model analysis showed that among the three agronomic factors, rice yield was affected the most significantly by net nitrogen ratio, followed by transplanting density and the seedling age had the least significant impact. Additionally, due to the interacting effect between variables x_1x_2, the planting density should be reduced for younger seedling, or vice versa, it should be increased for older seedlings (Table 2).

Table 2. The effect of interaction of seedling age with transplanting density on rice yield

x_1	x_2				
	−1.682	−1.000	0.000	1.000	1.682
1.682	11167.20	11441.85	11844.75	12247.65	12522.30
1.000	11333.85	11541.00	11844.75	12148.50	12355.65
0.000	11578.20	11686.35	11844.75	12003.15	12111.30
−1.000	11822.55	11831.55	11844.75	11857.95	11866.80
−1.682	11989.20	11930.70	11844.75	11758.80	11700.30

2.2   Effect of the Agronomic Factors on the Profit of Rice Production

In this study, prices for different items were estimated at 2.20 yuan/kg rice grains, 1.50 yuan/thousand seedlings, 2.10 yuan/kg urea, 2.70 yuan/kg potassium chloride, and 0.50 yuan/kg and calcium superphosphate. Only material cost was included in the production cost. The net profit from each experimental plot was converted into profit per hectare. A regression model was generated for the correlation between the three agronomic factors and the profit. After removal of the variables at the non-significant level ($\alpha = 0.25$), a simplified regression model for profit was developed as follows:

$\hat{y}_2 = 24664.05 + 292.35x_2 + 1478.55x_3 - 1047.60x_3^2 + 319.65x_1x_2$, in which the constrain interval is at $-1.682 \le x_i \le 1.682$, $i = 1, 2, 3$. Test of the significant difference confirmed that the profit regression model behaved basically the same as the yield model, x_3, x_3^2 both were at the extremely significant level, and x_2, x_1x_2 at a significant level ($\alpha = 0.25$). Therefore, the regression model for profit was also at the extremely significant level as a whole.

Model analysis found that the effect of the three agronomic factors on profit was very similar to that of yield. Net nitrogen ratio had the greatest effect on rice profit, followed by transplanting density, and the seedling age had the least significant impact. Furthermore, there was a significant impact from the interacting effect from variables of x_1 and x_2, therefore the transplanting density should be reduced for younger seedlings whereas it should be increased for older seedlings for high profit rice production. Although the interacting effect from x_1 and x_2 would have a similar influence on both yield and profit, the two factors must be adjusted appropriately according to local conditions in any agronomic scheme.

3 Computer Simulations for the Optimization of Agronomic Scheme

3.1 Optimization of High Yield Agronomic Schemes

A computer simulation was performed using the yield regression model within the constraint interval of $-1.682 \leq x_i \leq 1.682$, to screen for high yield agronomic schemes [9–11]. From the 125 sets of agronomic schemes, 33 sets were selected to produce a yield above 12000 kg/hm^2 (Table 3). Within 95 % confidence intervals, the selected values for x_1, x_2, x_3 were −0.301–0.525 (22–27 days), 0.276–1.024 (175350–203400 clusters/hm^2) and 0.783–1.143 (184.95–211.95 kg/hm^2), respectively.

Table 3. Computer simulation of the agronomic scheme for high yield rice production

Treatment levels	x_1	Frequency	x_2	Frequency	x_3	Frequency
−1.6818	6	0.1818	3	0.0909	0	0.0000
−1.0000	5	0.1515	3	0.0909	0	0.0000
0.0000	8	0.2424	5	0.1515	6	0.1818
1.0000	7	0.2121	11	0.3333	20	0.6061
1.6818	7	0.2121	11	0.3333	7	0.2121

Table 4. Computer simulation of agronomic schemes for high profit rice production

Treatment level	x_1	Frequency	x_2	Frequency	x_3	Frequency
−1.6818	6	0.1935	3	0.0968	0	0.0000
−1.0000	5	0.1613	4	0.1290	0	0.0000
0.0000	6	0.1935	5	0.1613	8	0.2581
1.0000	7	0.2258	10	0.3226	19	0.6129
1.6818	7	0.2258	9	0.2903	4	0.1290

3.2 Optimization of High Profit Agronomic Scheme

Computer simulation of the profit regression model using the constraint intervals of $-1.682 \leq x_i \leq 1.682$, was performed to select for the high efficiency agronomic schemes. From 125 schemes, 31 schemes would produce a profit above 24750 yuan/hm^2 (Table 4), where the selected values for x_1, x_2, x_3 value (within the 95 %

confidence intervals) were −0.321–0.558 (22–27 days), 0.121–0.917 (169545–199395 clusters/hm^2) and 0.641–1.019 (174.30–202.65 kg/hm^2), respectively. It can be seen that the effects of the three agronomic factors followed the same trend for improving profit as well as for increasing yield.

4   Conclusion and Discussion

In this study, regression mathematical models were constructed to determine the effect of agronomic factors including seedling age, transplanting density and net nitrogen fertilizer rate on the yield and profit of rice 'Yongyou No. 12'. An optimized agronomic scheme was developed for single-cropping of the late hybrid rice cultivar. The three agronomic factors affected yield or net profit of rice production in a similar manner. Among the three factors, net nitrogen ratio had the greatest effect on yield and net profit, followed by transplanting density, and seedling age had the least significant effect. By using modeling analysis in combination with the real field condition, the optimized agronomic scheme for the highest yield and highest profit for rice 'Yongyou No. 12' was projected to use seedlings of the 22–25 days old, at a transplanting density of c.a. 172500 clusters/hm2 (no less than 165000 clusters/hm2), and using a net nitrogen ratio at 180–195 kg/hm2 when growing the crop under the same field conditions used in this study.

However, all the single-cropping late rice farms had higher yield in 2011 compared to previous years, which led to a higher projected yield and net profit increases in the optimized agronomic scheme developed in this study. It remains to be tested if such high yield and profit will be stably achieved in future years.

References

1. Liu, W.M., Wu, L.H.: Effect of planting density and fertilizer application technology on productivity and economic benefit of sweet potato Zheshu 132. Seed **27**(6), 101–102 (2008)
2. Liu, W.M., Ji, Z.X., Cheng, X.S.: Test for optimizing main cultural techniques of a new sweet potato cultivar "Xinxiang". Acta Agriculturae Shanghai **24**(4), 48–50 (2008)
3. Liu, W.M., Wu, L.H.: A study of optimization in agronomic practices for a new sweet potato cultivar 'Zheshu 13'. Seed **26**(4), 73–75 (2007)
4. Liu, W.M., et al.: Comprehensive agronomic optimization for Zheqiu Soybean 3. Bull. Sci. Technol. **23**(3), 382–385 (2007)
5. Lu, X.J.: Effects of N, P, K combined application on the yield and quality of Cassava. Chin. J. Trop. Crops **34**(12), 2331–2335 (2013)
6. Liu, W.M., Wang, R.Z.: Influences of planting density and fertilization technique on the yield and benefit of annual Dioscorea. J. Anhui Agric. Sci. **36**(4), 1396–1397 (2008)
7. Liu, W.M.: The application of computer in the optimization studies of crop intercropping. Agric. Netw. Inf. **3**, 7–9 (2005)
8. Mao, S.S., Yu, Y., Zhou, J.X.: Regression Analysis and Experimental Design, pp. 201–211. East-China Normal University Press, Shanghai (1981)

9. Liu, W.M., Yu, Z.Y.: Application of rotary regression experimental design in agronomy optimization experiment. Appl. Probab. Stat. **3**(3), 274–275 (1987)
10. Liu, W.M.: Application of regression design in the study of intercropping system of crops. Stat. Manag. **25**(5), 1–5 (2005)
11. Liu, W.M., Xu, J.: A study on transplanting density and nitrogen fertilizer of in single-cropping late hybrid rice 'Zhongzheyou 1'. Hybrid Rice **21**(5), 50–51 (2006)

The Milk Somatic Cell Image Segmentation Method Based on Dimension Reduction and Fusion

Jie Bai, Heru Xue[✉], and Yanqing Zhou

College of Computer and Information Engineering,
Inner Mongolia Agricultural University, Huhhot 010010, China
baijie9554@163.com, xuehr@imau.edu.cn,
974864134@qq.com

Abstract. Milk somatic cell image segmentation has a very important effect on milk quality analysis. In this paper, a new segmentation method is proposed for color milk somatic cell image. Aiming at the milk somatic cell image, a new method for image segmentation is proposed. First of all, selecting the appropriate color components by reducing the dimension of RGB color image; Then using the k-means algorithm to segment the low dimension image; Finally, the segmented images are fused by the region splitting and merging process. The experimental results show that the proposed method is better than the original three-dimensional (3D) color space segmentation method. The method improves in the performance and running time, the correct segmentation rate reached 98.4 %. Therefore, this method has certain feasibility.

Keywords: Milk somatic cell image · Dimension reduction · k-means algorithm · Image fusion

1 Introduction

With the development of computer application technology, graphics and image processing has penetrated into all walks of life. the research on somatic cell bovine milk is a very important research field. The number of milk somatic cells is an important index to judge the quality of milk [1]. The more somatic cell counts, the higher the incidence of mastitis is. But the image segmentation is the first step of the milk somatic cell counting, which has very important significance to improve the quality of milk cow mastitis detection and diagnosis.

At present, domestic and foreign scholars for the segmentation of cell image are also put forward a series of effective methods, such as Threshold segmentation method, Edge detection method and Region growing method. Threshold segmentation method has the advantages of simple algorithm and is easy to realize, which can use boundary definition closed and connected overlapping regions. The disadvantage is difficult to obtain the accurate object boundary for the small difference between object and background [2]. The advantage of edge detection method is precise contour location, and the disadvantage is the inability to guarantee the closed contour [3]. Region growing method can get a closed contour, but it is hard to determine the growth of

D. Li and Z. Li (Eds.): CCTA 2015, Part I, IFIP AICT 478, pp. 580–586, 2016.
DOI: 10.1007/978-3-319-48357-3_55

termination conditions [4]. In biological cells, the cell shape varies, so there is no universal method for all cell images.

By comparing the existing algorithms, selecting reasonable low dimensional space and using K-means algorithm and image fusion technology to segment the milk somatic cell images can get better segmentation results, and it provides a basis for the further analysis of milk somatic cell image.

The remaining of the paper is organized as follows. Section 2 describes how to select the appropriate color components and implement the segmentation in the low dimensional space. Section 3 presents a fusion of these segmentation results and gets a final segmentation. Section 4 evaluates the segmentation results. conclusion is given in Sect. 5.

2 The Milk Somatic Cell Image Segmentation Base on the Low Dimensional Space

The color image contains more information than the gray image, so in many applications of computer vision and pattern recognition, the color image can obtain better results [5]. This paper uses the color milk somatic cell image. The processed color image will produce the problem of large computation and slow processing speed. So in order to overcome the above problems, it is necessary to reduce the dimensionality of 3D color space. In this paper, the 3D color space is projected to the 2D color space [6], and the k-means segmentation is implemented in 2D color space.

2.1 Select the Low Dimensional Space for the Milk Somatic Cell

In this paper, the milk somatic cell image is RGB color image. Therefore, it is necessary to analyze its color components, In the process of analysis, through calculating the standard deviation and the histogram of each color component, to select the spectral component with the most abundant informations. In the final analysis, the R component of the standard deviation is 0.1865, G is 0.1469, B is 0.0216. The histogram of different color components is shown in Fig. 1.

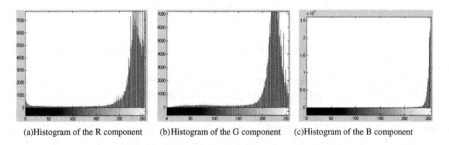

(a)Histogram of the R component   (b)Histogram of the G component   (c)Histogram of the B component

Fig. 1. Histograms of different color components (Color figure online)

By observing the color component histogram and the compute results of the standard deviation, the standard deviation of R is the largest, B is the smaller. It can be concluded that the distinguishing effect is from strong to weak is R, G, B. It also can be say that the R and G components have the significant difference in the nucleus, cytoplasm and the background. In this paper, the segmentation is performed in RG channel and RB channel by using k-means method.

2.2   Using K-Means to Segment the Milk Somatic Cell in Low Dimensional Space

Using the k-means algorithm, first of all, selecting the initial cluster centers, and then classifying all data points. In the end, the average value is calculated to adjust the cluster center for the each cluster, and continuous loop iteration, which makes the evaluation of clustering criterion function to achieve optimal performance, in order to reach the compactness within class and the independence in class of each cluster [7]. The image segmentation steps are described as follows:

1. Select the k sample points in the sample data sets N, and the k sample values are assigned to the initial clustering center $(\mu_1^{(1)}, \mu_2^{(1)}, \ldots, \mu_k^{(1)})$;
2. When the j iteration, to all points in the sample points $P_t(t = 1, \ldots, n)$, followed by calculating the Euclidean distance d (t, i) of each cluster center $\mu_i^{(j)}$;

$$d(t, i) = \sqrt{(P_t - \mu_i^{(j)})^2} \tag{1}$$

3. Find out the minimum distance of P_t and $\mu_i^{(j)}$, put P_t into the minimum distance cluster of $\mu_i^{(j)}$;
4. Update the clustering center of each cluster

$$\mu_i^{(j+1)} = \frac{1}{n_i} \sum_{t=1}^{n_i} P_{it} \tag{2}$$

5. Calculate the square error E_i of all the points in the data sets N, and compared with previous error E_{i-1}

$$E_i = \sum_{i=1}^{k} \sum_{t=1}^{n_i} |P_{it} - \mu_i^{j+1}|^2 \tag{3}$$

If $|E_{i+1} - E_i| < \delta$, the algorithm ends, or go to (2) iterative again. The segmentation results are shown in Fig. 2.

Figure 2 shows that it has different segmentation results by using k-means algorithm in the RG space, RB space and the original three-dimensional space. And the original image segmentation results (Fig. 2(f)) compared to in RG space segmentation results (Fig. 2(b)) the background is misclassification into cytoplasm, in RB space segmentation results (Fig. 2(d)) the nucleus is misclassification into cytoplasm. It can be seen that the effect of segmentation in the two two-dimensional subspaces is not the best, which is due to loss a color component. Therefore, In order to overcome the above problems, it is need to use the image fusion technology to fusion the segmentation results of subspace, and obtain the more accurate segmentation results.

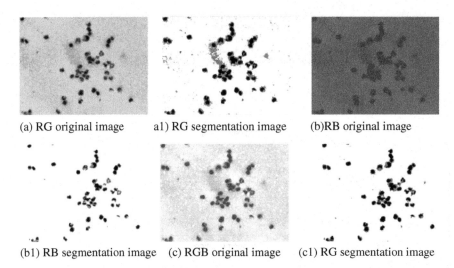

(a) RG original image   a1) RG segmentation image   (b)RB original image

(b1) RB segmentation image   (c) RGB original image   (c1) RG segmentation image

Fig. 2. Segmentation results of milk somatic cell image by k-means algorithm

3   Fusion Results of Low Dimensional Space Segmentation

Image fusion is the spatial matching of image data from different sources, and using complementary organic combination and producing new image data. This new data has a description of the research object and better representation of information, compared with the single information source, it can reduce or inhibit the ambiguity, incompleteness, uncertainty and error to explain the perceived object or environment and maximize the use of the information provided by the various sources of information [8, 9].

The fusion of milk somatic cell image is divided into the following steps. The first step is to apply the k-means segmentation process in two 2D subspaces, and then use the region splitting and merging process to segment again. Finally it gets a better segmentation image. The specific segmentation process as shown in Fig. 3.

The region splitting and merging process is briefly described as follows:

Inputting the 3D color space image I, Extracting it 2D subspace and implementing k-means algorithm in two 2D subspaces, the segmentation image are denoted by I_{RG} and I_{RB}. During the region splitting, The image I is divided into class L in the RG subspace, we denote by $I_{RG}^{(l)}$, $1 = 1, 2, \ldots, L_{RG}$ and $I_{RG}^{(1)} \cup I_{RG}^{(2)} \cup \ldots \cup I_{RG}^{(1)} = I_{RG}$, $I_{RG}^{(1)} \cap I_{RG}^{(t)} = \varphi(l \neq t)$, The RB subspace can also be obtained in segmentation result and expression of symbols for $\forall l \in \{1, 2, \ldots, L_{RG}\}$ and $t \in \{t_1, t_2, \ldots, t_k\} \subset \{1, 2, \ldots, L_{RB}\}$, The region $I_{RG}^{(1)}$ is subdivided into $I_{RG}^{(1)} \cap I_{RB}^{(t_1)}$, $I_{RG}^{(1)} \cap I_{RB}^{(t_2)}$, $\ldots$, $I_{RG}^{(1)} \cap I_{RB}^{(t_k)}$ and $I_{RG}^{(l)} - I_{RG}^{(t)} \cap \left(I_{RB}^{t_1} \cup I_{RB}^{t_2} \cup \ldots \cup I_{RB}^{t_k} \right)$. Above is the splitting process, this process has realized the secondary segmentation of I_{RG} relative to I_{RB}, the RG subspace image is further divided into many subclasses, the results of segmentation is denoted by I_{RG+RB}. Computing the respective regional of the average pixel components R, G, B which is reclassified, to

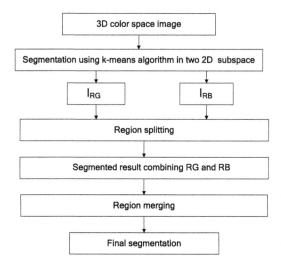

Fig. 3. The flow chart of milk somatic cell image segmentation

compute the average pixel components is assigned to all the pixels belonging to this class. At the end of the splitting process, the number of sub class may be larger than L. So it is necessary to carry out the above segmentation image region merging process. In this paper, by comparing the two color points of Euclidean distance $D = \sqrt{(r1 - r2)^2 + (g1 - g2)^2 + (b1 - b2)^2}$, the smallest color distance of two classes are first merged and the mean color values for new regions are updated. This process is iterative, until the image to achieve a specified class number date. In this experiment, the milk somatic cell image should be divided into three classes, namely the nucleus, cytoplasm and background.

The experiment is performed in Matlab2011b environment to achieve the above algorithm. This paper uses the milk somatic cell image that comes from the Inner Mongolia Agricultural University laboratory animal pathology. The image using

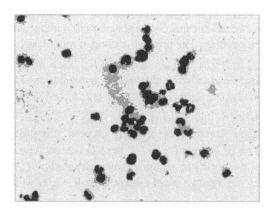

Fig. 4. The results of subspace images segmentation and fusion

Newman's staining and it is magnified by 40× oil lens, which is stored as a 24 bit color image. In this experiment, it can be divided into three categories, namely the nucleus, cytoplasm and background. The final result using k-means and fusion approach in two 2D subspaces as depicted by Fig. 4.

4   Experimentation and Results

In this paper, through computing the image segmentation correct rate, the quantitative results are evaluated. Firstly, a standard and stable method is used to segment the image as a reference image segmentation. The FCM [10] method is used to segment the milk somatic in the experiment, then the comparison calculation of segmented image and the reference image are made. If the nucleus is divided into the pixel numbers of nucleus is a, the cytoplasm is divided into the pixel numbers of cytoplasm is b, and the background is divided into the pixel numbers of background is c. The accuracy will be calculated: $(a + b+c)/n$, and n is the total number of the image.

Table 1 is the time comparison with the milk somatic cell image by k-means method in different color spaces. Tables 2 and 3 is the accuracy of the segmentation which is used the paper method and directly segment in 3D space using k-means method. As Table 2 shows, the segmentation accuracy is 98.4 %, and the Table 3 shows that the segmentation accuracy is 95.4 %. Through many times of experimenting, we can know that both segmentation methods in 3D color space and in 2D subspace can get the correct results. However the correct rate of 2D method is higher than using 3D method directly and the running time is shorter. So the method proposed in this paper has some maneuverability and practicability.

Table 1. Comparison between different methods

	3D method	2D method
Running time(s)	7.4 s	3.9 s

Table 2. Segmentation accuracy using 2D method

2D method						
Reference method	Classes	Nucleus	Cytoplasm	Background	Total	Accuracy
	Nucleus	15060	0	0	15060	(15060 + 18632 + 278358)/
	Cytoplasm	0	18632	5150	23782	317200 = 98.4 %
	Background	0	0	278358	278358	

Table 3. Segmentation accuracy using 3D method

3D method						
Reference method	Classes	Nucleus	Cytoplasm	Background	Total	Accuracy
	Nucleus	12960	2100	0	15060	(15060 + 11306 + 278358)/
	Cytoplasm	0	11306	12376	23782	317200 = 95.4 %
	Background	0	0	278358	278358	

5 Conclusions

Milk Somatic cell image segmentation is an important part of milk somatic cell image processing, due to color image with large amounts of data, segmentation directly in 3D color space will produce the problem of slow processing speed. However in this paper, we have presented an efficient method for the milk somatic cell image segmentation. Firstly, through analyzing the effect of each color component, the 3D color space is projected onto the low dimensional space. Secondly, the k-means algorithm is implemented in the low dimensional space to segment. Finally using region split and merge methods of the image fusion to obtain the final segmentation result. By comparison the method, the method proposed in the paper not only overcomes the disadvantages of the large amount of calculation and lower speed in 3D color space segment directly, but also improves the accuracy of segmentation. Therefore, the proposed method is feasible and has certain theoretical and practical significance.

Acknowledgment. Funds for this research was provided by the National Natural Science Foundation of China (Grant No. 61461041)

References

1. Hai-xia, Liu: Screening test of somatic cell count in bovine milk. China Dairy Ind. **06**, 31–34 (2004)
2. Han, S., Wang, L.: A survey of thresholding methods for image segmentation. Syst. Eng. Electron. **06**, 91–94 + 102 (2002)
3. Mohamed, H.T., Bouhlel, S., Derbel, N., Kamoun, L.: A Surrey and Evaluation of Edge Detection Operators Application to Medical Images. IEEE (2002)
4. Chang, Y.L., Li, X.B.: Adaptive image region-growing. IEEE-IP **3**(6), 868–872 (1994)
5. Xue, H., Ma, S., Pei, X.: Color image segmentation based on mathematical morphology and fusion. J. Image Graph. **11**(12), 1764–1767 (2006)
6. Kurugollu, F., Sankur, B., Harmanci, A.E.: Color image segmentation using histogram multithresholding and fusion. J. Image Vis. Comput. **19**(13), 915–928 (2001)
7. Zhou, A., Yu, Y.: The research about clustering algorithm of K-means. Comput. Technol. Dev. **21**(2), 2 (2011)
8. Xue, H.R., Pei, X.C.: Color image segmentation using fuzzy sets and fusion. In: The Sixth International Conference on Electronic Measurement & Instruments. Taiyuan ISTP (2003)
9. Zhao, X.: Modern digital image processing technology to improve and applications comments (MATLAB version), vol. 04, pp. 113–114. Bei Hang University Press (2012)
10. Fang, M.: The Application of FUZZY C-Means Algorithm in the Segmentation of Milk Somatic Cell Color Image. Inner Mongolia Agriculture University (2008)

Quantitative Detection of Pesticides Based on SERS and Gold Colloid

Yande Liu[(⊠)], Yuxiang Zhang, Haiyang Wang, and Bingbing He

Institute of Optics-Mechanics-Electronics Technology and Application,
East China Jiaotong University, Nanchang 330013, Jiangxi, China
jxliuyd@163.com

Abstract. The detection of pesticide residued in fruit is an important concern for consumers. Surface enhanced Raman spectroscopy (SERS) coupled with gold colloid was applied to analyze two kinds of pesticides (phosmet, chlorpyrifos) which were mainly used on the navel orange. The concentration of the phosmet samples of range from 3 to 33 mg/L and chlorpyrifos samples of range from 4 to 34 mg/L. Using Partial least squares (PLS) regression and the different preprocessing method for the spectral data analyses, and different pretreatment methods such as the Savitzky-Golay were compared. The optimal model of phosmet pesticide and chlorpyrifos pesticide were set up. The prediction correlation coefficient (R) and the root mean square error of prediction (RMSEP) of phosmet pesticide were 0.924 and 4.293 mg/L; The R and RMSEP of chlorpyrifos pesticide were 0.715 and 6.646 mg/L. It indicated that SERS technology is a effective method in the field of pesticide residue detection in fruit.

Keywords: SERS · Gold colloid · Pesticide residue

1 Introduction

SERS is an ultrasensitive testing technique, and It has achieved notable achievements in the last years in some areas such as food safety inspection [1, 2], environmental monitoring [3], materials science, surface science, biomolecular Sensing [4] and analytical chemistry. With the photons as probes, SERS has the advantages of high sensitivity, good selectivity, non-destructiveness, and not require vacuum conditions. Now, SERS has become a widely used method of analysis. Organophosphosphate(OP) pesticides has been used for decades to protect the growth of crops, and made a great contribution to ensure the agricultural production and famers' income [5]. However, with the increasing scale of pesticide use, agricultural products with the pesticide residues have a negative impact on human health.

The common methods used to detect the pesticides residues in agricultural products are mature, such as Gas Chromatography (GC) [6, 7], High Performance Liquid Chromatography (HPLC) [8–10], and etc. Although these methods have good accuracy, their sample pretreatment process is complex, time-consuming and testing cost is high. SERS has the advantages of faster, convenience, high sensitivity and nondestructive analytical method for determination of fruit pesticides. And the Raman analyses could be used in liquids [5, 11, 12]. For the past few years, SERS, which is a

© IFIP International Federation for Information Processing 2016
Published by Springer International Publishing AG 2016. All Rights Reserved
D. Li and Z. Li (Eds.): CCTA 2015, Part I, IFIP AICT 478, pp. 587–596, 2016.
DOI: 10.1007/978-3-319-48357-3_56

promising technology for the residues detection, has attracted much interest by researchers from the entire world. The theory of SERS is putting probed molecules onto the roughened surface of transition metals, enhancing the Raman signals more than one million times because of the chemical enhancement and electromagnetic field enhancement [13]. The strength of SERS depends on the interaction between analyte and Nanostructures substrate surface, and the most typical substrate is gold (Au), silver (Ag) and copper (Cu) [14].

Guerrini et al. [15] obtained the SERS spectra of dimethoate (DMT) and omethoate (OMT) by aggregated Ag hydrosols, and the result showed that the detection limit of DMT was 10^{-5} mol/L. Dhakal et al. [16] collected the Raman spectroscopy of organophosphorus (chlorpyrifos) pesticide. Detection limit for the chlorpyrifos residue in apple surface of the developed system was 6.69 mg/kg.

The purpose of this study was to discuss the feasibility of choosing Gold Nanostructures as SERS substrates for qualitative and quantitative analysis of fruit residues. Ganan navel orange was selected as the object of research. Choose phosmet and chlorpyrifos as the experimental subject. Spectral data collected were qualitative and quantitative analyzed. Nowadays, there are many reports on the application of using Gold Nanostructures as SERS Substrates in the field of pesticide residues, and the goal was to detect and characterize pesticide residues using Gold Nanostructures as SERS substrates.

2 Experiments

2.1 Materials

Two OP pesticides (phosmet and chlorpyrifos) were purchased from AccuStandard, Inc. $HAuCl_4 \cdot 4H_2O$ was purchased from Sinopharm Chemical Reagent Co., Ltd. NaCl, acetonitrile, methanol and sodium citrate were analytically pure, and purchased from Nanchang huake Instrument Co. Ltd.

2.2 Sample Preparation

Pure pesticide solutions: 5000 mg/L of phosmet and chlorpyrifos solutions were prepared by methanol and H_2O (1:1, v/v).

Oranges were cleaned repeatedly with ultrapure, and dry naturally. Clean the orange skin and Cut it into square pieces (2 cm × 2 cm), and a certain amount of standard pesticide solution of phosmet and chlorpyrifos were added on it. After waiting pesticides air-dry, the skin was grated and added to an acetonitrile solution. We could get the solution of pesticide after filter it. The 31 different concentrations of phosmet range from 3 to 33 mg/L at 1 mg/L interval. The 31 different concentrations of chlorpyrifos range from to 34 mg/L at 0.5 mg/L interval.

2.3   SERS Measurement

The SERS substrates of Gold Nanostructures were prepared by sodium citrate reduction method. The pesticide samples, colloidal gold and sodium chloride were mixed based on a certain volume ratio (10:3:3), shocked evenly on the shock tester. Mixed solution was put on the quartz plate, and collected SERS of it.

2.4   Data Analysis

The data analysis used Unscrambler v8.0 software. Pesticide original spectral data were processed by partial least squares (PLS) and the different pretreatment methods. The prediction model was established between the Raman spectrum and pesticide content using different modeling algorithms (PLS, PCR), and the effect of models was analyzed and evaluated using the following indicators:

$$R = \sqrt{\frac{\sum\limits_{i=1}^{n} (\hat{y}_i - y_i)^2}{\sum\limits_{i=1}^{n} (\hat{y}_i - \bar{y})^2}}$$

$$RMSEC = \sqrt{\frac{\sum\limits_{i=1}^{n} (y_i - \hat{y}_i)^2}{n-1}}$$

$$RMSEP = \sqrt{\frac{\sum\limits_{i=1}^{n} (y_i - \hat{y}_i)^2}{n-1}}$$

In these equations, n is the number of samples, y_i is the reference pesticide concentration (mg/L), $\hat{y}_i$ is the predicted pesticide concentration (mg/L), $\bar{y}$ is the average of the reference pesticide concentration (mg/L).

The R, RMSEC and RMSEP were used to evaluate the model. In the same concentration range, if the R value is higher and the RMSEP value is lower, the better predictability the model will get.

3   Results and Discussion

3.1   Characteristic Raman Shift of Phosmet, Chlorpyrifos

The chemical structures of phosmet and chlorpyrifos are quite similar (Fig. 1). Raman spectra of the phosmet powder were obtained, as shown in Fig. 2. The very strong peak at 650 cm^{-1} is the P=S stretching vibration; the strong peak at around 501 cm^{-1}, 605 cm^{-1}, 1189 cm^{-1}, 1774 cm^{-1} is attributed to CH$_3$ torsional vibration, C=O in-plane deformation, P–O–CH$_3$ out-of plane deformation and C=O stretch, respectively; the weak peak at around 1013 is attributed to P–O–C deformation [17, 18].

Raman spectra of the chlorpyrifos powder were obtained, as shown in Fig. 3. The very strong peak at 631 and 678 cm^{-1} is the P=S stretching vibration [19]. The peak at 1569 cm^{-1} is the C=C stretching. The peaks at 160 and 1240 cm^{-1} is the P-O vibration and ring breathing [20].

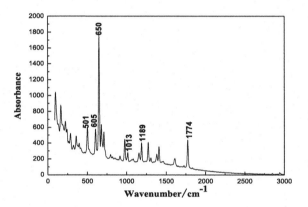

Fig. 1. Chemical structures of phosmet and chlorpyrifos

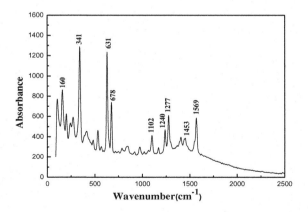

Fig. 2. Raman spectra of phosmet powder

Fig. 3. Raman spectra of chlorpyrifos powder

3.2   SERS Spectra of Phosmet and Chlorpyrifos

Raman spectrum has fingerprint characteristics, namely the spectrum of each substance has a specific peak. Its characteristic peak should be identified before the quantitative analysis about this two pesticides. With the increase of phosmet content in solution, the corresponding peak intensity was rise accordingly as shown in Fig. 4. In addition to differences in signal intensity, the peak shape and peak value remained the same. According to the peak position, the main characteristic peaks of phosmet was at 501 cm^{-1}, 605 cm^{-1}, 798 cm^{-1}, 1013 cm^{-1}, 1189 cm^{-1}, 1774 cm^{-1}. Compared with the Raman spectra of phosmet solid, the SERS spectra of phosmet solution were not obvious peak appeared at 650 cm^{-1}, and the intensity of peak vibration at 1013 cm^{-1}, 1189 cm^{-1}, 1774 cm^{-1} also significantly decreased in phosmet solution.

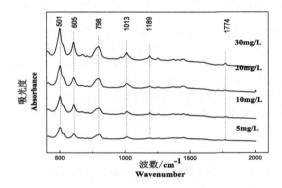

Fig. 4. Contrast of phosmet in different concentrations

Using gold colloid as SERS substrates, enhanced spectrum of chlorpyrifos in orange skin was collected, the average SERS spectrum as shown in Fig. 5, and the spectral range was 400 cm^{-1} ~ 1800 cm^{-1}. Compared with SERS spectra of chlorpyrifos and the Raman spectra of chlorpyrifos solid, as shown in Fig. 6, there were

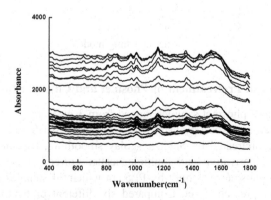

Fig. 5. Average Raman spectra of chlorpyrifos sample

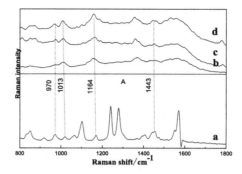

Fig. 6. Raman spectra of chlorpyrifos standard sample and SERS of chlorpyrifos solutions, 5 (b), 10 (c) and 20 mg/L (d)

enhanced effect in multiple bands using gold colloid. The intensity of peak vibration at 970 cm^{-1}, 1013 cm^{-1}, 1164 cm^{-1}, 1443 cm^{-1} as shown in Fig. 6, and the spectrum peaks belonging as shown in Table 1. With the increase of chlorpyrifos content in solution, the corresponding peak intensity was rise accordingly. Some peaks displacement of SERS spectra of solution changed in chlorpyrifos pesticide residues solutions, but the drift range was less than 10 cm^{-1} and not affect the accuracy of the model.

Table 1. Band assignments of major peak in Raman spectra acquired from chlorpyrifos

Band (cm^{-1})	Assignment
160	P–O vibration
341	N-cyclopropyl bending vibration
411	C–Cl stretch
631	P=S
678	P=S
1102	P–O–C stretch
1240	Ring mode
1277	Ring mode
1453	C–H deformation
1569	Ring stretching mode

3.3   SERS Spectral Pretreatment of Phosmet and Chlorpyrifos

The SERS spectra of phosmet samples ranged from 3 mg/L to 33 mg/L was collected by confocal Raman spectroscopy using gold colloid as SERS substrate. The original spectra data were pretreated by Savitzky-golay smoothing, baseline processing, first derivative, second derivative and multiple scatter correction etc. The quantitative analysis mode of phosmet was built by PLS, in which 24 samples for calibration set and 7 samples for prediction set. Compared six different pretreatment methods, the

Table 2. Comparison results for phosmet solution

Preprocessing method	Calibration Set		Prediction Set	
	R_C	RMSEC (mg/L)	R_P	RMSEP (mg/L)
Savitzky-golay smoothing	0.897	3.726	0.835	4.705
Base line	0.898	3.706	0.827	4.738
1st derivatives	0.968	2.120	0.873	4.754
2nd derivatives	0.966	2.173	0.883	4.662
MSC	0.890	3.853	0.665	6.017
S. G smoothing + 2nd	**0.988**	**1.254**	**0.924**	**4.293**

models were evaluated by Rp and RMSEP as shown in Table 2. The best prediction model was achieved with Rp of 0.924 and RMSEP of 4.293 mg/L, with the Savitzky-golay smoothing and 2nd derivative data preprocessing.

3.4 Building Model of Phosmet

The original spectral data were processed by different modeling algorithms combined with the savitzky gold smoothing and second derivatives data preprocessing. The analysis model of Raman spectroscopy and phosmet content was established using PLS and principal component regression(PCR), the prediction effect of the model was evaluated by R_P and RMSEP, as shown in Table 3. The value of factor lower resulted in that the information of model was not complete, and the predictive ablity of model was lower; on the other hand, the value of factor higher resulted in that the model was too complex, and fitting phenomenon appeared in the training. As shown in Fig. 7,

Table 3. Modeling comparison results for phosmet pesticide residues

Modeling algorithm	PCs	Prediction set	
		R_P	RMSEP (mg/L)
PCR	8	0.893	4.527
PLS	**6**	**0.924**	**4.293**

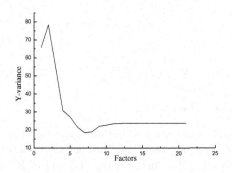

Fig. 7. Determination of principal factor number for detection of phosmet content

when the factor is 6, the RMSEP of PLS model is minimum. The PLS model predicted values of chlorpyrifos concentrations were showed in Fig. 8.

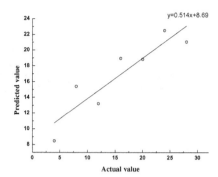

Fig. 8. Validation results of phosmet pesticide residues

3.5 SERS Spectral Pretreatment of Chlorpyrifos

The SERS spectra of chlorpyrifos samples ranged from 4 mg/L to 34 mg/L was collected by confocal Raman spectroscopy using gold colloid as SERS substrate. The quantitative analysis mode of chlorpyrifos was established by PLS, in which 23 samples for calibration set and 8 samples for prediction set. Table 4 shows the results of the models with different preprocessing methods obtained by PLS regression. The optimal model of chlorpyrifos pesticide was Rp of 0.715 and RMSEP of 6.646 mg/L, with the baseline and MSC data preprocessing.

Table 4. Comparison results for chlorpyrifos solution

Pretreatment method	PC	Calibration set		Prediction set	
		R_C	RMSEC (mg/L)	R_P	RMSEP (mg/L)
Origin	3	0.837	4.829	0.640	7.066
smoothing	3	0.837	4.827	0.640	7.067
MSC	7	0.974	1.980	0.714	6.743
Baseline	5	0.922	3.409	0.682	7.031
1st	5	0.990	1.192	0.686	7.216
2nd	4	0.987	1.380	0.797	7.918
Baseline + MSC	**8**	**0.989**	**1.273**	**0.715**	**6.646**

3.6 Building Model of Chlorpyrifos

The original spectral data were processed by different modeling algorithms combined with baseline and MSC data preprocessing. The analysis model of Raman spectroscopy and chlorpyrifos content was established using PLS and PCR, the prediction effect of the model was evaluated by Rp and RMSEP, as shown in Table 5. As shown in Fig. 9,

Table 5. Modeling comparison results for chlorpyrifos pesticide residues

Modeling algorithms	PC	R_C	RMSEC (mg/L)	R_P	RMSEP (mg/L)
PLS	8	0.989	1.273	0.715	6.646
PCR	13	0.941	2.981	0.780	6.006

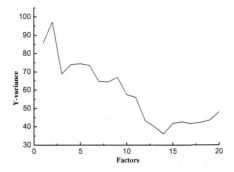

Fig. 9. Determination of principal factor number for detection of chlorpyrifos content

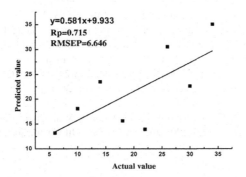

Fig. 10. Validation results of chlorpyrifos pesticide residues

when the factor is 13, the RMSEP of PLS model is minmum. The PLS model predicted values of chlorpyrifos concentrations were showed in Fig. 10.

4   Conclusions

Two kinds of pesticides (phosmet, chlorpyrifos) can be quantitatively measured and distinguished using SERS coupled with gold colloid. Pesticide residues on the surface of orange could be measured and the model has a good effect in the qualitative analysis of their SERS spectral data. It indicated that determination of pesticide residues using SERS coupled with gold colloid is feasible, compared the SERS method and other chemical detection methods such as GC and HPLC. But two kinds of pesticides are not enough for determination of pesticide residues, so we need to rich the kinds of pesticides.

Acknowledgments. This project is supported by the National Natural Science Foundation of China (31160250, 61178036), the National High Technology Research and Development Program of China (863 program) (no. 2012aa101906), East China Jiaotong University Legislation Research Fund (14JD01), and Key Laboratory of Carrying Tools and Equipment in Jiangxi.

References

1. Craig, A.P., Franca, A.S., Irudayaraj, J.: Annu. Rev. Food Sci. Technol. **4**(3), 369–380 (2013)
2. Huang, S., Hu, J., Guo, P., et al.: Anal Methods **7**, 4334–4339 (2015)
3. Vikesland, H.P.J.: Environ. Sci. Technol. **44**(20), 7749–7755 (2010)
4. Alvarez-Puebla, R.A., Liz-Marzán, L.M.: Small **6**(5), 604–610 (2010)
5. Pan, L., Dong, R., Wu, Y., et al.: Talanta **127**, 269–275 (2014)
6. Masoum, S., Mehran, M., Ghaheri, S.: J. Sep. Sci. **38**(3), 410–417 (2015)
7. Lisec, J., Schauer, N., Kopka, J., et al.: Nat. Protoc. **10**(9) (2015)
8. Kochansky, J.: J. Apic. Res. **43**(2), 60–64 (2015)
9. Tan, G., Yang, T., Miao, H., et al.: J. Chromatogr. Sci. (2015)
10. Rong, S., Zou, L., Zhang, Y., et al.: Food Chem. **170**, 303–307 (2015)
11. Qu, L.L., Liu, Y.Y., He, S.H., et al.: Biosens. Bioelectron. **77**, 292–298 (2016)
12. Lei, Z., Shu, J.Z., Xiao, Y.W., et al.: Talanta **148**, 308–312 (2016)
13. Kneipp, K., Kneipp, H., Itzkan, I., et al.: J. Phys.: Condens. Matter **14**(18), 597–624 (2002)
14. Li, J.F., Huang, Y.F., Ding, Y., et al.: Nature **464**(7287), 392–395 (2010)
15. Guerrini, L., Sanche-Corte, S., Cruz, V.L., et al.: J. Raman Spectrosc. **42**, 980–985 (2011)
16. Dhakal, S., Li, Y., Peng, Y., et al.: J. Food Eng. **123**(2), 94–103 (2014)
17. He, Q., Li, S., Guenter, S.: Spectrosc. Spectral Anal. **30**(12), 3249–3253 (2010)
18. Lee, P.C., Meisel, D.: J. Phys. Chem. **86**(17), 3391–3395 (1982)
19. Fang-Ying, J.I., Si, L.I., Dan-Ni, Y.U., et al.: Chin. J. Anal. Chem. **38**(8), 1127–1132 (2010)
20. El-Abassy, R.M., Donfack, P., Materny, A.: J. Am. Oil Chem. Soc. **86**(6), 507–511 (2009)

Research on Freshwater Fish Information Service Mode for Modern Production and Circulation in the Internet + Era

Xinping Fang[✉]

College of Economics and Management,
China Agricultural University, Beijing 100083, China
candyfxp@sohu.com

Abstract. In the traditional information service for freshwater fish, there existed several crucial problems involving in the information classification and content organization, which was not clear and specific to serve market and consultancy etc., information refreshing not timely, and information sharing difficult to take, and so on. These problems seriously restricted the development of the freshwater fish and the other relative industries. The "Internet + Strategy", which has been advocated and carried out in China since 2015, has promoted the formulation of new ideas for information service model to construct and enhance modern production and circulation of the freshwater fish. Based on the analysis of the core and characteristics of information service in the production and circulation for the modern freshwater fish, this study constructed a new information service model consisting of production, sale, transaction and circulation. In the development of the freshwater fish industry for the future, it is necessary and beneficial to accelerate the construction of the information service platform, integrate the information service as well as the improvement of information resources with the merchandise production and circulation, and strengthen the brand building of enterprises, especially the leading enterprises.

Keywords: Internet plus · Freshwater fish · Modern production and circulation · Information service

1 Introduction

The articles specific to the information service model of the freshwater fish are reported few until now, but the information service model is similar with the agriculture industry, which is well developed and could be referenced.

The information service models of the agriculture industry are well developed and widely reported. In 2005, it was suggested by Zheng and Hu that there are three models: traditional model, web information service model and blended model. Dang and Cen pointed out that there are three types: web service platform, mobile newspaper SDI, and mobile Internet for the agriculture information service model [1]. Zhong and Wan stated that the agriculture information service model is a multiple model consisting of many systems such as network system, sound system, video, short message, remote video, etc. [2]. Sun et al. explored the new models of integrating modern

D. Li and Z. Li (Eds.): CCTA 2015, Part I, IFIP AICT 478, pp. 597–603, 2016.
DOI: 10.1007/978-3-319-48357-3_57

information technology with professional cooperative information service [3]. It was suggested by Jiang that there existed several types of information service model – "government + farmer", "government + rural cooperative organization + farmer", "government + IT enterprise (communication enterprise) + farmer", "government + agriculture-related enterprise + rural cooperative organization + farmer" and "rural cooperative organization + farmer" [4]. Wang et al. advocated the models of "government + farmer", "government + association + farmer", "government + communication enterprise + association + farmer", "government + enterprise + association + farmer", "agricultural leading enterprise + farmer", and so on [5].

In another hand, the information service models of the agricultural product and the extension of agriculture technology are well studied, too. Hu proposed three types of agricultural technology extension model including the base type, the industry type and the universal type [7]. Zheng believed that there existed five kinds of information service models according to the service suppliers: the government, the industrial organization, the regional organization, the exporting market and the domestic market [8]. Wu made a deep analysis on the problems in freshwater fish production and circulation chain in China and its possible reasons, and proposed the strategies to upgrade the freshwater fish circulation chain [9].

At present, the existing and running information service models for agriculture, agricultural products, and the freshwater fish all were constructed and improved before the strategy of "Internet +", and are marked by the traditional information concepts, which were government-oriented, mid-scale organization-oriented, enterprise-oriented. These information service models often failed to clearly and specifically classify and organize the necessary information, to refresh information timely, and share the information widely, and so on. In the era of "Internet +" and the competitive domestic and international market economy, it is urgent and necessary to refresh the old or develop new information service models to serve and keep up with the rapid development of modern production and circulation in freshwater fish in China.

2 The Core Elements and Features of the Information Service Model of Modern Production and Circulation in Freshwater Fish

All the procedures in the freshwater fish production and circulation, including production information, sale information, transaction information and circulation information, should be accomplished by leading enterprises of large-scale fishery industry. Besides, production enterprises should build an information platform about production, sale, transaction and circulation information so that all the information can be acquired on the platform. In this way, information can be achieved timely and accurately at any time and intermediate links can be reduced. As all the intermediate links can be displayed on the information platform, the direct link of "freshwater fish to consumer" can be achieved in that freshwater fish can be delivered from the place of production to consumers quickly and directly, hence the freshness of freshwater fish can be guaranteed.

The most fundamental requirements of freshwater fish production and circulation are rapidity and freshness, which means that the rapidity of circulation and the freshness of the product are the core goals of freshwater fish production and circulation procedure. While the Internet plus Era has made these two requirements become possible. To make the core goals of "freshwater fish to consumer" come true, the following two aspects need to be fulfilled at least:

First, the network information platform must be built to present all the relevant information including production information in the production process, supply and price information of enterprises' sales, demand and preference information of consumers, payment and trade information of the transactions. In addition, the third party represented by government, industry associations and research institutes will also provide information of the industry, policies and researches to enrich the information platform and build up the big data net of freshwater fish modern production and circulation.

Second, all the intermediate procedures need to be fulfilled on the information platform so that product flow can be cut down greatly. The information flow in the platform will be clearer and more easily to access, at the same time, production enterprises and consumers can get all kinds of relevant information by web searching in real time. Therefore, all the intermediate procedures can be completed based on network information. Consumers can get production and sale information to decide whether to close a deal, if so, the transaction can be completed directly with trade and payment information. Once the production enterprises gain the information, the

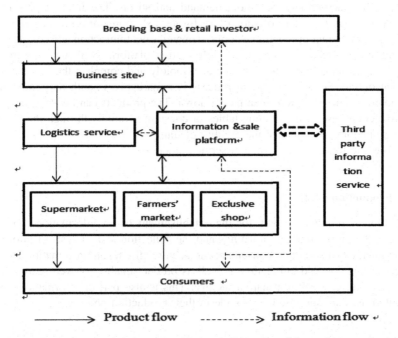

Fig. 1. Production flow and information flow of freshwater fish modern production and circulation in the Internet plus era

products will be delivered immediately. The breeding base and retail investors can bring the products together to the point of sale and transportation based on the directions from the information platform, and the logistics service will deliver the products to the supermarkets, farmers' markets and exclusive shops immediately, then the fish will be butchered based on the demand information of consumers from the platform and delivered to consumers' home through cold-chain logistics within one kilometer. As a result, all the procedures including sale, transaction, payment, supply and demand can be completed online to achieve the goal of linking production enterprises to consumers directly. The whole product flow chain covers production base, logistics sites, intermediate logistics, retail stores and consumers. Compared with traditional ways, some intermediate steps can be cut down thus the whole process is reduced greatly and the circulation efficiency is also improved significantly to ensure the rapidity and freshness of fresh fish. While all the intermediate procedures are based on the information monitoring of the Internet of Things around the clock so that consumers can get relevant information all time (Fig. 1).

3   The Construction of the Information Service Model of Modern Production and Circulation in Freshwater Fish

The nucleus of freshwater fish modern production and circulation lies in the rapid transfer from production to consumers, including the production, sale and circulation procedures, alongside with the relevant information of freshwater fish production, quality, sale, transaction, circulation, demand and so on. The Internet plus Era has made the delivery and transportation of freshwater fish more convenient as all kinds of information including production, supply, demand, sale, transaction and circulation can be achieved immediately through the information platform, which means that consumers can decide whether to make the deal rapidly and complete the transaction and payment online. So the transaction process is remarkably smooth and coherent and other intermediate procedures can be cut down. The products can be transported from the production base to the dinning-tables of consumers rapidly through the most efficient circulations to guarantee the rapidity of transportation and the freshness of fish. Therefore, the following information service model can be built (Fig. 2):

3.1   Production Link

Every link of production calls for the establishment of an information monitoring system based on the Internet of Things and the collection and analysis of information via the Internet. In so doing, consumers can acquire all relevant information regarding product production and circulation as well as product quality by web searching, QR code scanning, or APP information inquiry. This is also part of information services that enterprises are supposed to provide in their production process.

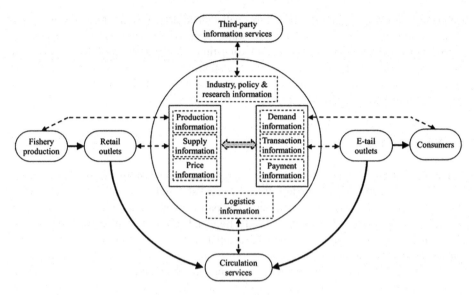

Fig. 2. Information service model of freshwater fish modern production and circulation in the Internet plus era

3.2   Transaction Link

Transaction entails transaction in both retail outlets and e-tail outlets. Enterprises build up sales information networks via Internet information platforms. Enterprises provide information concerning production, supply, and price by means of APP or online shops, and consumers offer such demand information as consumer preferences and purchase habits. Enterprises can recommend right products to consumers based on their demand information while consumers can make purchase decisions on the basis of supply information and product information provided by enterprises. If consumers decide to purchase, they can instantly accomplish online payment and transaction on the Internet. After receiving instant transaction information, enterprises deliver fish products directly through logistics to ensure the efficiency of delivery. The entire process of transaction and marketing is done based on enterprises' information platforms. Both consumers and enterprises can convey information via network platforms, thus getting transaction done at one go and saving the time it takes to go through intermediate steps.

3.3   Circulation Link

Logistics is central to freshwater fish production and circulation. The difficulty of freshwater fish production and circulation lies in the failure to realize long-distance and long-time circulation. In the Internet plus era, all intermediate transaction steps are done on the Internet, except that only logistics cannot be completely replaced by the Internet. While offering cold-chain logistics, third-party logistics enterprises (represented by SF Express) are incapable of transporting freshwater fish. Future enterprises should establish a logistics system featured by self-owned logistics by learning from the logistics model of Heshishuichan, providing specialized services, and striving to innovatively achieve long-distance and long-time transportation of freshwater fish products. Meantime, during logistics and transportation, an Internet-based monitoring system should be set up to achieve real-time monitoring and guarantee information services during the process of logistics.

4 Suggestions

In the Internet plus era, the production and circulation model of freshwater fish changed completely. It becomes possible to sell freshwater fish quickly and alive, thus creating a direct link from enterprises to consumers. The paper constructed a modern information service model to serve the modern production and circulation in freshwater fish in China and suggested several important ideas related to improve the information service for the production and circulation.

Firstly, efforts should be exerted to accelerate the construction of information service platforms. The information platforms are the key to a modern information service for modern production and circulation of freshwater fish. According to the various scales of fishery enterprises, different types of enterprise platforms could be established:

① Large-enterprise platforms for large-scale fishery enterprises. With adequate human and financial resources and relatively comprehensive systems of production, circulation and marketing, large enterprises could take a lead in constructing the internet-based information service platforms integrated with marketing, production and circulation.

② Enterprise union(or alliances) platform for small or mid-scale fishery enterprises. Due to limited human and financial resources, small fishery enterprises could imitate large enterprises to establish flexible joint fishery information and marketing platforms in joint unions. At the same time, small enterprises could initiate collaboration in other fields including production and circulation to reduce cost and increase profits.

Second, the integration of freshwater fishery between the production and circulation and the information service resources should be promoted. The first aspect of integration is about the production resources. Individual producers could integrate their production resources with other producers or enterprises through cooperatives or enterprise production bases to improve product quality, monitoring product information, and reduce production cost. Small scale enterprises could cooperate as producer teams to form an industry alliance via collaboration to build production bases for scale and standardized production. The second aspect of integration is about the integration of information resources from marketing, production and circulation, resource purchase and product sales as well as post-sale information feedback to coordinate every procedure in a whole enterprise and the whole industry. Small-scale or mid-scale enterprises could coordinate their production and sales based on their established joint information platforms suggested in the last paragraph to compete with large enterprises.

Thirdly, the brand building of enterprises should be enhanced, especially leading enterprises. With the developed production bases and the scale and standardized production, product quality will be guaranteed and brands could be set up, and the strategy of branding could be applied for in the market. Tongwei-labeled fish products of Tongwei fish are from the branded Tongwei Group, which is built in the long-term brand strategy. Therefore, while enterprises emphasize on quality during production, the marketing in product sale should be emphasized to take the brand strategy to build the long-term constant brand in the market.

References

1. Dang, Y.-l., Cen, J.-J.: Research of agricultural information service mode in big data era. J. Libr. Inf. Sci. Agric. **27**(6), 152–154 (2015)
2. Zhong, J., Wan, X.: Study on the application of multi-pattern mode in agricultural information service in Jinan. Hubei Agric. Sci. **52**(8), 1950–1952, 1958 (2013)
3. Sun, Y., et al.: Agricultural information service situation and countermeasures in Henan. J. Henan Agric. Sci. **42**(12), 158–161 (2013)
4. Jiang, Y.: Practice and exploration for the service of grass-root agricultural information taking the construction of agricultural information service in Langzhong city as a perspective. Hubei Agric. Sci. **48**(11), 2903–2906 (2009)
5. Wang, B.-J.: Model research of agricultural literature and information resources serving the new socialist countryside construction. J. Anhui Agric. Sci. **40**(10), 6276–6278 (2012)
6. Zi, W.-C., Liao, X.-G.: Research on the information service patterns of agricultural products' circulation based on supply chain management. Logistics Sci-Tech **34**(5), 19–20 (2011)
7. HU, C.-Q.: Patterns of current agricultural technology dissemination and application. Rural Technol. Dev. **5**, 42 (1997)
8. Zheng, Q.-M., Guo, X.-Y., Li, D., Shao, F.-W.: Socialized agricultural technological service system: pattern framework and implementation strategy. Agric. Econ. **1999**(1), 40–41
9. Wu, H.-M.: Research on equipment and technology integrating and mode optimizing of the freshwater-fish modern circulation. Ph.D. dissertation, China Agricultural University (2014)

Interactive Pruning Simulation of Apple Tree

Lili Yang[1], JiaFeng Chen[1], Jing Hua[2], MengZhen Kang[2],
and QiaoXue Dong[1(✉)]

[1] College of Information and Electric Engineering,
China Agriculture University, Beijing, China
dongqiaoxue@163.com
[2] State Key Laboratory of Management and Control for Complex Systems,
Institute of Automation, Chinese Academy of Science, Beijing, China

Abstract. Pruning treatment is one of the important management practices of perennial fruit tree cultivation. Combining fruit tree growth theory with mathematics model and software development is an effective way in agricultural research. Virtual apple tree is built as a case to study pruning treatment in this paper. First, architectural development of apple tree was analyzed and stored in xml file. Then, the simulation software of apple tree pruning is build based on the Qt framework and OpenGL graphics library. The interactive pruning and automatic pruning with setting conditions are realized in the software. Reaction laws of apple tree pruning are extracted from the analysis on the experiment data. Stochastic 3D apple tree architectures after being pruned also can be simulated. Result indicates that simulation is efficient, accurate and timely judgment of pruning reaction is possible. This work is the foundation of future research, which will simulate the apple tree development of architecture and biomass after pruning treatment over time.

Keywords: Apple tree · Pruning · Interactive · Simulation

1 Introduction

Tree architecture plays a key role in foliage distribution and consequently in light interception and carbon acquisition, which in turn strongly affect the reproductive growth of fruit trees. Pruning treatment is one of common measures for architecture study in perennial fruit tree cultivation. Reasonable pruning plays a key role in tree fruit quantity and quality [1]. It had been found that removal of flowering shoots at a young stage of growth with less pruning of old branches tends to stimulate growth of the remaining shoots and fruit set of adjacent inflorescences [2–4], Jean Stephan studied the effect of pruning strategies on shoot demography and development during 2 years based on experiment and demographic approach [5]. A biological model of apple tree production and a biomechanical model of apple tree development using mixed statistics are built successively [6, 7]. The perennial nature of apple tree make modeling complex, but the simplifications and codes description of tree topology presented by Costes make apple tree model possible [8, 9]. Many models are focus on the apple tree growth and development under the natural state rather than after pruning [10, 11]. Pruning research based on virtual apple tree model can be meaningful and efficient. However,

D. Li and Z. Li (Eds.): CCTA 2015, Part I, IFIP AICT 478, pp. 604–611, 2016.
DOI: 10.1007/978-3-319-48357-3_58

few interactive virtual platform of apple tree pruning are found [12, 13]. The work in this paper include: (a) Build a virtual apple tree model based on architectural development of apple trees. (b) Design and realize 3D interactive pruning simulation on general PC. Pruning strategies commonly used in real planting are considered in this platform. (c) Stochastic 3D architectures of virtual apple tree were modeled after pruning. The work is the foundation of further work, which focuses on the relationship between structure change which is reduced by pruning of apple tree and function change regarding as physiological character of fruit tree.

2 Overall Design

The technical framework of apple tree pruning simulation is established on visual apple tree model combined with computer graphics technology. First, the hierarchical data organization which describes apple tree's morphological structure is formed based on the geometry structure of the apple tree. Second, the three-dimensional model of apple tree is build based on OpenGL. Finally, an interactive interface of apple tree pruning simulation based on QT framework is developed. The work is showed in Fig. 1.

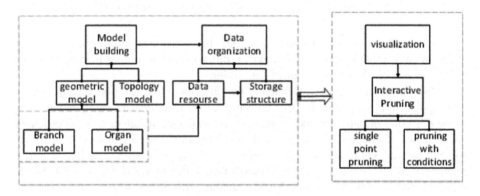

Fig. 1. Workflow of pruning simulation

3 Interactive Pruning Simulation

3.1 The Model

It is commonly accepted that plants are modular organisms that develop by the repetition of elementary botanical entities [14]. Here, apple tree architecture is described in Growth Units (GU) level, which were divided into latent bud, medium shoot (5–15 cm), short shoot (<5 cm) and long shoot (>15 cm) according to the length. Topology structure of apple tree, which describes the connection relationship among all units, is modeled using the dual-scale automata approach [15–17]. In this way, apple tree is divided into hierarchy according with self-similarity and branching order during apple tree growth. Different years of germination branches reflect difference vigor, which is called physiology age (PA). Except the trunk, primary branch on trunk is defined as PA1, and secondary branch on primary branch is defined as PA2 and so on. Each branch belongs to a certain GU according to the branch length. In commercial orchards apple tree usually are pruned short and less branch level. According to dual-scale automata model, four macro states are applied to simulate a young apple tree topology development (Fig. 2).

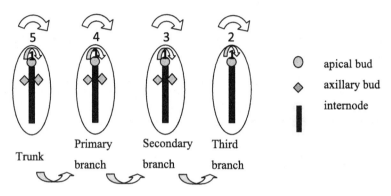

Fig. 2. Dual-scale automata model of a 5 years old apple tree

Position of structure (POS) unit describes the linked relationship between units. Different node rank on branch will develop different GU according to certain probability obtained from field experiment data fitting.

A local-to-global approach is taken in building apple tree 3D model. Apple tree is decomposed into four level structure unit: node, internode, branch, organ. Node refers to the key point of apple tree skeleton. Node information, including space coordinates and distance to internode surface, is calculated by extracting a certain rule based on some measured data. The adjacent nodes constitute an internode and a series of internodes constitute a branch. Organ refers to apple tree leaves and apple fruit. Single organ geometric structure is built by 3d scanning. All organs, whose space position, orientation and size are simulated by using the information of measured data in special coordination system, are assumed only having differences in position, angle and size for same type.

3.2   Data Organization

Two categories of data are included in apple tree model: one is node information, another is organ information. Both are saved as a standard XML file with different suffix. Only the key node information of tree skeleton is stored in model, not including the surface node information of the branches in order to reduce the data storage.

The information of function topology is added into geometry topology based on the three-dimensional model structure of apple tree mentioned in 3.1. Taking the branch for example, the model data include PA, POS and all the node information on the branch. The coordinates of the pruning point located the branch where the pruning point is on, then, the function topology information of selected branch like PA is obtained from the node information XML file. POS is designed as a sequence which stores the position order from trunk to branch or organ in an even number. For example, the list <10,1,8,3>, means a piece of leaf or a shoot born on the 3th small branch of the 8th node, which is on the first branch of the 10th node of the trunk. A branch and an organ data are organized as below. All nodes are described by 3D space coordinates x,y,z and the diameter of the node to the branch. The organ is described by a transfer matrix that can generate different size and direction organ by organ transformation.

```
</Branch>
<Branch PA="4" POS="193, 1, 55, 1, 13, 1">
        <Node>-50.1586, -6.02841, 307.927, 0.47941</Node>
        <Node>-48.5204, -6.78654, 307.927, 0.47941</Node>
        <Node>-46.883, -7.54428, 307.869, 0.47941</Node>
        <Node>-45.2471, -8.30138, 307.774, 0.481215</Node>
        <Node>-43.599, -9.06413, 307.676, 0.481215</Node>
        <Node>-41.9508, -9.82688, 307.578, 0.481215</Node>
        <Node>-40.3027, -10.5896, 307.48, 0.481215</Node>
</Branch>

<Organ ID="10" PA="1" POS="353, 1">
        <Matrix>0, -4.08864, -5.23322, 0</Matrix>
        <Matrix>0, 5.23322, -4.08864, 0</Matrix>
        <Matrix>6.64105, 0, 0, 924.37</Matrix>
</Organ>
```

Vector is used to store the data of apple tree model by using the C++ programming language. The node vector corresponds to a node which contains the node coordinates and diameter of the branch which the node is on. The branch vector takes a single node vector as a basic item, and the whole apple tree vector takes a single branch vector as a basic item. The branch position vector consists of POS of each branch, and so as branch age vector.

3.3   Interactive Pruning Simulation

Interactive pruning includes deterministic point pruning and combined conditional pruning. Deterministic point pruning means pruning the branch from the mouse clicked point. The following steps are needed: (1) Coordinates transformation. It refers to transform the pruning point coordinates given by mouse click to the coordinates in apple tree model space. In OpenGL. First, transform 2d coordinates on screen to 3d view-part coordinates. Then, transform view-part coordinates to the model coordinates. (2) Considering that the coordinates processed in (1) is on the surface of the branch and may not be contained in the file of node information, a method based on distance is applied to establish a mapping between clicked-point and points in the apple tree model. The method tries to find the closest node to the pruning point and then get the POS character of the branch with pruning point. (3) The last step is to find all the related branches based on the POS character of matching point, and deletes point data and organ data of corresponding branch.

There are many pruning treatments in orchard management, not only including shape pruning and conditional pruning determined by branch or organ types, angles or PA, but also including different level pruning based on branch length, branch ratio and so on. Combined pruning refers to interactive pruning of apple tree in the way that combines mouse clicking with pruning parameters input.

For example, make the apple tree spindle shape and cut off 1/3 of the branches whose physiology age is 2, the steps of interactive pruning are as follows: (1) Click the key point on the screen to define spindle shape. (2) Expand 2d spindle shape to 3d spindle body. (3) Pruning judgment. First, delete the related nodes which are not in the scope of 3d spindle body. Second, find all the branches whose physiology age is 2 and then find the nearest nodes of the point at a third of the branch; finally delete the related information based on the found node coordinates. (4) Redraw the apple tree model based on the updated data information.

4   Result

4.1   Running Time of Interactive Pruning Simulation

Interactive pruning simulation on apple tree is running on an ordinary computer (INTEL CORE i5-2410 M 2.3 GHz/4 GB/visual studio 2010/OpenGL 2.1/QT 5.1). 359 branches with 4411 sample nodes are included in node files. The program running time in different pruning modes are listed in the following Table 1.

Table 1.   Running time of different pruning modes

	Options	time/ms
Defined point pruning	① Select a branch of PA = 2 and cut it off near the midpoint	1
	② Select a branch of PA = 3 and cut it off near the midpoint	2
Pruning with conditions	③ Prune to make the tree look like sphere shape	1
	④ Cut off all the branches of PA = 3	3
	⑤ Cut off all the branches whose length < 0.1 m	1
	⑥ Cut off 1/3 of the branches of PA = 2	2
	⑦ ④ + ⑤	3
	⑧ ④ + ⑥	3
	⑨ ④ + ⑤+⑥	3

The results show that the pruning system makes it easy for users to select the branches need to be cut by mouse clicking or keyboard input with a single condition or several conditions combined. The simulation time with the average response time is ms level, which is independent of the complexity of pruning. The simulation of natural pruning could be achieved on ordinary PC hardware. The efficiency of the simulation could be probably improved if the GPU technology is applied.

4.2   3D Results of Interactive Pruning Simulation

The system comes into the view mode in default after the data of the tree model loaded. In this mode, mouse and keyboard are used to move, scale or rotate the visual tree model, then the exact position of pruning is easily to locate. The branches and organs are selected to be cut off by different ways in pruning mode after mode switching. Pruning is done before growth resumed in the spring when the leaves have not come out yet. Figure 3b, c clearly show the rendering of two different pruning cases. The pruning effect can be clearly observed.

(a)Without pruning   (b) Defined point pruning   (c) conditional pruning

Fig. 3. Renderings of two pruning cases with interactive interface.

4.3   Stochastic 3D Apple Tree Architectures After Being Pruned

Branching patterns of apple tree branches after all branches (PA = 1) being pruned were simulated on the foundation of experimental data fitting. Lateral buds in the following years will develop to short shoot, medium shoot, long shoot or stay the original status, stochastically. The reason of different development for same pruned treatment is complex. Internal and external factors such as environment, morphology, and assimilate competition all can result to different branch. Different 3D architecture of virtual apple tree can reflect different branching patterns.

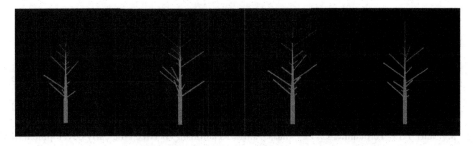

Fig. 4. Stochastic architecture of pruned tree

5 Discussion

Apple tree geometric model is established by the method that dividing the model into different structural units and levels, which reduces the complexity of the apple tree three-dimensional modeling. The topological relationship between the branches is well documented by reasonable way of data organization so that it is easily to realize the pruning operation that all the child branches also are cut off when the mother branch is cut off, which simplify pruning realization and improve simulation efficiency.

Both shape pruning and combined conditional pruning are considered in the view of actual planting. Two interactive ways of pruning, i.e. by mouse or keyboard and by parameters input, are provided for the convenient user operation. Stochastic architecture after pruned also is simulated according experiment data. If we can get more nodes information on pruned branch from base to top, discrete-time and discrete-state-space stochastic processes can be used to illustrate the branching architecture for any pruning measure. Now stochastic branch is limited to batch pruning for same physiology age branch.

The method of interactive pruning on visual apple tree in this paper can provide a reference for pruning simulation of other species. It also can be used in entertainment, education or tree research. Meanwhile, this work provides an intuitive, flexible auxiliary tool for the further research on the tree pruning effect combined with the function model of the apple tree. The stochastic architecture showed in Fig. 4 cannot be well explained, if tree structure model is considered independently. Pruning causes apple tree architecture changed. New organs will create, which change apple tree production (source) and demand (sink) ability. New local and global source-sink ratio is closely related to pruning bud fate. Function-structural plant models (FSPM) have more advantages to explain it [18–21]. With more field experiment and model developed, different virtual pruning treatment can be done to minimize the vegetative branch. Optimal pruning strategy can be got by analyzing source-sink ratio of reproductive branch. Optimization of simulation algorithm, the extensibility of the program simulation and verification of the simulation can be further studied.

Acknowledgements. This study was supported by the National Natural Science Foundation of China (#31200543).

References

1. Lauri, P.E., Trottier, C.: Patterns of size and fate relationships of contiguous organs in the apple (Malus domestica) crown. New Phytol. **163**, 533–546 (2004)
2. Lauri, P.E.: From tree architecture to tree training-an overview of recent concepts developed in apple in France. J. Korean Soc. Hortic. Sci. **43**, 782–788 (2002)
3. Lauri, P.E., Terouanne, E., Lespinasse, J.M.: Relationship between the early development of apple fruiting branches and the regularity of bearing-an approach to the strategies of various cultivars. J. Hortic. Sci. **72**, 519–530 (1997)

4. Lauri, P.E., Terouanne, E., Lespinasse, J.M.: Vegetative growth and reproductive strategies in apple fruiting branches-an investigation into various cultivars. Acta Hortic. **451**, 717–724 (1997)
5. Stephan, J., Lauri, P., Dones, N., et al.: Architecture of the pruned tree: impact of contrasted pruning procedures over 2 years on shoot demography and spatial distribution of leaf area in apple (Malus domestica). Ann. Bot. **99**, 1055–1065 (2007)
6. Susie, H., Oscar, C.: A biological model of apple tree production, 1091–1096
7. Smith, C., Godin, C., Yann, G., et al.: Simulation of apple tree development using mixed statistical and biomechanical models. 31–34
8. Costes, E., Sinoquet, H., Kelner, J.J., Godin, C.: Exploring within-tree architectural development of two apple tree cultivars over 6 years. Ann. Bot> **91**, 91–104 (2003)
9. Costes, E., Lauri, P.-E., Regnard, J.-L.: Tree architecture and production. Hortic. Rev. **32**, 1–60 (2006)
10. Costes, E., Guedon, Y.: Modelling branching patterns on 1-year-old trunks of six apple cultivars. Ann. Bot. **89**, 513–524 (2002)
11. Seleznyova, A., White, M., Tustin, S., Costes, E.: Application of Markovian models to study root/interstock effects on flowering of young apple tree. In: Godin et al. (eds.) Proceeding of the 4th International Workshop on Functional-Structural Plant Models, 7–11 June 2004, Montpellier, France, pp. 311–314 (2004)
12. Xia, N., Hu, B.G.: Evaluation of importance of pruning on branching structures in nectarine tree using the model of hidden Semi-Markov chain. In: The 3rd International Symposium on Intelligent Information technology in Agriculture (2005)
13. Atkins, T.A., O'Hagan, T.A., Rogers, W.J., et al.: Virtual reality in horticulture education, pruning simulated fruit trees in a virtual environment. Acta Horti. **416**, 243–246 (1996)
14. Barthelemy, D., Caraglio, Y.: Plant architecture: a dynamic, multilevel and comprehensive approach to plant form, structure and ontogeny. Ann. Bot. **99**, 375–407 (2007)
15. Xing, Z., de Reddye, P., Xiong, F.-L., et al.: Dual-scale automation model for virtual plant development. Chin. J. Comput. **24**(6), 608–1807 (2001)
16. Hong, G., Xiangdong, L., Jun, D.: A review of functional-structural tree model. World For. Res. **2**(23), 55–60 (2010)
17. Ferraro, Pascal: Toward a quantification of set-similarity in plants. Fractals **13**(2), 91–109 (2005)
18. Lin, Y., Kang, M., Hua, J.: Fitting a functional structural plant model based on global sensitivity analysis. In: 2012 IEEE International Conference in Automation Sciences and Engineering (CASE), Seoul, Korea, 20–24 August 2012
19. Wang, H., Kang, M., Hua, J.: Simulating plant plasticity under light environment: a source-sink approach. In: IEEE the Fourth International Symposium on Plant Growth Modeling, Simulation, Visualization and Applications, 31 October–04 November, 2012, Shanghai, China, pp. 431–438 (2012)
20. de Reffye, P., Kang, M.-Z., Hua, J., Auclair, D.: Stochastic modelling of tree annual shoot dynamics. Ann. For. Sci. **69**, 153–165 (2012)
21. Vos, J., Evers, J.B., Buch-Sorlin, G.H., et al.: Functional-structural plant modeling: a new versatile tool in crop science. J. Exp. Bot. **61**, 2101–2115 (2010)

Research on Key Technology of Grid Cell Division Method in Rural Community

Chunlei Shi and Bo Peng[(⊠)]

College of Information and Electrical Engineering,
China Agricultural University, Beijing 100083, People's Republic of China
pengbo_cau@126.com

Abstract. Village and town are the basic unit of our society, there vigorously development in social management and innovative plays an important role in promoting economic growth, maintaining social stability, improving people's living environment and other aspects, but the traditional government management mode can not meet the claim of the social development of the villages. Based on the idea of grid management model, this paper presents the principles and method of grid cell division in the process of the gird management in the village community, and to analyze the division method so as to provide evidence for the scientific, rational division of community grid unit.

Keywords: Management · Grid cell division · Village community

1 Introduction

As the market economy continues to develop, village society presents a diversified development trend, along with the development of the towns, village members appeared stratified society, dispatch phenomenon, which leads to differences complicate the values and interests of the main interests of the village community sources of. The village society of pluralism, diversity is bound to induce complex social needs, social conflicts and problems, the improper disposal may cause discontent among the members of the village community, and even leads to mass incidents [1]. For China, a vast country with a large agricultural population, the role of towns and villages in the county of socio-economic development can not be ignored. Facing the new situation and the new problems of rural society, rural grass-roots government's public service still follow the tradition of fragmentation, the fragmented nature of resource allocation and administrative work, this kind of government management model is clearly not meet the requirements of the village social development. Particularly in industrialized, urban reform orientation of the development process, long-term adherence to the "urban bias" in public policy-oriented, leading to a serious shortage of village public services.

How in the new rural governance system effectively extend the management and service functions of government to farmers, covering the entire villages is a new topic in the new situation of grass-roots social management innovation. Thus, according to the new characteristics of rural community development, the research on the method of new village community governance haves important theoretical significance and practical significance [2].

© IFIP International Federation for Information Processing 2016
Published by Springer International Publishing AG 2016. All Rights Reserved
D. Li and Z. Li (Eds.): CCTA 2015, Part I, IFIP AICT 478, pp. 612–618, 2016.
DOI: 10.1007/978-3-319-48357-3_59

Grid management is a new management model, which was worked out in the construction of China's urban grass-roots social management in recent years, it mainly refers to in accordance with certain criteria the community are divided into a plurality of mesh, as the basis of the IT and grid coordination mechanisms, the sharing of resources between the grid and efficient exchange of information, and ultimately achieve the optimal allocation of resources, improve the efficiency of the new management thinking for the community management [3], to solve community information due to the work and information integration across sectors shared problems brought [4]. Grid management centralized dynamic management, fine management and flat management, the use of digital management technology, from the perspective of social management, to integration of social wealth and resources, eliminate information silos, to achieve information sharing and coordination office, can greatly improve the efficiency of community management, saving management costs, laid the basis for innovative community management model [5, 6]. Therefore, exploring for grid management model for village communities has great significance in urban and rural dual economic structure and build a new socialist countryside construction.

2   Grid Cell Division

Grid Cell division is the basis of geographical network applications. It has a relatively large workload. The first step in constructing the grid management system is to divide each grid cell, there must be some criteria for the classification constructed so that for standard each grid cell's construction and management, in order to facilitate the implementation of grid management.

Village space meshing has rules division and irregular division. Village spatial data is complexity and variety, space meshing will directly affect the entire village grid system and the management based on this. Geography meshing should give full consideration to the workload of each mesh grid administrators, and should be divided according to the village space form, and should consider the integrated grid area, the number of households within the grid, the number of towns and villages within the grid and other parts of factors. In this paper, three division methods are proposed according to the administrative divisions, road blocks and regular grid.

2.1   Principle of Grid Cell Division and Geographical Elements

The division of the grid cell should follow the principle of a statutory basis, territorial management, geographical distribution, the convenience of management, space and relative stability. Space grid with rectangular shape is appropriate, and the ratio of length to width should not be too large, meanwhile, it should avoid the appearance of "7" shape or "concave" shape of the space grid.

The division of the grid cell should be in accordance with the layout of the village lanes, courtyards and open spaces, rivers, mountains, lakes and other natural geography. References to the standard of the relevant grid division recomm ended by the construction department can give the following method for the division of geographical elements. Geographical elements mentioned here may not be comprehensive, in the implementation should be flexible to grasp.

(1) Boundary. Spatial grid can not cross the township, village, administrative boundaries.

(2) Settlement places. In general, considering the ownership relations, put the buildings which have a wall or have unified label on the topographic map in a spatial grid. A building which is inconsistent with the ownership should be based on the adjacent relationship, as far as possible subdivided into a grid with rules shape.

(3) Hilly area. According to its boundary (the slope bottom or the bottom of the slope), in principle, hilly area is divided into an independent space grid. If there is a high degree of highway in the middle of the mountain, it can be divided into several mountainous areas in accordance with the road. In the boundary of the mountain division, first consider the river and road at the bottom of the slope, if existing rivers and roads divide the mountain and other according to the roads and the rivers. Otherwise, to divide the sideline according to the steep ridge or the fence, etc. In addition, lakes, reservoirs and rivers in the mountains and small area residents and other features are generally no longer to be divided, and as a part of the mountain space grid [7].

(4) Waters. It mainly includes linear rivers and lakes, reservoirs, etc. To make the central line of the river as the boundary of the space grid (If the river is the boundary of the administrative division, treatment by administrative divisions), and according to the grid object distribution, the river can be divided into the appropriate space grid. For relatively large reservoirs and lakes generally divided separately, if the area is lesser, which can consider a merger with the surrounding terrain for the same space grid.

2.2   Grid Cell Division Method

2.2.1   According to the Administrative Divisions

Administrative divisions are in the area of the state to facilitate the administration and hierarchically divided region, also known as administrative region. In the villages and towns, the village is primary mass self-government organizations of the villagers elected by the villagers, it is the township (town) the next level of administrative division, setting up a villagers' committee, and accepting the guidance of the administrative authority at the next higher level. Space grid division, which is according to the township (town) - the village of the administrative rank level division. The first step is to meet the needs of villages and towns management departments, do the division of "rough" class level. Township and village boundaries are certain legal boundaries, and once determined it is difficult to be amended, unless the authorities to modify the boundary line. The second step is to meet the needs of rural management and other functional departments, do the division of "fine" class level. Comprehensive analysis is needed to determine the size of the smallest unit grid, which is division of rural space grid by the comprehensive analysis of the survey report, multiple data source graphics overlay, contrast, and field survey. In the area of the villages and towns, the geographical elements are diverse. The division of the region is actually a partition of the geographical elements.

The basic idea of the geographic grid according to the administrative division is that village as the unit. According to the territorial management, geographical distribution, the status quo management, convenient management, integrity management objects, etc. The town is divided into a number of geographic grid units. The geographic grid divided according to the administration should follow the following principles.

(1) The division of the grid cell should be carried out according to the statutory topographic survey data, the scale should generally be with 1/500 advisable, but should not be less than 1/2000. When dividing grid cell, the larger the scale, the grid line to be more accurate, and management object is more accurate positioning.
(2) The maximum boundary of the unit grid is the villages boundary, and one village has a grid at least.
(3) The division of the grid should be adapted to the current situation, and not split more than ten thousand square meters of independent courtyard, division of the units of the independent courtyard. In the actual process of separating, the area of the unit grid may be reached tens of thousands of square meters.
(4) The division of the grid must enable service behavior to be easily realized, in the process of information service, whether considering the traffic tools, or consideration to the problem of rural politics of rapid processing, etc., all requirements can be easily arrived at the scene.
(5) Building can't be split. Otherwise the description of the location of the management object will be ambiguous. Villages and towns management object can't be split. Otherwise you will produce the phenomenon such as cross or management absence. Unit within the grid management object number is roughly balanced, for the implementation of the management behavior, evaluation and so on, are very necessary.
(6) The splicing between the elements of the grid must be seamless splicing, and can not overlap, otherwise it will inevitably produce the phenomenon such as cross or management absence.

2.2.2   According to the Road Blocks

In the villages and towns, block plays a very vital role in traffic. It also divides the village into different areas. Block itself is a kind of irregular geographic grid, therefore, it can be divided into geographical grid based on block elements. The basic idea of dividing geographic grid according to the road block division is that in the block formed between the main road as the basic unit in the towns and villages, in each block unit which is divided into irregular geographic grid according to the actual geographical distribution and non-trunk, the roads should be divided as an independent unit grid. The geographic grid separated according to the road blocks should follow the following principles.

(1) The division of the unit grid should be carried out according to the statutory topographic survey data, the scale should generally with 1/500 advisable, but should not be less than 1/2000. 1/500 1/1000 and 1/2000 are three kinds of scale topographic survey data can be used to divide the unit grid.

(2) The maximum boundary of the unit grid is the boundary of the road block, and one block has at least one unit grid.

(3) The division of the grid unit to be carried out in accordance with the layout of the village lanes, courtyards and public green space, square, bridge, open space, rivers, mountains, lakes and other natural geography.

(4) The division of the grid should maintain the integrity of the independent court-yard, and do not split the larger detached courtyard. In the actual process of dividing, the area of the unit grid may be reached tens of thousands of square meters.

(5) The division of the grid must enable service behavior to be easily realize, in the process of information service, whether considering the traffic tools, or consideration to the problem of rural politics of rapid processing, etc., all requirements can be easily arrived at the scene.

(6) Building can't be split, otherwise the description of the location of the management objective will be ambiguous;Villages and towns management object can't be split, otherwise you will produce the phenomenon such as cross or management absence. Unit within the grid management object number is approximately balanced, for the implementation of the management behavior, evaluation and so on, are very necessary.

(7) The splicing between the elements of the grid must be seamless splicing, and can not overlap, otherwise it will inevitably produce the phenomenon such as cross or management absence.

(8) The road as an independent unit grid to be divided.

2.2.3 According to the Regular Grid

Regular grid square is a kind of geographic data model, namely, can express geographic information as a series of grid cells, which are of the same size and align by rows and cols. The first step of dividing geographic grid is to divide the large scale map into regular grid square by with a certain distance. These grid squares are composed of a regular geographic grid. Each geographic grid determines its location by the coordinates of its center point. If necessary, geographic grid can be subdivided according to the actual surface intensity, such as sparse objects without the need of mesh subdivision, but intensive surface features can be subdivided into geographic grid according to the needs, from high level to low level of different thickness.

Geographic grid, divided by regular grid square, whose properties should be determined, because of their different properties contribute several algorithms in the following:

(1) Central Point Method, Central point method is the most simple method of properties discrimination. The core idea of Central point method is taking the attribute values of the central point in source regions as the attribute value of the entire grid area for each grid. Central point method is applicable to the

geographical elements with continuous distribution characteristics. In the practical application, there may be the center point of the grid area falls on the boundary of many source regions and can't determine the position ownership of the center point. Then require combined with other methods, or directly by analysts, according to the experience and the need to clarify the properties.

(2) Preponderant Area Method, Preponderant area method takes the attribute value of the source region, which is the largest component of the grid region, determines the value of the entire grid area. It can be applied to the case of a finer classification and smaller grid [8].

(3) Weightiness Method, Weightiness method according to the importance of the non-homologous region in the area of the grid, the most significant source region is chosen to decide the attribute value of the grid area. It is suitable for the region with special meaning and less area.

3 Analysis and Comparison

According to the division of administrative division the geographic grid, makes the information of towns and villages in the space forming three levels, and clearly determine the responsibility of village management at each level. The first level is townships (towns). The responsible person in the village management is the township (town) government. The second level is the village, the responsible person is the village committee; The third level is the geographical grid unit, the responsible person is a special geographic grid unit management personnel. Geographical grid encoding is the same with administrative divisions encoding. There is no phenomenon road and road intersection because of the management cross or management defects. The disadvantage is that the data is not easy to manage, and the flexibility is not high.

According to road blocks divide geographical grid, blocks network on the map has an obvious boundary, thereby dividing the grid cells relatively easy, and take the road as a separate grid may avoid road cross management or management vacancy phenomenon. The disadvantage is that according to the district road network divide geographical grid may be conflict with the administrative divisions. For example, two different villages or groups (teams) in the region are likely to be divided into a geographic grid unit, resulting in conflict with administrative divisions. Correspondingly, the geographical grid encoding and the grid components encoding conflict with the administrative divisions encoding. The intersection of the road can cause the phenomenon of the management across or management absence. The road unit grid and the grid of the adjacent units need to be equipped with the management personnel, which will cause the waste of human resources.

Geographic grid divided in accordance with the square grid is a regular grid, is contributing to store the geographic space information, management and spatial analysis, but it can't guarantee the surface features' integrity. Many surface features are split in different geographic grids, even the relatively larger independent courtyard will be split in several unit grids. That will easily cause the management across or management absence. And the properties in a single grid are not sole. There may be a number of different properties in the grid.

4 Conclusions

The new model of rural grid management is still at the stage of theoretical exploring and researching. This paper discusses the grid partition technology in the village and town grid management, and gives out the specific principles and method of grid division. In the implementation of the actual grid partition should according to the characteristics of local topography and the village or town planning layout, combined with the advantages and disadvantages of various divide method, considering each management subject of convergence after grid partition to avoid the fracture of management services rights, finally archive success of covering the whole management and full service in village management grid construction and operation process.

References

1. Lu, F.: Cooperative service: innovation mode of rural social management. Study Explor. **12**(1), 64–65 (2012)
2. Huang, A.: Governance of new rural communitese in the context of urbanization. J. GuangZhou Open Univ. GuangZhou. **11**(4), 56–59 (2011)
3. Gao, M.: The study of China's urban community grid management. Henan University, HeNan, pp. 1–3 (2013)
4. Fountain, J.E.: Building the Virtual State Information Technology and Institutional Change. Brookings Institution Press, Washington (2001)
5. Chen, M.: The research on Lucheng town community grid management. Anhui University, AnHui, 2-4.3-10 (2013)
6. Box, R.C.: Citizen Governance: Leading American Communities into the 21st Century. SAGE Publications Inc., New York (1998)
7. Li, L.: Plotting of city spatial grid based on field mode. Geospatial Inf. **6**(2), 96–99 (2008)
8. Zhu, L.: GIS study on the gridding of vector polygon in GIS. Geogr. Geo-Inf. Sci. **20**(1), 12–15 (2004)

A Review on Optical Measurement Method of Chemical Oxygen Demand in Water Bodies

Fei Liu[1,2], Peichao Zheng[1(✉)], Baichuan Huang[2], Xiande Zhao[2],
Leizi Jiao[2], and Daming Dong[2(✉)]

[1] Key Laboratory of Optical Fiber Communication Technology,
Chongqing University of Posts and Telecommunications,
Chongqing 400065, China
liufei8901@163.com, zhengpc@cqupt.edu.cn
[2] Beijing Research Center of Intelligent Equipment for Agriculture,
Beijing Academy of Agriculture and Forestry Sciences, Beijing 100097, China

Abstract. Water quality monitoring technology based on optical method is the trend for modern water environmental monitoring. Compared with the traditional monitoring methods, Spectroscopy is a more simple, a small amount of reagent consumption, good repeatability, high accuracy and rapid detection of significant advantages, which is very suitable for rapid and on-line monitoring determination of environment water samples COD. This paper summarized the status and research progress of optical methods for monitoring of COD in water. The basic principle of traditional analysis methods and optical methods for measuring COD in water were brief described, and compared to the characteristic of different waveband of the detection of COD. The principles and applications of spectroscopic methods commonly used spectral preprocessing methods and calibration methods were listed, and also introduced the progress of optical sensors. Finally, the future research focus and direction of spectroscopic methods were prospected.

Keywords: Spectroscopic · Water · COD

1 Introduction

Water body refers to the complex of rivers, lakes, ground water and other natural water, which includes water and the living organism and substances present in it. Water is an important material for human survival. It is also the necessary resources to human life. With the development of the social economy and the occupation and extension for the natural resources constantly by the human activities, the water quality encountered serious pollution which will threaten the security of water source, the water resource suffered waste that will make the fresh water shortage [1]. The global water source is severely polluted, so that the water cannot reach drinking water quality standards, which exacerbated the shortage of water. In 2012, over ten percent of 972 sections are worse than Grade V, where National Monitoring section of the Yangtze River and Yellow River and other trunk watersheds of China [2].

D. Li and Z. Li (Eds.): CCTA 2015, Part I, IFIP AICT 478, pp. 619–636, 2016.
DOI: 10.1007/978-3-319-48357-3_60

Water pollution means some chemical or physical parameters of the water do not meet the water quality standards. Industrial wastewater and domestic sewage are important sources of water pollution. Serious water pollution is toxic to aquatic plants and animals and even lead to death, while the harm to human health through the food chain. To ensure the quality and safety and good environmental protection, water quality monitoring work has been increasingly important. COD is one of the main monitoring items of surface water, domestic sewage and industrial waste water, and is an important parameter for water quality monitoring. Both types of industrial waste water or sewage must be treated to meet COD value of environmental emission standards allowed to discharge [3]. Therefore, measurement of COD index is particularly important in today's increasingly pay attention to environmental protection. Simultaneously, COD is one of the wastewater pollution indexes of pollutant discharge total amount control engineering [4].

Chemical Oxygen Demand refers to a substance that can be oxidized in water chemical oxidation under certain conditions, which can be consumed amount of oxidant to oxygen mg/L to indicate, it reflects the degree of water and nitrite, ferrous salts, sulfides, etc. So to some extent, COD is one of the important indicators of overall performance characterization of organic pollutants in water [5].

Organic pollutants in the microbial oxidation decomposition requires large amounts of dissolved oxygen in water, and water quality black stinking, choking water creatures, which worsened the aquatic environment and upset the balance of the ecosystem [6], huge losses to aquaculture. Therefore, timely and accurate monitoring water COD concentration is important for the control of pollutant emissions and monitoring water quality.

The standard method for determination of COD is permanganate index and potassium dichromate oxidation method. The former applies to groundwater and relatively clean surface water, drinking water analysis. The latter used for industrial wastewater and analysis of sewage [7]. The basic principle is potassium dichromate in strongly acidic solution, with an excess of potassium dichromate to reducing substances in water samples, and take ferroin solution as indicator, back titration with sulfuric acid ferrous ammonium to calculate the amount of reducing substances by the amount of consumed ferrous sulfate of ammonia [8].

This law applies to all types of containing COD value greater than 30 mg/L of water samples, and the upper limit of the undiluted water samples is 700 mg/L. Determination of the optimum range of 50–500 mg/L [5]. This method can detect water, sewage treatment plant effluent and moderately polluted waste water. Be of simple detection equipment and accurate measurement results [2]. However, this method has to be consumed expensive silver sulfate and a lot of concentrated sulfuric acid, in order to eliminate the interference of chloride ions [9]. It has the defect of long analysis period, heavy workload, reagent consumption, high energy consumption [10]. If the test solution contains chloride ions, nitrite ions, which will react with digestive agent or catalyst cause large deviation of measurement [11], For a large number of samples in the area of water quality, water quality monitoring and sewage treatment plants in the water treatment process of the determination, the standard method has great limitations [12]. There are some method such as standard improved method, polarography [13] and other methods for the detection of water COD, but still cannot solve the traditional method of secondary pollution, measuring long time, higher testing costs disadvantages.

With advances in optics manufacturing technology and equipment, optical methods and chemometrics method for the analysis of complex system of quantitative and qualitative sample has gradually been recognized by people. That has been successfully applied in many fields. Deepa and Ganesh [14] proposed a method to measure dissolved oxygen in water, using an optical measurement range 2-9 mg/L, with high efficiency, low cost, simple operations and other advantages. Kelley et al. [15] put forward a method based on the use of optical fiber sensor technology developed economy turbidity, which can be realized without reagents, measuring water turbidity and particle concentration with high sensitivity. Optical metrology method combined chemometrics algorithm can be applied to the detection of chlorophyll. O'Connell et al. [16] proposed a novel optical fiber sensors for real-time monitoring of the marine environment, and be able to identify the type of water chlorophyll. Chen et al. [17] presented an optical sensor combined ECRSA algorithm to detect the concentration of chlorophyll in turbid water. Azema et al. [18] studied the optical method to detect water suspended solids in water body. Optical methods can no regents, high sensitivity, fast measurement of water quality. although in situ, precise sensing device is still in the research stage [15, 19], but with the development of computer technology and modern scientific instruments, advanced water quality monitoring equipment will gradually be widely used.

Optical methods have also been applied to the measurement of water COD, there are visible spectroscopy [20], visible-near infrared spectroscopy [21], near infrared spectroscopy [22], dual-wavelength spectroscopy [23], UV spectroscopy [24], photochemical luminescence method [25] and other methods, optical detection method is fast, easy operation, ideal for water COD online real-time monitoring. In this paper, different bands spectroscopic methods to detect water COD research status and characteristics were reviewed, and elaborated the chemometrics algorithm which frequently used for optical method measuring water COD. And then introduced some online COD optical sensors, in order to provide a reference for the research on optical method of monitoring water COD.

2 Different Bands Spectroscopy Method for Measuring Water COD

Based on spectral analysis of water quality monitoring technology is an important development direction of modern water environment monitoring, compared with traditional methods, sensitive, rapid, simple and many other advantages, to meet the environmental requirements of the water environment monitoring, showing the broad application prospects. Therefore, launching the research of spectroscopy monitoring water COD is important. Currently, based on UV/UV-visible spectrum and near infrared spectra method research more extensive, other band spectroscopy method for measuring COD in water research literature failed to mention, pursuant to the above article only describes two spectroscopic methods.

2.1 UV Spectroscopy

In recent years, International and domestic academics carry out a lot of research about the UV absorption spectroscopy detection of COD technology. Utilizing the relationship between wastewater in UV spectral region absorbance and water COD calculate the COD value, which directly according to the organic matter has a very sensitive optical absorption principle. UV-visible spectroscopy can be measured the concentration of organic pollutants in water directly or indirectly [26, 27], and the analysis of its components [28], with advantages of sensitive, fast and simple. For the first time by using ultraviolet absorption spectrometry water COD is Japanese scholars [29]. In 1965, Ogura studied the relationship between water quality of rainwater, rivers and other natural water bodies and UV absorption, published a UV absorption spectrum of substances which found organic matter at 220 nm exists a certain correlation between the absorbance and water COD. But in order to masking ultraviolet absorption interference for the water containing a large amount of nitrate, 250 nm wavelength was a more suitable point for organic compounds measured. UV method has been included in the Japanese Industrial Standard K-0807, require determination of sewage absorbance at the wavelength of 253.7 nm, estimated COD value by the standard relationship between pre-established absorbance and COD. The Japanese government had made the official water quality monitoring indicators to evaluate the organic matter content of the water. Europe had been monitoring indicators to monitor water treatment effect of organic matter [30]. Langergraber et al. [31] applied UV-visible spectrometer to obtain the absorption spectra of paper mill wastewater, Compared with single wavelength and full band analysis model is established, the experimental results prove that the model based on full-band COD values are better prediction. Kong and Wu [32] studied UV spectrometer to detect the COD of industrial wastewater and dyeing wastewater, with fast, accurate, and other advantages. Su et al. [33] utilized UV photolysis and photochemical method to achieve a rapid, sensitive, on-line detection of COD, reached a very low detection limit. Roig and Thomas [34] measured a variety of water quality monitoring parameters by UV spectrum instrument, analysis showed that UV spectroscopy was a powerful tool for the detection of water COD and other water quality parameters.

Chen [35] measured the COD of tap water which have been treatment close to pure water at 220 nm, the results showed that such water sample of UV spectroscopy obtained very good correlation between acidic potassium permanganate method, but the UV method is more simple and fast. Song et al. [36] found the UV absorbance value and COD existed good linear correlation, which the sample collected from the SBR wastewater treatment plant secondary effluent filtered through a little suspended solids. Hu et al. [37] monitored the water COD by UV spectroscopy using a new nozzle, the experimental results showed that the relative error of experiment was less than ten percent comparing with standard method. Chen et al. [38] made liner regression about the detecting data of UV absorbance and COD in industrial wastewater, sewage and surface water. A strong correlation between UV absorbance and COD, in certain conditions UV absorbance under the water body can calculate the results of COD.

Wu et al. [39] found that the integrated area scanned by UV spectroscopy and scanning water samples COD had a good linear relationship. Xing [40] determined by UV spectroscopy and COD compared with potassium dichromate method, it had no secondary pollution and high measuring speed. Wang et al. [41] studied the relationship UV absorption in wastewater after biochemical treatment of the WISCO coking plant and COD value, the regression equation is established. The results showed that the relative errors are less than 6.4 % when wastewater COD values in the range of 0–100 ppm. That leading to a rapid predictive biological treatment plant effluent quality. Zhao et al. [30] designed and implemented online testing water COD instrument, but these principle are single-wavelength spectral data reflect organic pollutants in water, which demand high precision of the instrument. There were also using UV spectral measurements combined algorithm to detect water COD. Be of the regression model, neural network method. Feng et al. [24] studied the spectral analysis of water quality, used the least squares method combined full-band ultraviolet absorbance establish calibration model to estimate the water COD. When processing the water samples because water turbidity and other related factors [42], general used visible light absorbance to amend UV254. Charef et al. [43] established the correction model with a wavelength range of 200-800 nm UV-visible absorption combined with neural network algorithm, the estimated value and measured value had a high correlation. Şahin et al. [44] get the UV-visible absorption spectrum of the dyeing wastewater, through the whole spectrum method combined with neural network established calibration model had good prediction accuracy. Vaillant et al. [45] studied the ultraviolet spectrometry for qualitative and quantitative detection of the urban water quality, and described the development process of UV spectroscopy to detect the quality of water, the tendency is from the single-wavelength spectroscopy to the whole spectrum method for measure the absorbance of the water body. Directly determination of COD UV spectroscopy development process as Fig. 1 shown:

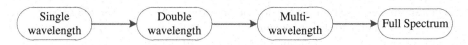

Fig. 1. UV spectroscopy development trends

Currently no chemical reagents detection technology has begun to apply in foreign countries, Europe and the United States free of chemical reagents Instrument's market share is substantial growth [46]. Simple single/dual wavelength photometer structure only applies to measure a single component of the water COD, but determine the actual water samples COD will be disturbed by many factors, and the different components of organic matter in water, the maximum absorption peak is not all at 254 nm, therefore, only at 254 nm spectral data to reflect the concentration of organic matter is very difficult. Most organics have ultraviolet absorption at the range of 200–400 nm, organic compounds by measuring absorbance in the ultraviolet band can indirectly reflect the content of organic matter in water, and thus optical measurement method is widely used in water and organic matter in the qualitative and quantitative determination.

The development of modern spectroscopy allows the full spectrum measurement becomes a reality, once detected in the water samples can be obtained by UV-visible absorbance curve over a wide spectral range, capable of a more comprehensive reflect water pollution, despite the many advantages of direct determination of COD by UV absorption spectroscopy, But at the moment for the combination of different water bodies of the organic components, and ultraviolet absorption spectrum and the COD value relationship exist shortage of research. For the particular wastewater, the regression equation should be determined according to the measured results or calibration curve. Due to the different water samples interfering factors specific makes calibration model no universality. UV absorption spectra measured COD directly, eliminating interfere of suspended solids, non-soluble colloidal substances and certain inorganic ions in water samples still need further study. Although the method is not perfect, but with the development of computer technology and chemometrics, research and development of new sensing devices, these problems will gradually be resolved. UV absorption spectroscopy monitoring the water quality will be widely available in the future work.

2.2   Near Infrared Spectroscopy

NIR spectroscopy is a new technology to quickly detect the chemical composition of the sample content, and the fastest growing spectroscopy since the 1990s, the principle is using of chemical substances in the optical properties of the near-infrared spectral region. It has been widely used in the agriculture, food, petroleum, chemical industry and other fields.

In recent years, domestic and foreign scholars have carried out the research of detect the water COD, that base on near-infrared spectroscopy and made great progress. Sarraguça et al. [47] obtained the activated sludge reaction tank effluent absorbance at the wavelength range of 900–1400 nm near-infrared band, utilizing PLS established calibration model based on the absorption spectrum data after SNV and MNCN processed, the model correlation coefficient was 23 %. Pan et al. [48] utilized S-G smoothing and PLS method through the different NIR spectrum band of sugar refinery wastewater established model, and ultimately selected 808-964 nm for the optimal information band, the predictive of validation sample set reached 96 %.

He et al. [49] studied the feasibility of near infrared spectroscopy to measure the COD, based on the varied divisions for the COD standard solution and wastewater samples, a regression model was established. The correlation coefficient of the standard theory and predicted COD values in standard solution was 0.9999. The correlation coefficient of measured and predicted COD values in sewage treatment plant effluent was 0.9453. It showed that the near-infrared spectroscopy applied for the determination of COD in sugar refinery waster was feasible. Furthermore, Xu [50] measured the COD in river water, combined with PLS algorithm based on the first derivative NIR spectrum established COD analysis model, the correlation coefficient of the predicted model reached 0.991, the Root Mean Square Deviation Correction equivalent 4.3 % of average COD value in water samples, met the needs of daily monitoring, the RMSEC and RMSEP was closely. The analysis showed that the model had better applicability.

NIR spectrometry for the determination of water COD had a stability result, which for the rapid determination of organic pollution in water bodies provided a viable analysis techniques. Ji et al. [22] measured the COD of sugar refinery wastewater, based on the FTIR/ATR spectroscopy, combined with the PLS method and SG smoothing built regression model, Prediction correlation coefficient reached 96.8 %.

These studies demonstrated the feasibility of measuring water indicators by near-infrared spectroscopy, but to establish a functional COD prediction model of near-infrared spectroscopy, that need to collect a sufficient number of representative samples of wastewater, so that the useful information and background information should be contained wide enough in the sample set. Eliminating interfere of temperature, pH, turbidity in still need further study, thereby reduced the SEC and SEP of regression model. Although near infrared spectroscopy has many advantages, but it also exists weak absorption intensity, low sensitivity and other defects. Material in the near infrared absorption coefficient generally small, which make the measured components should be greater than 0.1 % in general. In addition, the spectrum of water samples easily suffered influence by the state of the sample, measurement conditions and other external conditions [51]. But with the development and application of new chemometrics methods and new devices, these problems will gradually be resolved. Near-infrared light have excellent transmission performance in optical fibers, making near-infrared spectroscopy technology can achieve remote applications through fiber optic technology, which brings the industry on-line analysis and in situ measurements a great possibilities.

NIR spectroscopy has many advantages, such as no chemical reagents, simple analytical process, low cost analysis, etc., and NIR spectroscopy can fast, safe and operate easily for the quantitative determination of COD in wastewater. And can be combined with the near-infrared optical fiber technology to build large-scale online monitoring system, which has important practical significance for control the pollution sources of water and monitoring pollution.

2.3 Comparison of Several Spectroscopic Methods for COD Measurement

Using the NIR spectrometer was possible to measure COD in water on-line and in-situ, without sample pretreatment and consumes no chemicals, almost all of the main structure and organic matter composed of spectral information, which can be found in near infrared spectroscopy signal and spectrum stable. UV spectroscopy is use of the relationship between water sample absorbance and COD concentration obtained the COD values indirectly. From the absorbance intensity can determine the extent of organic pollution in water bodies, based on most of the organic matter in the UV band has the absorption peak. Using UV-visible spectroscopy and near-infrared spectroscopy to monitor reactor wastewater sludge wastewater treatment process.

Sarraguca et al. [47] utilized UV-visible spectroscopy and near-infrared spectroscopy to monitor wastewater in the treatment process of activated sludge reactor. The results showed the UV-VIS modeling results were more accurate than the results obtained with NIR. Analysis believed the presence of water had a huge impact on the near-infrared spectrum, masking some spectral characteristics that could be important for calculating the COD. Wu et al. [52] utilized UV-visible spectroscopy and near-infrared spectroscopy to monitor the standard solution of potassium hydrogen phthalate. After different pretreatments established UV-VIS modeling and NIR modeling by PLS, found that the UV absorption method had a higher correlation than the infrared transmission method. The prediction accuracy was slightly lower than the latter. However, these results were based on a standard solution for the object, did not have extensive, but also did not take into account the actual measurement of turbidity, pH effects on measurement accuracy.

To accommodate the technology development requirements of automated, real-time COD monitoring, although UV spectroscopy can achieve real-time on-line COD measurement. But there are some problems due to water absorption have a larger coefficient in the UV spectral region, make the higher requirements of sample pretreatment, so the tested water samples should be clear and transparent, the suspended solids that contribute to the COD values are removed after pretreated, which make the estimated COD inaccurately. Furthermore, many organic have no absorption spectrum in the UV spectrum region, it will cause the estimated value lower than actual [49]. Table 1 gave the data for the optical method detection of the COD in water.

Table 1. Spectroscopic methods based on different bands detected COD in water

Optical method	Range of measurement (mg/L)	LOD (mg/l)	Accuracy R^2	Water sample	Instrument complexity	Papers
UV spectroscopy	0–500	1.98	0.940	Wastewater	Simple, no pollution	Suryani et al. [53]
	0.2–20	0.08	0.989	River lake	Detect fast, no chemicals, high sensitivity, simple structure	Su et al. [33]
	115–427	137.37	0.88	municipal wastewater	Simple structure, no pollution	Charef et al. [43]
NIR spectroscopy	41–416	–	0.976	Sugar refinery wastewater	Simple structure, no pollution, fast	Pan et al. [48]
	30–500	33.9	0.9436	municipal wastewater	No pollution, Simple structure	He et al. [49]

Suryani et al. [53] combined with UV spectral deconvolution to build regression model, the LOD of COD reached 1.98 mg/L, the correlation coefficient of predicted and detection values was 0.94. Su et al. [33] utilized UV spectroscopy achieved rapid, highly sensitive on-line monitoring water COD, the LOD reached 0.08 mg/L in the range of 0.2-20 m/L, the correlation coefficient was 0.989. Charef et al. [43] detected the COD of municipal wastewater combined with ANN algorithm and UV modeling, tested COD value in the range of 115–427 mg/L, the LOD was 134.25 mg/L, the correlation coefficient of model between the estimated and measured COD was 0.88. Pan et al. [48] utilized NIR spectroscopy combined with the SG smoothing and MWPLS method built regression model, the determination of COD in the range of 41 to 416 mg/L, the correlation coefficient was 0.976. He et al. [49] utilized NIR spectroscopy combined with PLS built model, the correlation coefficient of predicted between detection values was 0.94, the LOD of COD in the range of 30 to 500 mg/L was 33.9 mg/L.

3   Chemometrics Methods for Optical Measurement of Water COD

3.1   Spectral Preprocessing Methods

When using an optical method for measuring water COD, the spectra obtained except contains the chemical information of water sample itself, also contains the circuit noise, the sample background, and other stray light noise and other unwanted information. Therefore, in order to improve the accuracy of the regression model built by chemometrics methods, eliminating noise and unwanted spectral data interference information becomes necessary, spectral preprocessing methods can be largely solved these problems. There are many commonly used methods such as Mean Centering, Normalization, Smoothing, Derivative method, SNV and detrending algorithm, Multiplicative Scatter Correction and other methods.

SNV pre-processing method for processing a spectrum, using the data points on the spectrum curve subtract the mean of spectral curve, and then divided by the standard deviation, which is mainly used to remove the interference of solid particle size, surface scattering, and optical path changes. Wu et al. [27] utilized UV spectroscopy to detect the COD of wastewater combined with different calibration methods and SNV, the model had better prediction accuracy, but less effective compared to other pretreatment methods, the experimental results showed that the spectral preprocessing not necessarily possible to improve the accuracy of the model unless suitable.

Derivative pretreatment is essentially intensity distribution derivation of spectral signal. The principle is to choose a few points of the original spectrum constitutes a window and obtained derivative spectra in the window by derivation. Derivative spectra can effectively eliminate the baseline and other background interference, and improve the resolution and sensitivity. But the derivative spectrum will also introduce noise information and reduce noise ratio. The first derivative and second derivative spectra were more commonly used. Xu et al. [50] detected the COD of river water by NIR spectroscopy, and used the original spectrum, utilized the first derivative and second derivative spectra of the water samples established regression model, the first derivative of NIR spectrum modeling predicted the best results after comprehensive comparison.

Smooth pretreatment is to choose an odd number of the wavelength points on original spectrum as the window, and take on average or fitting of wavelength points within the window. Complete smoothing of all the points by moving the window from left to right. Where the width of the window is a key smoothing parameter, if the window width is too small, it will lead to poor smoothing effects, if the spectral width of the window oversize the general will cause signal distortion. Moving average smoothing and SG convolution smoothing is the denoising smoothing algorithm that commonly used. Pan et al. [48] detected the COD in wastewater by NIR spectroscopy combined with SG smoothing method, the modeling results were better than the original spectrum modeling. Ji et al. [22] utilized the PLS algorithm combined with SG smoothed data established prediction model, built relationship between FTIR/ATR spectrum and COD values, prediction correlation coefficient reached 0.968, the SG modeling were better than the PLS modeling without SG smoothing.

Spectral deconvolution means the original spectrum is decomposed from a few number of features of spectra and a commonly tool for water COD determination combined with UV spectroscopy. Nam et al. [54] utilized the NV spectral deconvolution detected the nitrite content in frozen spinach, the correlation coefficient reached 0.9843. Martins et al. [55] used the spectral deconvolution method to estimate the COD value of hospital discharge wastewater, the UV absorbance and COD values were well correlated. Thomas et al. [56] measured the COD and TOC of wastewater by UV spectral deconvolution, the correlation coefficient of the estimated between measured values were 0.94, 0.92, which achieved a very good prediction.

3.2   Regression Approach

Research on the use of spectroscopy measurement data for COD value prediction has been further developed, the key technology of optical analysis method is to build a mathematical model of spectral data, which between the concentration of organic pollutants, as well as to improve the extrapolate ability of the model, Namely established a stable functional relationship between the spectral information and the nature of components, the stability and prediction accuracy of the calibration mode depend on the regression approach is very important. Established method calibration model is divided into linear and nonlinear correction method, which used multivariate calibration method based on linear regression have MLR, PCR and PLS, based on the non-linear regression methods have Artificial Neural Network, Support Vector Machine, Kernel Partial Least Squares. The following describes the above-described regression method used for the measurement of COD.

In spectral analysis, linear regression was developed earlier, which used to evaluate the correlation of the spectra prediction predicting outcomes between reference method for the determination results in a set of sample. Wu et al. [52] utilized the linear regression method established regression model based on the UV absorbance of the potassium hydrogen phthalate standard solution, the correlation coefficient was 0.986, standard deviation was 15.26344. Wang et al. [41] detected the coking wastewater COD value by UV spectro-photometry, the measured absorbance and COD value were better reliability and practicality through the establishment of linear regression line equation. A linear regression can only use one variable, modeling of spectral data using only a single wavelength point absorbance as a reference data, but when using spectroscopy measured water COD actually, the temperature of water sample will affect the measurement accuracy. Mu et al. [57] utilized the UV spectroscopy measured the water COD, through the experiment determined the data based on the ultraviolet absorbance of the standard solution changes with the temperature. And obtained the relationship between the two by multivariate regression method to fit the data, thereby eliminated the influence of the temperature and improved the model predictive power. Fogelman et al. [58] measured the COD values by UV spectroscopy combined with MLR algorithm, the correlation coefficient of the predicted and measured values was 0.85.

PCA (Principal Component Regression) for the main purpose is to exclude the overlapping information of the spectral data. The algorithm idea is to transform the original variable of the spectral matrix into a small number of linear combinations of several variables, the new variables should characterized the spectral features of the original variable as much as possible, and then select the main combinations for multiple linear regression. PCR overcomes the problem of multicollinearity between MLR input variables, and can improve the predictive ability of the mathematical model. The method can be used for more complex analysis system. It can be more accurately estimate the content of the component under test without knowing which interfering components. But the operating rate of PCR is lower than MLR. If you ignore the useful component in the selection of principal components retained noise, it will make a large deviation of the regression model prediction. Lourenço et al. [59] proved UV spectroscopy combined with PCA can be used to identify contaminants. Şahin et al. [44] analyzed the PCR modeling and PLS modeling predicted the COD of dyeing wastewater based on UV spectroscopy, PCR method was faster but slightly lower than the accuracy of PLS algorithm.

PLS can establish the relationship between the spectral data and composition and using the full spectrum and partial spectral data modeling. The decomposition of data matrix and regression interaction combined into one step, and the resulting characteristics vector are directly related to the measured components, therefore, it is a popularly multivariate calibration method at present [21]. The PLS combined the predict method of modeling with data analysis methods, and can simultaneously achieve regression modeling, simplify data structure and correlation analysis between the two groups of variables, which gave the multidimensional analysis of complex systems brought a great convenience. PLS is a multivariate analysis algorithms based on PCA. The principle is that decomposition treatment the spectral matrix X and density matrix Y at the same time, and taking into account the relationship between the two in the decomposition, overcoming the limitations of PCR performed only on the spectral matrix X decomposition. PLS method is a good combination of multiple linear regression and principal component analysis. Its basic step is to eliminate the multi-collinearity of spectral data, and then reuse the potential factor for egressions. The number of PLS factors which are the most critical parameters, If the number of factors introduced too little in modeling, the regression model prediction accuracy will be lower, while if introduced too many modeling factor, it will introduce noise impact of the predictive power of the regression model. Şahin et al. [44] used UV spectro-photometry combined with the PLS established the model based on the spectrum of textile waste water and COD, which improved the prediction accuracy of the model. Platikanov et al. [60] achieved multi-parameter water quality monitoring system by UV spectroscopy combined with PLS modeling, for the detection and analysis of sewage treatment plant effluent treatment conditions. And the prediction error is less than 20 %.

ANN (Artificial Neural Network) is proposed based on the simulation of the human brain structure, which the information storage and computing simultaneously in nerve cells. So to some extent, the neural network can simulate the brain nervous system active process, ANN possess self-learning, self-organizing, adaptive fault tolerance capabilities, distributed storage, highly nonlinear expression ability and parallel processing of information, which is the other traditional multivariate calibration methods are not available. More and more researchers began using ANN analysis method to solve the issue of Analytical Chemistry. In many neural networks, the most widely used is Back Propagation-artificial Neural Network, which is a typical feed forward artificial neural network model and has excellent non-linear approximation ability.

The essence of BP algorithm is transformed a set of input and output problem into a nonlinear mapping problem, and through the gradient descent algorithm iterative to solve the weights. BP algorithm is divided into the net input forward calculation and the error back-propagation of two processes. When Network training, the two processes appear alternately until the total error reaches the preset accuracy of the network, this part of the work called network training process, the weights are no longer change in the process of network training. For each given input, network through the forward calculation shows the output response, the work of this section shall be using the training to derive the network for predict or classify. BP network structure is shown in Fig. 2.

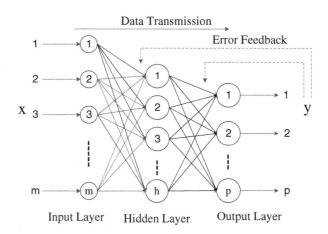

Fig. 2. The BP neural network topology

Kato et al. [61] detected the COD by UV-VIS spectroscopy combined with neural network algorithm, the regression models obtained a higher prediction accuracy. Fogelman et al. [58] utilized UV-VIS spectroscopy combined with different algorithms to measure wastewater COD values, the prediction accuracy of ANN modeling is higher than the MLR modeling. Charef et al. [43] used a novel smart sensing equipment to monitor water quality, the use of UV-visible spectroscopy combined with ANN calibration model established regression model, the correlation coefficient of estimated values of the training set and measured values was 0.952, the correlation coefficient of estimated values of the validation set and measured values was 0.88.

In recent stage, optical methods for water quality measurement commonly used partial least squares (PLS), principal component regression (PCR) and artificial neural network (ANN) and so on. PLS and PCR can significantly compress high-dimensional data and effectively remove the multicollinearity, so they widely used in spectral analysis, but the accuracy of the model that established by the PLS and PCR depends on the number of principal components, Principal component number of too little or too much will lead to the model of training sample fitting insufficient or excessive, which will reduce the accuracy of model predictions. And currently the method used to select the number of principal components is cross-validation, which has a larger calculated and prone to over-fitting. Although ANN method has a strong ability of approximation, but it needs a lot of experiments training to determine the appropriate neural network model and its application effect depends on the user's experience. Therefore, it is adverse to real-time spectrum analysis. Du et al. [62] established the model of COD and UV absorption spectrum based on SVM method, the experiments showed that the predictive power of the SVM modeling is better than PLS modeling. Wu et al. [63] proposed an iterative regression modeling algorithm based on Boosting theory, compared to the PLS, PCR, ANN algorithm had improved the prediction accuracy of TOC.

At present, using optical method combined with a variety of chemometrics methods to establish a mathematical prediction model for monitoring COD is the development trend. It can solve various problems, such as suspensions, PH, inorganic ions and other influence factor, which make the measurement of COD easier and practical. Chemometrics in dealing with chemical measurements of the experimental design, data processing, signal analysis and distinction, decision of chemical classification and prediction, solved a large number of the complex problem that traditional chemical methods are difficult to solve, and showed a strong vitality.

4 COD On-Line Optical Sensors

Optical sensor should meet the on-line analysis requirements of the analytical instrument miniature, high efficiency, rapidly detection and detection of COD in a wider spectral range, so that could improving the accuracy of on-line analyzers. With the development of optical fiber and optical fiber communication technology, Optical fiber sensing technology is a new rapidly developed sensing technology with light as a carrier, optical fiber as medium, perception and transmission of external signals. Compared with traditional sensors, fiber optic sensor has a high sensitivity, the flexural optical path, anti-electromagnetic interference, ease of connection with the computer, etc. [64]. Gu et al. [62, 65] developed an online submersible water quality monitoring system for the measurement of COD based UV spectroscopy, the instrument can be put directly into the water samples, according to the set time intervals and measured parameters at the same water samples at 200–720 nm spectral data, Determination the spectral data of water samples at the range of 200 to 700 nm under the set time intervals and measured parameters, and then automatically transmitted, storage and analysis the data of COD, achieved a real-time, online, in situ water quality testing. Langergraber et al. [31] detected the water quality of paper mill wastewater by UV-visible spectrometer, obtained a more accurate real-time spectral data, achieved the measurement of without reagents, no sample preparation, in-situ. Van den Broeke et al. [66] utilized insitu, on-line Invasive water quality monitor based on the UV-visible spectroscopy measured a variety of water samples, from sewage treatment plant aeration tank effluent and Vienna municipal wastewater to drinking water. The instrument was provided by Scan company. Different standards apply to the spectrometer detector has brought great challenges, but through adjusting the optical path of spectral absorption cell and the optimization of algorithms, and ultimately get a better measurement results, the spectrometer structure and application examples of Austrian Scan company shown in Figs. 1 and 2 in that paper.

In order to meet water quality monitoring data in real time, reliable higher standards, combined with spectral monitoring technology and wireless network technology to achieve monitoring a large area of water body is the inevitable trend, which will achieve monitoring the water quality changes in the water resources and grasp the overall message, providing an important reference for the comprehensive analysis of the water quality in the water basin.

The development of modern optical sensors for water quality monitoring has opened up a new field, and greatly expanded the application range of water quality monitoring instrument. It can be used in the wastewater treatment monitoring, leak detection in industrial production, monitoring environment summary hydrocarbon compounds and organic matter in drinking water, etc. Although some products have been put into market application in the world, the research and development of optical sensor lags fall behind the computer and communication technology, which greatly limits the water quality monitoring system of precision and test efficiency. In situ and real-time wide spectroscopy measurement is restricted by the optical sensors. It is the development requirement of water quality monitoring equipment. In order to get the breakthrough of key technology, we should strengthen on the technology research and development of the optical sensor, and overcome the technical bottleneck to speed up the development of water environment monitoring technology.

5 Conclusion

This article describes an optical method for measuring COD in water research progress and status in domestic and foreign. Briefly explained the chemometrics method that commonly used in the analysis of spectral data, the principle of optical method for measuring water COD is built on the regression model between spectral data and COD value, the extrapolation of model determines the measurement accuracy. So there are important significance for the research of spectral preprocessing and calibration methods for optical methods.

Water quality monitoring method based on optical technology is an important development direction of modern environmental monitoring, compared with traditional chemical analysis, electrochemical analysis and chromatographic analysis methods,

Spectroscopy is more simple operation, a small amount consumption of reagent, good repeatability, high accuracy and rapid detection of obvious advantages, which is very suitable for environmental water samples quick on-line monitoring.

Combination with optical methods for monitoring water COD technology and wireless sensor network technology to achieve a large area of water is the development trend. This will timely grasp the total information of the water quality monitoring in the area of the water resources. Developing based on integrated micro-spectrometer miniaturization water quality monitoring equipment, implementing intelligent, network-based, real-time in-situ monitoring of the water quality monitoring is the requirement of development. With the rapid development of sensor technology and computer technology, we believe that the direct determination of optical methods for monitoring water COD will get more and more widely application.

Acknowledgments. This work was supported by National High Technology Research and Development Program of China (No. 2013AA10230202).

References

1. Methods, E.O.S.W.: Water and Wastewater Monitoring and Analysis Methods. China Environmental Science Press (2002)
2. Guo-bing, L.: A review on detection methods of chemical oxygen demand in water bodies. Rock Minera Anal. **06**, 860–874 (2013)
3. Gui-bao, L., Hong-wu, T., Yao, Z.: The status quo of China's water environmental pollutant emission standards. China Stand. **09**, 47–48 (2002)
4. Von Sperling, M., de Lemos Chernicharo, C.A.: Urban wastewater treatment technologies and the implementation of discharge standards in developing countries. Urban Water **4**(1), 105–114 (2002)
5. Jun, W., et al.: Analysis of potassium dichromate method for determining COD in water. Henan Sci. (09), 1217–1219 (2012)
6. Fu-ling, G.: Improvement and research progress of COD determination methods. Sichuan Environ. **31**(1), 109–113 (2012)
7. Feng, Y.X., Hong, D.D.: The status and recent development for determination of COD. Chongqing Environ. Sci. **04**, 57–61 (1997)
8. Li, S.: Analysis of potassium dichromate method for determining COD in water. Metall. Power **02**, 73–75 (2011)
9. Yin-tao, S., et al.: The effect and elimination of chlorion interference in the determination of CODCr of sewage water. J. Wuhan Univ. Sci. Eng. **17**(4), 20–24 (2004)
10. Yan, Z., et al.: Status and recent development for determination of COD. Environ. Appl. Chem. **18**(05), 455–458 (2006)
11. Xi-hui, H., Liang-wan, L., Al, E.: Recent advance in determination methods of COD. J. Xihua Univ. Nat. Sci. **25**(03), 51–54, 115 (2006)
12. Zhang-zhong, L., et al.: Review on the determination of chemical oxygen demand. Ind. Water Treat. **20**(10), 9–11, 20 (2000)
13. Xue-ming, H., et al.: New-style chemical oxygen demand detection research with UV absorption spectroscopy. J. Atmosphering Environ. Opt. **1**(02), 26–28 (2006)

14. Deepa, N., Ganesh, A.: Sol-gel based portable optical sensor for simultaneous and minimal invasive measurement of pH and dissolved oxygen. Measurement **59**, 337–343 (2015)
15. Kelley, C., et al.: An affordable open-source turbidimeter. Sensors **14**(4), 7142–7155 (2014)
16. O'Connell, E., et al.: A mote interface for fiber optic spectral sensing with real-time monitoring of the marine environment. IEEE Sens. J. **13**(7), 2619–2625 (2013)
17. Chen, J., Sheng, H., Sun, J.: An empirical algorithm for hyperspectral remote sensing of chlorophyll-a in turbid waters: a case study on hyperion sensor. Sensor Lett. **11**(4), 623–631 (2013)
18. Azema, N., et al.: Wastewater suspended solids study by optical methods. Colloids Surf. A **204**(1), 131–140 (2002)
19. Bin Omar, A.F., Bin MatJafri, M.Z.: Turbidimeter design and analysis: a review on optical fiber sensors for the measurement of water turbidity. Sensors **9**(10), 8311–8335 (2009)
20. Yan-feng, L., et al.: Rapid determination of chemical oxygen demand by sulfuric-acid digestion and visible spectrometry. J. Gansu Sci. **16**(01), 39–44 (2004)
21. Haiyan, S., et al.: Approach to rapid detection of chemical oxygen demand in livestock wastewater based on spectroscopy technology. Trans. CSAE **22**(06), 148–151 (2006)
22. Ji, Q., Tao, P., Xin-yan, L.: Rapid determination of wastewater COD for sugar refinery by TIR/ATR spectroscopy. J. Jiamusi Univ. (Nat. Sci. Ed.), (03): 365–366, 369 (2011)
23. Ran, J., et al.: A dual-wavelength spectroscopic method for the low chemica oxygen demand determination. Spectrosc. Spectral Anal. **07**, 2007–2010 (2011)
24. Feng, W., et al.: On-line monitoring technology for chemical oxygen demand based on full-spectrum analysis. Acta Photonica Sin. **41**(08), 883–887 (2012)
25. Zhao, H., et al.: Development of a direct photoelectrochemical method for determination of chemical oxygen demand. Anal. Chem. **76**(1), 155–160 (2004)
26. Rui-xia, H., et al.: Ultraviolet absorption spectrum characterization approach for quantitative analysis of dissolved organic contaminants in sewage. J. Beijing Univ. Technol. **32**(12), 1062–1066 (2006)
27. Wu, Y-q, Shu-xin, D.U., Yun, Y.: Ultraviolet spectrum analysis methods for detecting the concentration of organic pollutants in water. Spectrosc. Spectral Anal. **31**(01), 233–237 (2011)
28. Xiaokun, Z.: UV absorption spectroscopy applied in structure elucidation of organic compounds. Inner Mongolia Petrochem. Ind. **11**, 171–173 (2007)
29. Ogura, N.: Ultraviolet absorbing materials in natural water. Nippon Kagaku Zasshi **86**(12), 1286–1288 (1965)
30. Zhao, Y., et al.: Novel method for on-line water COD determination using UV spectrum technology. Chin. J. Sci. Instrum. **31**(09), 1927–1932 (2010)
31. Langergraber, G., et al.: Monitoring of a paper mill wastewater treatment plant using UV/VIS spectroscopy. Water Sci. Technol. **49**(1), 9–14 (2004)
32. Kong, H., Wu, H.: A rapid determination method of chemical oxygen demand in printing and dyeing wastewater using ultraviolet spectroscopy. Water Environ. Res. **81**(11), 2381–2386 (2009)
33. Su, Y., et al.: Rapid, sensitive and on-line measurement of chemical oxygen demand by novel optical method based on UV photolysis and chemiluminescence. Microchem. J. **87**(1), 56–61 (2007)
34. Roig, B., Thomas, O.: UV spectrophotometry: a powerful tool for environmental measurement. Manag. Environ. Qual.: Intl. J. **14**(3), 398–404 (2003)
35. Chen, W.: Determination of COD in RO water by ultraviolet absorption spectrometry. Technol. Water Treat. **24**(06), 25–27 (1998)
36. Song, L., et al.: UV spectrophotometric method for the determination of sewage plant effluent COD. China Water Wastewater **18**(12), 85–86 (2002)

37. Hu, X-m, et al.: New-style chemical oxygen demand detection research with UV absorption spectroscopy. J. Atmos. Environ. Opt. **05**, 121–124 (2006)
38. Chen, G., et al.: Correlations between ultraviolet absorbance degree and COD in water. Manag. Technol. Environ. Monit. **17**(06), 11–13 (2005)
39. Wu, T-h, et al.: Rapid determination of the chemical oxygen demand of the sewage through continuous UV scanning spectrophotometry. Environ. Monit. China **25**(04), 57–60 (2009)
40. Xing, Z.: Research on the determination of chemical oxygen demand by UV absorption spectrum. Chem. Ind. Guangzhou (24) 94–95, 105 (2011)
41. Wang, G., et al.: Research on COD in waste water by Rapid Determination of Ultraviolet Spectrophtometry. Fuel Chem. Process. **31**(01), 31–33 (2000)
42. Feng, W., Wei, S.H., Fu, F.S.: Study on miniaturization and robustness of dispersion system for UV-VIS spectrophotometer. Chin. J. Sci. Instrum. **27**(11), 1437–1440 (2006)
43. Charef, A., et al.: Water quality monitoring using a smart sensing system. Measurement **28** (3), 219–224 (2000)
44. Şahin, S., Demir, C., Güçer, Ş.: Simultaneous UV-vis spectrophotometric determination of disperse dyes in textile wastewater by partial least squares and principal component regression. Dyes Pigm. **73**(3), 368–376 (2007)
45. Vaillant, S., Pouet, M.F., Thomas, O.: Basic handling of UV spectra for urban water quality monitoring. Urban Water **4**(3), 273–281 (2002)
46. van den Broeke, J., Langergraber, G., Weingartner, A.: On-line and in-situ UV/vis spectroscopy for multi-parameter measurements: a brief review. Spectrosc. Eur. **18**(4), 15–18 (2006)
47. Sarraguça, M.C., et al.: Quantitative monitoring of an activated sludge reactor using on-line UV-visible and near-infrared spectroscopy. Anal. Bioanal. Chem. **395**(4), 1159–1166 (2009)
48. Pan, T., et al.: Near-infrared spectroscopy with waveband selection stability for the determination of COD in sugar refinery wastewater. Anal. Methods **4**(4), 1046–1052 (2012)
49. He, J-c, et al.: Determination of chemical oxygen demand in wastewater by near-infrared spectroscopy. J. Zhejiang Univ. (Eng. Sci.) **41**(05), 752–755 (2007)
50. Xu, L.H.: Near infrared spectroscopy for the analysis of organic pollutants water research. Chin. J. Anal. Lab. **27**(z1), 448–450 (2008)
51. Zhigang, X., Liyan, M., Shijiia, L.: Application of near-infrared spectroscopy in water quality monitoring. Guizhou Agric. Sci. 41(8) (2013)
52. Wu, G-q, et al.: Determination of chemical oxygen demand in water using near infrared transmission and UV absorbance method. Spectrosc. Spectral Anal. **31**(06), 1486–1489 (2011)
53. Suryani, S., Theraulaz, F., Thomas, O.: Deterministic resolution of molecular absorption spectra of aqueous solutions: environmental applications. Trends Anal. Chem. **14**(9), 457–463 (1995)
54. Nam, P.H., et al.: A new quantitative and low-cost determination method of nitrate in vegetables, based on deconvolution of UV spectra. Talanta **76**(4), 936–940 (2008)
55. Martins, A.F., et al.: COD evaluation of hospital effluent by means of UV-spectral deconvolution. CLEAN–Soil, Air, Water **36**(10–11), 875–878 (2008)
56. Thomas, O., et al.: UV spectral deconvolution: a valuable tool for waste water quality determination. Environ. Technol. **14**(12), 1187–1192 (1993)
57. Mu, X., et al.: Influence of temperature change on COD measurement with UV spectrophotometry. Anal. Instrum. (06) 53–55 (2008)
58. Fogelman, S., Blumenstein, M., Zhao, H.: Estimation of chemical oxygen demand by ultraviolet spectroscopic profiling and artificial neural networks. Neural Comput. Appl. **15** (3–4), 197–203 (2006)

59. Lourenço, N.D., et al.: UV spectra analysis for water quality monitoring in a fuel park wastewater treatment plant. Chemosphere **65**(5), 786–791 (2006)
60. Platikanov, S., et al.: Chemometrics quality assessment of wastewater treatment plant effluents using physicochemical parameters and UV absorption measurements. J. Environ. Manag. **140**, 33–44 (2014)
61. Kato, Y., et al.: The prediction of Chemical Oxygen Demand (COD) in waste water by UV-visible absorption spectrum-neural network. Orient. J. Chem. **17**(1), 01–08 (2001)
62. Du, S., Xiaoli, W., Tiejun, W.: Support vector machine for ultraviolet spectroscopic water quality analyzers. Chin. J. Anal. Chem. **32**(9), 1227–1230 (2004)
63. Wu, L., Yangjun, L., Tiejun, W.: A boosting-partial least squares method for ultraviolet spectroscopic analysis of water quality. Chin. J. Anal. Chem. **34**(08), 1091–1095 (2006)
64. Sheng He, S.: Development trend of modern sensor. J. Electron. Measur. Instrum. **23**(1), 1–10 (2009)
65. Gu, J., et al.: On submersible water-quality monitoring systems based on spectral ultraviolet method. J. Safety Environ. **12**(6), 98–102 (2012)
66. van den Broeke, J., et al.: Use of in-situ UV/Vis spectrometry in water monitoring in Vienna. In: Proceedings IWA World Water Congress and Exhibition (2008)

Analysis of Changes in Agronomic Parameters and Disease Index of Rapeseed Leaf Leukoplakia Based on Spectra

Kunya Fu[1,2], Hongxin Cao[2(✉)], Wenyu Zhang[2], Weixin Zhang[2], Daokuo Ge[2], Yan Liu[2], Chunhuan Feng[2], and Weitao Chen[2]

[1] College of Agronomy, Nanjing Agricultural University,
Nanjing 210095, China
921186907@qq.com
[2] Institute of Agricultural Economy and Information/Engineering Research
Center for Digital Agriculture, Jiangsu Academy of Agricultural Sciences,
Nanjing 210014, China
caohongxin@hotmail.com, aunote@163.com,
nkyzwx@126.com, gedk@sina.com, liuyan0203@aliyun.com,
{921186907,1286234727,1303079141}@qq.com

Abstract. The non-destructive, rapid and accurate monitoring diagnosis of rapeseed diseases is of significance for sustainable development of rapeseed production and environment protection. The spectrum data of rapeseed leaf leukoplakia were collected in the experimental Farm of Jiangsu Academy of Agricultural Science in 2013 and 2014. Firstly, the common distinctive bands of the disease and the health were found by comparing reflectance spectrum of leaves in the field and under black background. The results showed that with the progress of the growth period, reflectance of disease leaves decreased earlier than healthy leaves. It was the best period to identify rapeseed leukoplakia from 11 days after early anthesis to 9 days after finish flowering to identify rapeseed leukoplakia in the field due to during this period the reflectance of healthy leaves remained at 35 % while the disease diseased to 30 %. The sensitive band was in the range of 760–1080 nm. The correlation among disease index (DI), agronomic parameters, and the reflectance of the disease samples were analysed, and the results showed that there were high correlations between DI, and agronomic parameters and reflectance, e.g., the correlation between the leaf moisture content and the reflectance in 460 nm, 550 nm, 650 nm, 710 nm, 760 nm, 1480 nm, and 1600 nm, between the leaf nitrogen content and the reflectance in 810 nm, 870 nm, 1080 nm, 1280 nm, 1320 nm, 1540 nm, 1600 nm, 1650 nm, and 1700 nm, and between the SPAD value and the reflectance in 1200 nm, 1280 nm, and 1540 nm had significance with $p < 0.01$. The quantitative models of agronomic parameters based on reflectance were developed by stepwise regression, principal component analysis, and curve fitting. The data of rapeseed leukoplakia in 2013 and rapeseed virus in 2014 were used to test. The results showed that in the same disease test, the quantitative models of moisture content based on reflectance were fit well. In the different disease test, the quantitative models were fit badly except the model of moisture content. The model on moisture content performed reasonably well, though performance of precision could probably be improved by further analysis, and the paper would provide a basis for spectrum-based identifying of rapeseed leaf leukoplakia.

© IFIP International Federation for Information Processing 2016
Published by Springer International Publishing AG 2016. All Rights Reserved
D. Li and Z. Li (Eds.): CCTA 2015, Part I, IFIP AICT 478, pp. 637–653, 2016.
DOI: 10.1007/978-3-319-48357-3_61

Keywords: Rapeseed leukoplakia · Spectrum reflectance · Disease index · Agronomic parameters

1 Introduction

Rapeseed is one of very important oilseed crops in the world, and its plant area in normal year is about 18–30 million ha. The plant area of rapeseed in China is about 6–7 million ha, and its total yield is about 10–13 million tons, which ranks the fifth place in crop production in China [1]. However, the control for diseases and insect pests is the key to crop production with high yielding, good quality, better benefit, ecology, and safety, and the traditional control for diseases and insect pests depends on manned or machined application of pesticide, which has some obvious disadvantages such as bigger work intensity, high cost, and environmental problems, etc. [2–5].

The spectrum technology is a new subject that detects changes in properties and shapes of object as non-touch manner through using physics theories, mathematics methods, and geography rules [6]. In recent years, studies on crop spectrum technology have made rapid progress. Some researchers pointed out that vegetation reflective spectrum can denote its health status, and some relevant physiological biochemistry indices, e.g. Xue *et al.* [7, 8], Zhang *et al.* [9], Li *et al.* [10], and Hou *et al.* [11] developed the reflective spectrum-based models of plant nitrogen content for rice, wheat, cucumber, orange, and cotton, respectively; Tian *et al.* [12, 13], Zhang *et al.* [14], Mao *et al.* [15], and Fu *et al.* [16] proposed the reflective spectrum-based quantitative models of plant water content for rice, wheat, rapeseed, lettuce, and alfalfa, respectively; and Xie *et al.* [17], Zheng *et al.* [18], Chen *et al.* [19], Zou *et al.* [20], and Li *et al.* [21] set up the reflective spectrum-based quantitative models of leaf SPAD value for rice, wheat, corn, tea plant, and apple. However, few studies have focused on the spectrum-based quantitative models for agronomic parameters of crop disease sample.

The objectives of this research were to find characteristic spectrum band using comparison of the reflective spectrum of health and disease leaf for rapeseed, to reveal the quantitative relationships between the characteristic spectrum and agronomic parameters of disease leaf for rapeseed, and to provide a basis for spectrum-based identifying of rapeseed leaf leukoplakia.

2 Materials and Methods

We conducted field experiments at the Jiangsu Academy of Agricultural Sciences farm, Nanjing, China (32°03′N) during the 2012 to 2013, and 2013 to 2014 growing seasons. The soil type of the experimental area is a hydragric anthrosol. Soil test results indicated the following: organic carbon, 13.8 g kg^{-1}; whole nitrogen, 2.03 g kg^{-1}; available phosphorus, 20.30 mg kg^{-1}; available potassium, 139.00 mg kg^{-1}; and pH, 7.31.

2.1   Materials

This study used 3 rapeseed cultivars breed by Institute of Economical Crops of Jiangsu Academy of Agricultural Sciences: (V1) Ningyou18 (normal), (V2) Ningyou16 (normal), and (V3) Ningza19 (hybrid).

2.2   Methods

2.2.1   Experimental Conditions and Design

The experiments on cultivars were a randomized complete design with 3 varieties (listed above) and 3 replications. The sowing date was Oct. 8 and the transplanting date was Nov. 9 for 2012 to 2013, and Oct. 3 and Nov. 1 for 2013 to 2014, respectively. The 5-leaf seedlings (30, and 33 seedling days for 2012 to 2013, and 2013 to 2014) were transplanted into 60- by 30-cm spacings in 7.98- by 3.50-m subplots. The fertilizer included 90 kg N ha^{-1} (as urea), 90 kg P$_2$O$_5$ ha^{-1}, and 90 kg K$_2$O ha^{-1}. The basal: overwinter: elongation fertilizer ratio was 5:3:2. The 15 kg borax ha^{-1} was sprayed on foliage after elongation for 2013 to 2014. Other management activities followed local production practices.

2.2.2   Measurements

Spectrum: we used CropScan MS16R Multispectral Radiometer with 16 wave bands (460, 550, 650, 710, 760, 810, 870, 1080, 1200, 1280, 1320, 1480, 1540, 1600, 1650, and 1700 nm) to collect reflective spectrum of healthy and leukoplakia disease leaf at 31.1° view field angle under field canopy background and under black cloth.

(1)   Reflectance spectrum of leaves in the field: we randomly selected 5 to 10 healthy, and leukoplakia disease leaf, and investigated their reflective spectrum during 10:00 to 12:00 AM under clearness or few cloud every 7 days for Mar. 15 to May 3 in 2013, and for Apr. 9 to May 6 in 2014. At measuring, the spectrum sensor should be plumbed with ground, and hold out about 30 cm high over target leaf. The data of each leaf were measured in 5 repeats, and their average values were computed.

(2)   Reflectance spectrum of leaves under black background: after finishing measurement in the field, the target leaves were taken from the plant, and lain on the black background to examine their reflective spectrum taking nature light as light source with the same method in (1).

Leaf water content: leaf fresh weights were weighed before reflectance spectrum of leaves was determined under black background, and after finishing measurement under black background, put in oven, and dried in 30 min. at 105°C, then at 80°C until reaching a stable weight, where dry weight were weighed using a 0.001 g electro-level. Leaf water content can be computed.

Leaf SPAD value: we used SPAD-502 Chlorophyll meter to measure SPAD value of upside, middle part, and basal for the healthy and the leukoplakia disease leaves, and their average values were computed.

Disease index (DI): we acquired leukoplakia disease leaves images using Nikon P5000 numeral camera under black background and 1 cm^2 reference, gained pixels and numbers of leukoplakia disease spot, and leaf area in Photoshop CS4.0, and the DI was computed.

All experiment data were analysed using Microsoft EXCEL.2003 and SPSS ver. 19.0.

2.2.3 Model Validation

The models developed in this paper were validated by calculating the correlation (R), the average absolute difference (d_a), the ratio of d_a to the average observation (d_{ap}), the standard error of absolute difference (S_d, like the root mean squared errors, RMSE (Evers *et al.* 2005)) [22, 23], and 1:1 figure of simulated and observed properties, which were defined as:

$$d_i = X_{Oi} - X_{Si}$$

$$d_a = \frac{1}{n} \sum_{i=1}^{n} d_i$$

$$\bar{X}_O = \frac{1}{n} \sum_{i=1}^{n} X_{Oi}$$

$$d_{ap}(\%) = d_a / \bar{X}_O \times 100$$

$$S_d = \sqrt{\frac{\sum_{i=1}^{n} (d_i - d_a)^2}{n - 1}}$$

with i = sample number, n = total number of measurements, $n - 1 = n$ when $n \geq 30$, X_{Si} = simulated value, and X_{Oi} = observed value.

3 Results

3.1 Changes in Reflectance Spectrum of Healthy and Leukoplakia Disease Leaves in Different Growth Periods for Rapeseed

3.1.1 Reflectance Spectrum of Leaf in Field

The field experiment data in 2013 showed that the changes in spectrum reflectance of rapeseed leaf in field were decreasing trend with growth period duration, and the biggest spectrum reflectance (around 810 nm) from early anthesis to the 17th day after end anthesis was reduced from 45 % to 30 %, in that the decline for spectrum reflectance of leukoplakia disease leaves was earlier than that of the healthy; The biggest spectrum reflectance from the 9th day after early anthesis to the 9th day after end anthesis was kept at around 35 % for healthy leaves, and reduced at around 30 % for leukoplakia disease leaves. Thus, this period was the optimum time for identifying rapeseed leukoplakia disease in field based on reflectance spectrum, and the average spectrum reflectance was about range of 30 % to 35 % for leukoplakia disease leaves; In total, the spectrum reflectance in near infrared band (760–1080 nm) for healthy leaves was higher than that of the leukoplakia disease, while there was a reverse status in short-wave infrared band (1080–1700 nm) except for around 1500 nm moisture absorb peak, and there was no obvious difference in others; In addition, because of no obvious difference for color of leukoplakia disease and healthy leaves in field, there was no obvious difference for the spectrum reflectance in visible light band (460–760 nm) for leukoplakia disease and healthy leaves (shown in Figs. 1, 2, 3 and 4).

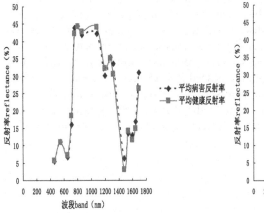

Fig. 1. Reflectance spectrum of *Brassica napus* leaves under the background of field groups in early anthesis (4.2)

Fig. 2. Reflectance spectrum of *Brassica napus* leaves under the background of field groups 11 days after anthesis (4.12)

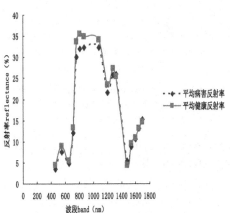

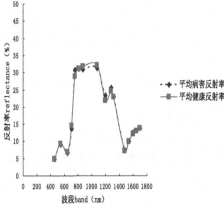

Fig. 3. Reflectance spectrum of Brassica napus leaves under the background of field groups 9 days after finish flowering (4.25)

Fig. 4. Reflectance spectrum of Brassica napus leaves under the background of field groups 17 days after finish flowering (5.3)

3.1.2   Reflectance Spectrum of Leaf in Black Background

In order to eliminate the effects of soil environmental factors in field on reflectance spectrum of leaf, Observations of reflectance spectrum for leaf at various growth period were made under black background (shown in Figs. 5, 6 and 7), the results showed that the spectrum reflectance in visible light band (460–760 nm) for healthy leaves less than that of the leukoplakia disease, and the reasons of that were that chlorophyll was destroyed when the leukoplakia disease occurred, absorption to light from coloring matter was reduced, and this made the spectrum reflectance in visible light band increase; In near infrared band (760–1080 nm), because the rapeseed leaf cell structure was broken when the leukoplakia disease occurred, the number of cell layer was diminished, this made the light transmission increase, and the spectrum reflectance was weaken correspondingly; Although the spectrum reflectance in near infrared band for healthy leaves was slight decreasing with rapeseed growth period duration, its highest value was from 70 % to around 65 %, there was more obvious diminishing for that of the leukoplakia disease leaves, and in total, the distance of spectrum reflectance between the healthy and the leukoplakia disease leaves was gradually increased in advantages to leukoplakia disease identifying.

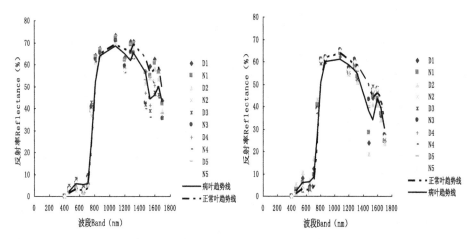

Fig. 5. Leaf reflectance under natural light in early anthesis (4.2)

Fig. 6. Leaf reflectance under natural light 1 day after finish flowering (4.17)

Note: D1, D2, D3, D4, and D5 denote the disease leaf reflectance, and N1, N2, N3, N4, and N5 denote the normal leaf reflectance, respectively.

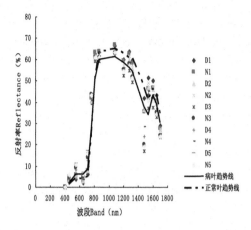

Fig. 7. Leaf reflectance under natural light 9 days after finish flowering (4.25)

3.2 Correlation Between the Reflectance and the Disease Index (DI), and Agronomic Parameter

3.2.1 Correlation Between the Reflectance and the DI, and Agronomic Parameter

The correlation between the reflectance and the disease index (DI), and agronomic parameter for disease leaves was shown in Table 1. We saw that the correlation between the reflectance and the DI at 1200 nm, 1480 nm, and 1600 nm was significant with $p < 0.05$, that of between the reflectance and the disease leaf moisture content in short-wave infrared band 1320 nm to 1650 nm was significant with $p < 0.01$, in which there was the highest negative correlation, -0.810, at 1540 nm, that of between the reflectance and the disease leaf nitrogen content in band 810 nm to 1320 nm, and 1540 nm to 1700 nm was significant with $p < 0.01$, in which there was correlation larger than 0.70 at band 810 nm, 870 nm, and 1080 nm, and that of between the reflectance and the disease leaf chlorophyll SPAD value in band 1200 nm, 1280 nm, and 1540 nm was significant with $p < 0.01$.

Table 1. Correlation coefficients between reflectance and DI, agronomic parameters (n = 22)

Band	Disease index (DI)	Moisture content	Nitrogen content	SPAD value
460	−0.173	0.471*	0.140	−0.063
550	−0.216	0.471*	0.171	−0.173
650	−0.113	0.299	0.153	−0.362
710	−0.191	0.449*	0.204	−0.171
760	−0.293	0.398	0.034	−0.030
810	0.344	0.199	−0.743**	−0.195
870	0.123	0.203	−0.730**	−0.191
1080	−0.018	0.170	−0.719**	−0.230
1200	0.470*	−0.465*	−0.593**	−0.640**

(*continued*)

Table 1. (*continued*)

Band	Disease index (DI)	Moisture content	Nitrogen content	SPAD value
1280	0.279	−0.471*	−0.660**	−0.593**
1320	0.316	−0.657**	−0.566**	−0.506*
1480	0.512*	−0.732**	−0.339	−0.496*
1540	0.198	−0.810**	−0.679**	−0.555**
1600	0.425*	−0.777**	−0.556**	−0.462*
1650	0.303	−0.750**	−0.657**	0.116
1700	0.014	−0.400	−0.670**	−0.339

Note: The asterisk * denote significant correlation at $p < 0.05$, and the asterisk ** denote significant correlation at $p < 0.01$; $r_{(20, 0.05)} = 0.423$, and $r_{(20, 0.01)} = 0.537$.

3.2.2 Changes in Agronomic Parameters with the DI

The field experiment data in 2013 showed that there were declines in the moisture content, the nitrogen content, and the SPAD value for rapeseed disease leaf (shown in Figs. 8, 9 and 10) with the DI increase, which revealed that some leaf physiological indices were in response to the effects of leukoplakia disease on rapeseed leaf, of that there was better correlation between the SPAD value and DI. However, we known that the SPAD value for rapeseed disease leaf increased with its nitrogen content increase through correlation analysis (Fig. 11), and the studies in literature [24–26] also indicated that there was higher correlation between the SPAD value and nitrogen content. Therefore, we considered that the SPAD value can be as one of index indicated severity degree of leukoplakia disease and leaf nitrogen content for rapeseed.

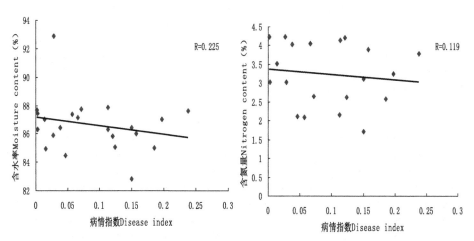

Fig. 8. Relationship between moisture content and DI

Fig. 9. Relationship between nitrogen content and DI

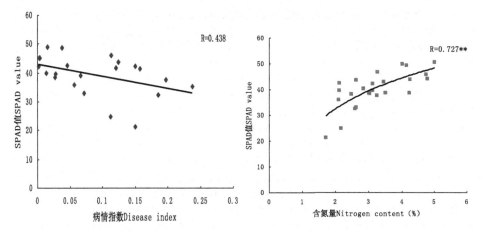

Fig. 10. Relationship between SPAD value and DI

Fig. 11. Relationship between nitrogen content and SPAD value

3.3   Spectrum-Based Quantitative Analysis of Agronomic Parameters

3.3.1   Reflectance-Based Quantitative Analysis of Agronomic Parameters

Stepwise Regression. In order to reflectance-based quantify some leaf agronomic parameters, it need to develop the regression models between the multi-spectrum variables and leaf agronomic parameters, such as the moisture content, the nitrogen content, and the SPAD value of disease leaf. The correlation analysis indicated that there was higher correlation of reflectance for various bands, and a serious issue with multicollinearity (Table 2). Thus, we found regression functions between the agronomic parameters and the reflectance using multiple stepwise regression and principal component analysis, of that the quantitative model for the disease leaf moisture content was taking the reflectance at 1540 nm (Eq. 1 in Table 3), for the disease leaf nitrogen content taking the reflectance at 810 nm (Eq. 2 in Table 3), and for the disease leaf nitrogen content taking the reflectance at 1280 nm, 1080 nm, and 650 nm as independent variable (Eq. 3 in Table 3), respectively, and the variable inflate factor (VIF) of these regression functions were less than 10, which revealed that there were higher tolerance among reflectance in various bands. The statistics test for the models was shown in Table 4.

Curve Fitting. Some studies in literature [27, 28] showed that there was larger difference when various models were fitted using same independent variable, thus, we fitted the reflectance-based equation on moisture content, nitrogen content for disease leaf chosen linearity ($y = ax + b$), logarithm ($y = \log_a bx$), quadratic ($y = ax^2 + bx + c$), and exponential equation ($y = a^{bx}$), and the decision coefficient was shown in Table 5. Of that the decision coefficients (R^2) of the reflectance-based quadratic

Table 2. Correlation coefficients between the reflectance for various bands

	460	550	650	710	760	810	870	1080	1200	1280	1320	1480	1540	1600	1650	1700
460	1.00															
550	0.96	1.00														
650	0.93	0.98	1.00													
710	0.86	0.95	0.97	1.00												
760	0.8	0.87	0.81	0.84	1.00											
810	−0.41	−0.41	−0.42	−0.42	−0.12	1.00										
870	−0.47	−0.42	−0.44	−0.40	−0.09	0.92	1.00									
1080	−0.45	−0.38	−0.40	−0.34	−0.05	0.82	0.98	1.00								
1200	−0.69	−0.78	−0.75	−0.79	−0.65	0.79	0.67	0.56	1.00							
1280	−0.73	−0.76	−0.76	−0.76	−0.54	0.85	0.86	0.81	0.93	1.00						
1320	−0.72	−0.74	−0.75	−0.75	−0.52	0.80	0.87	0.84	0.87	0.99	1.00					
1480	−0.76	−0.85	−0.84	−0.89	−0.73	0.71	0.66	0.58	0.96	0.94	0.92	1.00				
1540	−0.72	−0.81	−0.75	−0.79	−0.81	0.57	0.40	0.28	0.93	0.77	0.70	0.89	1.00			
1600	−0.75	−0.83	−0.79	−0.82	−0.74	0.71	0.63	0.55	0.98	0.92	0.89	0.97	0.94	1.00		
1650	−0.73	−0.79	−0.77	−0.79	−0.63	0.78	0.78	0.72	0.95	0.98	0.97	0.97	0.84	0.97	1.00	
1700	−0.58	−0.57	−0.59	−0.55	−0.31	0.78	0.91	0.92	0.71	0.91	0.94	0.76	0.49	0.72	0.87	1.00

Table 3. Stepwise regression of reflectance and parameters (n = 20)

Parameter	Model	F	Sig.	R^2	VIF	Eq
Moisture content	$Y = 112.7 - 0.6R_{1540}$	36.334	0.000	0.657	1.000	1
Nitrogen content	$Y = 20.805 - 0.28R_{810}$	22.119	0.000	0.551	1.000	2
SPAD	$Y = -143.658 - 4.475R_{1280} + 6.849R_{1080} - 2.302R_{650}$	15.914	0.000	0.737	3.688	3

Note: R_{650}, R_{810}, R_{1080}, R_{1280}, and R_{1540} denote the disease leaf reflectance at 650 nm, 810 nm, 1080 nm, 1280 nm, and 1540 nm, respectively.

Table 4. Coefficient of Equations and t-test

Parameter	Unstandardized coefficients	Corresponding variable	t	Sig.
Moisture content	0.005	R_{760}	4.298	0.000
	0.638	constant	11.972	0.000
Nitrogen content	−0.28	R_{810}	−4.703	0.000
	20.805	constant	5.563	0.000
SPAD	−4.475	R_{1280}	−6.261	0.000
	6.849	R_{1080}	4.524	0.000
	2.302	R_{650}	−3.520	0.003
	−143.685	constant	−1.975	0.065

equation on moisture content, and the exponential equation on nitrogen content were the largest, and the t-test on parameters in these two equations was shown in Table 6. We saw that all parameters in the quadratic equation on moisture content, $Y = 2.251 - 0.071R_{760} + 0.001R_{760}^2$, had significant with $p < 0.05$, and the parameter a in the exponential equation on nitrogen content had no significant in the t-test.

Table 5. Curve estimation of moisture content, nitrogen content and chosen bands

Parameter	Independent variable	Models	R^2	F	Sig.
Moisture content	R_{760}	linearity	0.480	18.472	0.000
		logarithm	0.460	17.069	0.001
		quadratic	0.620	15.487	0.000
		exponential	0.476	18.172	0.000
Nitrogen content	R_{810}	linearity	0.551	22.119	0.000
		logarithm	0.550	21.998	0.000
		quadratic	0.553	10.498	0.001
		exponential	0.571	23.974	0.000

Table 6. Coefficient of equations and t-test

Equation	Unstandardized coefficients	Parameter	t	Sig.
Moisture content	−0.071	b	−2.451	0.024
	0.001	a	2.642	0.016
	2.251	c	3.675	0.002
Nitrogen content	−0.096	b	−4.896	0.000
	1282.838	a	0.813	0.427

Principal Components Analysis. We selected the higher correlation bands with moisture content (460 nm, 550 nm, 650 nm, 710 nm, 760 nm, 1480 nm, and 1600 nm), nitrogen content (810 nm, 870 nm, 1080 nm, 1280 nm, 1540 nm, 1650 nm, and 1700 nm), and the SPAD value (1200 nm, 1280 nm, 1320 nm, and 1540 nm) for the disease leaf, made dimension reduction, and found principal components. Then, regression analysis was made selecting the first principal components as independent variable.

Because almost the parameters in regression equation with the highest decision coefficient had no significant in t-test (Table 7), and the decision coefficients were less than that of the stepwise regression, it was seen that principal components regression was not suitable for regression of multi-band combination and agronomic parameters.

Table 7. Coefficient of principal components regression equation and t-test

Model	Unstandardized coefficients	Corresponding variable	t	Sig.
Moisture content	−0.001	c	−1.768	0.095
	0.005	b	−1.233	0.234
	0.007	a	−0.665	0.515
	0.861	d	90.769	0.000
Nitrogen content	−0.734	c	−2.342	0.032
	−0.093	b	−0.716	0.485
	0.069	a	0.531	0.603
	3.353	d	17.752	0.000
SPAD	−2.866	c	−0.535	0.600
	−6.698	b	−1.874	0.078

(*continued*)

Table 7. (*continued*)

Model	Unstandardized coefficients	Corresponding variable	t	Sig.
	−3.619	a	−1.139	0.271
	31.400	d	9.906	0.000

3.3.2   Model Validation

The stepwise regression models developed in this paper were validated by using independent experiment data in 2013, and 2014, and the results showed that the average absolute difference (da), the ratio of da to the average observation (dap) of simulation and observation in moisture content were less than 5 %, and that of the nitrogen content, and the SPAD value were larger than 20 % except for da for the nitrogen content (Tables 8 and 9). The 1:1 figure on simulation and observation in the moisture content, the nitrogen content, and the SPAD value are represented in Figs. 12, 13, 14, 15, and 16. As illustrated by comparisons of 1:1 map, the model on moisture content performed reasonably well, though performance of precision could probably be improved by further analysis.

Table 8. Stepwise regression model test of the same disease samples

Parameters	N	d_a	$d_{ap}(\%)$	r	t	Sig.
SPAD	5	22.684	52.901	−0.512	7.247	0.002
moisture content	10	3.477 %	4.138	0.810	−1.137	0.269
nitrogen content	10	3.424 %	85.465	0.748	0.520	0.616

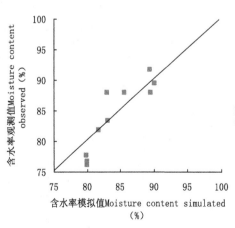

Fig. 12. Comparison of observation and simulation in moisture content of the same disease

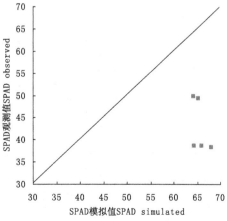

Fig. 13. Comparison of observation and simulation in SPAD value of the same disease

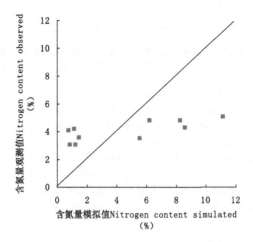

Fig. 14. Comparison of observation and simulation in nitrogen content of the same disease

Table 9. Stepwise regression model test of different diseases samples

Parameters	N	d_a	$d_{ap}(\%)$	r	t	Sig.
SPAD	21	20.260	76.849.	0.775	−12.934	0.000
moisture content	21	5.268 %	6.083	0.537	0.281	0.781

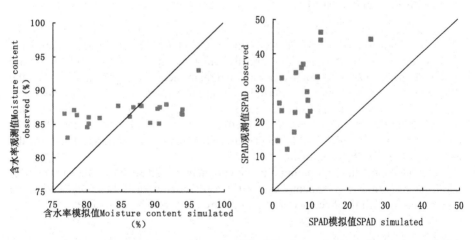

Fig. 15. Comparison of observation and simulation in moisture content of different diseases

Fig. 16. Comparison of observation and simulation in SPAD value of different diseases

4   Discussion

4.1   The Difference of Reflectance Spectrum Between the Normal and Disease Leaf Under Field Nature Condition Was Reduced

Because there was large difference of the severity degree of disease leaves collected in various growth periods, of that the severity degree of partial disease leaves was less, whose leaf color had no difference for the normal and disease leaves basically, and its chlorophyll structure was not destroyed, the correlation between the reflectance in visible bands (determined by chlorophyll absorption) and the DI was not high, and there was a significant correlation only in near infrared band (1480 nm), which indicated that there were large effects of leaf moisture content on disease leaf reflectance. This was in accordance with the low correlation between the reflectance in visible bands and the SPAD value.

4.2   There Were Serious Multicollinearity Among Bands

Because there was serious multicollinearity among bands, the independent variable with higher correlation was exclude using the stepwise regression, the multiple linear regression equations between the DI, the moisture content, the nitrogen content, and the SPAD value and reflectance were set up.

4.3   The Effects of Leaf Thickness on SPAD Value Were in Accordance with the Former Studies

Due to low correlation between the SPAD value and reflectance in visible light bands, and high correlation at 760 nm, and 810 nm (Table 1), leaf cell structure was principal factor of effect on reflectance in above bands, which was in accordance with the former studies. Bauerle et al. [30] pointed out that although there was significant between the SPAD value measured and leaf transmission and absorption, there was no high correlation between the SPAD value and reflectance. Yang et al. [29] also indicated that the leaf thickness was important factor of effect on SPAD estimating model. Thus, we thought about not only visible light band, but also near infrared band in the reflectance-based SPAD estimating.

In addition, the nitrogen is important component of cell structure, thus changes in the disease leaf nitrogen content can affect its cell structure, and these changes can be reflected in spectrum reflectance, further in correlation with SPAD value.

5   Conclusions

This study screened out the reflectance-based bands with large difference between the disease and the normal samples, found the identifying bands with the same character through comparison of the reflectance for canopy in field and the reflectance for leaf in black background, the optimum growth period for reflectance-based rapeseed leukoplakia identifying was form the 11th day after early anthesis to the 9th day after end anthesis, and 760–1080 nm was sensitive band in field identifying.

The correlation between the DI, the moisture content, the nitrogen content, and the SPAD value and reflectance in many character bands were analysed, and the multiple linear regression equations between the DI, the moisture content, the nitrogen content, and the SPAD value and reflectance were set up.

The same disease samples, rapeseed leukoplakia, and different disease samples, rapeseed virus, were used to validation of stepwise regression models, which made disease identifying precision raise.

Funding Information. This work was supported by National Natural Science Foundation of China [31171455, 31471415,31201127], Jiangsu Province Agricultural Scientific Technology Innovation Fund [CX (14)2114]; Jiangsu Academy of Agricultural Sciences Basic Scientific Research Work Special Fund [ZX(15)2008].

References

1. National Bureau of Statistics of China (2009). http://www.stats.gov.cn/tjsj/qtsj/gjsj/. Accessed 26 Sept 2013. (in Chinese with English Abstract)
2. Zhang, L.: The pollution of agricultural environment with chemical pesticide and its control. J. Nanjing Agric. Technol. Coll. **17**(4), 36–38, 43 (2001). (in Chinese with English Abstract)
3. State Environmental Protection Administration, China's Pesticide Pollution and Its Being Problems and Suggestions, Environment Protection, vol. 6, pp. 23–24 (2001). (in Chinese with English Abstract)
4. Zhu, M.Y., Zhang, J.X., Lu, Z.H., Wu, J.F.: Controlling the environmental pollution of chemical pesticide carrying out consecutions protection. Jiangxi Plant Prot. **26**(3), 29–31 (2003). (in Chinese with English Abstract)
5. Tu, Y.Q.: Thinking of problems of agriculture chemicals and circumstance. Reclaiming Rice Cultiv. **1**, 58–60 (2003). (in Chinese with English Abstract)
6. Wang, W.H.: Review for space optical remote sensing technology. Inf. Syst. Eng. (1), 159–141 (2012). (in Chinese with English Abstract)
7. Xue, L.H., Cao, W.X., Luo, W.H., Jiang, D., Meng, Y.L., Zhu, Y.: Diagnosis of nitrogen status in rice leaves with the canopy spectral reflectance. Scientia Agricultura Sinica **36**(7), 807–812 (2003). (in Chinese with English Abstract)
8. Xue, L.H., Cao, W.X., Luo, W.H., Zhang, X.: Correlation between leaf nitrogen status and canopy spectral characteristics in wheat. Acta Phytoecologica Sinica **28**(2), 172–177 (2004). (in Chinese with English Abstract)
9. Zhang, X.J., Li, M.Z., Zhang, Y.E., Zhao, P., Zhang, J.P.: Estimating nitrogen content of cucumber leaf based on solar irradiance spectral reflectance in greenhouse. Trans. Chin. Soc. Agric. Eng. **20**(6), 11–14 (2004). (in Chinese with English Abstract)
10. Li, J.M., Ye, X.J., Wang, Q.L., Zhang, C., He, Y.: Development of prediction models for determining N content in citrus leaves based on hyperspectral imaging technology. Spectrosc. Spectr. Anal. **34**(1), 212–216 (2014). (in Chinese with English Abstract)
11. Hou, X.J., Jiang, G.Y., Bai, L., Liu, S.J., Li, F.: Relationships between leaf nitrogen content and canopy reflected spectral in cotton. Xinjiang Agric. Sci. **45**(5), 776–781 (2008). (in Chinese with English Abstract)

12. Tian, Y.C., Cao, W.X., Jiang, D., Zhu, Y., Xue, L.H.: Relationship between canopy reflectance and plant water content in rice under different soil water and nitrogen conditions. Acta Phytoecologica Sinica **29**(2), 318–323 (2005). (in Chinese with English Abstract)

13. Tian, Y.C., Zhu, Y., Cao, W.X., Dai, T.B.: Relationship between canopy reflectance and plant water status of wheat. Chin. J. Appl. Ecol. **15**(11), 2072–2076 (2004). (in Chinese with English Abstract)

14. Zhang, X.D., Mao, H.P., Zuo, Z.Y., Gao, H.Y., Zhou, Y.: Study on estimation model for rape moisture content under water stress based on hyperspectral remote sensing. J. Anhui Agric. Sci. **39**(30), 18451–18452+18487 (2011). (in Chinese with English Abstract)

15. Mao, H.P., Zhang, X.D., Li, X., Zhang, Y.: Model establishment for grape leaves dry-basis moisture content based on spectral signature. J. Jiangsu Univ.: Nat. Sci. Ed. **29**(5), 369–372 (2008). (in Chinese with English Abstract)

16. Fu, Y.B., Fan, Y.M., Sheng, J.D., Li, N., Wu, H.Q., Li, M.T., Li, L., Zhao, Y.: Study on relationship between alfalfa canopy spectral reflectance and leaf water content. Spectrosc. Spectr. Anal. **33**(3), 766–769 (2013). (in Chinese with English Abstract)

17. Xie, X.J., Shen, S.H., Li, Y.X., Li, B.B., Cheng, G.F., Yang, S.B.: Red edge characteristics and monitoring SPAD and LAI for rice with high temperature stress. Trans. Chin. Soc. Agric. Eng. **26**(3), 183–190 (2010). (in Chinese with English Abstract)

18. Zheng, F., Yue, J.Q., Shao, Y.H., Wang, Q.C., Zhang, G.T., Li, C.H.: Relationship between NDVI values and SPAD readings in different spring wheat varieties. J. Triticeae Crops **28**(2), 291–294 (2008). (in Chinese with English Abstract)

19. Chen, Z.Q., Wang, L., Bai, Y.L., Yang, L.P., Lu, Y.L., Wang, H., Wang, Z.Y.: Hyperspectral prediction model for maize leaf SPAD in the whole growth period. Spectrosc. Spectr. Anal. **33**(10), 2838–2842 (2013). (in Chinese with English Abstract)

20. Zou, H.Y., Ding, L.X.: Study on estimation model of tea leaf SPAD value based on reflective spectra data. Remote Sens. Inf. **5**, 71–75 (2011). (in Chinese with English Abstract)

21. Li, M.X., Zhang, L.S., Li, B.Z., Zhang, H.Y., Guo, W.: Relationship between spectral reflectance feature and their chlorophyll contentrations and SPAD value of apple leaves. J. Northwest Forest. Univ. **25**(2), 35–39 (2010). (in Chinese with English Abstract)

22. Cao, H.X., Zhang, C.L., Li, G.M., Zhang, B.J., Zhao, S.L., Wang, B.Q., Jin, Z.Q.: Researches of simulation models of rape (Brassica napus L.) growth and development. Acta Agronomica Sinica **32**(10), 1530–1536 (2006). (in Chinese with English Abstract)

23. Evers, J.B., Vos, J., Fournier, C., Andrieu, B., Chelle, M., Struik, P.C.: An architectural model of spring wheat: evaluation of the effects of population density and shading on model parameterization and performance. Ecol. Model. **200**, 308–320 (2007)

24. Zhang, R.M.: Changes in leaf dynamic after early anthesis for Brassica Napus. Tillage Cultiv. **5**, 59–60 (1988). (in Chinese with English Abstract)

25. Wang, J., Han, D.W., Ren, G., Guo, J.Q., Zhang, Y.S., Wei, C.Z., Song, Y.M.: A study on relation between SPAD value, chlorophyll nitrogen content cotton. Xinjiang Agric. Sci. **43**(3), 167–170 (2006). (in Chinese with English Abstract)

26. Yu, Y.L., Jia. W.K., Wang, C.H., Jiang, B.W.: Study on the relationship of SPAD readings to nitrogen content and to yield in spring maize. J. Maize Sci. 19(4), 89–92+97 (2011). (in Chinese with English Abstract)

27. Hu, H., Bai, Y.L., Yang, L.P., Lu, Y.L., Wang, L., Wang, H., Wang, Z.Y.: Diagnosis of nitrogen nutrition in winter wheat (Triticum aestivum) via SPAD-502 and GreenSeeker. Chin. J. Eco-Agric. **18**(4), 748–752 (2010). (in Chinese with English Abstract)

28. Jiang, J.B., Chen, Y.H., Huang, W.J.: Study on hyperspectra estimation of pigment contents in canopy leaves of winter wheat under disease stress. Spectrosc. Spectr. Anal. **27**(7), 1363–1367 (2007)

29. Yang, H.Q., Yao, J.S., He, Y.: SPAD prediction of leave based on reflection spectroscopy. Spectrosc. Spectr. Anal. **29**(6), 1607–1610 (2009). (in Chinese with English Abstract)
30. Bauerle, W.L., Weston, D.J., Bowden, J.D.: Leaf absorptance of photosynthetically active radiation in relation to chlorophyll meter estimates among woody plant species. Sci. Hortic. **101**, 169–178 (2004). (in Chinese with English Abstract)

Author Index

An, Xiaofei II-117

Bai, Jie I-580
Bao, Zuda I-574
Bu, Yunlong I-425

Cai, Lixia I-45, I-564
Cao, Chengfu II-68
Cao, Hongxin I-459, I-502, I-637, II-133
Cao, Liying I-564, II-76, II-100
Cao, Yongqi I-556
Chang, Ruokui II-33, II-514
Chen, Ailian I-335
Chen, Fang-yi II-277
Chen, Guifen I-45, I-153, I-239, I-564, II-76,
    II-100, II-417
Chen, Guoxing I-356
Chen, Hang I-45, I-564
Chen, Huan II-68
Chen, Huiling I-8
Chen, JiaFeng I-604
Chen, Jianmei I-231
Chen, Liping II-194, II-202, II-237
Chen, Lu II-392
Chen, Shien II-143
Chen, Tian'en II-60, II-277, II-287, II-310,
    II-320
Chen, Weitao I-459, I-502, I-637
Chen, Xiaofeng II-92
Chen, Yingyi I-64, I-72
Chen, Yu I-317, I-485, I-536, II-330
Chen, Zixiao I-399
Cong, Yue I-528
Cui, Yongjie I-399

Deng, Lei II-586
Deng, Qiao II-496
Diao, Zhihua I-190
Dong, Daming I-619
Dong, Gang II-548
Dong, Jianjun I-348, II-117
Dong, Jing II-60, II-287
Dong, Lin I-546

Dong, QiaoXue I-19, I-604
Dong, Yansheng II-163
Du, Keshuang II-522
Du, Meng I-94
Du, Shangfeng I-19
Du, Shizhou II-68
Duan, Lingfeng I-356, I-390
Duan, Yanqing II-357

Fan, Jingchao I-275
Fan, Shuxiang II-194, II-202, II-237
Fang, Xinping I-597
Feng, Chunhuan I-459, I-502, I-637
Feng, Haikuan II-259
Feng, Jinkui II-522
Feng, Wenjie II-557, II-563, II-570
Fu, Kunya I-459, I-637
Fu, Licheng II-300
Fu, Liting II-496
Fu, Longsheng I-399
Fu, Shaolong I-446
Fu, Weiqiang I-528

Gao, Bingbo II-225
Gao, Jianmin I-399
Gao, Li II-310
Gao, Liangliang I-72
Gao, Nana I-348, I-528
Gao, Yang I-125
Gao, Zhen I-246
Ge, Daokuo I-459, I-502, I-637
Gong, Zhiyuan II-445
Guo, Jianhua II-117
Guo, Jiao I-246
Guo, Lei I-94
Guo, Qingzeng I-231
Guo, Xiuming I-275

Han, Lei II-539
Han, Xue I-325, II-41
Hao, Mingming II-15
Hao, Xingyao II-225
Hao, Yunpeng I-317

He, Bingbing I-587
He, Peng II-259
He, Yajuan I-161, I-325
Hou, Jinjian II-505
Hu, Hongyan I-536
Hu, Lin I-275
Hu, Yanan II-357
Hu, Yating I-564
Hua, Jing I-604
Huang, Baichuan I-619
Huang, Chenglong I-356, I-390
Huang, Linsheng I-300
Huang, Linya I-246
Huang, Wenqian II-194, II-202, II-237
Huang, Wuji II-269
Huang, Yanbo II-404, II-429

Ji, Wei I-485
Jia, Weikuan I-485, II-330
Jia, Wenshen II-505
Jia, Xuesha I-27
Jiang, Li II-310
Jiang, Lixia I-292
Jiang, Qiuxiang I-556
Jiang, Shuwen II-60, II-287
Jiao, Leizi I-619
Jiao, Weijie I-107, I-325, II-41
Jiao, Yang II-170
Jin, Xiuliang II-259
Jin, Zhiqing II-217
Jing, Qin II-479, II-487

Kang, MengZhen I-604

Li, Chao II-287
Li, Chenghua I-473
Li, Daoliang I-8, I-115, I-519, II-15
Li, Guisen II-300
Li, Han I-425
Li, Jian I-317, II-404
Li, Jinlei I-255
Li, Jizhou II-539
Li, Pingping I-208
Li, Shanghong I-173, II-530
Li, Shaoming I-446
Li, Shasha I-285
Li, Wei II-68
Li, Weifang I-366
Li, Weiwei I-153, I-239

Li, Xiaolan II-225
Li, Yifan II-445
Li, Yiyang I-82
Li, You I-528
Li, Zhenbo I-519
Li, Zhenhai II-259
Li, Zhixiao I-94
Liang, Dong I-300
Liang, Fengchen II-522
Lianjun, Yu II-455
Liao, Qinhong II-68
Liu, BingGuo I-546
Liu, Chen II-237
Liu, Chunhong II-496
Liu, Fei I-619
Liu, Hongman I-143
Liu, Hua II-269
Liu, Lingbo I-390
Liu, Minjuan II-392
Liu, Shanmei I-266
Liu, Sheng I-546
Liu, Shuangyin I-8
Liu, Weiming I-574
Liu, Xing I-143
Liu, Xueyan I-8
Liu, Yan I-459, I-502, I-637
Liu, Yande I-292, I-587, II-445
Liu, Yang II-1, II-33, II-133, II-217
Liu, Yanzhong I-64, I-72, II-570
Liu, Yijun I-375
Liu, Yuechen I-107, I-366
Liu, YuJie I-546
Liu, Yunling I-197
Liu, Zhe I-446
Liu, Zhihong II-185
Lu, Caiyun I-348
Lu, Jian II-76, II-100
Luo, Dan I-143
Luo, Zongqiang II-1
Lv, Changhuai II-300

Ma, AiSheng I-546
Ma, Jian II-346, II-369
Ma, Li II-100
Ma, Shangjie I-161
Ma, Wei II-429, II-437
Ma, Xinming I-53
Ma, Zhijie I-64, II-539
Ma, Zhiping I-161

Ma, Zhongren   II-143
Meng, Lumin   II-163
Meng, Ying   II-76, II-100
Meng, Zhijun   I-348, I-528, II-117
Mu, Weisong   II-586
Mu, Yuanjie   II-548

Nengfu, Xie   I-437
Niu, Chong   I-425
Niu, Luyan   II-570
Niu, Yuguang   I-425

Ouyang, Jihong   II-417

Pan, Ligang   II-505
Pan, Yuchun   II-225
Pei, Zhiyuan   I-161, I-335, I-366
Peng, Bo   I-612
Peng, Hui   I-266
Peng, Yuli   I-208

Qi, Lijun   II-429, II-437
Qi, Yan   I-173, II-530
Qian, Man   II-194, II-202
Qiao, Xi   I-519
Qiao, Yuqiang   II-68
Qin, Jing   II-539
Qin, Xu   II-522
Qiu, Yun   I-275
Qu, Mei   I-19

Ren, Yanna   I-53
Ruan, Chengzhi   I-485
Ruan, Huaijun   II-548, II-563

Sa, Liangbing   II-269
Shang, Minghua   II-548
Shao, ChangYong   I-546, II-522
Shao, MingXi   I-546
Shen, Kejian   I-366, II-41
Shen, Tian   II-330
Shi, Binjie   II-269
Shi, Chunlei   I-612
Shi, Chunlin   II-1, II-133, II-217
Shi, Qinglan   I-19, I-197
Shi, Xiaohui   II-320
Shi, Yongle   II-1
Shu, Jing   II-522
Si, Xiuli   I-153, I-239

Song, Chaoyu   I-231
Song, Chuwei   I-459, I-502
Song, Yu'e   II-92
Su, Baofeng   I-246, I-399
Sui, Yuanyuan   II-155, II-178, II-455
Sun, Fangli   I-231
Sun, Guannan   I-366
Sun, Jiabo   II-570
Sun, Li   I-325
Sun, Lijuan   II-185
Sun, Longqing   I-82
Sun, Min   I-115
Sun, Ming   II-111
Sun, Xinxin   I-82
Sun, Yonghong   I-231
Sunwei   I-437

Tang, Shuping   I-485
Tang, Xin   II-522
Teng, Ling   I-300
Teng, Xiaowei   II-163
Thiravong, Sisavath   II-479, II-487
Tian, Hongwu   I-255
Tian, Ruya   I-375
Tian, Xiaojing   II-143
Tian, Xuedong   I-27
Tong, Ling   I-473
Tu, Xingyue   I-64

Wang, Bin   I-399
Wang, Bingbing   II-185
Wang, Chaopeng   II-202
Wang, Chengguo   II-92
Wang, Cong   II-287
Wang, Dan   II-346
Wang, Danqiong   I-325
Wang, Danyang   I-473
Wang, Dengwei   II-310
Wang, Dong   II-505
Wang, Fei   I-325, I-366
Wang, Fengyun   II-579
Wang, Fujun   II-548
Wang, Guowei   I-317, I-536, II-155, II-178,
    II-455
Wang, Haijun   I-325, II-41
Wang, Haiyang   I-292, I-587
Wang, Hui   I-231
Wang, Jian   II-127
Wang, Jianfei   II-557

Wang, Jizhang   I-208
Wang, Jun   II-143
Wang, Lei   II-563, II-579
Wang, Lianlin   I-161
Wang, Linlin   II-155
Wang, Mingfei   I-255
Wang, Qingyan   II-194, II-237
Wang, Ruimei   II-586
Wang, Ting   II-392
Wang, Wei   II-277
Wang, Weiqing   I-1, II-248
Wang, Wenli   I-216
Wang, Xiang   I-94
Wang, Xin   II-469
Wang, Xiu   I-348, II-429, II-437
Wang, Xu   II-417
Wang, Yaru   I-94
Wang, Yuanhong   II-33, II-514
Wang, Yuefei   II-392
Wang, Yuepeng   I-143
Wei, Ruijiang   II-469
Wei, Xiufang   II-1
Wei, Xueli   I-528
Wei, Yong   II-33, II-269, II-514
Wei, Yuquan   I-190
Wu, Baoguo   II-530
Wu, Beibei   I-190
Wu, Dingfeng   II-127
Wu, Guangwei   I-348, II-117
Wu, Haiyun   II-514
Wu, Lei   I-375
Wu, Qiulan   II-185
Wu, Quan   I-325, I-366
Wu, Yuanyuan   I-190
Wu, Zihan   II-111

Xi, Lei   I-53, I-412
Xia, Xue   I-275
Xiaochun, Zhong   I-437
Xie, Xinhua   I-300
Xinning, Hao   I-437
Xiong, Lizhong   I-356, I-390
Xu, Jing   I-64, I-72
Xu, Jingrong   II-479, II-487
Xu, Xin   I-412
Xu, Yan   I-125, I-143
Xuan, Shouli   II-1, II-133, II-217
Xue, Changying   II-133
Xue, Heru   I-580

Xue, Jiani   I-64
Xuefu, Zhang   I-437

Yan, Yun   II-392
Yang, Dandan   II-185
Yang, Fang   I-27
Yang, Fuqin   II-259
Yang, Guijun   II-259
Yang, Jiao   II-155
Yang, Jie   II-300
Yang, Jingjing   II-202
Yang, Jutian   II-143
Yang, Lili   I-19, I-604
Yang, Sisi   I-564
Yang, Songqiang   I-27
Yang, Wanneng   I-356, I-390
Yang, Xiaorong   II-346
Yang, Xinbin   I-8
Yang, Yujian   II-48
Yang, Yuqin   I-45
Yao, Rujing   II-404
You, Jiong   I-366
Yu, Haiyang   II-259
Yu, Haiye   II-155, II-178, II-455
Yu, Helong   II-15
Yu, Hua   I-412
Yu, Huihui   I-64
Yu, Lianjun   II-155, II-170, II-178
Yu, Lu   I-216
Yuan, Xue   II-392
Yuan, Xuexia   II-48
Yuan, Xuyin   II-539

Zeng, Lusheng   I-231
Zhai, Dekun   II-185
Zhai, Ruifang   I-266
Zhan, Shaobin   I-8
Zhang, Baohua   II-194, II-202, II-237
Zhang, Benhua   I-473
Zhang, Caiyu   II-300
Zhang, Chi   II-310
Zhang, Chong   I-27
Zhang, Chuanhong   II-170
Zhang, Fangming   II-300
Zhang, Fengrong   I-125
Zhang, Hang   I-519
Zhang, Hao   I-412, I-536
Zhang, Jiayu   I-173
Zhang, Jingjing   II-463

Zhang, Jinheng I-231
Zhang, Lei II-155, II-178, II-455
Zhang, Lili II-522
Zhang, Liping II-346, II-382
Zhang, Qiyu I-72, II-92
Zhang, Rui I-94
Zhang, ShouSheng I-546
Zhang, Weixin I-459, I-502, I-637
Zhang, Wen I-197
Zhang, Wenyu I-459, I-502, I-637, II-133
Zhang, Xiangqian II-68
Zhang, Xiaodong I-446
Zhang, Xiaoqian I-366
Zhang, Xiaoyan II-570
Zhang, Xin I-216
Zhang, XiuMei I-546
Zhang, Xuefu I-375
Zhang, Yong II-48
Zhang, Yuanyuan II-404
Zhang, Yun II-357
Zhang, Yuxiang I-292, I-587
Zhang, Zhenhong I-446
Zhao, Dean I-485, II-330
Zhao, Dongling I-94
Zhao, Guogang II-155, II-178, II-455
Zhao, Hu I-335
Zhao, Hua II-127
Zhao, Jia II-557, II-563
Zhao, Jingjie II-586
Zhao, Jingling I-300
Zhao, Ke I-556

Zhao, LiJing I-546
Zhao, Lijing II-522
Zhao, Peng-fei II-277
Zhao, Wenyong I-8
Zhao, Xiande I-619
Zhao, Xiaoyu I-125
Zhao, Xin II-170
Zhao, Xuehua I-8
Zhao, Yanhua II-133
Zhao, Yun I-64, II-539
Zhao, Zhu II-68
Zhao, Zizhu II-33, II-514
Zhen, Zhumi I-64, I-72
Zheng, Guang I-53, I-412
Zheng, Huamei II-522
Zheng, Jiye II-579
Zheng, Peichao I-619
Zheng, Wengang I-216, I-425
Zhou, Guomin I-275, II-127
Zhou, Jian I-125
Zhou, Yanbing II-225
Zhou, Yanqing I-580
Zhou, Zengchan I-425
Zhou, Zhimei I-556
Zhu, Dehai I-446
Zhu, Huiqin II-469
Zhu, Ling I-72, II-92
Zhu, Yiming I-285
Zou, Ling II-111
Zou, Wei II-437
Zuo, Yan I-300

Printed in the United States
By Bookmasters